Programmed Cell Death in Cancer Progression and Therapy

ADVANCES IN EXPERIMENTAL MEDICINE AND BIOLOGY

Recent Volumes in this Series

Volume 607
EUKARYOTIC MEMBRANES AND CYTOSKELETON: ORIGINS AND EVOLUTION
Edited by Gáspár Jékely

Volumes 608
BREAST CANCER CHEMOSENSITIVITY
Edited by Dihua Yu and Mien-Chie Hung

Volume 609
HOT TOPICS IN INFECTION AND IMMUNITY IN CHILDREN VI
Edited by Adam Finn and Andrew J. Pollard

Volume 610
TARGET THERAPIES IN CANCER
Edited by Francesco Colotta and Alberto Mantovani

Volume 611
PETIDES FOR YOUTH
Edited by Susan Del Valle, Emanuel Escher, and William D. Lubell

Volume 612
RELAXIN AND RELATED PETIDES
Edited by Alexander I. Agoulnik

Volume 613
RECENT ADVANCES INTO RETINAL DEGENERATION
Edited by Joe G. Hollyfield, Matthew M. LaVail, and Robert E. Anderson

Volume 614
OXYGEN TRANSPORT TO TISSUE XXIX
Edited by Kyung A. Kang

Volume 615
PROGRAMMED CELL DEATH IN CANCER PROGRESSION AND THERAPY
Edited by Roya Khosravi-Far and Eileen White

Roya Khosravi-Far • Eileen White

Programmed Cell Death in Cancer Progression and Therapy

 Springer

Roya Khosravi-Far
Harvard Medical School
Boston, MA
USA

Eileen White
Rutgers University
Piscataway, NJ
USA

ISBN 978-1-4020-6553-8   e-ISBN 978-1-4020-6554-5

Library of Congress Control Number: 2007937291

Printed on acid-free paper.

9 8 7 6 5 4 3 2 1

springer.com

Dedication

We dedicate this book to three deserving groups. First, our families for their support and encouragement, Simin, Ghasem, Reza and Ali Khosravi-Far, and Greg, Jason and Melissa Diamond. Second, to our students, fellows and assistants, who with their hard work pave the road to discovery. Last but not least, to anyone who has been touched by cancer, as they are our motivation for this work and for our research.

Contents

Foreword... ix

Contributors... xi

Chapter 1   **Cell Death: History and Future** 1
            Zahra Zakeri and Richard A. Lockshin

Chapter 2   **Caspase Mechanisms** 13
            Guy S. Salvesen and Stefan J. Riedl

Chapter 3   **The Mitochondrial Death Pathway**..................... 25
            Anas Chalah and Roya Khosravi-Far

Chapter 4   **Apoptotic Pathways in Tumor Progression
            and Therapy** 47
            Armelle Melet, Keli Song, Octavian Bucur, Zainab Jagani,
            Alexandra R. Grassian, and Roya Khosravi-Far

Chapter 5   **Therapeutic Targeting of Death Pathways in Cancer:
            Mechanisms for Activating Cell Death in Cancer Cells** 81
            Ting-Ting Tan and Eileen White

Chapter 6   **Overcoming Resistance to Apoptosis in
            Cancer Therapy** 105
            Peter Hersey, Xu Dong Zhang, and Nizar Mhaidat

Chapter 7   **Trail Receptors: Targets for Cancer Therapy**............. 127
            Robin C. Humphreys and Wendy Halpern

Chapter 8   **Rational Design of Therapeutics Targeting the BCL-2
            Family: Are Some Cancer Cells Primed for Death
            but Waiting for a Final Push?**........................ 159
            Victoria Del Gaizo Moore and Anthony Letai

Chapter 9   **Autophagy and Tumor Suppression: Recent Advances in Understanding the Link between Autophagic Cell Death Pathways and Tumor Development** 177
Shani Bialik and Adi Kimchi

Chapter 10   **Regulation of Programmed Cell Death by the P53 Pathway** 201
Kageaki Kuribayashi and Wafik S. El-Deiry

Chapter 11   **Regulation of Programmed Cell Death by NF-κB and its Role in Tumorigenesis and Therapy** 223
Yongjun Fan, Jui Dutta, Nupur Gupta, Gaofeng Fan, and Céline Gélinas

Chapter 12   **Targeting Proteasomes as Therapy in Multiple Myeloma** 251
Dharminder Chauhan, Teru Hideshima, and Kenneth C. Anderson

Chapter 13   **Histone Deacetylase Inhibitors: Mechanisms and Clinical Significance in Cancer: HDAC Inhibitor-Induced Apoptosis** 261
Sharmila Shankar and Rakesh K. Srivastava

Chapter 14   **RNA Interference and Cancer: Endogenous Pathways and Therapeutic Approaches** 299
Derek M. Dykxhoorn, Dipanjan Chowdhury, and Judy Lieberman

Chapter 15   **Cancer Stem Cells and Impaired Apoptosis** 331
Zainab Jagani and Roya Khosravi-Far

Index 345

Foreword

Apoptosis is a tightly regulated cell-suicide program that plays an essential role in development and maintenance of tissue homeostasis by eliminating unnecessary or harmful cells. Impairment of this native defense mechanism of the cell promotes uncontrolled growth and frequently confers chemoresistance to tumor cells. Substantial progress has been made in the elucidation of several of the underlying mechanisms of apoptotic signaling and their dysregulation in cancer. These advances have facilitated the identification of new drug targets for promising apoptosis-inducing therapeutic strategies. Several of the novel therapeutic agents directed against these targets demonstrate enhanced apoptotic killing and sensitize resistant cancer cells to antineoplastic agents. As a number of these agents have entered the clinic and more are in the pipeline, this is an exciting time for reaping the benefits of years of basic science discoveries through their translation into cancer therapies.

In this book, the regulation of apoptotic signaling in normal cells and the means by which this protective response is suppressed in cancer cells will be discussed. In addition, the novel apoptosis-inducing therapeutic strategies will be summarized. We hope that this book will be a useful source for scientists and clinicians.

Acknowledgments

We acknowledge the editorial support of Susan Glueck, Jessica Platti, Lydia Gregg and Tami Sharkey. We also recognize the assistance of Cristina Alves dos Santos and Melania Ruiz of Springer.

Contributors

Kenneth C. Anderson
The Jerome Lipper Multiple Myeloma Center, Department of Medical Oncology,
Dana Farber Cancer Institute, Harvard Medical School, Boston, MA 02115, USA

Shani Bialik
Department of Molecular Genetics, Weizmann Institute of Science, Rehovot,
Israel 76100

Octavian Bucur
Department of Pathology, Harvard Medical School, Beth Israel Deaconess
Medical Center, 99 Brookline Avenue, Boston, MA 02215, USA

Anas Chalah
Department of Pathology, Harvard Medical School, Beth Israel Deaconess
Medical Center, 99 Brookline Avenue, Boston, MA 02215, USA

Dharminder Chauhan
The Jerome Lipper Multiple Myeloma Center, Department of Medical Oncology,
Dana Farber Cancer Institute, Harvard Medical School, Boston, MA 02115, USA

Dipanjan Chowdhury
CBR Institute for Biomedical Research and Department of Pediatrics,
Harvard Medical School, 200 Longwood Avenue, Boston, MA 02115, USA

Victoria Del Gaizo Moore
Medical Oncology, Dana-Farber Cancer Institute, Boston, MA 02115, USA

Jui Dutta
Center for Advanced Biotechnology and Medicine; Graduate Program in
Biochemistry and Molecular Biology, Robert Wood Johnson Medical School,
University of Medicine and Dentistry of New Jersey, Piscataway, NJ 08854-5638,
USA

Derek M. Dykxhoorn
CBR Institute for Biomedical Research and Department of Pediatrics,
Harvard Medical School, 200 Longwood Avenue, Boston, MA 02115, USA

Wafik S. El-Deiry
Laboratory of Molecular Oncology and Cell Cycle Regulation, Departments of
Medicine (Hematology/Oncology), Genetics, and Pharmacology, The Institute
for Translational Medicine and Therapeutics and the Abramson Comprehensive
Cancer Center, University of Pennsylvania School of Medicine, Philadelphia, PA

Gaofeng Fan
Center for Advanced Biotechnology and Medicine; Graduate Program in
Biochemistry and Molecular Biology, Robert Wood Johnson Medical School,
University of Medicine and Dentistry of New Jersey, Piscataway,
NJ 08854-5638, USA

Yongjun Fan
Center for Advanced Biotechnology and Medicine, Robert Wood Johnson
Medical School, University of Medicine and Dentistry of New Jersey,
Piscataway, NJ 08854-5638, USA

Céline Gélinas
Center for Advanced Biotechnology and Medicine; Department of Biochemistry
and Cancer Institute of New Jersey, Robert Wood Johnson Medical School,
University of Medicine and Dentistry of New Jersey, Piscataway,
NJ 08854-5638, USA

Alexandra R. Grassian
Department of Pathology, Harvard Medical School, Beth Israel Deaconess
Medical Center, 99 Brookline Avenue, Boston, MA 02215, USA

Nupur Gupta
Center for Advanced Biotechnology and Medicine; Graduate Program in
Biochemistry and Molecular Biology, Robert Wood Johnson Medical School,
University of Medicine and Dentistry of New Jersey, Piscataway,
NJ 08854-5638, USA

Wendy Halpern
Human Genome Sciences, Pathology, Pharmacokinetics and Toxicology,
Rockville, MD 20850, USA

Peter Hersey
Immunology and Oncology Unit, Newcastle Mater Hospital, Newcastle,
NSW 2300, Australia

Teru Hideshima
The Jerome Lipper Multiple Myeloma Center, Department of Medical Oncology,
Dana Farber Cancer Institute, Harvard Medical School, Boston, MA 02115, USA

Robin C. Humphreys
Human Genome Sciences, Development Sciences and Research, Rockville,
MD 20850, USA

Zainab Jagani
Novartis Institutes for Biomedical Research, Novartis Oncology,
250 Massachussetts Avenue, Cambridge MA 02139

Roya Khosravi-Far
Department of Pathology, Harvard Medical School, Beth Israel Deaconess
Medical Center, 99 Brookline Avenue, Boston, MA 02215, USA

Adi Kimchi
Department of Molecular Genetics, Weizmann Institute of Science, Rehovot,
Israel 76100

Kageaki Kuribayashi
Laboratory of Molecular Oncology and Cell Cycle Regulation, Departments of
Medicine (Hematology/Oncology), Genetics, and Pharmacology, The Institute
for Translational Medicine and Therapeutics and the Abramson Comprehensive
Cancer Center, University of Pennsylvania School of Medicine, Philadelphia, PA

Anthony Letai
Medical Oncology, Dana-Farber Cancer Institute, Boston, MA 02115, USA

Judy Lieberman
CBR Institute for Biomedical Research and Department of Pediatrics,
Harvard Medical School, 200 Longwood Avenue, Boston, MA 02115, USA

Richard A. Lockshin
Queens College and the Graduate Center of the City University of New York,
St. John's University, New York, USA

Armelle Melet
Department of Pathology, Harvard Medical School, Beth Israel Deaconess
Medical Center, 99 Brookline Avenue, Boston, MA 02215, USA

Nizar Mhaidat
Immunology and Oncology Unit, Newcastle Mater Hospital, Newcastle,
NSW 2300, Australia

Stefan J. Riedl
Program on Apoptosis and Cell Death, Burnham Institute for Medical Research,
10901 N. Torrey Pines Road, La Jolla, CA 92037, USA. Tel.: 858.646.3114;
fax: 858.646.3197; e-mail: gsalvesen@burnham.org

Guy S. Salvesen
Program on Apoptosis and Cell Death, Burnham Institute for Medical Research,
10901 N. Torrey Pines Road, La Jolla, CA 92037, USA. Tel.: 858.646.3114;
fax: 858.646.3197; e-mail: gsalvesen@burnham.org

Sharmila Shankar
Department of Biochemistry, The University of Texas Health Center at Tyler,
11937 US Highway 271, Tyler, TX 75708-3154, USA

Keli Song
Department of Pathology, Harvard Medical School, Beth Israel Deaconess
Medical Center, 99 Brookline Avenue, Boston, MA 02215, USA

Rakesh K. Srivastava
Department of Biochemistry, The University of Texas Health Center at Tyler,
11937 US Highway 271, Tyler, TX 75708-3154, USA. Tel.: 903-877-7559;
fax: 903-877-5320; e-mail: rakesh.srivastava@uthct.edu

Ting-Ting Tan
Center for Advanced Biotechnology and Medicine; Rutgers University,
Piscataway, NJ 08854, USA, New Brunswick, NJ 08901, USA

Eileen White
Center for Advanced Biotechnology and Medicine; Department of Molecular
Biology and Biochemistry, 679 Hoes Lane, Room 140; Rutgers University,
Piscataway, NJ 08854, USA; Cancer Institute of New Jersey, New Brunswick,
NJ 08901, USA

Zahra Zakeri
Queens College and the Graduate Center of the City University of New York,
St. John's University, New York, USA

Xu Dong Zhang
Immunology and Oncology Unit, Newcastle Mater Hospital, Newcastle,
NSW 2300, Australia

Chapter 1
Cell Death: History and Future

Zahra Zakeri* and Richard A. Lockshin

Abstract Cell death was observed and understood since the 19th century, but there was no experimental examination until the mid-20th century. Beginning in the 1960s, several laboratories demonstrated that cell death was biologically controlled (programmed) and that the morphology was common and not readily explained (apoptosis). By 1990, the genetic basis of programmed cell death had been established, and the first components of the cell death machinery (caspase 3, bcl-2, and Fas) had been identified, sequenced, and recognized as highly conserved in evolution. The rapid development of the field has given us substantial understanding of how cell death is achieved. However, this knowledge has made it possible for us to understand that there are multiple pathways to death and that the commitment to die is not the same as execution. A cell that has passed the commitment stage but is blocked from undergoing apoptosis will die by another route. We still must learn much more about how a cell commits to death and what makes it choose a path to die.

Keywords apoptosis, autophagy, autophagic cell death, history, lysosome

1 Cell Death has Long been Recognized as an Important Biological Problem

Cell death was seen and reported as early as 1842 by Carl Vogt (see Clarke and Clarke, 1996) although at that time it was not called cell death. Cell death was recognized almost as soon as the normal form of a living cell was understood, i.e., by the middle of the 19th century. If a living being can die, it is reasonable to watch a cell

Zahra Zakeri
Department of Biology, Queens College and the Graduate Center of the City University of New York, 65–30 Kissena Blvd., Flushing, New York, 11361 USA,

Richard A. Lockshin
St. John's University, Queens, New York, USA

* To whom correspondence should be addressed: e-mail: zahra_zakeri@hotmail.com.

R. Khosravi-Far and E. White (eds.), *Programmed Cell Death in Cancer Progression and Therapy*.
© Springer 2008

die. Thus, the first histologists recognized dying cells. They even recognized morphologies of death that one would today describe as apoptosis. Much of the work in the 19th century relied on the histological identification of dying cells with limited recognition of its importance and regulation. This history is well described by Clarke and other authors in a number of reviews (Clarke and Clarke, 1995, 1996; Häcker and Vaux, 1997; Lockshin and Zakeri, 2001). What is more interesting is to see when scientists began to appreciate that death was not an accident. In other words, at a certain time the phrase "cells that die are replaced by mitosis" was transformed into "cells die, and they are then replaced by mitosis" (the emphasis changed: it is no longer that an accident is repaired by an organized act, but that the organized step, i.e., cell suicide, is followed by a repair process). With the change in emphasis also came the observation that cell death could not be an abnormal event.

It became obvious that the death of cells seen during metamorphosis of amphibians or insects could not be considered abnormal (Terre, 1889; Janet, 1907; Pérez, 1910). However, the idea that the death was under some sort of control came much later. The great insect physiologist V. B. Wigglesworth well understood that the growth and disappearance of muscles in the blood-sucking insect *Rhodnius prolixus* depended on molting and thus on molting hormones (Wigglesworth, 1972). But the importance and generalization of that idea dates only from the middle of the 20th century.

2 The Mechanism of Cell Death Becomes a Question of Interest

Much of the realization and generalization of the appearance of cell death was reported by the developmental biologists. Starting in the 1950s, A. Glücksmann assembled a long list of instances of cell death, which he classified according to their function (Glücksmann, 1951, 1965). His classifications were heavily teleological, based on the presumed value to the organism (elimination of vestigial organs, scaffolding or basis for construction of a secondary organ, metamorphic loss of structure, etc.). The value of these reports was that they established the commonness and reproducibility of cell death as a biological activity. Certainly, the implication was that the deaths derived in some manner from the organization of the animal, but he did not specifically argue a physiology of cell death. It was John Saunders who really began the experimental phase. His experiments were very simple but revealing. From his transplantation experiments using the wing of a chick embryo, he concluded that the cells in the designated area were condemned to die but were neither dead nor moribund: hence he noted: "The death clock is ticking" (Saunders, 1966).

At the same time, Richard Lockshin, working in the laboratory of Carroll M. Williams, on the disappearance of muscles in the large American silkmoths, noted the activation of lysosomes (at the time recently discovered) just before the death of the muscles and the dependency of both the death and the activation of lysosomes on the action of hormones (Lockshin and Williams, 1964, 1965a–d). It was evident that the death of the muscles followed a biological plan and they described the plan as a program and the process, programmed cell death. Later, following the lead of Jamshed Tata (1966), who studied pieces of tadpole tail in culture, they observed a need for protein synthesis for the execution of death (Lockshin, 1969), and Jacques Beaulaton and Richard Lockshin described the morphology of dying cells and of those protected from death (Beaulaton and Lockshin, 1977, 1978; Lockshin and Beaulaton, 1974a, b, 1979). The phrase programmed cell death was very much in style, and was readily accepted by the handful of biologists who worked on the subject.

A few years after the establishment of the term programmed cell death, John Kerr, an Australian pathologist working with A.R. Currie, and a postdoctoral fellow named Andrew Wyllie noted that dying cells in tadpole tail, epidermis, thymus, tumors, and other tissues resembled each other: they were rounded, dense, with blebs, and with rounded or fragmented nuclei, and the chromatin was very condensed and pushed against the nuclear membrane. The curiosity was striking for not only did they resemble each other, but also their morphology was difficult to explain. (Kerr, 1971; Kerr et al., 1972; Kerr and Harmon, 1991).

A necrotic cell (to use today's terminology) is easily explicable: without oxygen or energy, the cell ferments. Lactate accumulates in the cell and draws in water by osmotic pressure. Soon the cell explodes. It is much more difficult to explain how a cell shrinks. The shrinkage presumes either a loss of osmoles or an expulsion of water by hydrostatic pressure (Lockshin and Beaulaton, 1981). Later, it was calculated that the force exerted by the cytoskeleton was not enough to expel the water, and John Cidlowski explained the loss of osmolality (Bortner and Cidlowski, 2002). Kerr, Wyllie, and Currie chose the name "apoptosis" for that generalized form of death, thus indicating three things: (1) the form of death was general and common; (2) it suggested a very interesting physiology of death; and (3) death perhaps followed a ritual as well disciplined as birth, i.e., mitosis (Kerr et al., 1972). Still the field remained quiet. What catapulted the field of studying cell death was the recognition of its role and the morphology of apoptosis in cancer, which is the focus of this book. After 36 years, we readily accept that, for homeostasis to function, it is certainly necessary for both birth and death to be regulated.

It is not worth the trouble to insist too much on the distinction between "programmed cell death" and "apoptosis." At first, "programmed cell death" described a process, whereas "apoptosis" described a morphological conformation. The former term was used primarily in development, whereas the latter often referred to pathological situations. The implication of a requirement for synthesis of protein or mRNA for programmed cell death was vehemently argued for the case of apoptosis. Today, however, both terms are used in an essentially interchangeable fashion, and to insist like a scholastic on the purity of the terms no longer makes sense. It is also true that to explain a phenomenon by saying "the loss of cells is by apoptosis" does not say anything more than "the loss of cells is by cell death." As described below, we now understand that apoptosis is perhaps the most efficient means of cell destruction, but there are others; not all deaths are apoptotic, and if apoptosis is blocked, the cell may default to an alternative pathway. Today it is not only acceptable but also fashionable to describe some deaths as nonapoptotic or to use other names to describe cell death.

It is therefore more important to explain how apoptosis occurs and the trigger that launches it. For that story we are indebted to Robert Horvitz and his little worms.

3   The Genetics of Cell Death Reveals the Physiology

The first big step was taken when Brenner, Sulston, and others traced the embryonic descent of each cell in the nematode worm *Caenorhabditis elegans*. Horvitz and colleagues, among others, showed that all the cell deaths in the embryo (13% of the cells, 131 total) were under the control of a handful of genes, which they named ced (cell death defective, after the mutant phenotype). The activity of the *ced* genes was regulated by *ces* (cell death selection) genes. These results were very interesting, but the discovery, shortly thereafter, that one gene encoded a type of restriction protease, a CASPase (cysteinyl protease cleaving at the carboxyl side of an aspartic acid) was earthshaking. First, one now had the first mechanism of death; and second, the gene was conserved from worm to mammal (Horvitz, 2003). Its function was therefore obviously important. This recognition was quickly followed by the identification of substrates and homologous genes in mammals and the realization that mutations of these and other cell death genes were at the origin of different cancers. Thus began the excitement that led by mid-2006 to over 180,000 publications. But the explosion of interest is also due to the realization that cell death is an important component of diseases such as neurodegenerative diseases (Lang-Rollin et al., 2003; Tolkovsky et al., 2004), acquired immune deficiency syndrome (AIDS) (Ameisen and Capron, 1991), cancers (Yonish-Rouach et al., 1993), and immunologic diseases (Golstein et al., 1995a, b; Nagata and Golstein, 1995; Golstein, 1997).

4   The Activation of Cell Death is a Decision by a Cell, but the Decision is made Based on the Type of Cell; Activities by its Neighboring Cells; Nutrients, Kinins, Growth Factors, and Other Components of its Environment; and the Past History of the Cell

We know today the events taking place during the process of apoptosis, and the sequence of the activation of enzymes and the molecular partners that encourage or block apoptosis. These stories are worth telling and constitute a major section of this book. However, they are far from the entire story, because in many situations either the components do not change in amount or only a fraction of supposedly identical cells die in response to challenge. The decision to commit suicide is always very delicate and depends on many factors beyond the machinery itself. B-cell lymphoma can be caused by the transposition and activation of the antiapoptotic gene *bcl-2*, but in the lymphoma the pre-caspases are still present and can be activated by more intense challenge. This is also true for the p53-type cancers: the cancerous cells persist certainly not because of the failure of apoptosis, but because apoptosis is invoked only at a very high threshold of challenge. We have much to learn about the establishment of the threshold. By the same token, in cell culture, when cells are subjected to a modest challenge, the destiny of sister cells is always very variable as a function of time or sensitivity to dose of toxin or other inducer of cell death. As the 19th-century physiologist Claude Bernard remarked: "Life is the result of contact between the organism and the milieu; we cannot comprehend it by the organism itself, no more than by the environment alone." The activation of the death pathway is a type of positive feedback in which the threshold is vigorously defended, but once passed, death progresses without recourse. This threshold may be marked by the activation of caspases or of caspase-3. Thanks to many investigators, we know the partners of the apoptosome and the competition that determines their assembly and activation, but what do we still not know? What is the role of adenosine triphosphate (ATP), guanidine triphosphate (GTP), ceramide, NO, prostaglandin, and other resources of metabolism? How do the more sensitive and less sensitive cells differ? What makes in the same environment one cell more sensitive and the other more resistant? Why is it that much of our effort to block cell death only delays the event and that 100% block of cell death is less attainable?

5 It is Easy to Categorize, but Harder to Live with Intermediates or Ambiguity

We human beings love names and categories. A child who asks five times per minute "What is that?" is generally satisfied if one gives him or her the name of the object. We readily classify everything: "Is it a boy or a girl? Animal or vegetable?" "Are his politics to the left or the right?" We tolerate ambiguity very poorly. Thus, we had apoptosis vs necrosis. Then we found that there are intermediates, and more complex deaths. There have been several efforts to reduce confusion by describing and defining intermediate or alternative forms of cell death (see, e.g., Jaattela, 2004; Sperandio et al., 2004; Kroemer et al., 2005). While these efforts have some value in clarifying concepts, occasionally there is a sense that square pegs are being forced into round holes, and that the focus really needs to be on the process and the physiology. When all is said and done, what we are examining is more often than not a corpse, and this corpse may be the result of concurrent or sequential events, some of which may have been aborted or failed to conclude. One very common error is to suppose that there are only one, two, or three forms of death. But one does not need an instruction manual to die. If caspases are inhibited when cells are exposed to a very strong toxin, the cells will die, but the death demonstrates neither the existence of a second pathway to death nor the absence of a caspase pathway. If a toxin blocks, for example, all sources of energy, or the possibility of protein synthesis, the cell is going to die. For example, there are many reports that an embryo, before the point at which it begins to synthesize its own RNA, resists apoptosis (Hensey and Gautier, 1997, 1999; Negron and Lockshin, 2004). We find that such an embryo, exposed to cycloheximide, does not display a single morphological sign of apoptosis. Nevertheless, it activates caspase 3 at the same time as an older embryo would, though the cells of an older embryo would become apoptotic. The difference is that the cells of the younger embryo burst, necrotic, immediately after the activation of caspase 3, whereas the older cells survive another 90 min and thus have the time to transform themselves into apoptotic cells. It is therefore a weakness (of uncertain origin) of the younger cells so that they cannot reach the stage of apoptosis (Negron and Lockshin, 2004).

What is possible is that the most efficient pathway is the one preferred by the menaced cell and is determined by the nature of the cell, the agent that induces cell death, the environment and history of the cell, and much more. We see cells that consume the bulk of their cytoplasm before dying, in a form of death that is called "autophagic cell death" but we do not know if the autophagy in this situation differs in character from the autophagy seen in a starving cell (a protective autophagy, not necessarily fatal). We also do not know if death by autophagy is not more correctly described as an autophagy that continues without resolution, perhaps terminated by an apoptotic death. In the case of insect metamorphosis, it is possible that an activated autophagic process terminates by apoptosis. For example, the metamorphic death of labial glands or salivary glands (homologous organs in moths and flies, respectively) is well known as an autophagic type II cell death (Zakeri et al., 1993). At the beginning of metamorphosis, there is an activation of lysosomes and an expansion of the autophagic compartment. The bulk of the cytoplasm is eliminated without intervention of phagocytes, and without indication of any sign of apoptosis – no DNA fragmentation, no coalescence or margination of chromatin, no exteriorization of phosphatidylserine, no activation of caspases (even though the cells contain caspase genes). The death is therefore purely manifested by increased activity of autophagy, for instance, the death of a cell of mammary epithelium (Zakeri et al., 1995). But at the end (4th of 5 days for disappearance of the labial gland of *Manduca sexta*; 12th of 13.5 h for disappearance of the labial gland of Drosophila) one sees: cleavage of DNA, exteriorization of phosphatidylserine, coalescence and margination of chromatin, cleavage of caspase substrates. It would seem as if, for bulky, cytoplasm-rich cells, outside of the mitotic cycle, the elimination of cytoplasm is the priority, and this occurs by autophagy. For the moment, we do not know if this autophagy differs from an autophagy provoked by the lack of nourishing substrates in the milieu. In this case, autophagic death would not be death by autophagy, but autophagy activated by an unknown failure of the cell, with apoptosis being activated only when the autophagy had carried the cell beyond the point at which it could survive. For instance, in sympathetic neurons deprived of nerve growth factor, the autophagy and the threatened death of the cell is reversible until the mitochondria have been consumed (Xue et al., 1999, 2001; Zakeri and Lockshin,

2002, 2004; Lockshin and Zakeri, 2004a–c; C.O. B. Facey and R.A. Lockshin, in preparation). This argument would be consistent with observations of others emphasizing the protective role of autophagy: that inactivation of autophagy genes disrupts the formation of dauer larvae in *Caenorhabditis* (Melendez et al., 2003); the neonatal death of mice that lack Atg5, during the time that they must switch from placental nourishment to milk (Kuma et al., 2004); and the importance of autophagy in prolonging the life of cells deprived of growth factors (Boya et al., 2005; Lum et al., 2005). If such cells are not rescued, they die with a morphology typically described as autophagic cell death, with DNA fragmentation occurring very late if at all (Okada and Mak, 2004; Kroemer et al., 2005). One draws from this discussion the following arguments. First, apoptosis or programmed cell death is a very important and well-regulated process; it is not the event once considered to be passive. Second, in an acute situation a temporary protection against cell death, or in which one wishes to kill certain cells, interference with apoptosis promises a good outcome. But in more chronic situations, such as neurodegeneration, diseases such as AIDS or autoimmune disease, blocking apoptosis only allows alternative pathways to be exposed, and at the end of the day the cell will die. What is threatening the cell, and the limit to which it can be pushed before it invokes the death sequence, are questions that still must be resolved. Similarly, when one wishes to activate cell death as in cancer, the cell resources become a major consideration and one must examine very carefully the issue of specific targeting.

6 Where are We Now? Do We Know Where We are Going?

The aim of this chapter is not to review the over 180,000 publications on cell death, but to emphasize that several of our most favored arguments are based on somewhat tenuous ground and that we should not avoid the ambiguities. The value of this book is that the several authors confront these ambiguities and the options that we perceive today. The longer one works as a scientist, the more suspicious one becomes of predictions: there are always surprises, and we all remember instances such as very renowned professors being oblivious to about-to-break ideas such as clonal selection. Thus, it seems inappropriate if not self-destructive to attempt to imagine the next 5 or 10 years. Nevertheless, there are several themes that can be recognized as important for current research:

- The regulation of cell death is an important factor in disease. An important and even determining factor in many cancers is the reluctance of the affected cells to die on schedule, usually by mutation of genes in the cell death pathways; and in other diseases, the pathology is exacerbated or caused by the suicide of cells that appear to be capable of surviving.
- One cannot address these deaths simply by focusing on the direct apoptosis pathway. In many situations such as the p53- and bcl-2-driven cancers, the effectors (initiator and effector caspases) are in place but are not activated at

the appropriate time. In other situations, the affected cell is in dire straits and blocking caspases simply allows the cell to die by other means, or to persist in a nonfunctional zombie-like state, alive but incapable of fulfilling its physiological role.

- It is important therefore to consider apoptosis, or cell death in general, as more of a symptom or result than as a process, and to learn more about what in the environment or the history of the cell initiates the process. After all, in a community, murder and suicide rates are statistics, but they are only statistics revealing an underlying social pathology. Controlling access to guns may have value, but if this move simply diverts the pathology to knives and poison, the issue is not resolved. Similarly, blocking apoptosis, particularly in the acute situation, may provide real benefit, but more often the problem will persist. If, for instance, bystander cells cross the threshold to suicide in AIDS, will more aggressive positive support such as lymphokines help keep them below the threshold? Can one disrupt the Fas–FasL interaction that triggers receptor-mediated death? What effect will such disruption have on the physiology of the organism?

- Similarly, in situations such as neurodegenerative disease, cells are clearly agonizing over extended periods of time before they ultimately fail. Almost certainly, if one blocks the immediate executioner, the cells will still be agonizing and will probably die using other pathways, or remain alive in a weakened and poorly functional state. The question is far more to disrupt the process: to recognize the causes of stress on the cell, and to relieve the stress or support the cell so that it can better resist the stress.

- Activation of apoptosis as an oncolytic intervention will require considerable subtlety. Certainly, specific activation of apoptosis in cancer, especially disseminated cancer, is theoretically very interesting, because it promises to be considerably less toxic than systemic antimetabolites or antimitotics. However, most cancer cells are not dangerous because they have lost the death effector machinery; for one reason or another, they have increased their threshold to activating it. Thus, any efforts to address these cancers will have to reach the malignant cells in a highly specific and targeted manner.

- Large, postmitotic, cytoplasm-rich cells including postweaning mammary epithelium and postcastration prostatic epithelium undergo substantial autophagy prior to dying, in what has been called autophagic cell death. We need to understand whether the autophagy is a death process or an agony, and what the threshold and point of no return are. If the autophagy represents an agony, we need to know why, for instance, insect larval tissue appears to be agonizing during metamorphosis, when the blood is filled with available nutrients. If we can recognize and relieve agony in threatened cells that it is desirable to maintain, we will have accomplished much without directly manipulating the cell death pathway.

- Perhaps the nicest element of the following chapters is that they emphasize process rather than cell death itself. This seems to be the pathway leading to the most important growth in the field, since we still have much to learn.

References

Ameisen, J. C. and Capron, A. (1991). Cell dysfunction and depletion in AIDS: the programmed cell death hypothesis. Immunol Today Dev 12, 102–105.

Beaulaton, J. and Lockshin, R. A. (1977). Ultrastructural study of the normal degeneration of the intersegmental muscles of *Antheraea polyphemus* and *Manduca sexta* (Insecta, Lepidoptera) with particular reference to cellular autophagy. J Morphol Dev 154, 39–58.

Beaulaton, J. and Lockshin, R. A. (1978). Ultrastructural study of neuromuscular relations during degeneration of the intersegmental muscles. Biol Cellulaire Dev 33, 169–174.

Bortner, C. D. and Cidlowski, J. A. (2002). Apoptotic volume decrease and the incredible shrinking cell. Cell Death Differ Dev 9, 1307–1310.

Boya, P., Gonzalez-Polo, R. A., Casares, N., Perfettini, J. L., Dessen, P., Larochette, N., Metivier, D., Meley, D., Souquere, S., Yoshimori, T., Pierron, G., Codogno, P., and Kroemer, G. (2005). Inhibition of macroautophagy triggers apoptosis. Mol Cell Biol Dev 25, 1025–1040.

Clarke, P. G. H. and Clarke, S. (1995). Historic apoptosis. Nat Dev 378, 230.

Clarke, P. G. H. and Clarke, S. (1996). Nineteenth century research on naturally occurring cell death and related phenomena. Anat Embryol Dev 193, 81–99.

Glücksmann, A. (1951). Cell deaths in normal vertebrate ontogeny. Biol. Rev. Cambridge Phil Soc Dev 26, 59–86.

Glücksmann, A. (1965). Cell death in normal development. Arch Biol (Liege) Dev 76, 419–437.

Golstein, P. (1997). Controlling cell death [comment]. Sci Dev 275, 1081–1082.

Golstein, P., Marguet, D., and Depraetere, V. (1995a). Fas bridging cell death and cytotoxicity: the reaper connection. Immunol Rev Dev 146, 45–56.

Golstein, P., Marguet, D., and Depraetere, V. (1995b). Homology between reaper and the cell death domains of Fas and TNFR1. Cell Dev 81, 185–186.

Häcker, G. and Vaux, D. L. (1997). A chronology of cell death. Apoptosis 2, 247–256.

Hensey, C. and Gautier, J. (1997). A developmental timer that regulates apoptosis at the onset of gastrulation. Mech Dev 69, 183–195.

Hensey, C. and Gautier, J. (1999). Developmental regulation of induced and programmed cell death in *Xenopus* embryos. In: Mechanisms of Cell Death, eds. Z. Zakeri, R. A. Lockshin, and L. Benitez-Bribiesca, New York Academy of Sciences, New York City, pp. 105–119.

Horvitz, H. R. (2003). Nobel lecture. Worms, life and death. Biosci Rep Dev 23, 239–303.

Janet, C. (1907). Anatomie du corselet et histolyse des muscles vibrateurs après le vol nuptial, chez la reine de la fourmi (*Lasius niger*). DuCourtieux et Gout, Limoges, pp. 1–150.

Jaattela, M. (2004). Multiple cell death pathways as regulators of tumour initiation and progression. Oncogene Dev 23, 2746–2756.

Kerr, J. F. R. (1971). Shrinkage necrosis: a distinct mode of cellular death. J Pathol Dev 105, 13–20.

Kerr, J. F. R. and Harmon, B. V. (1991). Definition and incidence of apoptosis: an historical perspective. In: Apopotosis: The Molecular Biology of Cell Death, eds. L. D. Tomei and F.O. Cope. Cold Spring Harbor Press, Cold Spring Harbor, NY, pp. 5–30.

Kerr, J. F. R., Wyllie, A. H., and Currie, A. R. (1972). Apoptosis: a basic biological phenomenon with wide-ranging implications in tissue kinetics. Br J Cancer Dev 26, 239–257.

Kroemer, G., El-Deiry, W. S., Golstein, P., Peter, M. E., Vaux, D., Vandenabeele, P., Zhivotovsky, B., Blagosklonny, M. V., Malorni, W., Knight, R. A., Piacentini, M., Nagata, S., and Melino, G. (2005). Classification of cell death: recommendations of the nomenclature committee on cell death. Cell Death Differ Dev 12 (Suppl 2), 1463–1467.

Kuma, A., Hatano, M., Matsui, M., Yamamoto, A., Nakaya, H., Yoshimori, T., Ohsumi, Y., Tokuhisa, T., and Mizushima, N. (2004). The role of autophagy during the early neonatal starvation period. Nat Dev 432, 1032–1036.

Lang-Rollin, I. C., Rideout, H. J., Noticewala, M., and Stefanis, L. (2003). Mechanisms of caspase-independent neuronal death: energy depletion and free radical generation. J Neurosci Dev 23, 11015–11025.

Lockshin, R. A. (1969). Programmed cell death. Activation of lysis of a mechanism involving the synthesis of protein. J Insect Physiol Dev 15, 1505–1516.

Lockshin, R. A. and Beaulaton, J. (1974a). Programmed cell death. Cytochemical evidence for lysosomes during the normal breakdown of the intersegmental muscles. J Ultrastruct Res Dev 46, 43–62.

Lockshin, R. A. and Beaulaton, J. (1974b). Programmed cell death. Life Sci Dev 15, 1549–1565.

Lockshin, R. A. and Beaulaton, J. (1979). Cytological studies of dying muscle fibers of known physiological parameters. Tissue Cell Dev 11, 803–819.

Lockshin, R. A. and Beaulaton, J. (1981). Cell death: questions for histochemists concerning the causes of the various cytological changes. Histochem J Dev 13, 659–666.

Lockshin, R. A. and Williams, C. M. (1964). Programmed cell death. II. Endocrine potentiation of the breakdown of the intersegmental muscles of silkmoths. J Insect Physiol Dev 10, 643–649.

Lockshin, R. A. and Williams, C. M. (1965a). Programmed cell death. III. Neural control of the breakdown of the intersegmental muscles. J Insect Physiol Dev 11, 605–610.

Lockshin, R. A. and Williams, C. M. (1965b). Programmed cell death. I. Cytology of the degeneration of the intersegmental muscles of the pernyi silkmoth. J Insect Physiol Dev 11, 123–133.

Lockshin, R. A. and Williams, C. M. (1965c). Programmed cell death. IV. The influence of drugs on the breakdown of the intersegmental muscles of silkmoths. J Insect Physiol Dev 11, 803–809.

Lockshin, R. A. and Williams, C. M. (1965d). Programmed cell death. V. Cytolytic enzymes in relation to the breakdown of the intersegmental muscles of silkmoths. J Insect Physiol Dev 11, 831–844.

Lockshin, R. A. and Zakeri, Z. (2001). Programmed cell death and apoptosis: origins of the theory. Nat Rev Mol Cell Biol Dev 2, 545–550.

Lockshin, R. A. and Zakeri, Z. (2004a). When Cells Die II. Wiley-Liss, New York.

Lockshin, R. A. and Zakeri, Z. (2004b). Apoptosis, autophagy, and more. Int J Biochem Cell Biol Dev 36, 2405–2419.

Lockshin, R. A. and Zakeri, Z. (2004c). Caspase-independent cell death? Oncogene Dev 23, 2766–2773.

Lum, J. J., Bauer, D. E., Kong, M., Harris, M. H., Li, C., Lindsten, T., and Thompson, C. B. (2005). Growth factor regulation of autophagy and cell survival in the absence of apoptosis. Cell Dev 120, 237–248.

Melendez, A., Talloczy, Z., Seaman, M., Eskelinen, E. L., Hall, D. H., and Levine, B. (2003). Autophagy genes are essential for dauer development and life-span extension in C. elegans. Sci Dev 301, 1387–1391.

Nagata, S. and Golstein, P. (1995). The fas death factor. Sci Dev 267, 1449–1456.

Negron, J. F. and Lockshin, R. A. (2004). Activation of apoptosis and caspase-3 in zebrafish early gastrulae. Dev Dyn Dev 231, 161–170.

Okada, H. and Mak, T. W. (2004). Pathways of apoptotic and non-apoptotic death in tumour cells. Nat Rev Cancer Dev 4, 592–603.

Pérez, C. (1910). Recherches histologiques sur la métamorphose des muscides (Calliphora erythrocephala Mg). Arch Zool Expér Gén 5e Série Dev 4, 1–274.

Saunders, J. W., Jr. (1966). Death in embryonic systems. Sci Dev 154, 604–612.

Sperandio, S., Poksay, K., de, B., Lafuente, I. M. J., Liu, B., Nasir, J., and Bredesen, D. E. (2004). Paraptosis: mediation by MAP kinases and inhibition by AIP-1/Alix. Cell Death Differ Dev 11, 1066–1075.

Tata, J. R. (1966). Requirement for RNA and protein synthesis for induced regression of tadpole tail in organ culture. Dev Biol Dev 13, 77–94.

Terre, L. (1889). Contribution á l'étude de l'histolyse et de l'histogénèse du tissu musculaire chez l'abeille. C R Soc Biol (IIe Série) Dev 51, 896–898.

Tolkovsky, A. M., Bampton, E. T. W., and Goemans, C. G. (2004). Cell death in neuronal development and maintenance. In: When Cells Die II, eds. Lockshin, R. A. and Zakeri, Z. Wiley-Liss, New York, pp. 175–200.

Wigglesworth, V. B. (1972). The Principles of Insect Physiology. Chapman & Hall, London.

Xue, L., Fletcher, G. C., and Tolkovsky, A. M. (1999). Autophagy is activated by apoptotic signalling in sympathetic neurons: an alternative mechanism of death execution. Mol Cell Neurosci Dev 14, 180–198.

Xue, L., Fletcher, G. C., and Tolkovsky, A. M. (2001). Mitochondria are selectively eliminated from eukaryotic cells after blockade of caspases during apoptosis. Curr Biol Dev 11, 361–365.

Yonish-Rouach, E., Grunwald, D., Wilder, S., Kimchi, A., May, E., Lawrence, J.-J., May, P., and Oren, M. (1993). p53-Mediated cell death: relationship to cell cycle control. Mol Cell Biol Dev 13, 1415–1423.

Zakeri, Z. and Lockshin, R. A. (2002). Cell death during development. J Immunol Methods Dev 265, 3–20.

Zakeri, Z. and Lockshin, R. A. (2004). Cell death: shaping an embryo. In: When Cells Die II, eds. Lockshin, R. A. and Zakeri, Z. Wiley-Liss, New York, pp. 27–58.

Zakeri, Z., Bursch, W., Tenniswood, M., and Lockshin, R. A. (1995). Cell death. Programmed, apoptosis, necrosis, or other. Cell Death Differ Dev 2, 87–96.

Zakeri, Z. F., Quaglino, D., Latham, T., and Lockshin, R. A. (1993). Delayed internucleosomal DNA fragmentation in programmed cell death. FASEB J Dev 7, 470–478.

Chapter 2
Caspase Mechanisms

Guy S. Salvesen* and Stefan J. Riedl

Abstract The main effectors of apoptosis encompass proteases from the caspase family, which reside as latent precursors in most nucleated animal cells. The apoptotic caspases constitute a minimal two-step signaling pathway. The apical (initiator) caspases are activated within oligomeric signaling complexes in response to apoptotic stimuli. Their mechanism of activation probably results from proximity-induced clustering to the dimeric active forms. Once activated, the apical caspases directly activate the executioner (effector) caspases by limited proteolytic cleavage. The distinct activation mechanisms explain how an apoptotic stimulus is converted to proteolytic activity, and how this activity is amplified to allow for limited proteolysis of the dozens of protein substrates whose cleavage is required for efficient apoptosis.

Keywords apoptosome, caspase, DISC, IAP, inhibition, protease, zymogen

1 Apoptosis and Limited Proteolysis

Apoptosis is a mechanism to regulate cell number, and is vital throughout the life of all metazoan animals. Although several different types of biochemical events have been recognized as important in apoptosis, perhaps the most fundamental is the participation of the caspases[1–3] – a family of proteases found in multicellular animals. The name caspase comes from cysteine-dependent aspartate-specific protease,[4] thus their enzymatic properties are governed by a stringent specificity for protein substrates containing Asp, and by the use of a Cys side chain for catalyzing peptide bond cleavage. The use of a Cys side chain as a nucleophile during peptide bond hydrolysis is common to several protease families. However, the primary

Guy S. Salvesen and Stefan J. Riedl
Program on Apoptosis and Cell Death, Burnham Institute for Medical Research,
10901 N. Torrey Pines Road, La Jolla, CA 92037, USA

*To whom correspondence should be addressed: Tel.: 858.646.3114; fax: 858.646.3197;
e-mail: gsalvesen@burnham.org

R. Khosravi-Far and E. White (eds.), *Programmed Cell Death in Cancer Progression and Therapy*.
© Springer 2008

13

specificity for Asp turns out to be very rare among proteases throughout biotic kingdoms. Of all known mammalian proteases only the caspase activator granzyme B, a serine protease, has the same primary specificity.[5, 6] Caspases cleave a number of cellular proteins,[7, 8] and the process is one of limited proteolysis where a small number of cuts, usually only one, are made. Sometimes cleavage results in activation of the protein, sometimes in inactivation, but presumably not in degradation since their substrate specificity distinguishes the caspases as among the most restricted of endopeptidases. This is an important distinction from the other cytosolic proteases such as the proteasome, which permits signaling by wholesale destruction of regulatory proteins such as IκB in NFκB signaling and PDS1 in anaphase promotion,[9] or calpains, whose specificity has pretty much defied analysis in vivo so far.[10]

2   Caspase Signaling

A consensus view of caspases places them in two main groups. First are the cytokine activators (inflammatory caspases) related to caspase-1, probably including mouse caspase-11 and its orthologs caspase-4 and caspase-5 in humans. Their role is to respond to infection by rapidly converting active cytokines (IL-1β, IL-18) from intracellular stores. Confirmation of the important roles of the caspases in the inflammatory cytokine response comes from gene ablation experiments in mice. Animals ablated in caspase-1 or caspase-11 are deficient in cytokine processing,[11, 12] but without any overt apoptotic phenotype. The second group constitutes the apoptotic caspases that transduce and execute death signals. The phenotypes of these knockouts are very gross, usually antiapoptotic, and vary from early embryonic lethality to perinatal lethality to relatively mild with defects in the process of normal oocyte ablation.[13, 14] Researchers in the area have placed the apoptotic caspases in two converging pathways, such that some are activated by others (Fig. 2.1). The core pathways probably represent the minimal apoptotic program, and certainly its simplicity is complicated by cell-specific additions that help to fine-tune individual cell fates. Nevertheless, the basic order and at least some of the essential functions and, importantly, the catalytic and activation mechanisms are known.

2.1   *Caspase Activation*

In common with most proteolytic enzymes, caspases reside as latent forms (zymogens) that are usually activated by limited proteolysis. It is relatively easy to imagine that the caspases operating at the bottom of the pathway are activated by ones above. But the question of how the first caspase in a pathway became activated, how the first death signal was generated, was initially perplexing.

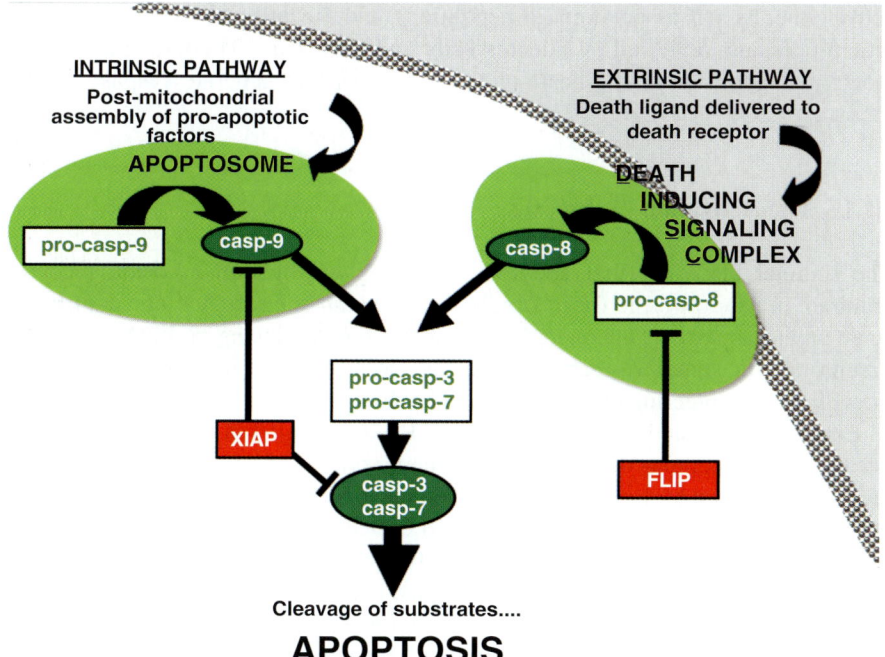

Fig. 2.1 The framework of apoptosis. Death may be signaled through ligand enforced clustering of receptors at the cell surface via the extrinsic pathway, which leads to the activation of apical caspase-8.[54] This caspase then directly activates the executioner caspase-3 and caspase-7 (and possibly 6), which are primarily responsible for the limited proteolysis that defines apoptotic dismantling of the cell. Irreparable damage to the genome caused by mutagens, pharmaceuticals that inhibit DNA repair, or ionizing radiation – transmitted by a mechanism thought to involve the release of cytochrome C from mitochondria via the intrinsic pathway – engages the same executioner caspases.[55] The latter events progress through the apical caspase-9 and its cofactor Apaf-1.[26] Activation of the extrinsic pathway is regulated by FLIP, which modulates the recruitment of caspase-8 to its adapters.[56] The execution phase is regulated through direct caspase inhibition by XIAP, which can also regulate the active form of caspase-9. In turn, the IAPs are under the influence of antagonist proteins that compete with caspases for IAPs.[44] Though other modulators may regulate the apoptotic pathway in a cell-specific manner, this framework is considered common to most mammalian cells

How exactly does a recruited zymogen become active? To understand this one must understand the unusual properties of caspase zymogens that set them apart from most other proteases. For, unlike most other proteases, simple ectopic expression of caspase zymogens in *Escherichia coli* usually results in their autolytic cleavage by limited proteolysis within a "linker segment" that separated the large (~20kDa) and small (~10kDa) subunits of the catalytic domain.[15, 16] This processing is a consequence of intrinsic proteolytic activity residing in the caspase zymogens. It is not due to *E. coli* proteases since catalytically disabled caspase mutants fail to undergo processing. In vitro, apical caspase zymogens can be induced to become active either by self-association (dimerization), and executioner caspases by proteo-lytic processing, reviewed by Fuentes-Prior and Salvesen.[17] These distinct require-ments activation are at the heart of the processes that generate caspase activity in vivo.

2.1.1   The Activation Complexes

The seminal discovery that apoptotic signaling via ligation of death receptors required, in its most basic form, simply a transmembrane receptor, an adapter mol-ecule and a caspase[18, 19] uncovered a solution to the perplexing problem of how the first proteolytic signal was generated during apoptosis, since it implicated a caspase directly in the triggering event, as outlined below.

Extracellular ligands such as FasL and TRAIL that bind in a conventional man-ner to the extracellular domains of transmembrane receptors trigger the extrinsic pathway, reviewed by LeBlanc and Ashkenazi.[20] The death signal is transmitted to the cytosol by receptor clustering followed by recruitment of the apical caspase-8 (Fig. 2.1). The caspase-8 paralog, caspase-10, is also an initiator in death-receptor-mediated cell death, at least in humans (mice apparently lack a caspase-10 gene), although there is controversy in the literature regarding the ability of caspase-10 to functionally substitute for caspase-8 in death receptor signaling.[21] Structural infor-mation on the conformation adopted by the receptors in this complex is very sketchy, but recent data on the adaptor protein FADD,[22] and homologues of cas-pase-8[23, 24] suggest that activation of caspase-8 occurs at the cytosolic face of the cell membrane by an induced proximity mechanism (see below). The exact process of ligand binding and receptor oligomerization may require receptor internalization in addition to clustering,[25] but it is clear that the death-inducing signaling complex (DISC) represents a common example of a typical ligand-dependent transmem-brane signaling receptor.

The receptor of the intrinsic pathway – the apoptosome – in contrast is not a typical transmembrane signaling receptor.[26] In the apoptosome, the cytosolic protein Apaf-1[27] senses the release of ligand, cytochrome C, which, upon binding to Apaf-1, triggers its oligomerization. As a "soluble" receptor, Apaf-1 lacks the two-dimensional arrange-ments of transmembrane domains spanning the cell membrane, and uses another mechanism to generate a two-dimensional surface, or platform, for signaling. This process has been reviewed,[28] and involves a mechanism-based oligomerization that uses the specialized AAA+ domain of Apaf-1 to generate a ring with sevenfold sym-metry for the recruitment of caspase-9, the apical caspase of the intrinsic pathway.

2.1.2   Apical Caspases: Induced Proximity

Having formed a seven-membered recruitment platform, the apoptosome must now activate pro-caspase-9. In common with other caspases, caspase-9 is a dimer in its active form.[29] However, pro-caspase-9 exists in cells at a concentration of ~20 nM,[30]

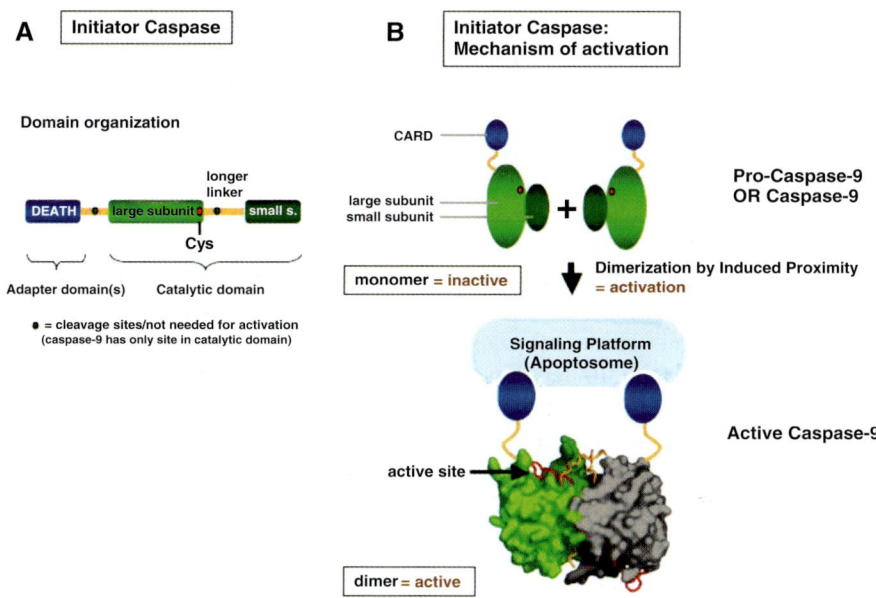

Fig. 2.2 Initiator caspases: architecture and activation. (A) Initiator caspases are expressed as single chains, comprising one or two adaptor domain(s) belonging to the DEATH domain family at the N-terminus followed by a catalytic domain, which can be divided into a large and small subunit and a relatively long loop region between the subunits. Although they can be cleaved (as revealed, for example, in their crystal structures) initiator caspases, such as caspase-9, show full activity in their uncleaved forms, which could be due to the long linker loops between subunits. (B) Their activity is regulated by dimerization, instead of by cleavage. Initiator caspases exist as inactive monomers (*top*). Binding to an oligomeric platform, such as the apoptosome in the case of caspase-9, occurs via adaptor domains (such as CARD, caspase recruitment domain) and results in an induced proximity of the catalytic domains of initiator caspases. Recent results suggest that this leads to dimerization, which allows for the formation of a productive active site as shown here in the structure of cleaved, dimeric caspase-9 (*bottom*). Interestingly, only one of the two sites adopts the active form in the crystal structure of caspase-9. PDB entry: caspase-9, 1JXQ

and the K_d for dimerization in buffers in the physiologic range is more than $50\,\mu M$ in vitro.[31] The zymogen therefore must exist as a monomer under normal physiologic conditions in vivo (Fig. 2.2). Following formation of the apoptosome, and uncovering of the caspase-9-binding site on Apaf-1, the zymogen can associate with the activator complex.

Two hypotheses have been put forward for the activation of caspase-9. The first was an "allosteric model" that postulated the activation of a monomer directly by the apoptosome.[32] The second, the "induced proximity model," postulated that the apoptosome provides a platform for caspase-9 dimerization.[33] The contrasting hypotheses have been reviewed extensively, including recent revisions of the models.[34, 35]

Significantly, unlike the executioner caspases 3 and 7, pro-caspase-9 does not need to be cleaved in the linker region to become active[30, 36] (Fig. 2.2A). Not only is cleavage unnecessary, but also it is insufficient to produce an active enzyme. Instead, caspase-9 is activated by small-scale rearrangements of surface loops that define the substrate cleft and catalytic residues.[29] In the induced proximity model, this is achieved by dimerization of caspase-9 monomers within the apoptosome,[37] with the dimer interface providing surfaces compatible with catalytic organization of the active site (Fig. 2.2B). Consistent with the induced proximity model is the finding that activation of caspase-9 by the apoptosome has bimolecular kinetics, and that a hybrid containing the catalytic domain of caspase-8 tagged onto the recruitment domain of caspase-9 is also activated by the apoptosome.[31] It is tempting to speculate that a similar dimerization mechanism activates the caspase-8 zymogen to trigger the extrinsic pathway, especially since clustering of adaptors is critical for caspase-8 activation.[22–24] Finally, we note that the induced proximity activation of apical caspases by dimerization may also explain the requirement for apoptosome-like structures, known as inflammasomes, to activate inflammatory caspase zymogens, reviewed in Refs.[28, 38]

2.1.3 Executioner Caspases: Activation by Cleavage

Once an apical caspase has become active ensuing activation of the executioners is more easily explained. At their cytosolic concentration in human cells, the caspase-3 and caspase-7 zymogens are already dimers, but they are not active (Fig. 2.3). Cleavage within their respective linker segments is required for activation.[39, 40] Caspase-6 is not as widely studied as caspase-3 and caspase-7, but is classified as an executioner caspase based on its lack of a long pro-domain and its cleavage downstream of the initiators. The crystal structures of zymogen caspase-7, active caspase-7, and inhibitor-bound caspase-7 serve as models with which to rationalize the apparent conflict between the cleavage mechanism for executioner caspase activation and the dimerization mechanism for apical caspase activation.[39, 40, 57] When cleaved and uncleaved caspase-7 structures are compared, a similar reordering of catalytic and substrate binding residues occurs as seen in caspase-9, so the fundamental mechanism of zymogen activation is equivalent (Fig. 2.3B). Only the driving forces are distinct. Most importantly, the linker segment of pro-caspase-7 blocks ordering of the catalytic residues, and requires cleavage to allow a productive active site. The new N- and C-terminal sequences so generated aid in active site stabilization. The property that allows the very different driving forces of dimerization and cleavage to converge on the same activation mechanism seems to be the unusual mobility of the residues that together constitute the caspase active site, which are mainly placed on flexible loops and not ordered secondary structure.

Progress in understanding caspase structures and mechanisms now allows us to answer the question of why the executioner caspase zymogens are dimeric whereas the apical caspase zymogens are monomeric at physiologic concentrations.

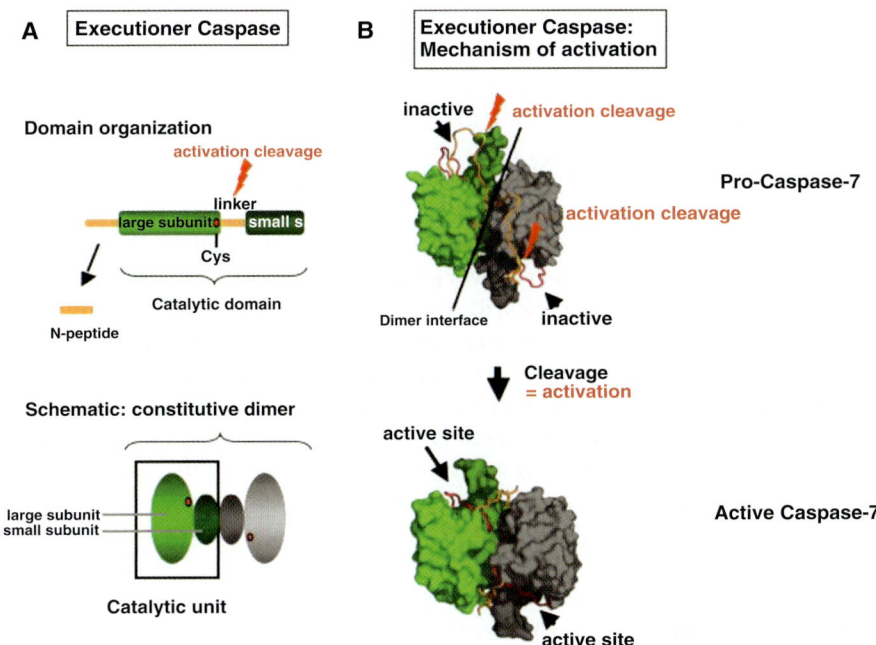

Fig. 2.3 Executioner caspases: architecture and activation. (A) Activation of executioner caspases. Caspases are initially expressed as single-chain proteins that undergo an activation cleavage. An executioner caspase is typically cleaved twice, ultimately leading to the release of a short N-terminal peptide. The actual activation cleavage divides the catalytic unit into a large and small subunit. The position of the active-site cysteine residue is indicated in red. Bottom: schematic illustrating that executioner caspases are constitutive dimers of two catalytic units. (B) Surface rendering of an executioner caspase (caspase-7) preactivation and postactivation cleavage. The same color code as in panel A is used and important loop regions are displayed as ribbons. Cleavage releases strains on surface loops (red and orange) and the chains rearrange. The newly formed termini of large and small subunit (orange) interact with each other across the other catalytic unit, and with the red loops to nicely align the substrate-binding pockets at the bottom of the active-site cleft. This results in a highly active caspase (*bottom*). PDB entries: caspase-7, 1F1J; pro-caspase-7, 1GQF

Much of the reason for this lies the relatively weak hydrophobic character of the dimer interface in caspase-8 and caspase-9, strongly contrasting with a more hydrophobic nature of the dimer interface in caspase-3 and caspase-7. Specifically, the K_d for caspase-3 dimerization is less than 50 nM,[41] which is more than three orders of magnitude tighter than that for caspase-8 ~~50 μM).[42] Interestingly, caspase-3, like apical caspases, can also be activated by experimentally induced proximity employing hybrids that possess engineered dimerization domains (see, e.g., Mallet et al.[43]). In this context, the dimerization domains, which must be introduced as multiple tandem copies, likely recruit high local concentrations of preformed pro-caspase-3 dimers leading to proteolytic activation *in trans* of the type seen in high-level expression in *E. coli*.

2.2   IAPs: Caspase Inhibitors

The best-characterized endogenous caspase inhibitor is the X-linked IAP (XIAP), a member of the IAP family, also known as BIRC proteins.[44] The family is broadly distributed and, as the name implies, the founding members are capable of blocking apoptosis, having initially been identified in baculoviruses, reviewed by Verhagen et al.[45] Eight distinct IAPs have been identified in humans, and despite initial reports, it seems now that only XIAP is capable of directly inhibiting caspases,[46] having been found by multiple research groups to be a potent but restricted inhibitor targeting caspases 3, 7, and 9, reviewed by Salvesen and Duckett[44] and Deveraux and Reed.[47] IAPs have functions in addition to caspase inhibition because they have been found in organisms such as yeast which neither contain caspases nor undergo apoptosis.[48]

XIAP contains three baculovirus IAP repeat (BIR) domains, which represent the defining characteristic of the family. Currently, there is no known function for BIR1, but the second BIR domain (BIR2) of XIAP specifically target caspases 3 and 7 ($K_i \approx 0.1$–1 nM), and the third BIR domain (BIR3) specifically target caspase 9 ($K_i \approx 10$ nM). This led to the general assumption that the BIR domain itself was important for caspase inhibition. Surprisingly, the structures of BIR2 in complex with caspase-3 and caspase-7 have revealed that the BIR domain has a secondary role in the inhibitory mechanism, and that the main inhibitory contacts are made by the flexible region preceding the BIR domain.[49–51] Interestingly, the mechanism of inhibition of caspase-9 by the BIR3 domain requires cleavage in the inter subunit linker to generate the new sequence NH_2-ATPF.[36] In part, this explains the cleavage of caspase-9 during apoptosis, which as described above is not required for its activation. Paradoxically, it is required for its inactivation by XIAP. Another surprise was in store for researchers when the structure of the BIR3–caspase-9 complex was solved.[52] Here, there was no interaction of the BIR3 domain with the active site, but instead the BIR was found associated with the dimer interface of caspase-9. Essentially, BIR3 had monomerized the caspase thus reversing the activation mechanism.

Together, the structures of BIR2 bound to the executioner caspase-3 and caspase-7, and BIR3 bound to the apical caspase-9 complete our understanding of caspase activation, at least at the structural level. Each domain has found a solution matched to the special properties of their targets. BIR2 binds to active caspase-3 and caspase-7 in their dimeric active forms, with a very specific and somewhat unusual geometry blocking the catalytic site. BIR3 subverts the dimer/monomer transition of caspase-9, and is thus totally selective for this protease. The BIR domains of XIAP represent extraordinary mechanisms that are unique in the field of protease inhibitors, achieving tight binding and stringent specificity.

3 Conclusions

Stemming from the original observation that executioner caspases are activated by proteolysis, but that apical caspases are not, has come from the current understanding of the controls placed on caspase activity. So far as we know, caspase zymogens reside mainly as soluble cytosolic proteins. Upon ligation of death receptors, or formation of the apoptosome, apical caspase zymogen monomers are recruited to their cognate activation complexes where they are activated, most likely by proximity-induced dimerization. Thus, the first protease in each pathway gains catalytic activity. Following this, the zymogens of executioner caspases are activated by a direct proteolytic attack of the apical caspases. The executioners now orchestrate the demise of the cell by cleaving a large number of cellular proteins. There is no appropriate evidence to suggest why two steps are required for apoptosis. For example, in *C. elegans* a single apoptotic caspase, CED3, seems to be able to orchestrate apoptosis on its own. Possibly, advanced animals incorporated the executioner caspases as a mechanism to provide additional regulation, or to allow the apical apoptotic caspases to function in additional, non-death, roles. Indeed, the proposed non-death roles of apoptotic caspases, reviewed by Lamkanfi et al.,[53] provides a fertile field of investigation now that the fundamental mechanisms of caspase activation have been elucidated.

References

1. Salvesen, G. S. and Dixit, V. M. (1997). Caspases: intracellular signaling by proteolysis. Cell 91, 443–446.
2. Cohen, G. M. (1997). Caspases: the executioners of apoptosis. Biochem J. 326, 1–16.
3. Thornberry, N. A. and Lazebnik, Y. (1998). Caspases: enemies within. Science 281, 1312–1316.
4. Alnemri, E. S. et al. (1996). Human ICE/CED-3 protease nomenclature. Cell 87, 171.
5. Odake, S. et al. (1991). Human and murine cytotoxic T lymphocyte serine proteases: subsite mapping with peptide thioester substrates and inhibition of enzyme activity and cytolysis by isocoumarins. Biochemistry 30, 2217–2227.
6. Harris, J. L. et al. (2000). Rapid and general profiling of protease specificity by using combinatorial fluorogenic substrate libraries. Proc Natl Acad Sci USA 97, 7754–7759.
7. Nicholson, D. W. (1999). Caspase structure, proteolytic substrates, and function during apoptotic cell death. Cell Death Differ 6, 1028–1042.
8. Timmer, J. C. and Salvesen, G. S. (2007). Caspase substrates. Cell Death Differ 14, 66–72.
9. Gutierrez, G. J. and Ronai, Z. (2006). Ubiquitin and SUMO systems in the regulation of mitotic checkpoints. Trends Biochem Sci 31, 324–332.
10. Suzuki, K., Hata, S., Kawabata, Y., and Sorimachi, H. (2004). Structure, activation, and biology of calpain. Diabetes 53 (Suppl 1), S12– S18.
11. Kuida, K. et al. (1995). Altered cytokine export and apoptosis in mice deficient in interleukin-1-beta converting enzyme. Science 267, 2000–2003.
12. Wang, S., Miura, M., Jung, Y.-K., Zhu, H., and Yuan, J. (1998). Murine caspase-11, an ICE-interacting protease, is essential for the activation of ICE. Cell 92, 501–509.
13. Zheng, T. S., Hunot, S., Kuida, K., and Flavell, R. A. (1999). Caspase knockouts: matters of life and death. Cell Death Differ 6, 1043–1053.

14. Oppenheim, R. W. et al. (2001). Programmed cell death of developing mammalian neurons after genetic deletion of caspases. J Neurosci 21, 4752–4760.

15. Orth, K., O'Rourke, K., Salvesen, G. S., and Dixit, V. M. (1996). Molecular ordering of apoptotic mammalian CED-3/ICE-like proteases. J Biol Chem 271, 20977–20980.

16. Stennicke, H. R. and Salvesen, G. S. (1997). Biochemical characteristics of caspases-3, -6, -7, and -8. J Biol Chem 272, 25719–25723.

17. Fuentes-Prior, P. and Salvesen, G. S. (2004). The protein structures that shape caspase activity, specificity, activation and inhibition. Biochem J 384, 201–232.

18. Boldin, M. P., Goncharov, T. M., Goltsev, Y. V., and Wallach, D. (1996). Involvement of MACH, a novel MORT1/FADD-interacting protease, in Fas/APO-1- and TNF receptor-induced cell death. Cell 85, 803–815.

19. Muzio, M., Stockwell, B. R., Stennicke, H. R., Salvesen, G. S., and Dixit, V. M. (1998). An induced proximity model for caspase-8 activation. J Biol Chem 273, 2926–2930.

20. LeBlanc, H. N. and Ashkenazi, A. (2003). Apo2L/TRAIL and its death and decoy receptors. Cell Death Differ 10, 66–75.

21. Kischkel, F. C. et al. (2001). Death receptor recruitment of endogenous caspase-10 and apoptosis initiation in the absence of caspase-8. J Biol Chem 276, 46639–46646.

22. Carrington, P. E. et al. (2006). The structure of FADD and its mode of interaction with pro-caspase-8. Mol Cell 22, 599–610.

23. Yang, J. K. et al. (2005). Crystal Structure of MC159 reveals molecular mechanism of DISC assembly and FLIP inhibition. Mol Cell 20, 939–949.

24. Li, F. Y., Jeffrey, P. D., Yu, J. W., and Shi, Y. (2006). Crystal structure of a viral FLIP: insights into FLIP-mediated inhibition of death receptor signaling. J Biol Chem 281, 2960–2968.

25. Lee, K. H. et al. (2006). The role of receptor internalization in CD95 signaling. EMBO J 25, 1009–1023.

26. Li, P. et al. (1997). Cytochrome c and dATP-dependent formation of Apaf-1/Caspase-9 complex initiates an apoptotic protease cascade. Cell 91, 479–489.

27. Zou, H., Henzel, W. J., Liu, X., Lutschg, A., and Wang, X. (1997). Apaf-1, a human protein homologous to C. elegans CED-4, participates in cytochrome c-dependent activation of caspase-3. Cell 90, 405–413.

28. Riedl, S. J. and Salvesen, G. S. (2007). The apoptosome: signalling platform of cell death. Nat Rev Mol Cell Biol 8, 405–413.

29. Renatus, M., Stennicke, H. R., Scott, F. L., Liddington, R. C., and Salvesen, G. S. (2001). Dimer formation drives the activation of the cell death protease caspase 9. Proc Natl Acad Sci USA 98, 14250–14255.

30. Stennicke, H. R. et al. (1999). Caspase-9 can be activated without proteolytic processing. J Biol Chem 274, 8359–8362.

31. Pop, C., Timmer, J., Sperandio, S., and Salvesen, G. S. (2006). The apoptosome activates caspase-9 by dimerization. Mol Cell 22, 269–275.

32. Rodriguez, J. and Lazebnik, Y. (1999). Caspase-9 and APAF-1 form an active holoenzyme. Genes Dev 13, 3179–3184.

33. Boatright, K. M. et al. (2003). A unified model for apical caspase activation. Mol Cell 11, 529–541.

34. Boatright, K. M. and Salvesen, G. S. (2003). Mechanisms of caspase activation. Curr Opin Cell Biol 15, 725–731.

35. Shi, Y. (2004). Caspase activation: revisiting the induced proximity model. Cell 117, 855–858.

36. Srinivasula, S. M. et al. (2001). A conserved XIAP-interaction motif in caspase-9 and Smac/DIABLO regulates caspase activity and apoptosis. Nature 410, 112–116.

37. Acehan, D. et al. (2002). Three-dimensional structure of the apoptosome: implications for assembly, procaspase-9 binding and activation. Mol Cell 9, 423–432.

38. Martinon, F. and Tschopp, J. (2004). Inflammatory caspases: linking an intracellular innate immune system to autoinflammatory diseases. Cell 117, 561–574.

39. Riedl, S. J. et al. (2001). Structural basis for the activation of human procaspase-7. Proc Natl Acad Sci USA 98, 14790–14795.
40. Chai, J. et al. (2001). Crystal structure of a procaspase-7 zymogen. Mechanisms of activation and substrate binding. Cell 107, 399–407.
41. Bose, K. and Clark, A. C. (2001). Dimeric procaspase-3 unfolds via a four-state equilibrium process. Biochemistry 40, 14236–14242.
42. Donepudi, M., Mac Sweeney, A., Briand, C., and Gruetter, M. G. (2003). Insights into the regulatory mechanism for caspase-8 activation. Mol Cell 11, 543–549.
43. Mallet, V. O. et al. (2002). Conditional cell ablation by tight control of caspase-3 dimerization in transgenic mice. Nat Biotechnol. 20, 1234–1239.
44. Salvesen, G. S. and Duckett, C. S. (2002). IAP proteins: blocking the road to death's door. Nat Rev Mol Cell Biol 3, 401–410.
45. Verhagen, A. M., Coulson, E. J., and Vaux, D. L. (2001). Inhibitor of apoptosis proteins and their relatives: IAPs and other BIRPs. Genome Biol 2, REVIEWS3009.
46. Eckelman, B. P., Salvesen, G. S., and Scott, F. L. (2006). Human inhibitor of apoptosis proteins: why XIAP is the black sheep of the family. EMBO Rep 7, 988–994.
47. Deveraux, Q. L. and Reed, J. C. (1999). IAP family proteins – suppressors of apoptosis. Genes Dev 13, 239–252.
48. Uren, A. G., Coulson, E. J., and Vaux, D. L. (1998). Conservation of baculovirus inhibitor of apoptosis repeat proteins (BIRPs) in viruses, nematodes, vertebrates and yeasts. Trends Biochem Sci 23, 159–162.
49. Chai, J. et al. (2001). Structural basis of caspase-7 inhibition by XIAP. Cell 104, 769–780.
50. Huang, Y. et al. (2001). Structural basis of caspase inhibition by XIAP: differential roles of the linker versus the BIR domain. Cell 104, 781–790.
51. Riedl, S. J. et al. (2001). Structural basis for the inhibition of caspase-3 by XIAP. Cell 104, 791–800.
52. Shiozaki, E. N. et al. (2003). Mechanism of XIAP-mediated inhibition of caspase-9. Mol Cell 11, 519–527.
53. Lamkanfi, M., Festjens, N., Declercq, W., Vanden Berghe, T., and Vandenabeele, P. (2007). Caspases in cell survival, proliferation and differentiation. Cell Death Differ 14, 44–55.
54. Ashkenazi, A. and Dixit, V. M. (1998). Death receptors: signaling and modulation. Science 281, 1305–1308.
55. Green, D. R. and Reed, J. C. (1998). Mitochondria and apoptosis. Science 281, 1309–1312.
56. Chang, D. W. et al. (2002). c-FLIP(L) is a dual function regulator for caspase-8 activation and CD95-mediated apoptosis. EMBO J 21, 3704–3714.
57. Wei, Y., et al. (2000). The structures of caspases-1, -3, -7 and -8 reveal the basis for substrate and inhibitor selectivity. Chem Biol 7, 423–432.

Chapter 3
The Mitochondrial Death Pathway

Anas Chalah and Roya Khosravi-Far*

Abstract Mitochondria have long been known to be critical for cell survival due to their role in energy metabolism. However, not until the mid-1990s did it become evident that mitochondria are also active participants in programmed cell death (PCD). This chapter focuses mainly on the role the mitochondria in mammalian cell death and cancer progression and therapy.

Keywords apoptosis, death receptors, mitochondria, bid, membranes, phospholipases, cardiolipin

1 Introduction

Apoptosis, or programmed cell death (PCD), is an evolutionarily conserved mechanism for the selective removal of aging, damaged or otherwise unwanted cells (Abe et al., 2000; Degli Esposti, 1999; Lawen, 2003; Ozoren and El-Deiry, 2003; Peter and Krammer, 1998; Strasser et al., 2000; Thorburn, 2004). It is an essential component of many normal physiological processes such as embryogenesis, normal tissue development, and the immune response (Vaux and Korsmeyer, 1999). Thus, regulation of apoptosis is critical for tissue homeostasis and its deregulation can lead to a variety of pathological conditions including carcinogenesis and chemoresistance (Burns and El-Deiry, 2003; Daniel et al., 2001; Green and Evan, 2002; Ozoren and El-Deiry, 2003; Sheikh and Huang, 2004; Thompson, 1995; Zornig et al., 2001).

Apoptosis is mediated primarily through the activation of specific proteases called caspases (cysteinyl, aspartate-specific proteases) (Algeciras-Schimnich et al., 2002;

Anas Chalah and Roya Khosravi-Far
Department of Pathology, Harvard Medical School, Beth Israel Deaconess Medical Center, 99 Brookline Avenue, Boston, MA 02215, USA

*To whom correspondence should be addressed: Tel.: (617) 667–8526; fax: (617) 667–3524; e-mail: rkhosrav@bidmc.harvard.edu

R. Khosravi-Far and E. White (eds.), *Programmed Cell Death in Cancer Progression and Therapy.*
© Springer 2008

Ozoren and El-Deiry, 2003; Salvesen and Dixit, 1997; Stegh and Peter, 2001; Thorburn, 2004). Caspases are effectors of cell suicide and cleave multiple substrates, leading to biochemical and morphological changes that are characteristic of apoptotic cells(Abe et al., 2000; Strasser et al., 2000). These alterations include: mitochondrial outer membrane permeabilization; cell membrane remodeling and blebbing; exposure of phosphatidylserine (PS) at the external surface of the cell; cell shrinkage with cytoskeletal rearrangements; nuclear condensation; and DNA fragmentation (Ashkenazi and Dixit, 1999; Green and Evan, 2002; Lawen, 2003; Peter and Krammer, 2003; Schulze-Osthoff et al., 1998; Thorburn, 2004). These morphological changes culminate in the formation of apoptotic bodies that are normally eliminated by phagocytosis (Geske and Gerschenson, 2001; Wallach, 1997). In mammalian systems, the extrinsic death receptor pathway and the intrinsic mitochondrial pathway are the two major signaling systems that result in the activation of the executioner/effector caspases and the consequent demise of the cell (Abe et al., 2000; Ozoren and El-Deiry, 2003; Peter and Krammer, 2003; Strasser et al., 2000; Thorburn, 2004). In many cell types, including cancer cells, activation of the extrinsic pathway also engages the mitochondrial pathway for full execution of cell death (Jaattela, 2004; Khosravi-Far and Esposti, 2004; Kroemer, 2003; Newmeyer and Ferguson-Miller, 2003; Thorburn, 2004). Thus, many apoptotic signals merge at the mitochondria, and thus mitochondria have been termed "gatekeepers" of the apoptotic machinery (Jaattela, 2004; Khosravi-Far and Esposti, 2004; Kroemer, 2003; Newmeyer and Ferguson-Miller, 2003; Thorburn, 2004).

As gatekeepers, the proteins comprising the intrinsic mitochondrial pathway are the major mediators of the cytotoxic effects of many chemotherapeutic agents and radiation therapy (Brenner et al., 2003; Costantini et al., 2000; Debatin et al., 2002; Hersey and Zhang, 2003). Cancer cells often evade this apoptosis and develop chemoresistance and radioresistance. Indeed, disruption of the mitochondrial apoptotic machinery has been observed in many tumors (Daniel et al., 2001; Morisaki and Katano, 2003). It is also likely that disruption of the mitochondrial machinery or mutations in the mitochondrial DNA could play a role in cancer initiation. Because of the central role of mitochondria in these processes, various components of the mitochondrial machinery can be targets for novel therapeutic strategies.

2 The Mitochondrial Pathway of Apoptosis

Mitochondria are thought to be the primary organelles involved in mediating most apoptotic pathways in mammalian cells (Green and Kroemer, 2004; Kroemer, 2003; Newmeyer and Ferguson-Miller, 2003; Ravagnan et al., 2002; Sorice et al., 2004; Zamzami and Kroemer, 2001). Mitochondria are engaged via the intrinsic pathway of cell death, which can be initiated by a variety of stress stimuli, including ultraviolet (UV) radiation, γ-irradiation, heat, DNA damage, the actions of some oncoproteins and tumor suppressor genes (i.e., P53), viral virulence factors, and most chemotherapeutic agents (Fig. 3.1) (Kroemer, 2003). These diverse forms

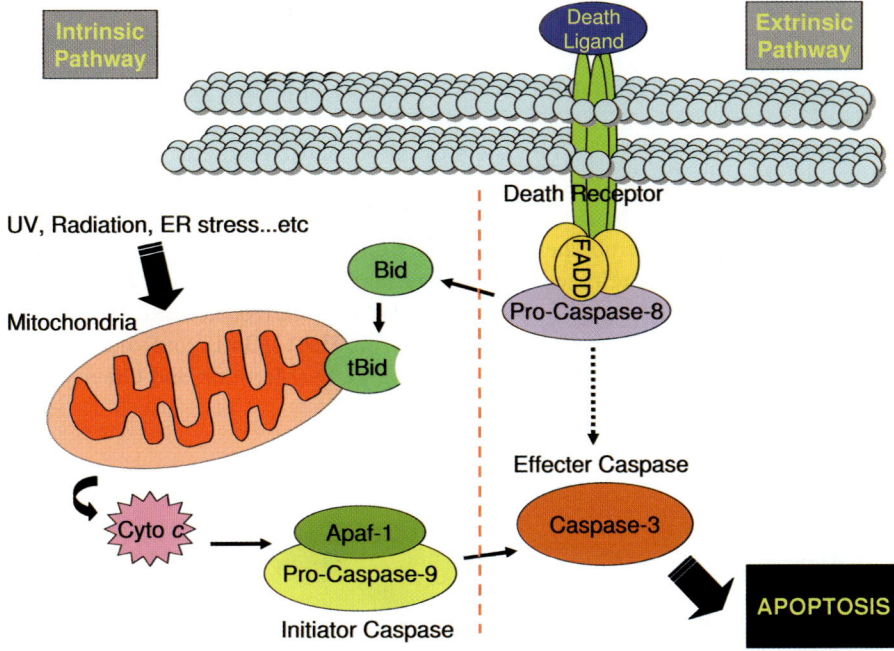

Fig. 3.1 Schematic representation of the intrinsic and extrinsic apoptotic pathways

of stress are sensed by multiple cytosolic or intraorganellar molecules. Transduction of these signals to the mitochondria ultimately results in alterations of the outer mitochondrial membrane (OM) (Esposti et al., 2003; Green and Kroemer, 2004; Kuwana et al., 2002; Newmeyer and Ferguson-Miller, 2003; Zamzami and Kroemer, 2001). These changes in the OM then lead to increased permeability to proteins that normally reside between the OM and the inner mitochondrial membrane (IM), enabling these proteins to escape the mitochondria and diffuse into the cytosol.

The mitochondrial pathway of apoptosis can also be activated in response to death ligands. In a majority of cells (type II cells), including tumor cells, extracellular death signals engage the mitochondria in a way that is equivalent to the intrinsic pathway (Abe et al., 2000; Algeciras-Schimnich et al., 2002; Ozoren and El-Deiry, 2002; Peter and Krammer, 1998). In these cells, signals originating from the death ligand-induced activation of caspase-8 and caspase-10 bifurcate into two arms, one of which directly engages mitochondria via a sequence of events causing activation of the effector caspases (i.e., caspase-3). The second arm promotes the cleavage of noncaspase substrates, such as Bid, inducing changes in the mitochondrial OM and the release of apoptogenic factors and activation of caspase-9, which then cooperates with the less-efficient activation of caspase-8 in these cells.

3   The Release of Proapoptotic Factors

Mitochondria contain and release many soluble proteins that are involved in the apoptotic cascade (Fig. 3.2) (Daniel et al., 2001; Debatin et al., 2002; Green and Kroemer, 2004; Reed, 2004). The variety of mitochondrial proteins participating in this pathway indicates the pivotal role of these organelles in determining cellular fates. Bcl-2 family members control apoptosis by regulating the permeabilization of the mitochondrial membrane (Chao and Korsmeyer, 1998; Cory et al., 2003; Daniel et al., 2001). The release of mitochondrial proteins, including cytochrome *c*, apoptosis-inducing factor (AIF), second mitochondria-derived activator of caspases (Smac/Diablo), high-temperature requirement A2 (HtrA2/Omi), and endonuclease G, is believed to play a pivotal role in inducing PCD (Martinou and Green, 2001; Zamzami and Kroemer, 2001).

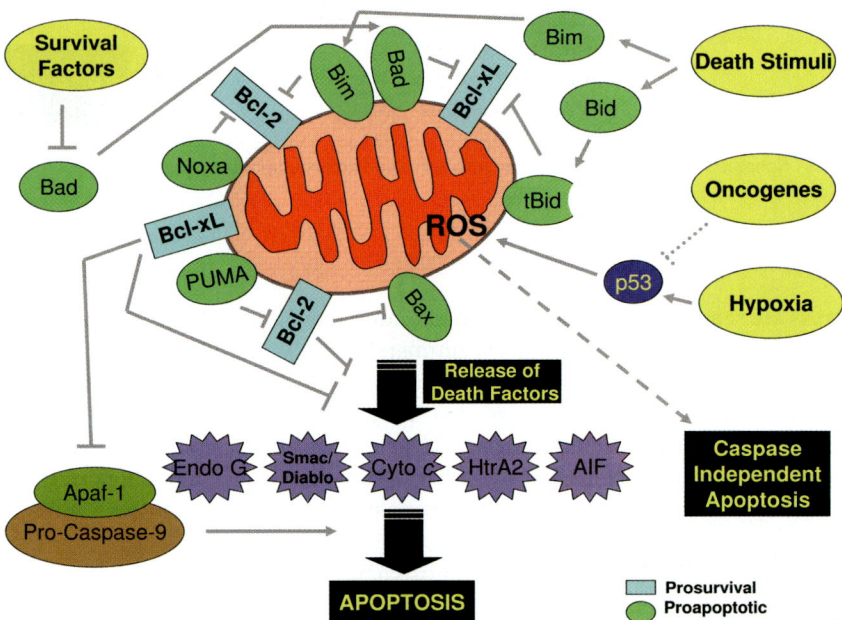

Fig. 3.2 Mitochondrial membrane permeabilization is regulated by an elegant balance of opposing actions of proapoptotic and antiapoptotic Bcl-2 family members. Bax, Bad, and Bak promote the release of cytochrome *c* and AIF through the formation of transmembrane channels across the mitochondrial outer membrane, while Bcl-2 and Bcl-X$_L$ delay this release and abort the apoptotic response, leading to cell survival. Besides the release of mitochondrial proapoptotic components, the loss of mitochondrial membrane integrity results in the loss of many essential biochemical cellular functions such as ATP synthesis and results in the generation of reactive oxygen species (ROS). The increased levels of ROS are directly linked to the oxidation of lipids, proteins, and nucleic acids

4   Cytochrome C

Cytochrome c (Cyt c), a small (13 kDa) nuclear encoded mitochondrial protein, was the first protein identified as being released from mitochondria upon apoptosis. It is considered a key regulator of apoptosis because once it is released from the mitochondrial intermembrane space (IMS), the cell is irreversibly committed to death (Green and Evan, 2002; Kluck et al., 1997; Zhivotovsky et al., 1998a; Zhivotovsky et al., 1998b) and Cyt c is synthesized in the cytosol and translocates to the mitochondria as an unfolded apoprotein through the TOM (translocase in the OMM) complex (Diekert et al., 2001). The driving force for translocation of apo-Cyt c into the IMS appears to be its interaction with the enzyme cytochrome c heme lyase (Dumont et al., 1991; Mayer et al., 1995).

The release of cytochrome c to the cytosol is considered among the major steps in the intrinsic death pathway (Kluck et al., 1997; Newmeyer and Ferguson-Miller, 2003; Zhivotovsky et al., 1998a). Once it escapes to the cytosol, it is captured by the apoptosis protease activating factor (APAF-1), a 130 kDa adaptor protein (Soengas et al., 1999; Zou et al., 1999). Prior to binding Cyt c, APAF-1 is virtually inactive. Once bound to Cyt c, the APAF-1 monomer goes through a cytochrome c-induced conformational change that promotes its activation. Further oligomerization occurs, resulting in a cartwheel-shaped heptameric structure containing seven Cyt c/APAF-1 complexes. This larger multiprotein complex is termed the apoptosome (Acehan et al., 2002; Adrain et al., 2001; Adrain et al., 1999; Srinivasula et al., 1999). Pro-caspase-9 is recruited to the apoptosome through its CARD domain, promoting its cleavage and converting it to an active protease (Adrain et al., 1999). Consequently, caspase-9 dissociates from the complex and goes on to activate effector caspases (3, 6, and 7) which collectively orchestrate the execution of apoptosis (Slee et al., 1999; Srinivasula et al., 1999; Zou et al., 1999).

5   Apoptosis-Inducing Factor

The precursor of the protein AIF is synthesized in the cytosol and imported into mitochondria (Susin et al., 1999). It contains an N-terminal mitochondrial localization sequence (MLS) which is cleaved upon its mitochondrial translocation to form the mature 57 kDa AIF (Susin et al., 1999). Under apoptosis-inducing conditions, AIF translocates through the permeabilized mitochondrial outer membrane to the cytosol (Cande et al., 2002; Susin et al., 1999). Subsequently, AIF is transported to the nucleus where it induces ATP-independent nuclear chromatin condensation, as well as large-scale DNA fragmentation (Cande et al., 2002; Susin et al., 1999). In contrast to cytochrome c, AIF acts in a caspase-independent fashion and does not require the presence of cytosolic factors to induce apoptotic features in the nuclei (Lorenzo et al., 1999; Miramar et al., 2001; Susin et al., 1999; Zamzami and Kroemer, 2001). Moreover, AIF translocation occurs in Apaf-1-null mice which fail to activate the executioner caspase (Cecconi et al., 1998). However, some studies indicate that crosstalk does occur between AIF and the apoptotic caspase cascade (Cande et al., 2002). For instance, AIF was observed to trigger the release of cytochrome c from isolated mitochondria (Susin et al., 1999). Additionally, AIF interacts with heat-shock protein 70 (Hsp70), a known protective factor and inhibitor of Apaf-1-dependent caspase activation (Ravagnan et al., 2002).

6   Smac/Diablo

Second mitochondria-derived activator of caspases (Smac) is a 22 kDa mitochondrial protein also known as direct IAP-associated binding protein with low pI (Diablo). Inhibitors of apoptosis (IAP) family members have the ability to interact and inhibit the enzymatic activity of caspases through their baculovirus inhibitor repeat (BIR) functional motif (Deveraux and Reed, 1999; Miller, 1999). Smac/Diablo was first identified as a mammalian IAP (Srinivasula et al., 1999; Verhagen and Vaux, 2002). Specifically, XIAP, c-IAP1, and c-IAP2 are proapoptotic factors regulated by Smac/ Diablo (Ekert et al., 2001; Srinivasula et al., 1999; Verhagen and Vaux, 2002). The Smac/Diablo precursor is synthesized in the cytosol, then imported to the mitochondria where it is cleaved and activated. A mature form of Smac/Diablo is released to the cytosol under apoptotic conditions. Unlike cytochrome c, which directly activates APAF-1 and caspase-9, Smac/Diablo binds to the BIR domains of multiple IAP members, antagonizing them and promoting indirect caspase activation (Ekert et al., 2001; Srinivasula et al., 1999; Verhagen and Vaux, 2002). Smac/Diablo and cytochrome c were found to be released from the mitochondria at around the same time. Moreover, the release was found to coincide with mitochondrial membrane potential depolarization (Rehm et al., 2003; Springs et al., 2002; Verhagen and Vaux, 2002). However, a recent study presented evidence suggesting that the release of Smac/Diablo may, in fact, depend on the release of cytochrome c (Hansen et al., 2006).

7   HtrA2/Omi

HtrA2, also referred to as Omi, is a mitochondrial protein that belongs to the family of serine proteases. This proapoptotic protein is expressed as a 50 kDa precursor that is cleaved at the N-terminal, upon translocation to the mitochondria, to generate the active 36 kDa protein (Hegde et al., 2002; Martins et al., 2002; Suzuki et al., 2001; Verhagen and Vaux, 2002). Similar to cytochrome c and Smac/Diablo, mature HtrA2/Omi localizes to the IMS (Hegde et al., 2002; Suzuki et al., 2004). Its release to the cytosol is stimulated by apoptotic triggers. Upon its release, HtrA2/Omi binds directly to the BIR domain of IAPs and inhibits their caspase-inhibitory activity (Suzuki et al., 2001). The first four N-terminal amino acids of the mature HtrA2 protein (AVPS) constitute the IAP-binding motif.

In addition to the proapoptotic effect of IAP binding and inhibition, Omi/HtrA2 appears to utilize its serine protease activity to induce an IAP inhibition-independent, caspase-independent apoptosis (Hegde et al., 2002; Suzuki et al., 2001). Recently, it was reported that the proapoptotic serine protease activity of HtrA2/Omi also plays a significant role in antagonizing IAPs. The observed HtrA2 cleavage of c-IAP produced significant caspase activation and sensitized cells to apoptosis (Yang et al., 2006).

8 Endonuclease G

As with most mitochondrial proteins, Endonuclease G is expressed as a precursor in the cytosol. Upon its translocation to the mitochondria, the 33 kDa protein is cleaved to a 28 kDa mature form (Cote and Ruiz-Carrillo, 1993). During apoptosis, endonuclease G is released from the mitochondrial IMS and translocates to the nucleus, where it causes oligonucleosomal DNA fragmentation (Li et al., 2001; van Loo et al., 2001). Endonuclease G release appears to be dependent on caspase activation downstream of mitochondria (Arnoult et al., 2003). Interestingly, endonuclease G-induced DNA degradation was observed to be caspase-independent (Li et al., 2001; Susin et al., 1999), suggesting an important role for endonuclease G in bringing about caspase-independent cell death.

9 Mitochondrial Proteins and Caspase Activation

Among the various proteins that leak out of mitochondria, a few, such as cytochrome c, play a major role in promoting caspase activation. (Kluck et al., 1999; Saelens et al., 2004) These apoptogenic factors are released in a hierarchical manner during cell death. Upon activation of the intrinsic pathway, cytochrome c, Htr2A/Omi and Smac/Diablo are released first, with similar kinetics (Saelens et al., 2004). The subsequent release of AIF and endonuclease G (Arnoult et al., 2003; Penninger and Kroemer, 2003) is associated with more severe damage to both the outer and inner membranes. Notably, cytochrome c has been shown to be directly involved in the mediation of cell death, as it is indispensable for the activation of Apaf-1 and subsequent formation of the apoptosome (Arnoult et al., 2003).

The apoptosome itself is a platform for recruiting and facilitating the autocatalytic activation of pro-caspase-9, the apical caspase of the intrinsic pathway of apoptosis (Adams and Cory, 2002; Baliga and Kumar, 2003; Cain et al., 2002; Chinnaiyan, 1999; Hill et al., 2003; Salvesen and Renatus, 2002; Shi, 2002). The activation of caspase-9 leads to the local accumulation of zymogens, promoting an autocatalytic process of downstream caspase activation (Adams and Cory, 2002; Baliga and Kumar, 2003; Cain et al., 2002; Chinnaiyan, 1999; Hill et al., 2003; Salvesen and Renatus, 2002; Shi, 2002). However, the apoptosome requires additional regulatory factors, including Smac/Diablo, for full activation of the caspase cascade. Smac/Diablo interacts with several IAPs to release them from their inhibitory interaction with pro-caspase-9 and other caspases (Adams and Cory, 2002; Baliga and Kumar, 2003; Cain et al., 2002; Shi, 2002). Smac/Diablo is also present in the mitochondria, where it is directly attached to the OM and is released upon alterations in the OM permeability (Cain et al., 2002; Saelens et al., 2004).

10   Mechanisms of Mitochondrial Protein Release

The exact mechanism by which mitochondrial proapoptotic components are released from the IMS is a matter of a long and ongoing debate. Currently, two general mechanisms are considered: nonspecific and specific release (Lim et al., 2001). The opening of the permeability transition pore (PTP) located in the mitochondrial IMS is proposed as the first possible mechanism. The permeability pore is comprised of three proteins: cyclophilin D, adenine nucleotide translocator (ANT), and voltage-dependent anion channel (VDAC), a matrix, an inner membrane, and an outer membrane protein, respectively (Crompton, 1999). The opening of the PTP triggers many processes, including (A) loss of the proton gradient produced by the electron transport machinery; (B) leakage of cellular water into the mitochondrial matrix, resulting in the gradual swelling of the IMS and the rupturing of the inflexible OM (Green and Kroemer, 2004); and (C) leakage of apoptotic factors from the IMS into the cytoplasm, which begins the cascade of proteolytic activities leading ultimately to nuclear damage and cell death (Brenner et al., 2000; Dejean et al., 2006; Kroemer, 2003; Marzo et al., 1998a; Marzo et al., 1998b). This mechanism represents a nonspecific release mode for proapoptotic mitochondrial mediators. However, the physical outer membrane disruption theory fails to explain the release of proapoptotic factors such as cytochrome c and AIF in the absence of any loss of outer membrane structural integrity (Dejean et al., 2006).

The second suggested mode of release involves the opening of large outer membrane channels that would allow cytochrome c and other IMS proteins to move into the cytosol. In contrast with the other scenarios, this model would leave the outer membrane largely intact. A benefit of this model is that there is no need for the mitochondrial matrix to swell. This better fits with the evidence that mitochondrial morphology remains the same in most cell death in vivo. Several outer membrane channels, including the VDAC and mitochondrial apoptosis-induced channel (MAC), have been targeted as possible specific regulators of mitochondrial release. Both provide aqueous pathways through the hydrophobic environment of the mitochondrial membrane.

VDAC is a 30 kDa highly conserved voltage-dependent, ion-selective, mitochondrial OM protein. The OM is densely packed with VDAC proteins which form barrel structures that enclose 3 nm internal diameter channels. VDAC can switch between two functional states, open and partially open. The "open" state is defined by large conductance and anion selectivity, while the "partially open" state is defined by lower conductance (about half that of the fully open state) and cation selectivity. The voltage-dependent change between these two states is widely attributed to structural rearrangements that lead to changes of size and charge distributions within the channel (Colombini et al., 1996; Mangan and Colombini, 1987; Thomas et al., 1993).

MAC was first identified in 2001. It is a mitochondrial outer membrane channel that, according to some reports, forms at early stages of the intrinsic apoptotic pathway (Dejean et al., 2006; Guo et al., 2004). Alternatively, other studies have reported the formation of MAC at late stages of the extrinsic apoptotic pathway (Guihard et al., 2004). MAC was found to be slightly cation-selective, and unlike VDAC, voltage-independent (Dejean et al., 2005; Guo et al., 2004). MAC activity was found to be induced by apoptosis and regulated by Bax, a proapoptotic Bcl-2 family protein. Bax translocation to the mitochondria was linked to MAC formation and cytochrome c release (Antonsson et al., 1997; Dejean et al., 2006; Guo et al., 2004; Saito et al., 2000; Schendel et al., 1997). Bax oligomerization is proposed to form MAC channels (Cheng et al., 2001; Dejean et al., 2006; Wei et al., 2001). The pore diameter of the MAC channel was measured to be ~4 nm, which is proposed to allow for the release of the ~3 nm diameter cytochrome c (Pavlov et al., 2001).

11   The Bcl-2 Family of Proteins and Regulation of the Mitochondrial Pathway to Cell Death

The process of mitochondrial release of proapoptotic factors such as cytochrome c is elegantly regulated through members of the Bcl-2 family (Fig. 3.2) (Antonsson et al., 1997; Cory et al., 2003; Danial and Korsmeyer, 2004; Green and Kroemer, 2004; Schendel et al., 1997). In mammals, the antiapoptotic members of this family include Bcl-2, Bcl-X_L, and Bcl-W, while the proapoptotic members include Bax, Bak, Bad, Bik, Bim, and Bid. The proapoptotic family members are further classified based on domain sequence homology into two groups: one that contains multiple BH domains and one that contains only the BH3 domain (Cheng et al., 2001; Fiers et al., 1999; Kuwana and Newmeyer, 2003; Wei et al., 2001). The fate of the cell depends to a great degree on the precious balance of function between these proapoptotic and antiapoptotic Bcl-2 proteins. Studies have shown that Bax, Bad, and Bak promote the release of AIF and cytochrome c, while Bcl-2 and Bcl-X_L delay the release and abort the apoptotic response, promoting cell survival (Cory and Adams, 2002; Yang et al., 1997).

It is believed that Bcl-2 family members regulate the apoptotic response by controlling mitochondrial membrane permeabilization (MMP) (Green and Kroemer, 2004). The proapoptotic proteins Bax and Bak have been shown to contribute to the formation of transmembrane channels across the mitochondrial OM, leading to the escape of AIFs (Dejean et al., 2005; Korsmeyer et al., 2000; Kuwana et al., 2002; Nechushtan et al., 2001; Wei et al., 2001). Bcl-2, Bcl-W, and Bcl-X_L are, on the other hand, believed to prevent pore formation and to inhibit the release of cytochrome c from the mitochondria (Kluck et al., 1997; Yang et al., 1997). Moreover, heterodimerization of Bax or Bad with Bcl-2 or Bcl-X_L is thought to inhibit their protective effect.

Bid is a potent proapoptotic protein that is normally located in the cytosol, but also shuttles through the surfaces of intracellular membranes due to its lipid-interacting capacity. Bid plays an important role in the mitochondrial pathway to apoptosis as it has been identified as the link between the death receptor signal and the release of cytochrome c. Activated caspase-8 engages the intrinsic apoptotic pathway through the truncation of Bid (Li et al., 1998; Luo et al., 1998). Upon death signaling, activated caspase-8 cleaves Bid (26 kDa) into two fragments: a C-terminus fragment (15 kDa) and an N-terminus fragment (11 kDa) (Luo et al., 1998). The 15 kDa fragment, which contains the BH3 domain, is termed truncated Bid or tBid. This functional fragment translocates to the mitochondria where it interacts with several proteins through its BH3 domain (Wang et al., 1996). There are two modes of Bid proapoptotic action. (1) In the BH3-dependent mode, Bid interacts with the antiapoptotic Bcl-X_L through its BH3 domain and prevents the formation of the Bcl-X_L/Apaf1 antiapoptotic complex. (2) In the BH3-independent mode, after truncation, Bid is proposed to form selective channels similar to BAX through its structural motifs (Chou et al., 1999; McDonnell et al., 1999). Moreover, tBid has been shown to induce the oligomerization of Bax and Bak, resulting in MAC formation and the subsequent release of proapoptotic cytochrome c (Eskes et al., 2000; Wei et al., 2000).

The mitochondrial receptor for caspase-cleaved Bid is thought to be cardiolipin (CL), a mitochondrial lipid (Esposti et al., 2003; Kuwana et al., 2002; Newmeyer and Ferguson-Miller, 2003; Sorice et al., 2004). CL is a glycerophospholipid that is synthesized and localized in the inner membrane of the mitochondria, making it one of its major constituents (Khosravi-Far and Esposti, 2004; McMillin and Dowhan, 2002; Schlame et al., 2000; Wright et al., 2004). This dimeric molecule apparently plays a significant role in controlling the mitochondrial membrane structure and function. Abnormal mitochondrial morphology and function have been observed in cells defective in the CL synthesis mechanism (Ohtsuka et al., 1993). It has been proposed that upon apoptotic stimulation, CL contributes to the apoptotic signal through the recruitment of cytosolic proteins such as tBid to the mitochondrial membrane. Additionally, it is thought that CL is involved in altering MMP, leading to the subsequent release of proapoptotic factors (Lutter et al., 2000).

12   Mitochondria and Oxidative Stress

Mitochondria are the sites of aerobic respiration. Energy is generated in mitochondria through the process of ATP synthesis via the oxidative phosphorylation pathway. This process, however, also results in the formation of single unpaired electrons, leading to reactive oxygen species (ROS). ROS such as hydrogen peroxide (H_2O_2), the superoxide anion (O_2^-), and hydroxyl radicals (OH) are highly reactive molecules generated and eliminated in a balanced process in normal cells. In particular, free radicals (superoxides) are byproducts of ATP generation by the mitochondrial respiratory chain (Andreyev et al., 2005; Beyer, 1992; Raha and Robinson, 2000). Cellular energy is usually liberated from ATP molecules through the removal of single phosphate-oxygen groups, producing adenosine diphosphate (ADP). ADP is recycled in the mitochondria where it is recharged through oxidative processes to reproduce ATP. Since ROS are harmful, the balance between energy supply and energy demand is extremely critical. Any shift in this balance would introduce excess ROS to cells and would result in oxidative stress.

The damaging effect of elevated levels of ROS is thought to be due to the highly reactive free electrons available to form stable chemical bonds. While H_2O_2 is free to escape the mitochondrion, both the superoxide anion and hydroxyl radicals have limited diffusion, and are more likely to contribute to inner membrane damage of mitochondria (Szeto, 2006). Several studies have demonstrated a direct relationship between mitochondrial ROS and the mitochondrial apoptotic pathway. For example, the release of cytochrome c to the cytosol has been linked to mitochondrial oxidation (Shidoji et al., 1999). It is believed that the release mechanism might involve the opening of mitochondrial PTPs (Vieira et al., 2001). Several antioxidant compounds, such as ascorbic acid (vitamin C), α-tocopherol (vitamin E), and ubiquinol are naturally present in the cell and act to protect against the effects of ROS (Sies and de Groot, 1992).

13   Mitochondria and Cancer

Given the important roles mitochondria play in cellular energy metabolism, free radical formation and PCD, defects in mitochondrial function are suspected to contribute to the development and progression of cancer and to resistance to therapy (Bettaieb et al., 2003; Brenner et al., 2003; Costantini et al., 2000; Debatin et al., 2002; Hersey and Zhang, 2003; Jaattela, 2004; Kasibhatla and Tseng, 2003; Kim et al., 2004). Defective apoptosis is one of the hallmarks of tumorigenicity and is implicated in multiple stages of tumor progression (Burns and El-Deiry, 2003; Hanahan and Weinberg, 2000; Ozoren and El-Deiry, 2003). Furthermore, the ability of tumor cells to escape apoptosis plays a key role in promoting resistance to conventional chemotherapy and radiation therapy (Abe et al., 2000; Barnhart et al., 2004; Daniel et al., 2001; El-Deiry, 1997; Thompson, 1995; Zornig et al., 2001).

A link between mitochondria and cancer progression was suggested over half a century ago when Warburg reported the role of mitochondria in cellular energy metabolism. This phenomenon was coined the "Warburg effect." The Warburg effect suggested that the development of an injury to the respiratory machinery is an important event in carcinogenesis (Warburg, 1951). This injury results in compensatory increases in glycolytic ATP production to fulfill the energy needs of tumor cells. Since then, preferential reliance on glycolysis over the oxidative metabolism has been shown to correlate with tumor progression in several types of cancer (Semenza et al., 2001). Since the initial report of the Warburg effect, a number of cancer-related mitochondrial defects have also been identified (Brenner et al., 2003; Carew and Huang, 2002; Debatin et al., 2002; Jaattela, 2004). These defects include altered expression and activity of respiratory chain subunits and glycolytic enzymes, changes in oxidation of NADH-linked substrates and mutations in mitochondrial DNA. Thus, the differences in energy metabolism between normal cells and cancer cells constitute a biochemical basis for the development of therapeutic strategies that might selectively kill cancer cells in their compromised respiratory state.

Furthermore, dysregulation of members of the Bcl-2 family has been detected in a variety of malignancies, especially hematological cancers. Bcl-2 itself was originally discovered as an oncogene in B cell lymphoma Danial and Korsmeyer, 2004. Additionally, overexpression of Bcl-2 has been detected in AML and non-Hodgkin's lymphomas. Dysregulation of other Bcl-2 family proteins have also been detected in other cancers; for example, increased expression of Mcl-1 has been detected in relapsed AML and multiple myeloma. Increased expression levels and mutations in the promoter of the *mcl-1* gene have also been observed in chronic lymphoblastic leukemias. These studies reiterate that changes to the mitochondrial-associated proteins, mainly members of the Bcl-2 family, are directly involved in tumor progression.

Additionally, there is some evidence that alterations in the mitochondrial DNA could also be involved in cancer progression. Besides hosting hundreds of nuclear encoded proteins, mitochondria have their own DNA that encodes 13 mitochondrial proteins (Schatz, 1995; Singh et al., 1999). Mutations in mtDNA could occur during oxidative phosphorylation involving ROS. Investigations of human bladder, lung, neck, and head primary tumors revealed a high percentage of mtDNA mutation (~50%) in these tumors (Fliss et al., 2000). These observations suggest a link between cancer development and mitochondrial dysfunction; however, they do not present a clear answer to whether mitochondrial DNA mutation is simply a result, or rather the cause, of alterations in PCD.

Mitochondria also play an important role in resistance to chemotherapy and radiation therapy. Since mitochondria are integrators of apoptotic signaling pathways, induction of apoptosis in many cell types leads to the induction of MMP (Brenner et al., 2003; Kroemer, 2003). MMP defines the point of no return in most PCD pathways and is regulated by pre-mitochondrial signal transduction pathways. These pathways involve caspase-dependent and caspase-independent mechanisms, members of the Bcl-2 family of proteins and changes in the composition of mitochondrial membranes (Bettaieb et al., 2003; Brenner et al., 2003; Green and Kroemer, 2004; Kim et al., 2004; Kroemer, 2003; Kuwana et al., 1998; Newmeyer and Ferguson-Miller, 2003; Peter and Krammer, 1998; Ravagnan et al., 2002; Sorice et al., 2004; Waterhouse et al., 2001; Zamzami and Kroemer, 2001). In response to MMP, proapoptotic factors are released into the cytosol to trigger the execution of cell death. This is likely due to the opening of protein channels such as the VDAC. Under pathological conditions, cancer cells escape from apoptosis and/or become resistant to treatment by affecting MMP (Bettaieb et al., 2003; Debatin et al., 2002; Hersey and Zhang, 2003; Kim et al.,

2004). Therefore, overcoming abnormalities in tumor cells that suppress MMP could lead to therapeutic targets by generating a potent proapoptotic stimulus. Additionally, since MMP is an early event in apoptosis, strategies to detect this process can be useful in assessing the response to chemotherapy.

Mutations in mtDNA have been implicated in the cellular response to chemotherapy. For example, Singh et al. (1999) examined the response of a tumor cell line lacking mitochondrial DNA to several anticancer drugs, including adriamycin (a DNA-interacting drug widely used in chemotherapy for its role in binding DNA and stopping the process of replication). Cancer cells lacking mtDNA showed great chemotherapy resistance, indicating an important role of the mitochondrial genome in regulating the cellular response to therapeutic agents. Similar findings were also reported in A549 non-small-cell lung cancer cell lines and their rho0 derivatives in which mitochondrial DNA has been eradicated (Lo et al., 2005). The parental cell line showed increased sensitivity to chemotherapy when compared with the mtDNA-compromised derivative cell line. Notably, the restoration of mtDNA restored chemosensitivity of the resistant cell line (Lo et al., 2005).

14 Targeting Mitochondria in Cancer Therapy

As mitochondria are gatekeepers of apoptotic signals, targeting mitochondria to induce apoptosis of malignant cells is an important therapeutic strategy. In the past several years, extensive research has focused on screening for chemical compounds, small molecules and peptides that could target the mitochondria. Therapeutic tactics have included strategies that involve the Bcl-2 family proteins, activation of PTPs, the respiratory chain, mitochondrial DNA depletion, and selective targeting of ROS-stressed malignant cells, as well as targeting inhibitors of apoptosis such as IAPs (Dias and Bailly, 2005). Targeting the antiapoptotic members of the Bcl-2 family, namely Bcl-2 and Bcl-X_L, and targeting the PTP are among the most studied mechanisms (Dias and Bailly, 2005; O'Neill et al., 2004; Shangary and Johnson, 2003; Walensky, 2006). Targeting of the Bcl-2 family of proteins is discussed in Chapter 8. Here, we will briefly describe strategies for targeting and activation of the PTP.

15 Targeting and Activation of the Permeability Transition Pore

The induction of proapoptotic protein release through increased PTP formation and opening has been explored in the recent years as a possible mechanism for cancer treatment. As a chemotherapeutic approach, this method involves perturbation of the mitochondrial membrane through direct targeting of the components of the membrane permeability transition pore complex (PTPC) (Brenner et al., 2003; Costantini et al., 2000; Debatin et al., 2002; Fantin and Leder, 2006; Galluzzi et al., 2006; Khosravi-Far and Esposti, 2004; Morisaki and Katano, 2003; Reed, 2004). Additionally, alterations in energy metabolism, such as depletions in ADP and ATP, can also facilitate formation of the PTPC.

In addition to therapeutic strategies that target Bcl-2 family members, several chemotherapeutic agents such as paclitaxel or etopiside have been shown to induce opening of the PTPC, albeit at high concentrations. Additionally, several experimental anticancer agents act directly on the components of the PTPC. For example, the synthetic retinoid CD437, arsenic acid and lonidamine are inhibitors of ANT (Debatin et al., 2002; Fantin and Leder, 2006; Galluzzi et al., 2006). Arsenic acid also inhibits the VDAC. Hexokinase, which is a component of the PTPC and a major player in maintaining the malignant state of transformed cells, is also inhibited by lonidamine (Debatin et al., 2002; Fantin and Leder, 2006; Galluzzi et al., 2006). Additionally, jasmonates are known to act selectively and directly on cancer cell mitochondria in a PTPC-mediated mechanism, resulting in membrane depolarization, swelling, and the release of cytochrome c (Rotem et al., 2005) leading to apoptosis of tumor cells. Similarly, lamellarins are another group of anticancer drugs that target mitochondria of cancer cells and induce permeability transition effects (Kluza et al., 2006).

16   Conclusions and Future Prospects

Mitochondria are the power generators of the cell due to their involvement in glucose metabolism, and they are "gatekeepers" of the cell involved in integrating apoptotic signals in majority of cells. Because tumor cells rely on glycolysis and since evasion of apoptosis is one of the hallmarks of cancer, mitochondria therefore play a central role in cancer cell biology. The intrinsic and extrinsic death pathways leading to changes in mitochondrial permeability; the components of the PTPC, including members of the Bcl-2 family; apoptogenic factors and their regulators, and mutations in mtDNA have been studied extensively in the past for their contributions to cancer progression or resistance to therapy. These constitute an extensive list of targets that could induce apoptosis, some with possible specificity for cancer cells. Therapeutic agents against many of these targets, including Bcl-2 family members and components of the PTP, are currently at various stages in the development pipeline. The ultimate goal of these studies is to generate novel mitotoxic agents that can selectively induce apoptosis of cancer cells and reduce the possibility of resistance.

Acknowledgments We thank Susan Glueck for assistance with preparation of the manuscript.

References

Abe, K., Kurakin, A., Mohseni-Maybodi, M., Kay, B., and Khosravi-Far, R. (2000). The complexity of TNF-related apoptosis-inducing ligand. Ann N Y Acad Sci 926, 52–63.

Acehan, D., Jiang, X., Morgan, D. G., Heuser, J. E., Wang, X., and Akey, C. W. (2002). Three-dimensional structure of the apoptosome: implications for assembly, procaspase-9 binding, and activation. Mol Cell 9, 423–432.

Adams, J. M. and Cory, S. (2002). Apoptosomes: engines for caspase activation. Curr Opin Cell Biol 14, 715–720.

Adrain, C., Slee, E. A., Harte, M. T., and Martin, S. J. (1999). Regulation of apoptotic protease activating factor-1 oligomerization and apoptosis by the WD-40 repeat region. J Biol Chem 274, 20855–20860.

Adrain, C., Creagh, E. M., and Martin, S. J. (2001). Apoptosis-associated release of Smac/DIABLO from mitochondria requires active caspases and is blocked by bcl-2. Embo J 20, 6627–6636.

Algeciras-Schimnich, A., Shen, L., Barnhart, B. C., Murmann, A. E., Burkhardt, J. K., and Peter, M. E. (2002). Molecular ordering of the initial signaling events of CD95. Mol Cell Biol 22, 207–220.

Andreyev, A. Y., Kushnareva, Y. E., and Starkov, A. A. (2005). Mitochondrial metabolism of reactive oxygen species. Biochemistry (Mosc) 70, 200–214.

Antonsson, B., Conti, F., Ciavatta, A., Montessuit, S., Lewis, S., Martinou, I., Bernasconi, L., Bernard, A., Mermod, J. J., Mazzei, G., et al. (1997). Inhibition of bax channel-forming activity by bcl-2. Science 277, 370–372.

Arnoult, D., Gaume, B., Karbowski, M., Sharpe, J. C., Cecconi, F., and Youle, R. J. (2003). Mitochondrial release of AIF and EndoG requires caspase activation downstream of bax/bak-mediated permeabilization. Embo J 22, 4385–4399.

Ashkenazi, A. and Dixit, V. M. (1999). Apoptosis control by death and decoy receptors. Curr Opin Cell Biol 11, 255–260.

Baliga, B. and Kumar, S. (2003). Apaf-1/cytochrome c apoptosome: an essential initiator of caspase activation or just a sideshow? Cell Death Differ 10, 16–18.

Barnhart, B. C., Legembre, P., Pietras, E., Bubici, C., Franzoso, G., and Peter, M. E. (2004). CD95 ligand induces motility and invasiveness of apoptosis-resistant tumor cells. Embo J 23, 3175–3185.

Bettaieb, A., Dubrez-Daloz, L., Launay, S., Plenchette, S., Rebe, C., Cathelin, S., and Solary, E. (2003). Bcl-2 proteins: targets and tools for chemosensitisation of tumor cells. Curr Med Chem Anti-Canc Agents 3, 307–318.

Beyer, R. E. (1992). An analysis of the role of coenzyme Q in free radical generation and as an antioxidant. Biochem Cell Biol 70, 390–403.

Brenner, C., Cadiou, H., Vieira, H. L., Zamzami, N., Marzo, I., Xie, Z., Leber, B., Andrews, D., Duclohier, H., Reed, J. C., and Kroemer, G. (2000). Bcl-2 and bax regulate the channel activity of the mitochondrial adenine nucleotide translocator. Oncogene 19, 329–336.

Brenner, C., Le Bras, M., and Kroemer, G. (2003). Insights into the mitochondrial signaling pathway: what lessons for chemotherapy? J Clin Immunol 23, 73–80.

Burns, T. F. and el-Deiry, W. S. (2003). Cell death signaling in malignancy. Cancer Treat Res 115, 319–343.

Cain, K., Bratton, S. B., and Cohen, G. M. (2002). The apaf-1 apoptosome: a large caspase-activating complex. Biochimie 84, 203–214.

Cande, C., Cecconi, F., Dessen, P., and Kroemer, G. (2002). Apoptosis-inducing factor (AIF): key to the conserved caspase-independent pathways of cell death? J Cell Sci 115, 4727–4734.

Carew, J. S. and Huang, P. (2002). Mitochondrial defects in cancer. Mol Cancer 1, 9.

Cecconi, F., Alvarez-Bolado, G., Meyer, B. I., Roth, K. A., and Gruss, P. (1998). Apaf1 (CED-4 homolog) regulates programmed cell death in mammalian development. Cell 94, 727–737.

Chao, D. T. and Korsmeyer, S. J. (1998). BCL-2 family: regulators of cell death. Annu Rev Immunol 16, 395–419.

Cheng, E. H., Wei, M. C., Weiler, S., Flavell, R. A., Mak, T. W., Lindsten, T., and Korsmeyer, S. J. (2001). BCL-2, BCL-X(L) sequester BH3 domain-only molecules preventing BAX- and BAK-mediated mitochondrial apoptosis. Mol Cell 8, 705–711.

Chinnaiyan, A. M. (1999). The apoptosome: heart and soul of the cell death machine. Neoplasia 1, 5–15.

Chou, J. J., Li, H., Salvesen, G. S., Yuan, J., and Wagner, G. (1999). Solution structure of BID, an intracellular amplifier of apoptotic signaling. Cell 96, 615–624.

Colombini, M., Blachly-Dyson, E., and Forte, M. (1996). VDAC, a channel in the outer mitochondrial membrane. Ion Channels 4, 169–202.

Cory, S. and Adams, J. M. (2002). The bcl2 family: regulators of the cellular life-or-death switch. Nat Rev Cancer 2, 647–656.

Cory, S., Huang, D. C., and Adams, J. M. (2003). The bcl-2 family: rolcs in cell survival and oncogenesis. Oncogene 22, 8590–8607.

Costantini, P., Jacotot, E., Decaudin, D., and Kroemer, G. (2000). Mitochondrion as a novel target of anticancer chemotherapy. J Natl Cancer Inst 92, 1042–1053.

Cote, J. and Ruiz-Carrillo, A. (1993). Primers for mitochondrial DNA replication generated by endonuclease G. Science 261, 765–769.

Crompton, M. (1999). The mitochondrial permeability transition pore and its role in cell death. Biochem J 341 (Pt 2), 233–249.

Danial, N. N. and Korsmeyer, S. J. (2004). Cell death: critical control points. Cell 116, 205–219.

Daniel, P. T., Wieder, T., Sturm, I., and Schulze-Osthoff, K. (2001). The kiss of death: promises and failures of death receptors and ligands in cancer therapy. Leukemia 15, 1022–1032.

Debatin, K. M., Poncet, D., and Kroemer, G. (2002). Chemotherapy: targeting the mitochondrial cell death pathway. Oncogene 21, 8786–8803.

Degli Esposti, M. (1999). To die or not to die–the quest of the TRAIL receptors. J Leukoc Biol 65, 535–542.

Dejean, L. M., Martinez-Caballero, S., Guo, L., Hughes, C., Teijido, O., Ducret, T., Ichas, F., Korsmeyer, S. J., Antonsson, B., Jonas, E. A., and Kinnally, K. W. (2005). Oligomeric bax is a component of the putative cytochrome c release channel MAC, mitochondrial apoptosis-induced channel. Mol Biol Cell 16, 2424–2432.

Dejean, L. M., Martinez-Caballero, S., and Kinnally, K. W. (2006). Is MAC the knife that cuts cytochrome c from mitochondria during apoptosis? Cell Death Differ 13, 1387–1395.

Deveraux, Q. L. and Reed, J. C. (1999). IAP family proteins – suppressors of apoptosis. Genes Dev 13, 239–252.

Dias, N. and Bailly, C. (2005). Drugs targeting mitochondrial functions to control tumor cell growth. Biochem Pharmacol 70, 1–12.

Diekert, K., de Kroon, A. I., Kispal, G., and Lill, R. (2001). Isolation and subfractionation of mitochondria from the yeast Saccharomyces cerevisiae. Methods Cell Biol 65, 37–51.

Dumont, M. E., Cardillo, T. S., Hayes, M. K., and Sherman, F. (1991). Role of cytochrome c heme lyase in mitochondrial import and accumulation of cytochrome c in Saccharomyces cerevisiae. Mol Cell Biol 11, 5487–5496.

Ekert, P. G., Silke, J., Hawkins, C. J., Verhagen, A. M., and Vaux, D. L. (2001). DIABLO promotes apoptosis by removing MIHA/XIAP from processed caspase 9. J Cell Biol 152, 483–490.

el-Deiry, W. S. (1997). Role of oncogenes in resistance and killing by cancer therapeutic agents. Curr Opin Oncol 9, 79–87.

Eskes, R., Desagher, S., Antonsson, B., and Martinou, J. C. (2000). Bid induces the oligomerization and insertion of bax into the outer mitochondrial membrane. Mol Cell Biol 20, 929–935.

Esposti, M. D., Cristea, I. M., Gaskell, S. J., Nakao, Y., and Dive, C. (2003). Proapoptotic bid binds to monolysocardiolipin, a new molecular connection between mitochondrial membranes and cell death. Cell Death Differ.

Fantin, V. R. and Leder, P. (2006). Mitochondriotoxic compounds for cancer therapy. Oncogene 25, 4787–4797.

Fiers, W., Beyaert, R., Declercq, W., and Vandenabeele, P. (1999). More than one way to die: apoptosis, necrosis and reactive oxygen damage. Oncogene 18, 7719–7730.

Fliss, M. S., Usadel, H., Caballero, O. L., Wu, L., Buta, M. R., Eleff, S. M., Jen, J., and Sidransky, D. (2000). Facile detection of mitochondrial DNA mutations in tumors and bodily fluids. Science 287, 2017–2019.

Galluzzi, L., Larochette, N., Zamzami, N., and Kroemer, G. (2006). Mitochondria as therapeutic targets for cancer chemotherapy. Oncogene 25, 4812–4830.

Geske, F. J. and Gerschenson, L. E. (2001). The biology of apoptosis. Hum Pathol 32, 1029–1038.

Green, D. R. and Evan, G. I. (2002). A matter of life and death. Cancer Cell 1, 19–30.

Green, D. R. and Kroemer, G. (2004). The pathophysiology of mitochondrial cell death. Science 305, 626–629.

Guihard, G., Bellot, G., Moreau, C., Pradal, G., Ferry, N., Thomy, R., Fichet, P., Meflah, K., and Vallette, F. M. (2004). The mitochondrial apoptosis-induced channel (MAC) corresponds to a late apoptotic event. J Biol Chem 279, 46542–46550.

Guo, L., Pietkiewicz, D., Pavlov, E. V., Grigoriev, S. M., Kasianowicz, J. J., Dejean, L. M., Korsmeyer, S. J., Antonsson, B., and Kinnally, K. W. (2004). Effects of cytochrome c on the mitochondrial apoptosis-induced channel MAC. Am J Physiol Cell Physiol 286, C1109–1117.

Hanahan, D. and Weinberg, R. A. (2000). The hallmarks of cancer. Cell 100, 57–70.

Hansen, T. M., Smith, D. J., and Nagley, P. (2006). Smac/DIABLO is not released from mitochondria during apoptotic signalling in cells deficient in cytochrome c. Cell Death Differ 13, 1181–1190.

Hegde, R., Srinivasula, S. M., Zhang, Z., Wassell, R., Mukattash, R., Cilenti, L., DuBois, G., Lazebnik, Y., Zervos, A. S., Fernandes-Alnemri, T., and Alnemri, E. S. (2002). Identification of Omi/HtrA2 as a mitochondrial apoptotic serine protease that disrupts inhibitor of apoptosis protein-caspase interaction. J Biol Chem 277, 432–438.

Hersey, P. and Zhang, X. D. (2003). Overcoming resistance of cancer cells to apoptosis. J Cell Physiol 196, 9–18.

Hill, M. M., Adrain, C., and Martin, S. J. (2003). Portrait of a killer: the mitochondrial apoptosome emerges from the shadows. Mol Interv 3, 19–26.

Jaattela, M. (2004). Multiple cell death pathways as regulators of tumour initiation and progression. Oncogene 23, 2746–2756.

Kasibhatla, S. and Tseng, B. (2003). Why target apoptosis in cancer treatment? Mol Cancer Ther 2, 573–580.

Khosravi-Far, R. and Esposti, M. D. (2004). Death receptor signals to mitochondria. Cancer Biol Ther 3, 1051–1057.

Kim, R., Emi, M., Tanabe, K., and Toge, T. (2004). Therapeutic potential of antisense Bcl-2 as a chemosensitizer for cancer therapy. Cancer 101, 2491–2502.

Kluck, R. M., Bossy-Wetzel, E., Green, D. R., and Newmeyer, D. D. (1997). The release of cytochrome c from mitochondria: a primary site for Bcl-2 regulation of apoptosis. Science 275, 1132–1136.

Kluck, R. M., Esposti, M. D., Perkins, G., Renken, C., Kuwana, T., Bossy-Wetzel, E., Goldberg, M., Allen, T., Barber, M. J., Green, D. R., and Newmeyer, D. D. (1999). The pro-apoptotic proteins, Bid and Bax, cause a limited permeabilization of the mitochondrial outer membrane that is enhanced by cytosol. J Cell Biol 147, 809–822.

Kluza, J., Gallego, M. A., Loyens, A., Beauvillain, J. C., Sousa-Faro, J. M., Cuevas, C., Marchetti, P., and Bailly, C. (2006). Cancer cell mitochondria are direct proapoptotic targets for the marine antitumor drug lamellarin D. Cancer Res 66, 3177–3187.

Korsmeyer, S. J., Wei, M. C., Saito, M., Weiler, S., Oh, K. J., and Schlesinger, P. H. (2000). Pro-apoptotic cascade activates BID, which oligomerizes BAK or BAX into pores that result in the release of cytochrome c. Cell Death Differ 7, 1166–1173.

Kroemer, G. (2003). Mitochondrial control of apoptosis: an introduction. Biochem Biophys Res Commun 304, 433–435.

Kuwana, T. and Newmeyer, D. D. (2003). Bcl-2-family proteins and the role of mitochondria in apoptosis. Curr Opin Cell Biol 15, 691–699.

Kuwana, T., Smith, J. J., Muzio, M., Dixit, V., Newmeyer, D. D., and Kornbluth, S. (1998). Apoptosis induction by caspase-8 is amplified through the mitochondrial release of cyto-chrome c. J Biol Chem 273, 16589–16594.

Kuwana, T., Mackey, M. R., Perkins, G., Ellisman, M. H., Latterich, M., Schneiter, R., Green, D. R., and Newmeyer, D. D. (2002). Bid, bax, and lipids cooperate to form supramolecular openings in the outer mitochondrial membrane. Cell 111, 331–342.

Lawen, A. (2003). Apoptosis-an introduction. Bioessays 25, 888–896.

Li, H., Zhu, H., Xu, C. J., and Yuan, J. (1998). Cleavage of BID by caspase 8 mediates the mitochondrial damage in the Fas pathway of apoptosis. Cell 94, 491–501.

Li, L. Y., Luo, X., and Wang, X. (2001). Endonuclease G is an apoptotic DNase when released from mitochondria. Nature 412, 95–99.

Lim, M. L., Minamikawa, T., and Nagley, P. (2001). The protonophore CCCP induces mitochondrial permeability transition without cytochrome c release in human osteosarcoma cells. FEBS Lett 503, 69–74.

Lo, S., Tolner, B., Taanman, J. W., Cooper, J. M., Gu, M., Hartley, J. A., Schapira, A. H., and Hochhauser, D. (2005). Assessment of the significance of mitochondrial DNA damage by chemotherapeutic agents. Int J Oncol 27, 337–344.

Lorenzo, H. K., Susin, S. A., Penninger, J., and Kroemer, G. (1999). Apoptosis inducing factor (AIF): a phylogenetically old, caspase-independent effector of cell death. Cell Death Differ 6, 516–524.

Luo, X., Budihardjo, I., Zou, H., Slaughter, C., and Wang, X. (1998). Bid, a Bcl2 interacting protein, mediates cytochrome c release from mitochondria in response to activation of cell surface death receptors. Cell 94, 481–490.

Lutter, M., Fang, M., Luo, X., Nishijima, M., Xie, X., and Wang, X. (2000). Cardiolipin provides specificity for targeting of tBid to mitochondria. Nat Cell Biol 2, 754–761.

Mangan, P. S. and Colombini, M. (1987). Ultrasteep voltage dependence in a membrane channel. Proc Natl Acad Sci USA 84, 4896–4900.

Martinou, J. C. and Green, D. R. (2001). Breaking the mitochondrial barrier. Nat Rev Mol Cell Biol 2, 63–67.

Martins, L. M., Iaccarino, I., Tenev, T., Gschmeissner, S., Totty, N. F., Lemoine, N. R., Savopoulos, J., Gray, C. W., Creasy, C. L., Dingwall, C., and Downward, J. (2002). The serine protease Omi/HtrA2 regulates apoptosis by binding XIAP through a reaper-like motif. J Biol Chem 277, 439–444.

Marzo, I., Brenner, C., and Kroemer, G. (1998a). The central role of the mitochondrial megachannel in apoptosis: evidence obtained with intact cells, isolated mitochondria, and purified protein complexes. Biomed Pharmacother 52, 248–251.

Marzo, I., Brenner, C., Zamzami, N., Jurgensmeier, J. M., Susin, S. A., Vieira, H. L., Prevost, M. C., Xie, Z., Matsuyama, S., Reed, J. C., and Kroemer, G. (1998b). Bax and adenine nucleotide translocator cooperate in the mitochondrial control of apoptosis. Science 281, 2027–2031.

Mayer, A., Neupert, W., and Lill, R. (1995). Translocation of apocytochrome c across the outer membrane of mitochondria. J Biol Chem 270, 12390–12397.

McDonnell, J. M., Fushman, D., Milliman, C. L., Korsmeyer, S. J., and Cowburn, D. (1999). Solution structure of the proapoptotic molecule BID: a structural basis for apoptotic agonists and antagonists. Cell 96, 625–634.

McMillin, J. B. and Dowhan, W. (2002). Cardiolipin and apoptosis. Biochim Biophys Acta 1585, 97–107.

Miller, L. K. (1999). An exegesis of IAPs: salvation and surprises from BIR motifs. Trends Cell Biol 9, 323–328.

Miramar, M. D., Costantini, P., Ravagnan, L., Saraiva, L. M., Haouzi, D., Brothers, G., Penninger, J. M., Peleato, M. L., Kroemer, G., and Susin, S. A. (2001). NADH oxidase activity of mitochondrial apoptosis-inducing factor. J Biol Chem 276, 16391–16398.

Morisaki, T. and Katano, M. (2003). Mitochondria-targeting therapeutic strategies for overcoming chemoresistance and progression of cancer. Curr Med Chem 10, 2517–2521.

Nechushtan, A., Smith, C. L., Lamensdorf, I., Yoon, S. H., and Youle, R. J. (2001). Bax and Bak coalesce into novel mitochondria-associated clusters during apoptosis. J Cell Biol 153, 1265–1276.

Newmeyer, D. D. and Ferguson-Miller, S. (2003). Mitochondria: releasing power for life and unleashing the machineries of death. Cell 112, 481–490.

O'Neill, J., Manion, M., Schwartz, P., and Hockenbery, D. M. (2004). Promises and challenges of targeting Bcl-2 anti-apoptotic proteins for cancer therapy. Biochim Biophys Acta 1705, 43–51.

Ohtsuka, T., Nishijima, M., Suzuki, K., and Akamatsu, Y. (1993). Mitochondrial dysfunction of a cultured Chinese hamster ovary cell mutant deficient in cardiolipin. J Biol Chem 268, 22914–22919.

Ozoren, N. and El-Deiry, W. S. (2002). Defining characteristics of types I and II apoptotic cells in response to TRAIL. Neoplasia 4, 551–557.

Ozoren, N. and El-Deiry, W. S. (2003). Cell surface Death Receptor signaling in normal and cancer cells. Semin Cancer Biol 13, 135–147.

Pavlov, E. V., Priault, M., Pietkiewicz, D., Cheng, E. H., Antonsson, B., Manon, S., Korsmeyer, S. J., Mannella, C. A., and Kinnally, K. W. (2001). A novel, high conductance channel of mitochondria linked to apoptosis in mammalian cells and Bax expression in yeast. J Cell Biol 155, 725–731.

Penninger, J. M. and Kroemer, G. (2003). Mitochondria, AIF and caspases – rivaling for cell death execution. Nat Cell Biol 5, 97–99.

Peter, M. E. and Krammer, P. H. (1998). Mechanisms of CD95 (APO-1/Fas)-mediated apoptosis. Curr Opin Immunol 10, 545–551.

Peter, M. E. and Krammer, P. H. (2003). The CD95(APO-1/Fas) DISC and beyond. Cell Death Differ 10, 26–35.

Raha, S. and Robinson, B. H. (2000). Mitochondria, oxygen free radicals, disease and ageing. Trends Biochem Sci 25, 502–508.

Ravagnan, L., Roumier, T., and Kroemer, G. (2002). Mitochondria, the killer organelles and their weapons. J Cell Physiol 192, 131–137.

Reed, J. C. (2004). Apoptosis mechanisms: implications for cancer drug discovery. Oncology (Williston Park) 18, 11–20.

Rehm, M., Dussmann, H., and Prehn, J. H. (2003). Real-time single cell analysis of Smac/DIABLO release during apoptosis. J Cell Biol 162, 1031–1043.

Rotem, R., Heyfets, A., Fingrut, O., Blickstein, D., Shaklai, M., and Flescher, E. (2005). Jasmonates: novel anticancer agents acting directly and selectively on human cancer cell mitochondria. Cancer Res 65, 1984–1993.

Saelens, X., Festjens, N., Vande Walle, L., van Gurp, M., van Loo, G., and Vandenabeele, P. (2004). Toxic proteins released from mitochondria in cell death. Oncogene 23, 2861–2874.

Saito, S., Hiroi, Y., Zou, Y., Aikawa, R., Toko, H., Shibasaki, F., Yazaki, Y., Nagai, R., and Komuro, I. (2000). beta-Adrenergic pathway induces apoptosis through calcineurin activation in cardiac myocytes. J Biol Chem 275, 34528–34533.

Salvesen, G. S. and Dixit, V. M. (1997). Caspases: intracellular signaling by proteolysis. Cell 91, 443–446.

Salvesen, G. S. and Renatus, M. (2002). Apoptosome: the seven-spoked death machine. Dev Cell 2, 256–257.

Schatz, G. (1995). Mitochondria: beyond oxidative phosphorylation. Biochim Biophys Acta 1271, 123–126.

Schendel, S. L., Xie, Z., Montal, M. O., Matsuyama, S., Montal, M., and Reed, J. C. (1997). Channel formation by antiapoptotic protein Bcl-2. Proc Natl Acad Sci USA 94, 5113–5118.

Schlame, M., Rua, D., and Greenberg, M. L. (2000). The biosynthesis and functional role of cardiolipin. Prog Lipid Res 39, 257–288.

Schulze-Osthoff, K., Ferrari, D., Los, M., Wesselborg, S., and Peter, M. E. (1998). Apoptosis signaling by death receptors. Eur J Biochem 254, 439–459.

Semenza, G. L., Artemov, D., Bedi, A., Bhujwalla, Z., Chiles, K., Feldser, D., Laughner, E., Ravi, R., Simons, J., Taghavi, P., and Zhong, H. (2001). The metabolism of tumours: 70 years later. Novartis Found Symp 240, 251–260; discussion 260–254.

Shangary, S. and Johnson, D. E. (2003). Recent advances in the development of anticancer agents targeting cell death inhibitors in the Bcl-2 protein family. Leukemia 17, 1470–1481.

Sheikh, M. S. and Huang, Y. (2004). Death receptors as targets of cancer therapeutics. Curr Cancer Drug Targets 4, 97–104.

Shi, Y. (2002). Apoptosome: the cellular engine for the activation of caspase-9. Structure (Camb) 10, 285–288.

Shidoji, Y., Hayashi, K., Komura, S., Ohishi, N., and Yagi, K. (1999). Loss of molecular interaction between cytochrome c and cardiolipin due to lipid peroxidation. Biochem Biophys Res Commun 264, 343–347.

Sies, H. and de Groot, H. (1992). Role of reactive oxygen species in cell toxicity. Toxicol Lett 64–65 Spec No, 547–551.

Singh, K. K., Russell, J., Sigala, B., Zhang, Y., Williams, J., and Keshav, K. F. (1999). Mitochondrial DNA determines the cellular response to cancer therapeutic agents. Oncogene 18, 6641–6646.

Slee, E. A., Adrain, C., and Martin, S. J. (1999). Serial killers: ordering caspase activation events in apoptosis. Cell Death Differ 6, 1067–1074.

Soengas, M. S., Alarcon, R. M., Yoshida, H., Giaccia, A. J., Hakem, R., Mak, T. W., and Lowe, S. W. (1999). Apaf-1 and caspase-9 in p53-dependent apoptosis and tumor inhibition. Science 284, 156–159.

Sorice, M., Circella, A., Cristea, I. M., Garofalo, T., Renzo, L. D., Alessandri, C., Valesini, G., and Esposti, M. D. (2004). Cardiolipin and its metabolites move from mitochondria to other cellular membranes during death receptor-mediated apoptosis. Cell Death Differ 11, 1133–1145.

Springs, S. L., Diavolitsis, V. M., Goodhouse, J., and McLendon, G. L. (2002). The kinetics of translocation of Smac/DIABLO from the mitochondria to the cytosol in HeLa cells. J Biol Chem 277, 45715–45718.

Srinivasula, S. M., Ahmad, M., Guo, Y., Zhan, Y., Lazebnik, Y., Fernandes-Alnemri, T., and Alnemri, E. S. (1999). Identification of an endogenous dominant-negative short isoform of caspase-9 that can regulate apoptosis. Cancer Res 59, 999–1002.

Stegh, A. H. and Peter, M. E. (2001). Apoptosis and caspases. Cardiol Clin 19, 13–29.

Strasser, A., O'Connor, L., and Dixit, V. M. (2000). Apoptosis signaling. Annu Rev Biochem 69, 217–245.

Susin, S. A., Lorenzo, H. K., Zamzami, N., Marzo, I., Snow, B. E., Brothers, G. M., Mangion, J., Jacotot, E., Costantini, P., Loeffler, M., et al. (1999). Molecular characterization of mitochondrial apoptosis-inducing factor. Nature 397, 441–446.

Suzuki, Y., Imai, Y., Nakayama, H., Takahashi, K., Takio, K., and Takahashi, R. (2001). A serine protease, HtrA2, is released from the mitochondria and interacts with XIAP, inducing cell death. Mol Cell 8, 613–621.

Suzuki, Y., Takahashi-Niki, K., Akagi, T., Hashikawa, T., and Takahashi, R. (2004). Mitochondrial protease Omi/HtrA2 enhances caspase activation through multiple pathways. Cell Death Differ 11, 208–216.

Szeto, H. H. (2006). Cell-permeable, mitochondrial-targeted, peptide antioxidants. Aaps J 8, E277–283.

Thomas, L., Blachly-Dyson, E., Colombini, M., and Forte, M. (1993). Mapping of residues forming the voltage sensor of the voltage-dependent anion-selective channel. Proc Natl Acad Sci USA 90, 5446–5449.

Thompson, C. B. (1995). Apoptosis in the pathogenesis and treatment of disease. Science 267, 1456–1462.

Thorburn, A. (2004). Death receptor-induced cell killing. Cell Signal 16, 139–144.

van Loo, G., Schotte, P., van Gurp, M., Demol, H., Hoorelbeke, B., Gevaert, K., Rodriguez, I., Ruiz-Carrillo, A., Vandekerckhove, J., Declercq, W., et al. (2001). Endonuclease G: a mitochondrial protein released in apoptosis and involved in caspase-independent DNA degradation. Cell Death Differ 8, 1136–1142.

Vaux, D. L. and Korsmeyer, S. J. (1999). Cell death in development. Cell 96, 245–254.

Verhagen, A. M. and Vaux, D. L. (2002). Cell death regulation by the mammalian IAP antagonist Diablo/Smac. Apoptosis 7, 163–166.

Vieira, H. L., Belzacq, A. S., Haouzi, D., Bernassola, F., Cohen, I., Jacotot, E., Ferri, K. F., El Hamel, C., Bartle, L. M., Melino, G., et al. (2001). The adenine nucleotide translocator: a target of nitric oxide, peroxynitrite, and 4-hydroxynonenal. Oncogene 20, 4305–4316.

Walensky, L. D. (2006). BCL-2 in the crosshairs: tipping the balance of life and death. Cell Death Differ 13, 1339–1350.

Wallach, D. (1997). Apoptosis. Placing death under control. Nature 388, 123, 125–126.

Wang, K., Yin, X. M., Chao, D. T., Milliman, C. L., and Korsmeyer, S. J. (1996). BID: a novel BH3 domain-only death agonist. Genes Dev 10, 2859–2869.

Warburg, E. (1951). Therapeutic imperative and evaluation of therapeutics. Ugeskr Laeger 113, 86–88.

Waterhouse, N. J., Goldstein, J. C., Kluck, R. M., Newmeyer, D. D., and Green, D. R. (2001). The (Holey) study of mitochondria in apoptosis. Methods Cell Biol 66, 365–391.

Wei, M. C., Lindsten, T., Mootha, V. K., Weiler, S., Gross, A., Ashiya, M., Thompson, C. B., and Korsmeyer, S. J. (2000). tBID, a membrane-targeted death ligand, oligomerizes BAK to release cytochrome c. Genes Dev 14, 2060–2071.

Wei, M. C., Zong, W. X., Cheng, E. H., Lindsten, T., Panoutsakopoulou, V., Ross, A. J., Roth, K. A., MacGregor, G. R., Thompson, C. B., and Korsmeyer, S. J. (2001). Proapoptotic BAX and BAK: a requisite gateway to mitochondrial dysfunction and death. Science 292, 727–730.

Wright, M. M., Howe, A. G., and Zaremberg, V. (2004). Cell membranes and apoptosis: role of cardiolipin, phosphatidylcholine, and anticancer lipid analogues. Biochem Cell Biol 82, 18–26.

Yang, J., Liu, X., Bhalla, K., Kim, C. N., Ibrado, A. M., Cai, J., Peng, T. I., Jones, D. P., and Wang, X. (1997). Prevention of apoptosis by Bcl-2: release of cytochrome c from mitochondria blocked. Science 275, 1129–1132.

Yang, X., Fraser, M., Moll, U. M., Basak, A., and Tsang, B. K. (2006). Akt-mediated cisplatin resistance in ovarian cancer: modulation of p53 action on caspase-dependent mitochondrial death pathway. Cancer Res 66, 3126–3136.

Zamzami, N. and Kroemer, G. (2001). The mitochondrion in apoptosis: how Pandora's box opens. Nat Rev Mol Cell Biol 2, 67–71.

Zhivotovsky, B., Hanson, K. P., and Orrenius, S. (1998a). Back to the future: the role of cytochrome c in cell death. Cell Death Differ 5, 459–460.

Zhivotovsky, B., Orrenius, S., Brustugun, O. T., and Doskeland, S. O. (1998b). Injected cytochrome c induces apoptosis. Nature 391, 449–450.

Zornig, M., Hueber, A., Baum, W., and Evan, G. (2001). Apoptosis regulators and their role in tumorigenesis. Biochim Biophys Acta 1551, F1–37.

Zou, H., Li, Y., Liu, X., and Wang, X. (1999). An APAF-1.cytochrome c multimeric complex is a functional apoptosome that activates procaspase-9. J Biol Chem 274, 11549–11556.

Chapter 4
Apoptotic Pathways in Tumor Progression and Therapy

Armelle Melet, Keli Song, Octavian Bucur, Zainab Jagani, Alexandra R. Grassian, and Roya Khosravi-Far*

Abstract Apoptosis is a cell suicide program that plays a critical role in development and tissue homeostasis. The ability of cancer cells to evade this programmed cell death (PCD) is a major characteristic that enables their uncontrolled growth. The efficiency of chemotherapy in killing such cells depends on the successful induction of apoptosis, since defects in apoptosis signaling are a major cause of drug resistance. Over the past decades, much progress has been made in our understanding of apoptotic signaling pathways and their dysregulation in cancer progression and therapy. These advances have provided new molecular targets for proapoptotic cancer therapies that have recently been used in drug development. While most of those therapies are still at the preclinical stage, some of them have shown much promise in the clinic. Here, we review our current knowledge of apoptosis regulation in cancer progression and therapy, as well as the new molecular targeted molecules that are being developed to reinstate cancer cell death.

Keywords apoptosis, cancer, therapy, inhibitors, signal transduction, oncogene, intrinsic, extrinsic

1 Introduction

Apoptosis, is an evolutionarily conserved mechanism for the selective removal of unwanted cells (Abe et al., 2000b; Degli Esposti, 1999; Lawen, 2003; Ozoren and El-Deiry, 2003; Peter and Krammer, 1998; Strasser et al., 2000; Thorburn, 2004). Regulation of apoptosis is critical for tissue homeostasis, therefore, its deregulation can lead to a variety of pathological conditions, including cancer. For this reason,

Armelle Melet, Keli Song, Octavian Bucur, Zainab Jagani, Alexandra R. Grassian,
and Roya Khosravi-Far

Department of Pathology, Harvard Medical School, Beth Israel Deaconess Medical Center,
99 Brookline Avenue, Boston, MA 02215, USA

*To whom correspondence should be addressed: Tel.: (617) 667-8526; fax: (617) 667-3524;
e-mail: rkhosrav@bidmc.harvard.edu

R. Khosravi-Far and E. White (eds.), *Programmed Cell Death in Cancer Progression
and Therapy*.
© Springer 2008

inhibition of apoptosis or the promotion of resistance to apoptosis contributes to carcinogenesis and chemoresistance (Burns and el-Deiry, 2003; Daniel et al., 2001; Green and Evan, 2002; Ozoren and El-Deiry, 2003; Sheikh and Huang, 2004; Thompson, 1995; Zornig et al., 2001).

Apoptosis is primarily mediated through the activation of specific proteases called caspases (cysteinyl, aspartate-specific proteases) (Algeciras-Schimnich et al., 2002; Ozoren and El-Deiry, 2003; Salvesen and Dixit, 1997; Stegh and Peter, 2001; Thorburn, 2004). Caspases, which are effectors of PCD, cleave multiple substrates, leading to biochemical and morphological changes that are characteristic of suicidal cells (Abe et al., 2000b; Bouillet et al., 2000). Cells undergoing apoptosis undergo cell membrane remodeling and blebbing; the exposure of phosphatidylserine (PS) at the external surface of the cell; cell shrinkage with cytoskeletal rearrangements; nuclear condensation; and DNA fragmentation (Ashkenazi and Dixit, 1999; Green and Evan, 2002; Lawen, 2003; Peter and Krammer, 2003; Schulze-Osthoff et al., 1998; Thorburn, 2004). These morphological changes culminate in the formation of apoptotic bodies that are normally eliminated by phagocytosis (Geske and Gerschenson, 2001; Wallach, 1997).

In this chapter, we introduce the major apoptotic machinery and discuss some recent insights into the involvement of apoptosis in cancer progression, cancer therapy, and resistance to therapy.

2   Apoptotic Machinery

In mammals, the two major signaling systems that result in the activation of caspases and the consequent induction of apoptosis are the *extrinsic* death receptor pathway and the *intrinsic* mitochondrial pathway (Fig. 4.1) (Abe et al., 2000b; Ozoren and El-Deiry, 2003; Peter and Krammer, 2003; Strasser et al., 2000; Thorburn, 2004). In the past few years, increasing evidence indicates that the death receptor and mitochondrial pathways are not isolated systems. Instead, significant cross talks and "biofeedbacks" regulate the apoptotic machinery (Abe et al., 2000b; Li and Yuan, 1999; Reed, 2000; Zornig et al., 2001).

2.1   *The Death Receptor Pathway of Apoptosis*

The extrinsic apoptotic pathway is activated upon the binding of cytokine ligands (i.e., FasL, tumor necrosis factor [TNF], and TNF-related apoptosis-inducing ligand [TRAIL]) to members of the TNFα receptor superfamily, which are usually called the death receptors (i.e., Fas, also called CD95/Apo-1; TNF receptors; and TRAIL receptors) (Abe et al., 2000b; Ashkenazi and Dixit, 1999; Ozoren and El-Deiry, 2003; Peter and Krammer, 2003; Schulze-Osthoff et al., 1998; Thorburn, 2004). Death receptors contain an intracellular globular interaction domain known

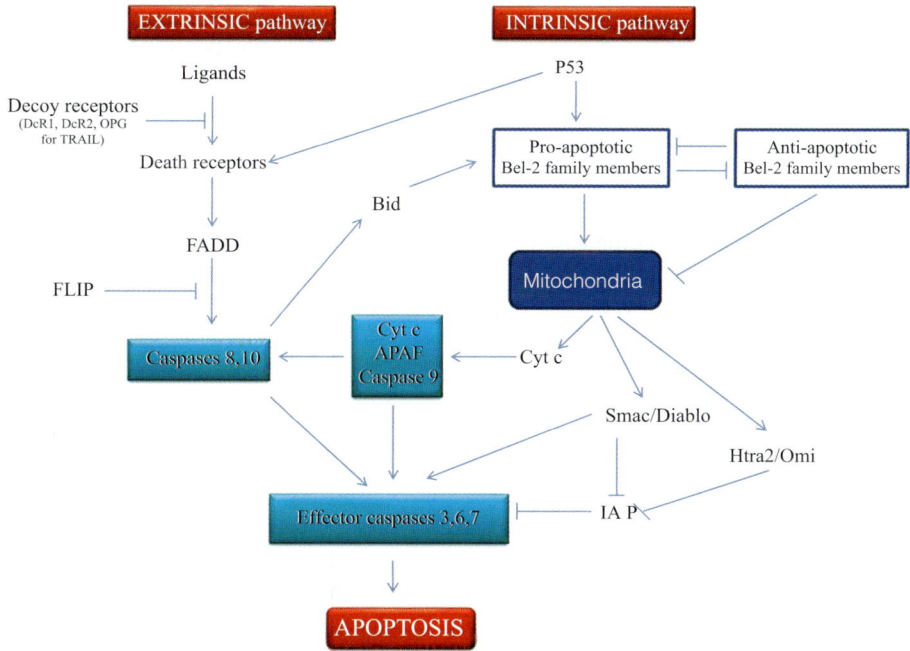

Fig. 4.1 Extrinsic and Intrinsic Apoptotic Machinery

as a death domain (DD). Upon ligand binding to their extracellular domains, death receptors aggregate at the cell surface and possibly form trimes. This results in the recruitment of adaptor molecules to the aggregated intracellular domains of the receptors. The Fas-associated death domain (FADD) is one of the major adaptors to be recruited to the death receptors. FADD possesses a DD that interacts either directly with the DD of death receptors, or indirectly through another adaptor molecule, TNF receptor-associated death domain (TRADD). FADD also contains a second protein interaction domain, known as the death effector domain (DED). This DED domain interacts with the DED of the weakly active zymogen pro-caspase-8, to form an intracellular multiprotein complex known as the death-inducing signaling complex (DISC) (Abe et al., 2000a; Ashkenazi and Dixit, 1998; Boatright et al., 2003; Cory and Adams, 2002; Wallach et al., 1999). Once formed, the DISC promotes the proximity-induced processing of caspase-8, which then proceeds to be further activated via an autoproteolysis mechanism (Salvesen and Dixit, 1999; Yang et al., 1998). Active caspase-8 subsequently activates executioner/effector caspases, such as caspase-3, leading to cell execution via degradation of the nucleus and other intracellular structures (Ashkenazi and Dixit, 1998; Cohen, 1997; Peter and Krammer, 2003; Scaffidi et al., 1998). This direct activation of caspase-dependent cell execution is thought to occur in certain cell types, including thymocytes, that are classified as type I cells (Boatright et al., 2003; Ozoren and El-Deiry, 2002; Scaffidi et al., 1998). These cells are able to efficiently activate caspase-8 so that it can cleave and consequently activate its primary targets, the executioner caspases including caspase-3. This simplified pathway of type I cells plays an important role in the immune response that is involved in the deletion of transformed cells (Hickman, 2002; Zornig et al., 2001) and resembles the linear pathway of developmental cell death established in genetic studies of *Caenorhabditis elegans* (Horvitz, 1999; Vaux, 2002). Nonetheless, PCD in *C. elegans* is distinct in that Bcl-2/Ced-9 is unable to block caspase activation following death receptor stimulation in type I cells (Peter and Krammer, 2003; Scaffidi et al., 1999).

2.2   *The Mitochondrial Pathway of Apoptosis*

Mitochondria are thought to be the central organelles involved in mediating most apoptotic pathways in mammalian cells (Green and Kroemer, 2004; Kroemer, 2003; Newmeyer and Ferguson-Miller, 2003; Ravagnan et al., 2002; Sorice et al., 2004; Zamzami and Kroemer, 2001). Mitochondria are engaged via the intrinsic pathway of cell death, which can be initiated by a variety of stress stimuli, including ultraviolet (UV) radiation, γ-irradiation, heat, DNA damage, the actions of some oncoproteins and tumor suppressor genes, viral virulence factors, and most chemotherapeutic agents (Kroemer, 2003). These diverse forms of stress are sensed or decoded by multiple cytosolic or intraorganellar molecules which then transduce the signals to mitochondria, resulting in alterations in the permeability of the outer mitochondrial membrane (OM) (Esposti et al., 2003; Green and Kroemer, 2004; Kuwana et al., 2002; Newmeyer and Ferguson-Miller, 2003; Zamzami and Kroemer, 2001). This leads to increased permeability to apoptotic proteins that are normally trapped between the OM and the inner mitochondrial membrane (IM), thus enabling these proteins to escape the mitochondria and diffuse into the cytosol.

The release of apoptotic factors leads to apoptosome-mediated activation of caspases (Fig. 4.1). The apoptosome works like a large platform for recruiting and facilitating the self-activation of the apical caspase of the intrinsic pathway of apoptosis, pro-caspase-9 (Adams and Cory, 2002; Baliga and Kumar, 2003; Cain et al., 2002; Chinnaiyan, 1999; Hill et al., 2003; Salvesen and Renatus, 2002; Shi, 2002). The apoptosome promotes the local accumulation of zymogens that initiate an autocatalytic activation of caspase-9 in a manner similar to the activation of caspase-8 at the DISC (Adams and Cory, 2002; Baliga and Kumar, 2003; Cain et al., 2002; Chinnaiyan, 1999; Hill et al., 2003; Salvesen and Renatus, 2002; Shi, 2002). The apoptosome, however, requires additional regulatory factors to fully activate the caspase cascade. These factors include Smac/Diablo, a protein that interacts with several inhibitor of apoptosis proteins (IAPs) and displaces them from their inhibitory interaction with pro-caspase-9 and other caspases.(Adams and Cory, 2002; Baliga and Kumar, 2003; Cain et al., 2002; Shi, 2002).

The mitochondrial pathway can also be activated in response to death ligands. In type II cells, selective cleavage of Bid by caspase-8 has been found to connect upstream signals from the DISC to the mitochondria (Gross et al., 1999; Li et al., 1998; Luo et al., 1998). Furthermore, genetic ablation of Bid reduces Fas-induced hepatotoxicity and mitochondrial damage (Zinkel et al., 2003). The caspase-cleaved form of Bid, tBid, migrates to the OM, where it cooperates with other Bcl-2 family proteins, such as Bak or Bax, to induce the release of mitochondrial proteins into the cytosol (Wei et al., 2001b).

3 Oncogene-Induced Evasion of Apoptosis: A Mechanism for Tumor Progression

It has become clear that a fundamental property of cancer cells is their ability to evade the apoptotic cellular death program (Hanahan and Weinberg, 2000). This not only promotes their unchecked growth, but also suggests a mechanism whereby they can be controlled. Investigating the mechanisms underlying this resistance of tumor cells to apoptosis remains of significant interest, since a desired goal of anticancer therapies is to selectively unleash the apoptotic potential of tumor cells.

In the normal cellular context, proliferation and death programs are tightly linked. Given this, cells harboring a single oncogenic mutation driving proliferation undergo a protective growth inhibitory response, appropriately resulting in apoptosis of the pre-neoplastic cell. In contrast, such as in the progression of tumors, oncogenes overcome these protective cellular responses by taking advantage of cooperating, additional mutations in apoptosis signaling molecules, resulting in the abnormal proliferation and survival/antiapoptosis of the tumor cell (Lowe et al., 2004). In a classic example, overexpression of the wild-type c-Myc oncoprotein can induce apoptosis and sensitize cells towards a host of apoptotic stimuli in certain cell types (Pelengaris et al., 2002). However, several events, including inactivation of p53, overexpression of Bcl-2, and loss of Bim are able to cooperate with Myc in inducing tumorigenesis (Pelengaris et al., 2002). In another strategy for tumorigenicity, fusion proteins such as Bcr-Abl can simultaneously activate multiple pathways, including those involved in cellular proliferation and in the promotion of survival and suppression of apoptosis.

3.1 Myc

C-Myc is a proto-oncogene first identified as the cellular homologue of the oncogene found in the avian myelocytomatosis retrovirus (Gonda et al., 1982). The other two Myc genes in mammals are MYCN and L-Myc. Myc, which is a transcription factor, can both activate and repress target genes. Recent estimates suggest that c-Myc could regulate as many as 15% of genes in genomes from flies to humans

(Fernandez et al., 2003; Orian et al., 2003). These target genes are involved in diverse functions including cell proliferation, differentiation, cell adhesion, metabolism, DNA repair, and apoptosis (Dang, 1999; Oster et al., 2002). Myc expression and its activity in normal cells are tightly regulated. However, Myc overexpression has been found in up to 50% of all human cancers (Alitalo and Schwab, 1986; Pompetti et al., 1996). Myc is thought to contribute to tumorigenesis through unrestrained cellular growth and proliferation and also exerts its effects on cellular processes such as cellular adhesion, angiogenesis and genomic instability (Calvisi et al., 2004; Felsher and Bishop, 1999; Ingvarsson, 1990; Knies-Bamforth et al., 2004). In addition to its established roles in promoting cellular growth and proliferation, Myc was also found to be an inducer of apoptosis (Evan and Vousden, 2001; Evan et al., 1992). It has been reported that Myc potentiates apoptosis through both p53-dependent and p53-independent mechanisms (Sakamuro et al., 1995). Even though the mechanisms by which Myc protein drives such disparate functions are still not well understood, it has been suggested that the ability of Myc to sensitize cells to apoptosis could be an intrinsic property of cells. Abrogation of this proapoptotic property profoundly contributes to cancer progression (Sakamuro et al., 1995). Some principles have also emerged from studies in cell culture and animal models to explain how Myc can promote cancer growth while acting as an inducer of apoptosis.

Modes of Myc dysregulation include chromosomal translocation and amplification, activation of upstream growth stimulatory signaling cascades, and increased protein stability (Oster et al., 2002). One of the important cellular processes caused by Myc dysregulation is genomic instability, which is prone to additional genomic mutations. Thus, activation of other oncogenes may follow in response to Myc deregulation. The antiapoptotic functions of some oncogenes can overcome the proapoptotic function of Myc. For example, in a conditional transgenic model of Myc-induced breast adenocarcinomas (Arvanitis and Felsher, 2006; Boxer et al., 2004; D'Cruz et al., 2001; Hutchinson and Muller, 2000), Myc inactivation results in tumor regression in about 50% of the tumors. Many of the tumors that initially regress subsequently relapse. Half of the tumors that do not regress and those that later relapse have active mutations in K-Ras or H-Ras. In those mice, mutant Ras appears to facilitate the ability of tumors to become independent of Myc.

Myc cooperation with other oncogenes is another important mechanism by which Myc promotes tumorigenesis. In mice, when the *C-Myc* transgene is coupled to the immunoglobulin heavy chain μ-enhancer, it leads to B-cell-specific overexpression of the *C-Myc* gene and development of lymphomas (Adams et al., 1985; Harris et al., 1988). This Eμ-Myc mouse is a model for the human disease Burkitt's lymphoma, where a reciprocal chromosomal translocation to the immunoglobulin locus leads to inappropriate expression of Myc in the B-cell compartment. The lymphomas that develop in the mouse model are consistently clonal, indicating that additional mutations are necessary to produce tumors. However, mice doubly transgenic for Eμ-Myc and Eμ-BCL2 mutations display a marked decrease in latency of disease, developing a leukemia of early progenitor cells (Strasser et al., 1990), rather than the lymphoma that develops with Eμ-Myc alone (Harris et al., 1988).

Studies in both cell culture and transgenic mice have shown that enforced Bcl2 expression was capable of blocking Myc-induced apoptosis and left the proliferation functions of Myc intact (Bissonnette et al., 1992; Fanidi et al., 1992; Letai et al., 2004).

Notably, the specific consequences of Myc inactivation appear to depend both on the type of cancer cells and the constellation of genetic events unique to a given tumor. Studies in conditional transgenic mouse model systems have shown that Myc inactivation results in the proliferation arrest, differentiation and/or apoptosis of tumor cells (Arvanitis and Felsher, 2006). Additionally, recent reports have suggested that targeted inactivation of Myc is a potential approach to cancer therapy, if used in conjunction with other anticancer treatments (Arvanitis and Felsher, 2006). To date, however, no drugs that target Myc have been identified for the treatment of humans with cancer (Arvanitis and Felsher, 2006).

3.2 Signaling Pathways Activated by Bcr-Abl and the Suppression of Apoptosis

The Bcr-Abl fusion protein activates multiple signaling pathways that lead to proliferation, reduced dependence on growth factors, apoptosis, and abnormal interactions with the extracellular matrix and stroma. Recent research suggests that one key mechanism by which Bcr-Abl facilitates the expansion of myeloid cells involves the suppression of apoptosis. Notably, the primary consequence of tyrosine kinase inhibition with imatinib in Bcr-Abl-transformed cells is the induction of apoptosis (Druker et al., 1996; Gambacorti-Passerini et al., 1997). Additionally, in growth factor-dependent hematopoietic cells, Bcr-Abl induces the survival and proliferation of cells that would otherwise undergo apoptotic cell death in response to growth factor withdrawal (Bedi et al., 1994). Furthermore, antisense oligonucleotide-mediated inhibition of Bcr-Abl expression in these transformed cells results in apoptosis without altering their cell cycle (Bedi et al., 1994). It has also been demonstrated that Bcr-Abl-positive cells are highly resistant to various apoptotic stimuli and become sensitized to drug treatment upon antisense inhibition of Bcr-Abl (McGahon et al., 1994). Additional evidence for the antiapoptotic effects of Bcr-Abl comes from experiments with temperature-sensitive Bcr-Abl kinase mutants, in which induction of Bcr-Abl kinase activity at the permissive temperature led to a significant decrease in apoptosis in the absence of growth factors (Carlesso et al., 1994; Kabarowski et al., 1994). In fact, studies in primary cells have revealed that chronic myelogenous leukemia (CML) progenitors show a normal proliferative response to growth factors and do not have a greater proliferative potential than normal progenitors (Emanuel et al., 1991). Furthermore, in the absence of serum and growth factors, neither normal nor CML progenitors proliferated, yet the latter maintained higher cell viability (Bedi et al., 1994).

As a result of its elevated tyrosine kinase activity, the Bcr-Abl fusion protein activates several signaling pathways, including Ras (Sawyers et al., 1995), PI3-K/Akt (Skorski et al., 1997; Varticovski et al., 1991), Stat (Carlesso et al., 1996; Ilaria and

Van Etten, 1996; Shuai et al., 1996), and NF-κB (Reuther et al., 1998) some of which may be crucial for its leukemogenic activity. In accordance with the ability of Bcr-Abl to substitute for the requirement of cytokines, many of these pathways are also activated by hematopoietic cytokines upon binding to their respective cytokine receptors. A functional consequence of the activation of these pathways involves changes in the activity and gene expression of key molecules, which have a direct impact on cellular survival, growth, and behavior. In particular, the Ras (Sawyers et al., 1995), PI3-K/Akt (Varticovski et al., 1991), Stat (Nieborowska-Skorska et al., 1999; Sillaber et al., 2000), NF-κB (Reuther et al., 1998), and FOXO (Ghaffari et al., 2003) pathways are capable of transmitting antiapoptotic signals, which could promote the evasion of Bcr-Abl-transformed cells from apoptosis. Thus, determining which of these antiapoptotic signals plays a role in Bcr-Abl-mediated evasion of apoptosis and promotion of leukemogenesis is of interest, especially since the cross talk between, and potential cooperation among, these pathways may be important in mediating the leukemogenic effects of Bcr-Abl.

4   Chemotherapeutic Drugs and Conventional Radiation-Induced Apoptosis in Tumor Cells

Aberrant cell proliferation, a major hallmark of cancer, has been exploited for anti-cancer drug development. Most existing chemotherapeutic drugs interfere with DNA synthesis and cell division, thereby preferentially killing rapidly dividing cells such as cancer cells (Schulze-Bergkamen and Krammer, 2004). These drugs include such diverse groups as antimetabolites, genotoxic/DNA-damaging agents (alkylating and intercalating agents, topoisomerase inhibitors) and mitotic inhibitors (vinca alkaloids and taxanes) (Luqmani, 2005). It is now well established that these cyto-toxic agents exert their antitumor activity mainly through induction of apoptosis and that defects in apoptotic pathways can lead to treatment failure (Johnstone et al., 2002; Kaufmann and Earnshaw, 2000; Lowe and Lin, 2000; Mesner et al., 1997).

Apoptosis induced by chemotherapeutic drugs primarily involves the mito-chondrial apoptotic pathway and, in some cases, the death receptor pathway and upregulation of death receptors and/or ligands (Bucur et al., 2006; Pommier et al., 2004). The relative contribution of each pathway to drug-induced apoptosis may depend on the cytotoxic drug, dose, kinetics, and cell type ((Fulda et al., 2001b), reviewed in (Debatin and Krammer, 2004)).

4.1   DNA-Damaging Agents and Induction of Apoptosis

Chemotherapeutic drugs damage DNA either directly or indirectly (antimetabolites) and subsequently initiate a DNA-damage response through both p53-dependent and p53-independent mechanisms (Waxman and Schwartz, 2003). Irradiation mainly induces direct DNA damage, but can also act indirectly, one example being the modulation of the epigenetic effectors in distant bystander tissue in vivo. X-ray exposure to one part of the animal body induces DNA strand breaks and causes an increase in levels of Rad51 in unexposed bystander tissue (Koturbash et al., 2006).

Drug- and radiation-induced DNA damage is first sensed by DNA-binding factors such as Rad17-RFC and the Rad9-1-1 supercomplex, BRCA1 and the Ku subunit of DNA-PK. The DNA damage signal is then transduced by activation of the PI3K family members DNA-PK, ATM, and ATR which, in turn, phosphorylate effector kinases such as the Ser/Thr kinases Chk1 and Chk2 and the tyrosine kinase c-Abl. These activated kinases then phosphorylate their downstream targets including the transcription factors p53, p73, and E2F, resulting in the transactivation of numerous genes involved in DNA repair, cell cycle arrest, and apoptosis.

The tumor suppressor p53 plays a key role in cellular response to stress and DNA damage (Meek, 2004) and has been implicated frequently in drug-induced apoptosis (Blagosklonny, 2002). Following DNA damage, p53 is induced by phosphorylation via ATM and Chk2. Phosphorylation of p53 not only enhances its DNA binding and transcriptional activity, but also stabilizes the protein by inhibiting its MDM2-mediated ubiquitination and its subsequent proteasomal degradation. The resulting increase in protein stability ultimately enhances the transcription of p53 target genes. p53 was shown to activate the mitochondrial apoptotic pathway by upregulating proapoptotic genes such as Bax, Bid, Noxa, and Puma, and down-regulating antiapoptotic proteins such as Bcl-2 and Mcl-1 (Michalak et al., 2005; Schuler and Green, 2005; Yu and Zhang, 2005). In addition, p53 can activate the extrinsic pathway by upregulating death receptors such as Fas, DR4, and DR5, although this pathway alone seems insufficient to induce apoptosis in some cancer cells (Reinke and Lozano, 1997). Recent evidence also suggests that p53 exerts proapoptotic functions independent of transcription, by translocating to the mitochondrion (Erster and Moll, 2005; Marchenko et al., 2000) and binding to Bcl-2 (Mihara et al., 2003; Tomita et al., 2006) and Bcl-XL (Mihara et al., 2003; Xu et al., 2006).

DNA-damaging agents can also induce apoptosis through p53-independent pathways involving, for instance, the transcription factor E2F (Lin et al., 2001). E2F exerts important proapoptotic activity in p53-deficient cells through transactivation of proapoptotic genes such as *Apaf-1* (Furukawa et al., 2002; Moroni et al., 2001), the caspase proenzymes (Nahle et al., 2002), *p73* (Irwin et al., 2003; Seelan et al., 2002; Stiewe and Putzer, 2000; Wang and Ki, 2001), and through repression of *Mcl-1* (Croxton et al., 2002). In certain cell types, radiation treatment, when used alone, may activate death receptors to execute the apoptotic program (Gong and Almasan, 2000). Finally, DNA-damaging drugs can engage a stress response via the stress-activated protein kinase/JNK pathway to activate the AP-1 and NF-κB-dependent transcription of *FasL* (Herr and Debatin, 2001; Kasibhatla et al., 1998).

4.2 *Targeting the Apoptotic Machinery Directly*

Targeting Bcl-2 family of proteins, death receptors, IAPs, caspases, and p53 are discussed in Chapter 8.

4.3   Microtubule Inhibitors and Induction of Apoptosis

Like DNA-damaging agents, microtubule inhibitors also lead to the phosphorylation
and stabilization of p53 as a mechanism for drug-induced apoptosis (Blagosklonny,
2002; Wang et al., 1999). However, in MCF-7 breast cancer cells, inactivation of p53
does not affect cellular sensitivity to paclitaxel killing. In those cells, p53 may act as a
survival factor by blocking them in the G2/M phase, rather than serving as an apoptotic
inducer. By contrast, the transcription factor FOXO3a has been shown to upregulate
the proapoptotic Bcl-2 family member, Bim, and contribute to paclitaxel-induced cell
death in MCF7 cells (Sunters et al., 2003). Similarly, another FOXO family member,
FOXO1, has been implicated in drug-induced apoptosis through the transcriptional
activation of the TNF-R1-associated protein TRADD (Rokudai et al., 2002).

4.4   Anticancer Therapeutics and Other Forms of Cell Death

In addition to classical apoptosis, anticancer drugs sometimes trigger autophagic
and necrotic modes of cell death (Gozuacik and Kimchi, 2004; Kim et al., 2006;
Kondo et al., 2005; Nelson and White, 2004). For instance, tamoxifen induces
autophagic cell death in MCF-7 cells (Bursch et al., 1996). Similarly, the
alkylating agent temozolomide kills malignant glioma cells through autophagy
rather than apoptosis (Kanzawa et al., 2004). Some reports also indicate that
paclitaxel and vinblastine induce both autophagic and apoptotic cell death (Broker
et al., 2004; Hirsimaki and Hirsimaki, 1984). Necrotic cell death (Proskuryakov
et al., 2003) has also been observed in vitro in resistant human ovarian carcinoma
cells exposed to HPMA copolymer-bound doxorubicin (Demoy et al., 2000) and
in vivo in p53/Bcl-2-deficient mice treated with DNA-alkylating agents (Zong
et al., 2004). These alternative modes of cell death have only recently been identified and their respective importance and possible cross talk in drug cytotoxic
action remain to be further defined. These forms of cell death together with
mitotic catastrophe are further discussed in Chapter 3.

5   Mechanisms of Radiation and Drug Resistance

Drug resistance can be classified as nononcogenic (impaired drug-target interaction)
and oncogenic (deregulation of apoptosis and the cell cycle). Principal mechanisms
of nononcogenic resistance include increased drug membrane export involving the
PgP protein product of the MDR gene; decreased drug activation; increased drug
degradation; enhanced DNA repair; and mutations of drug targets (Longley and
Johnston, 2005; Luqmani, 2005). In oncogenic resistance, the drug interacts with
its target, but downstream pathways of apoptosis and the cell cycle are altered

(Longley and Johnston, 2005; Luqmani, 2005). Intrinsic or acquired oncogenic resistance can result from multiple mechanisms, as outlined below.

5.1 Prosurvival Signaling (Mitogenic Kinases and NF-κB)

Mitogenic protein tyrosine kinases play a major role in drug resistance through their regulation of antiapoptotic signaling pathways (Blume-Jensen and Hunter, 2001). These include, for instance, members of the EGFR and Ras families, Bcr-Abl, and Akt. Overexpression of EGFR and Her-2 has been reported to increase resistance to chemotherapeutic drugs (Chevallier et al., 2004; Knuefermann et al., 2003; Mendelsohn and Fan, 1997; Nagane et al., 1998; Pegram et al., 1997). Activated Ras family members have also been shown to decrease cells' sensitivity to cytotoxic agents (Fan et al., 1997; Jansen et al., 1997). For example, several reports suggest that expression of Ras oncoproteins can contribute to cisplatin resistance by reducing drug uptake and increasing the degree of DNA repair (Dempke et al., 2000; Levy et al., 1994). Similarly, Bcr-Abl-expressing hematopoietic cell lines and various patient-derived CML cell lines are highly resistant to apoptotic induction by chemotherapy (Aichberger et al., 2005; Bedi et al., 1994; Cortez et al., 1996; Gesbert and Griffin, 2000; Keeshan et al., 2001; McGahon et al., 1994; Ray et al., 2004; Skorski, 2002; Underhill-Day et al., 2006).

The PI3K/Akt pathway, at the crossroads of multiple signaling networks, has been shown to be overactivated by upstream mitogenic kinases and oncogenic mutations in a wide range of tumors (Osaki et al., 2004). A number of studies have established that overexpression or activation of Akt increases chemoresistance both in cell lines (Page et al., 2000; Pommier et al., 2004) and in vivo (Kim et al., 2005; Martelli et al., 2005; McCormick, 2004; Wendel et al., 2004). Accordingly, inhibition of the PI3K/Akt pathway enhances the cytotoxic effects of a variety of chemotherapeutic agents (Hennessy et al., 2005; Nguyen et al., 2004; Nicholson et al., 2003; O'Gorman et al., 2000; Toretsky et al., 1999). The PTEN tumor suppressor is frequently mutated in human tumors. Loss of PTEN is associated with constitutive survival signaling through the PI3K/Akt pathway. Adenovirus-mediated expression of PTEN completely suppressed Akt activation in various cancer cell lines, such as the LNCaP prostate cancer cell line, and enhanced apoptosis induced by a broad range of apoptotic stimuli, including the chemotherapeutic agents mitoxantrone and etoposide, and death receptor-mediated treatments such as TRAIL, TNF-α, and agonistic antibodies against Fas (Yuan and Whang, 2002).

Finally, tumors with constitutive NF-κB activity are highly resistant to cytotoxic drugs (Arlt and Schafer, 2002; Baldwin, 2001). Accordingly, inhibition of NF-κB dramatically increases the sensitivity of these tumors to drugs by downregulation of antiapoptotic proteins (Nakanishi and Toi, 2005). Moreover, treatment with diverse cytotoxic drugs (including 5-FU, doxorubicin, paclitaxel, and cisplatin) can activate NF-κB, thereby blunting the ability of chemotherapy to induce cell death (Chuang et al., 2002).

5.2   Loss of p53 Function

Loss of p53 function is frequently encountered in human tumors and plays a critical role in resistance to chemotherapeutic drugs and conventional radiation (Levine et al., 2004; Lowe et al., 1994; Weller, 1998). Mechanisms responsible for p53 dysfunction include mutations or allelic loss in the p53 gene; upregulation of p53 inhibitors such as Mdm2; silencing of key p53 coactivators such as ARF; and altered upstream or downstream signaling (Vogelstein et al., 2000). For instance, lymphomas from mice deficient in p53 are markedly resistant to chemotherapy both in vitro and in vivo (Schmitt et al., 1999). p53 expression is predictive for response to chemotherapy in non-small-cell lung cancers (NSCLC) (Harada et al., 2003). Moreover, p53 mutations have been correlated with resistance to doxorubicin treatment and early relapse in patients with breast carcinomas (Aas et al., 1996). Cancers that retain wild-type p53 are more likely to respond to chemotherapy than other tumor types. However, many types of wt p53 tumors with defective apoptotic machinery do not undergo apoptosis despite genotoxic stress (Blagosklonny, 2001; Gudas et al., 1996).

5.3   Defective Apoptotic Machinery

5.3.1   Defective Mitochondrial Activation

The Bcl-2 protein family plays a pivotal role in the regulation of the mitochondrial apoptotic pathway and, consequently, its members serve as major regulators of tumor sensitivity to drugs (Kirkin et al., 2004; Kostanova-Poliakova and Sabova, 2005; Pommier et al., 2004). Overexpression of antiapoptotic Bcl-2 members such as Bcl-2, Bcl-XL, and Mcl-1, or deficiency of the proapoptotic members Bak and Bax, has been associated with drug resistance in cell lines, mouse models, and patients (Kirkin et al., 2004; Kostanova-Poliakova and Sabova, 2005; Pommier et al., 2004). Indeed, overexpression of Bcl-2 (Dole et al., 1994; Kamesaki et al., 1993; Miyashita and Reed, 1992; Walton et al., 1993) or Bcl-XL (Amundson et al., 2000) prevents apoptosis induced by most chemotherapeutic drugs in vitro. Some evidence also indicates similar effects with Mcl-1 overexpression (Song et al., 2005; Zhou et al., 1997). In concordance with overexpression data, downregulation of Bcl-XL or Bcl-2 has been shown to sensitize cancer cells to DNA damage-induced apoptosis. Fibroblasts deficient in both Bak and Bax are resistant to apoptosis induced by various agents (Wei et al., 2001a). While Bak deficiency renders Jurkat cells resistant to staurosporin, bleomycin, and cisplatin (Wang et al., 2001), loss of Bax expression is associated with acquired resistance to oxaliplatin (Gourdier et al., 2002) or resistance to 5-FU (Zhang et al., 2000) in colon carcinoma cell lines. Clinically, high expression of antiapoptotic Bcl-2 family members (Reed, 1996) and loss or inactivation of Bax (Ionov et al., 2000; Tai et al., 1998) has been correlated with poor response to chemotherapy in some types of malignancy, but not all kinds of tumors.

5.3.2 Impaired Activation of the Death Receptor Pathway

Alterations in activation of the death receptor pathway are also implicated in chemoresistance. For instance, downregulation of Fas/CD95 in lymphoid and solid tumors is often associated with resistance to drug-induced cell death (Debatin and Krammer, 2004; Friesen et al., 1997; Fulda et al., 1998a; Fulda et al., 1998b). Direct downstream signaling molecules such as FADD and c-FLIP are also involved. Of note, absence or low expression levels of FADD in acute myeloid leukemia cells predicts resistance to chemotherapy and poor outcome (Tourneur et al., 2004). c-FLIP silencing dramatically sensitizes colorectal cancer cells to the chemotherapeutic agents 5-fluorouracil, oxaliplatin, and irinotecan (Longley et al., 2006). In addition, decoy receptors seem to be also implicated in resistance of cancer cells to different treatments, like Apo2L/TRAIL. This ligand has five receptors, two of which have cytoplasmic DDs (DR4 and DR5) and three of which act as "decoys" (DcR1, DcR2, and osteoprotegerin [OPG]). DcR1 and OPG lack a cytosolic region and DcR2 has a truncated, nonfunctional cytoplasmic DD (Almasan and Ashkenazi, 2003).

5.3.3 Deregulation of Caspase Activation

Both the death receptor and mitochondrial pathways lead to the activation of caspases, the final effectors of apoptotic cell death. Deregulation in the expression of caspases or their regulators (Apaf-1 and IAPs) has been observed in tumors. Although caspase mutations occur at low frequency, caspase expression and function appears to be impaired frequently by epigenetic mechanisms in cancer cells (Teitz et al., 2000). Caspase-8 expression was found to be inactivated by hypermethylation in varied resistant tumors, including childhood neuroblastoma, Ewing and malignant brain tumors and melanoma (Teitz et al., 2000). Importantly, restoration of caspase-8 expression by gene transfer or by demethylation treatment sensitizes resistant tumor cells to drug-induced apoptosis (Fulda et al., 2001a; Teitz et al., 2001). Downregulation of caspase-3 has been proposed as a possible mechanism for breast cancer chemoresistance. Doxorubicin-induced apoptosis was restored by reconstitution of caspase-3 expression in caspase-3-deficient MCF-7 breast cancer cells (Devarajan et al., 2002).

Inhibition of caspase activity by members of the IAP family is also involved in chemotherapy resistance in some tumors. The X-linked inhibitor of apoptosis protein (XIAP) is a factor in chemoresistance of human androgen-insensitive DU145 prostate cancer cells, as its inhibition induces apoptosis and enhances sensitivity to chemotherapy (Amantana et al., 2004). Overexpression of many IAPs has been reported in multidrug-resistant HL-60 leukemia cells (Notarbartolo et al.,

2002). Survivin expression has been shown to inhibit paclitaxel-induced apoptosis in HeLa cells (Giodini et al., 2002) and to correlate with paclitaxel resistance in ovarian cancer (Zaffaroni et al., 2002). As a final example, loss of *Apaf-1* has been associated with chemoresistance in melanoma cells (Soengas et al., 2001).

6   Strategies to Overcome Chemotherapeutic Resistance

Conventional drugs target cancer cells preferentially, but not exclusively. As a result, they also kill high-proliferating normal cells from bone marrow and the gut, causing unwanted side effects. Moreover, the efficacy of therapy is limited by innate or acquired resistance to such agents. New targeted cancer therapies, though, aim at using drugs that interfere with specific defects of cancer cells to improve selectivity. Current therapies include the use of monoclonal antibodies, small molecules, RNAi, and adenovirus-based gene therapy. These methods are currently being developed and studied for use alone or in combination with conventional drugs to overcome resistance (Fesik, 2005). The major proapoptotic targeted therapies are outlined below.

6.1   Inhibition of Mitotic Kinases (RTK, Ras, Akt, and mTOR)

Targeting the mitotic kinases that are involved in the survival of cancer cells has become a potential strategy for the induction of apoptosis either as a single treatment or in combination with traditional therapies. Two major approaches are being considered for targeting these kinases: small-molecule inhibitors and blocking monoclonal antibodies.

Tyrosine kinase inhibitors have been designed to compete with and prevent the binding of ATP to the tyrosine kinase domain. One of the greatest advances in molecular targeted therapy in cancer involves the treatment of CML with a small-molecule inhibitor of the Bcr-Abl oncogenic kinase called imatinib-mesylate (Gleevec) (Deininger et al., 2005; Druker et al., 1996). Imatinib leads to unprecedented responses in the chronic phase, with 80% of newly diagnosed patients achieving complete hematological remission. Imatinib has been shown to eradicate Bcr-Abl-positive leukemia cells through the induction of apoptotic (le Coutre et al., 1999) or nonapoptotic caspase-independent cell death (Okada et al., 2004). Since imatinib also inhibits other kinases such as c-Kit and PDGFR, its application has been broadened to other types of cancer such as c-Kit-positive gastrointestinal stromal tumors (GIST) (Dagher et al., 2002). Other successful examples of targeted small molecules include two EGFR/HER1 tyrosine kinase inhibitors, gefitinib (Iressa), and erlotinib (Tarceva), both of which have recently been approved by the Food and Drug Administration (FDA) for the treatment of NSCLC (De Marinis et al., 2006).

Humanized monoclonal antibodies targeting the EGFR superfamily have also been developed to bind the extracellular domain of these receptors competitively and thus prevent tyrosine kinase activation. Trastuzumab (herceptin), a monoclonal antibody against the extracellular domain of Her-2, is a prime example (Emens, 2005). The drug, which is approved for the treatment of breast cancers overexpressing Her-2, is best used in combination with paclitaxel for first-line therapy, but may also be used as a single agent as second- and third-line therapy.

As Ras mutations have been found in a great majority of carcinomas, targeting of Ras or downstream effector pathways of Ras has been of great interest (Khosravi-Far et al., 1998; Khosravi-Far and Der, 1994; Wennerberg et al., 2005). Farnesyl transferase inhibitors have potently inhibited Ras in preclinical studies, but have exhibited rather disappointing results in clinics so far (Appels et al., 2005). Chemical inhibitors of the PI3K/Akt pathway have a potential use as suppressors of tumor growth and inducers of apoptosis (Hennessy et al., 2005). Although inhibition of the PI3K family members has been shown to inhibit growth of both cancer cells in vitro and tumors in animal models, these compounds so far lack selectivity. By contrast, rapamycin and its analogues, which inhibit the Akt downstream substrate mTOR, slow the growth of tumors in animal models without displaying significant toxicity (Morgensztern and McLeod, 2005) (Dudkin et al., 2001; Eng et al., 1984). These compounds are currently in clinical trials for the treatment of breast, colon, and lung cancers.

6.2 *Targeting of Transcription Factors*

Several transcription factors, including p53, members of the FOXO superfamily, and NF-κB, are involved in drug-induced cellular response and have therefore emerged as attractive targets for new apoptosis-inducing therapies (Kim et al., 2003). Restoring p53 activity in tumor cells has a therapeutic potential because p53 loss or dysfunction in many tumors is a major cause of drug resistance. Different approaches to restore p53 function include gene transfer of wt p53, chemical restoration of wt p53 activity, and inhibition of Mdm2–p53 interaction (Blagosklonny, 2002). Many clinical trials employing wt p53 gene transfer are ongoing in different types of p53-deficient cancers. p53 activity can also be restored by small molecules that modify mutant p53 back to wt (Bykov et al., 2003; Foster et al., 1999). CP-31398, a styrylquinazoline, restores a wt DNA-binding conformation to mutant p53 and is capable of suppressing tumor growth in vitro and in vivo (Luu et al., 2002). Blocking the interaction of Mdm2–p53 in order to inhibit p53 degradation has also been considered as a valuable strategy for cancer therapy. A small molecule inhibiting the p53 pocket of Mdm2 (IC50 = 100 nM) was recently discovered within a series of *cis*-imidazoline analogues called the nutlins (Vassilev et al., 2004). Dose-dependent antiproliferative and cytotoxic activities of nutlins were shown to be dependent on the p53 status of tumor cell lines. Nutlins inhibited the growth of tumors in xenograft models without causing significant toxicities (Vassilev et al., 2004). These results emphasize that small molecule inhibitors of Mdm2 could be valuable anticancer agents, especially for tumors retaining wt p53 but overexpressing Mdm2.

Recently, Hu et al. (2004) demonstrated that FOXO3a is inactivated by IKKβ in two thirds of breast cancer patients studied and that the presence of active FOXO3a correlates with improved patient survival. Additionally, FOXO3a has also been shown to be involved in paclitaxel-induced apoptosis in MCF-7 breast cancer cells. Notably, the FOXO family of transcription factors have been shown to regulate expression of proapoptotic genes such as Fas (Suhara et al., 2002), TRAIL (Ghaffari et al., 2003; Modur et al., 2002), and Bim (Gilley et al., 2003; Stahl et al., 2002). Taken together, these studies suggest that downregulation of FOXO transcription factors may be a key mechanism in tumorigenesis. As a proof of concept, chemical library screening identified a series of compounds that could target FOXO1 to the nucleus and that restored the induction of apoptosis in PTEN-null cells (Kau et al., 2003; Wang and El-Deiry, 2004).

Finally, an NF-κB inductive response to cytotoxic drugs can be targeted through its physiological inhibitor, IκB. Indeed, adenovirus-based inhibition of NF-κB elicited by gene delivery of an IκB superrepressor abrogates chemoresistance in some types of tumors such as androgen-independent prostate cancer cells and in glioma-derived cell lines (Orlowski and Baldwin, 2002). Targeting the IKK kinases that phosphorylate and promote the proteasomal degradation of IκB could be another approach, which is all the more appealing since FOXO3a is also regulated by IKKs in breast cancer cells. Moreover, conditional knockout of IKKβ in intestinal epithelial cells impedes irradiation-induced NF-κB activation and promotes the activation of p53 and apoptosis in those cells (Egan et al., 2004). Thus, as suggested by Finnberg and El-Deiry (2004), direct activation of FOXO3a, inhibition of NF-κB, and indirect activation of p53 by targeting IKKs could be an effective multifaceted anticancer therapy to inhibit cellular proliferation and promote cell death by multiple signaling pathways.

6.3   Direct Targeting of the Apoptotic Machinery

6.3.1   Activators of the Intrinsic/Mitochondrial Pathway

The mitochondria, as a major cell death checkpoint, constitute a prominent target for new anticancer therapies. The mitochondrial pathway can be selectively targeted by gene delivery of proapoptotic proteins such as Apaf-1 (Perkins et al., 2000) or Bax (Kagawa et al., 2000; Kaliberov et al., 2002). Alternatively, overexpressed antiapoptotic proteins such as Bcl-2, Bcl-XL, and XIAP can be downregulated. An antisense oligo against BCl-2, oblimersen, sensitizes patient-derived malignant melanoma cells to apoptosis induced by dacarbazine (Jansen et al., 2000) and has been recently approved by FDA for use in combination with this drug in advanced melanoma (Kim et al., 2004; Klasa et al., 2002). Phase II/III clinical trials are being carried out to assess the benefits of oblimersen in combination with conventional drugs in Acute myelogenous leukemia (AML) and Non-small Cell Lung Cancer (NSCLC). Varied designer ligands (peptidomimetics or organic small molecules) that bind to the BH3-binding pocket of BCL-2 and Bcl-XL have been shown to induce apoptosis in vitro (Yin et al., 2005; Degterev et al., 2001; Enyedy et al., 2001; Kutzki et al., 2002; Tzung et al., 2001; Walensky et al., 2004; Wang et al., 2000). To date, a stapled BH3 peptide has been reported to inhibit the growth of leukemia xenografts (Walensky et al., 2004) and a small-molecule inhibitor of the Bcl-2 family members, ABT-737 (Abbot Laboratories), has been shown to induce regression of solid tumors in vivo (Oltersdorf et al., 2005). Moreover, a dual Bcl-2/BclXL antagonist (GX15-070, GeminX Biotechnology) entered clinical trials last year.

Inhibition of XIAP and other IAPs is another mechanism considered to induce apoptosis in cancer cells (Huang et al., 2004; Schimmer et al., 2006). Knockdown of XIAP by antisense oligos or RNAi induces apoptosis in cancer cells (Adams, 2003; Lima et al., 2004; McManus et al., 2004). Peptidic and nonpeptidic inhibitors of XIAPs have also been reported (Huang et al., 2004; Schimmer et al., 2006). In particular, cell-permeable Smac peptidomimetics that inhibit IAPs potently induce caspase activation and apoptosis in cancer cells and inhibit tumor growth in xenograft mouse models (Fulda et al., 2002).

Finally, drugs which act directly on mitochondrial components are also being developed to enforce cell death in tumor cells in which upstream apoptotic pathways are disabled (reviewed in Dias and Bailly, 2005; Bouchier-Hayes et al., 2005; Costantini et al., 2000; Debatin et al., 2002). Betulinic acid, a natural pentacyclic triterpenoid which acts via the permeabilization transition pore, has been shown to exert antitumor effects against neuroblastodermal and malignant head and neck tumors irrespective of their p53 status (Fulda and Debatin, 2000; Pisha et al., 1995).

6.3.2 Activators of the Extrinsic/Death Receptors Pathway

There is significant interest in targeting the extrinsic pathway to circumvent drug resistance, since chemorefractory cells tend to have dysfunctional p53 and defects in their intrinsic pathway. Death receptor ligands such as Fas, TNF, and TRAIL can be strong inducers of apoptosis in tumor cells in vitro. Among these ligands, TRAIL emerges as the most promising antitumor agent due to its lack of toxicity (Abe et al., 2000b; Yagita et al., 2004). Unlike Fas and TNF, recombinant TRAIL induces tumor regression in preclinical models with little toxicity to normal tissues (Ashkenazi et al., 1999; Walczak et al., 1999) and is currently in phase I clinical trials for the treatment of solid tumors. Agonistic antibodies against DR4 and DR5 also induce apoptosis in cancer cells, but not in normal cells and slow the growth of tumors in xenograft tumor models with no apparent systemic toxicity (Yagita et al., 2004). Phases I and II clinical trials have been initiated for an agonistic antibody targeting DR4 and phase II clinical trials are ongoing for an antibody that targets DR5.

6.4   General Inhibitors (Proteasome, Hsp90, and HDAC)

Targeting more general cellular components such as the 26S proteasome (Adams, 2004), the molecular chaperone protein Hsp90 (Whitesell and Lindquist, 2005), and histone deacetylases (HDAC) (Minucci and Pelicci, 2006; Yoo and Jones, 2006) has led to some surprising success in specific anticancer therapy. These therapies, while not intended to induce apoptosis, preferentially kill cancer cells by exploiting their greater dependence on the targeted cellular processes than their normal counterparts.

The proteasome inhibitor Velcade has been approved for treatment of multiple myeloma and is under evaluation as a single agent or in combination chemotherapy for the treatment of other hematopoietic and solid cancers (Adams and Kauffman, 2004). Preclinical studies demonstrate that proteasome inhibition by Velcade potentiates the activity of other cancer therapeutics, in part by downregulating chemoresistance pathways such as NF-κB and by inducing proapoptotic proteins such as p53 or FOXO3a (Fujita et al., 2005; Ghaffari et al., 2003).

Hsp90 is a molecular chaperone protein required for the stability and function of multiple mutated chimeric and overexpressed signaling proteins. Hsp90 inhibitors have shown promising antitumor activity in preclinical model systems (Banerji et al., 2005) and a 17-AAG compound has reached phase II clinical trials (Heath et al., 2005).

HDAC inhibitors are novel anticancer agents in clinical development that target the family of HDAC enzymes responsible for deacetylating core nucleosomal histones and other proteins. The precise mechanisms resulting in the antiproliferative biological effects of these agents are not fully understood. Nevertheless, a phase I clinical trial of suberoylanilide hydroxamic acid (SAHA) has shown that it is well tolerated, and has antitumor activity in both solid and hematological tumors (Kelly et al., 2005).

6.5   Combined Treatment as a Strategy to Overcome Resistance to Conventional Radiation and Chemotherapeutic Drugs

Most chemotherapeutic agents utilize the apoptotic pathway to induce cancer cell death, as does radiation. To overcome resistance to apoptosis, combination therapies involving two or more treatments can be used. The efficiency is usually highest when these treatments act on different signaling pathways. Targeting apoptosis through both the intrinsic and extrinsic pathways has been shown to be a good strategy. For example, joint activation of the intrinsic pathway (via the Bcl-2 family of proteins) and the extrinsic pathway (using different ligands like Apo2L/TRAIL) has a synergistic effect in prostate cancer cell lines (Almasan and Ashkenazi, 2003). Also, resistance to death receptor-induced cell death (as in resistance to Apo2L/ TRAIL treatment) can be overcome by using a variety of therapeutic strategies, like the activation of the intrinsic pathway, inhibition of survival factors, metabolic inhibition (blocking protein synthesis), proteasome inhibition (bortezomib), and others (Bucur et al., 2006).

6.5.1 Conclusions and Future Directions

Apoptosis and its deregulation in cancer has been an intensive field of research over the past decades. A deeper understanding of the mechanisms involved in evasion of cancer cells from apoptosis, and its link to drug resistance, has enabled the recent development of molecular targeted proapoptotic therapies. These therapies have produced significant results in cancer treatment and, in the case of Gleevec in CML, have even exceeded expectations. However, as for conventional drugs, resistance has subsequently emerged, even to single targeted agents. To avoid resistance, combination therapies involving both an apoptosis inducer and a conventional drug appear to be the best approach to date, but different strategies can be also used.

When apoptosis is impaired, resistance can often be overcome by targeting both the extrinsic and intrinsic pathways of apoptosis. In addition, alternative modes of cell death, such as autophagy, mitotic catastrophe, or necrosis, might be activated. Involvement of these different types of cell death in drug-induced cytotoxicity raises the possibility of using these newly identified cellular pathways instead to treat chemoresistant cancers. A deeper understanding of these alternative modes of cell death and identification of the interplay and molecular switches between apoptosis–autophagy and necrosis might provide new therapeutic targets for cancer therapy.

Finally, it is becoming clearer that tumor cells are not homogeneous and that neither most conventional drugs nor even targeted agents such as Gleevec can eliminate the cancer stem cells from which the disease arises (Bhatia et al., 2003). Relapses could occur in part due to the failure of current therapies to target this specific and original cancer cell population. Therefore, future directions of cancer research must better decipher the different modes of cell death and their potential application in attacking cancer stem cells.

Acknowledgments We thank Susan Glueck for assistance with preparation of the manuscript.

References

Aas, T., Borresen, A. L., Geisler, S., Smith-Sorensen, B., Johnsen, H., Varhaug, J. E., Akslen, L. A., and Lonning, P. E. (1996). Specific P53 mutations are associated with de novo resistance to doxorubicin in breast cancer patients. Nat Med 2, 811–814.

Abe, K., Kurakin, A., Mohseni-Maybodi, M., Kay, B., and Khosravi-Far, R. (2000a). The complexity of TNF-related apoptosis-inducing ligand. Ann NY Acad Sci, 926, 52–63.

Abe, K., Kurakin, A., Mohseni-Maybodi, M., Kay, B., and Khosravi-Far, R. (2000b). The complexity of TNF-related apoptosis-inducing ligand. Ann N Y Acad Sci 926, 52–63.

Adams, J. (2003). The proteasome: structure, function, and role in the cell. Cancer Treat Rev 29 (Suppl 1), 3–9.

Adams, J. (2004). The proteasome: a suitable antineoplastic target. Nat Rev Cancer 4, 349–360.

Adams, J. and Kauffman, M. (2004). Development of the proteasome inhibitor Velcade (Bortezomib). Cancer Invest 22, 304–311.

Adams, J. M. and Cory, S. (2002). Apoptosomes: engines for caspase activation. Curr Opin Cell Biol 14, 715–720.

Adams, J. M., Harris, A. W., Pinkert, C. A., Corcoran, L. M., Alexander, W. S., Cory, S., Palmiter, R. D., and Brinster, R. L. (1985). The c-myc oncogene driven by immunoglobulin enhancers induces lymphoid malignancy in transgenic mice. Nature 318, 533–538.

Aichberger, K. J., Mayerhofer, M., Krauth, M. T., Skvara, H., Florian, S., Sonneck, K., Akgul, C., Derdak, S., Pickl, W. F., Wacheck, V., et al. (2005). Identification of mcl-1 as a BCR/ABL-dependent target in chronic myeloid leukemia (CML): evidence for cooperative antileukemic effects of imatinib and mcl-1 antisense oligonucleotides. Blood 105, 3303–3311.

Algeciras-Schimnich, A., Shen, L., Barnhart, B. C., Murmann, A. E., Burkhardt, J. K., and Peter, M. E. (2002). Molecular ordering of the initial signaling events of CD95. Mol Cell Biol 22, 207–220.

Alitalo, K. and Schwab, M. (1986). Oncogene amplification in tumor cells. Advances in Cancer Research 47, 235–281.

Almasan, A. and Ashkenazi, A. (2003). Apo2L/TRAIL: apoptosis signaling, biology, and potential for cancer therapy. Cytokine Growth Factor Rev 14, 337–348.

Amantana, A., London, C. A., Iversen, P. L., and Devi, G. R. (2004). X-linked inhibitor of apoptosis protein inhibition induces apoptosis and enhances chemotherapy sensitivity in human prostate cancer cells. Mol Cancer Ther 3, 699–707.

Amundson, S. A., Myers, T. G., Scudiero, D., Kitada, S., Reed, J. C., and Fornace, A. J., Jr. (2000). An informatics approach identifying markers of chemosensitivity in human cancer cell lines. Cancer Res 60, 6101–6110.

Appels, N. M., Beijnen, J. H., and Schellens, J. H. (2005). Development of farnesyl transferase inhibitors: a review. Oncologist 10, 565–578.

Arlt, A. and Schafer, H. (2002). NFkappaB-dependent chemoresistance in solid tumors. Int J Clin Pharmacol Ther 40, 336–347.

Arvanitis, C. and Felsher, D. W. (2006). Conditional transgenic models define how MYC initiates and maintains tumorigenesis. Semin Cancer Biol 16, 313–317.

Ashkenazi, A. and Dixit, V. M. (1998). Death receptors: signaling and modulation. Science 281, 1305–1308.

Ashkenazi, A. and Dixit, V. M. (1999). Apoptosis control by death and decoy receptors. Curr Opin Cell Biol 11, 255–260.

Ashkenazi, A., Pai, R. C., Fong, S., Leung, S., Lawrence, D. A., Marsters, S. A., Blackie, C., Chang, L., McMurtrey, A. E., Hebert, A., et al. (1999). Safety and antitumor activity of recombinant soluble Apo2 ligand. J Clin Invest 104, 155–162.

Baldwin, A. S. (2001). Control of oncogenesis and cancer therapy resistance by the transcription factor NF-kappaB. J Clin Invest 107, 241–246.

Baliga, B. and Kumar, S. (2003). Apaf-1/cytochrome c apoptosome: an essential initiator of caspase activation or just a sideshow? Cell Death Differ 10, 16–18.

Banerji, U., O'Donnell, A., Scurr, M., Pacey, S., Stapleton, S., Asad, Y., Simmons, L., Maloney, A., Raynaud, F., Campbell, M., et al. (2005). Phase I pharmacokinetic and pharmacodynamic study of 17-allylamino, 17-demethoxygeldanamycin in patients with advanced malignancies. J Clin Oncol 23, 4152–4161.

Bedi, A., Zehnbauer, B. A., Barber, J. P., Sharkis, S. J., and Jones, R. J. (1994). Inhibition of apoptosis by Bcr-Abl in chronic myeloid leukemia. Blood 83, 2038–2044.

Bhatia, R., Holtz, M., Niu, N., Gray, R., Snyder, D. S., Sawyers, C. L., Arber, D. A., Slovak, M. L., and Forman, S. J. (2003). Persistence of malignant hematopoietic progenitors in chronic myelogenous leukemia patients in complete cytogenetic remission following imatinib mesylate treatment. Blood 101, 4701–4707 (epub 2003 Feb 4706).

Bissonnette, R. P., Echeverri, F., Mahboubi, A., and Green, D. R. (1992). Apoptotic cell death induced by c-myc is inhibited by bcl-2. Nature 359, 552–554.

Blagosklonny, M. V. (2001). Paradox of Bcl-2 (and p53): why may apoptosis-regulating proteins be irrelevant to cell death? Bioessays 23, 947–953.

Blagosklonny, M. V. (2002). P53: an ubiquitous target of anticancer drugs. Int J Cancer 98, 161–166.

Blume-Jensen, P. and Hunter, T. (2001). Oncogenic kinase signalling. Nature 411, 355–365.

Boatright, K. M., Renatus, M., Scott, F. L., Sperandio, S., Shin, H., Pedersen, I. M., Ricci, J. E., Edris, W. A., Sutherlin, D. P., Green, D. R., and Salvesen, G. S. (2003). A unified model for apical caspase activation. Mol Cell 11, 529–541.

Bouchier-Hayes, L., Lartigue, L., and Newmeyer, D. D. (2005). Mitochondria: pharmacological manipulation of cell death. J Clin Invest 115, 2640–2647.

Bouillet, P., Huang, D. C., O'Reilly, L. A., Puthalakath, H., O'Connor, L., Cory, S., Adams, J. M., and Strasser, A. (2000). The role of the pro-apoptotic Bcl-2 family member bim in physiological cell death. Ann N Y Acad Sci 926, 83–89.

Boxer, R. B., Jang, J. W., Sintasath, L., and Chodosh, L. A. (2004). Lack of sustained regression of c-MYC-induced mammary adenocarcinomas following brief or prolonged MYC inactivation. Cancer Cell 6, 577–586.

Broker, L. E., Huisman, C., Span, S. W., Rodriguez, J. A., Kruyt, F. A., and Giaccone, G. (2004). Cathepsin B mediates caspase-independent cell death induced by microtubule stabilizing agents in non-small cell lung cancer cells. Cancer Res 64, 27–30.

Bucur, O., Ray, S., Bucur, M. C., and Almasan, A. (2006). APO2 ligand/tumor necrosis factor-related apoptosis-inducing ligand in prostate cancer therapy. Front Biosci 11, 1549–1568.

Burns, T. F. and el-Deiry, W. S. (2003). Cell death signaling in maligancy. Cancer Treat Res 115, 319–343.

Bursch, W., Ellinger, A., Kienzl, H., Torok, L., Pandey, S., Sikorska, M., Walker, R., and Hermann, R. S. (1996). Active cell death induced by the anti-estrogens tamoxifen and ICI 164 384 in human mammary carcinoma cells (MCF-7) in culture: the role of autophagy. Carcinogenesis 17, 1595–1607.

Bykov, V. J., Selivanova, G., and Wiman, K. G. (2003). Small molecules that reactivate mutant p53. Eur J Cancer 39, 1828–1834.

Cain, K., Bratton, S. B., and Cohen, G. M. (2002). The Apaf-1 apoptosome: a large caspase-activating complex. Biochimie 84, 203–214.

Calvisi, D. F., Ladu, S., Factor, V. M., and Thorgeirsson, S. S. (2004). Activation of beta-catenin provides proliferative and invasive advantages in c-myc/TGF-alpha hepatocarcinogenesis promoted by phenobarbital. Carcinogenesis 25, 901–908.

Carlesso, N., Frank, D. A., and Griffin, J. D. (1996). Tyrosyl phosphorylation and DNA binding activity of signal transducers and activators of transcription (STAT) proteins in hematopoietic cell lines transformed by Bcr/Abl. J Exp Med 183, 811–820.

Carlesso, N., Griffin, J. D., and Druker, B. J. (1994). Use of a temperature-sensitive mutant to define the biological effects of the p210Bcr-Abl tyrosine kinase on proliferation of a factor-dependent murine myeloid cell line. Oncogene 9, 149–156.

Chevallier, P., Robillard, N., Wuilleme-Toumi, S., Mechinaud, F., Harousseau, J. L., and Avet-Loiseau, H. (2004). Overexpression of Her2/neu is observed in one third of adult acute lymphoblastic leukemia patients and is associated with chemoresistance in these patients. Haematologica 89, 1399–1401.

Chinnaiyan, A. M. (1999). The apoptosome: heart and soul of the cell death machine. Neoplasia 1, 5–15.

Chuang, S. E., Yeh, P. Y., Lu, Y. S., Lai, G. M., Liao, C. M., Gao, M., and Cheng, A. L. (2002). Basal levels and patterns of anticancer drug-induced activation of nuclear factor-kappaB (NF-kappaB), and its attenuation by tamoxifen, dexamethasone, and curcumin in carcinoma cells. Biochem Pharmacol 63, 1709–1716.

Cohen, G. M. (1997). Caspases: the executioners of apoptosis. Biochem J 326 (Pt 1), 1–16.

Cortez, D., Stoica, G., Pierce, J. H., and Pendergast, A. M. (1996). The Bcr-Abl tyrosine kinase inhibits apoptosis by activating a Ras-dependent signaling pathway. Oncogene 13, 2589–2594.

Cory, S. and Adams, J. M. (2002). The Bcl2 family: regulators of the cellular life-or-death switch. Nat Rev Cancer 2, 647–656.

Costantini, P., Jacotot, E., Decaudin, D., and Kroemer, G. (2000). Mitochondrion as a novel target of anticancer chemotherapy. J Natl Cancer Inst 92, 1042–1053.

Croxton, R., Ma, Y., Song, L., Haura, E. B., and Cress, W. D. (2002). Direct repression of the Mcl-1 promoter by E2F1. Oncogene 21, 1359–1369.

D'Cruz, C. M., Gunther, E. J., Boxer, R. B., Hartman, J. L., Sintasath, L., Moody, S. E., Cox, J. D., Ha, S. I., Belka, G. K., Golant, A., et al. (2001). c-MYC induces mammary tumorigenesis by means of a preferred pathway involving spontaneous Kras2 mutations. Nature Medicine 7, 235–239.

Dagher, R., Cohen, M., Williams, G., Rothmann, M., Gobburu, J., Robbie, G., Rahman, A., Chen, G., Staten, A., Griebel, D., and Pazdur, R. (2002). Approval summary: imatinib mesylate in the treatment of metastatic and/or unresectable malignant gastrointestinal stromal tumors. Clin Cancer Res 8, 3034–3038.

Dang, C. V. (1999). c-Myc target genes involved in cell growth, apoptosis, and metabolism. Mol Cell Biol 19, 1–11.

Daniel, P. T., Wieder, T., Sturm, I., and Schulze-Osthoff, K. (2001). The kiss of death: promises and failures of death receptors and ligands in cancer therapy. Leukemia 15, 1022–1032.

De Marinis, F., De Santis, S., and De Petris, L. (2006). Second-line chemotherapy for non-small cell lung cancer. Ann Oncol 17 (Suppl 5), v68–v71.

Debatin, K. M. and Krammer, P. H. (2004). Death receptors in chemotherapy and cancer. Oncogene 23, 2950–2966.

Debatin, K. M., Poncet, D., and Kroemer, G. (2002). Chemotherapy: targeting the mitochondrial cell death pathway. Oncogene 21, 8786–8803.

Degli Esposti, M. (1999). To die or not to die–the quest of the TRAIL receptors. J Leukoc Biol 65, 535–542.

Degterev, A., Lugovskoy, A., Cardone, M., Mulley, B., Wagner, G., Mitchison, T., and Yuan, J. (2001). Identification of small-molecule inhibitors of interaction between the BH3 domain and Bcl-xL. Nat Cell Biol 3, 173–182.

Deininger, M., Buchdunger, E., and Druker, B. J. (2005). The development of imatinib as a therapeutic agent for chronic myeloid leukemia. Blood 105, 2640–2653 (epub 2004 Dec 2623).

Demoy, M., Minko, T., Kopeckova, P., and Kopecek, J. (2000). Time- and concentration-dependent apoptosis and necrosis induced by free and HPMA copolymer-bound doxorubicin in human ovarian carcinoma cells. J Control Release 69, 185–196.

Dempke, W., Voigt, W., Grothey, A., Hill, B. T., and Schmoll, H. J. (2000). Cisplatin resistance and oncogenes – a review. Anticancer Drugs 11, 225–236.

Devarajan, E., Sahin, A. A., Chen, J. S., Krishnamurthy, R. R., Aggarwal, N., Brun, A. M., Sapino, A., Zhang, F., Sharma, D., Yang, X. H., et al. (2002). Down-regulation of caspase 3 in breast cancer: a possible mechanism for chemoresistance. Oncogene 21, 8843–8851.

Dias, N. and Bailly, C. (2005). Drugs targeting mitochondrial functions to control tumor cell growth. Biochem Pharmacol 70, 1–12.

Dole, M., Nunez, G., Merchant, A. K., Maybaum, J., Rode, C. K., Bloch, C. A., and Castle, V. P. (1994). Bcl-2 inhibits chemotherapy-induced apoptosis in neuroblastoma. Cancer Res 54, 3253–3259.

Druker, B. J., Tamura, S., Buchdunger, E., Ohno, S., Segal, G. M., Fanning, S., Zimmermann, J., and Lydon, N. B. (1996). Effects of a selective inhibitor of the Abl tyrosine kinase on the growth of Bcr-Abl positive cells. Nat Med 2, 561–566.

Dudkin, L., Dilling, M. B., Cheshire, P. J., Harwood, F. C., Hollingshead, M., Arbuck, S. G., Travis, R., Sausville, E. A., and Houghton, P. J. (2001). Biochemical correlates of mTOR inhibition by the rapamycin ester CCI-779 and tumor growth inhibition. Clin Cancer Res 7, 1758–1764.

Egan, L. J., Eckmann, L., Greten, F. R., Chae, S., Li, Z. W., Myhre, G. M., Robine, S., Karin, M., and Kagnoff, M. F. (2004). IkappaB-kinasebeta-dependent NF-kappaB activation provides radioprotection to the intestinal epithelium. Proc Natl Acad Sci USA 101, 2452–2457.

Emanuel, P. D., Bates, L. J., Castleberry, R. P., Gualtieri, R. J., and Zuckerman, K. S. (1991). Selective hypersensitivity to granulocyte-macrophage colony-stimulating factor by juvenile chronic myeloid leukemia hematopoietic progenitors. Blood 77, 925–929.

Emens, L. A. (2005). Trastuzumab: targeted therapy for the management of HER-2/neu-overexpressing metastatic breast cancer. Am J Ther 12, 243–253.

Eng, C. P., Sehgal, S. N., and Vezina, C. (1984). Activity of rapamycin (AY-22,989) against transplanted tumors. J Antibiot (Tokyo) 37, 1231–1237.

Enyedy, I. J., Ling, Y., Nacro, K., Tomita, Y., Wu, X., Cao, Y., Guo, R., Li, B., Zhu, X., Huang, Y., et al. (2001). Discovery of small-molecule inhibitors of Bcl-2 through structure-based computer screening. J Med Chem 44, 4313–4324.

Erster, S. and Moll, U. M. (2005). Stress-induced p53 runs a transcription-independent death program. Biochem Biophys Res Commun 331, 843–850.

Esposti, M. D., Cristea, I. M., Gaskell, S. J., Nakao, Y., and Dive, C. (2003). Proapoptotic Bid binds to monolysocardiolipin, a new molecular connection between mitochondrial membranes and cell death. Cell Death Differ 10, 1300–1309.

Evan, G. I. and Vousden, K. H. (2001). Proliferation, cell cycle and apoptosis in cancer. Nature 411, 342–348.

Evan, G. I., Wyllie, A. H., Gilbert, C. S., Littlewood, T. D., Land, H., Brooks, M., Waters, C. M., Penn, L. Z., and Hancock, D. C. (1992). Induction of apoptosis in fibroblasts by c-myc protein. Cell 69, 119–128.

Fan, J., Banerjee, D., Stambrook, P. J., and Bertino, J. R. (1997). Modulation of cytotoxicity of chemotherapeutic drugs by activated H-ras. Biochem Pharmacol 53, 1203–1209.

Fanidi, A., Harrington, E. A., and Evan, G. I. (1992). Cooperative interaction between c-myc and bcl-2 proto-oncogenes. Nature 359, 554–556.

Felsher, D. W. and Bishop, J. M. (1999). Transient excess of MYC activity can elicit genomic instability and tumorigenesis. Proc Natl Acad Sci USA 96, 3940–3944.

Fernandez, P. C., Frank, S. R., Wang, L., Schroeder, M., Liu, S., Greene, J., Cocito, A., and Amati, B. (2003). Genomic targets of the human c-Myc protein. Genes Dev 17, 1115–1129.

Fesik, S. W. (2005). Promoting apoptosis as a strategy for cancer drug discovery. Nat Rev Cancer 5, 876–885.

Finnberg, N. and El-Deiry, W. S. (2004). Activating FOXO3a, NF-kappaB and p53 by targeting IKKs: an effective multi-faceted targeting of the tumor-cell phenotype? Cancer Biol Ther 3, 614–616 (epub 2004 July 2024).

Foster, B. A., Coffey, H. A., Morin, M. J., and Rastinejad, F. (1999). Pharmacological rescue of mutant p53 conformation and function. Science 286, 2507–2510.

Friesen, C., Fulda, S., and Debatin, K. M. (1997). Deficient activation of the CD95 (APO-1/Fas) system in drug-resistant cells. Leukemia 11, 1833–1841.

Fujita, T., Washio, K., Takabatake, D., Takahashi, H., Yoshitomi, S., Tsukuda, K., Ishibe, Y., Ogasawara, Y., Doihara, H., and Shimizu, N. (2005). Proteasome inhibitors can alter the signaling pathways and attenuate the P-glycoprotein-mediated multidrug resistance. Int J Cancer 117, 670–682.

Fulda, S. and Debatin, K. M. (2000). Betulinic acid induces apoptosis through a direct effect on mitochondria in neuroectodermal tumors. Med Pediatr Oncol 35, 616–618.

Fulda, S., Friesen, C., and Debatin, K. M. (1998a). Molecular determinants of apoptosis induced by cytotoxic drugs. Klin Padiatr 210, 148–152.

Fulda, S., Los, M., Friesen, C., and Debatin, K. M. (1998b). Chemosensitivity of solid tumor cells in vitro is related to activation of the CD95 system. Int J Cancer 76, 105–114.

Fulda, S., Kufer, M. U., Meyer, E., van Valen, F., Dockhorn-Dworniczak, B., and Debatin, K. M. (2001a). Sensitization for death receptor- or drug-induced apoptosis by re-expression of caspase-8 through demethylation or gene transfer. Oncogene 20, 5865–5877.

Fulda, S., Meyer, E., Friesen, C., Susin, S. A., Kroemer, G., and Debatin, K. M. (2001b). Cell type specific involvement of death receptor and mitochondrial pathways in drug-induced apoptosis. Oncogene 20, 1063–1075.

Fulda, S., Wick, W., Weller, M., and Debatin, K. M. (2002). Smac agonists sensitize for Apo2L/TRAIL- or anticancer drug-induced apoptosis and induce regression of malignant glioma in vivo. Nat Med 8, 808–815 (epub 2002 July 2015).

Furukawa, Y., Nishimura, N., Satoh, M., Endo, H., Iwase, S., Yamada, H., Matsuda, M., Kano, Y., and Nakamura, M. (2002). Apaf-1 is a mediator of E2F-1-induced apoptosis. J Biol Chem 277, 39760–39768 (epub 32002 July 39730).

Gambacorti-Passerini, C., le Coutre, P., Mologni, L., Fanelli, M., Bertazzoli, C., Marchesi, E., Di Nicola, M., Biondi, A., Corneo, G. M., Belotti, D., et al. (1997). Inhibition of the ABL kinase activity blocks the proliferation of BCR/ABL + leukemic cells and induces apoptosis. Blood Cells Mol Dis 23, 380–394.

Gesbert, F. and Griffin, J. D. (2000). Bcr/Abl activates transcription of the Bcl-X gene through STAT5. Blood 96, 2269–2276.

Geske, F. J. and Gerschenson, L. E. (2001). The biology of apoptosis. Hum Pathol 32, 1029–1038.

Ghaffari, S., Jagani, Z., Kitidis, C., Lodish, H. F., and Khosravi-Far, R. (2003). Cytokines and Bcr-Abl mediate suppression of TRAIL-induced apoptosis through inhibition of forkhead FOXO3a transcription factor. Proc Natl Acad Sci USA 100, 6523–6528.

Gilley, J., Coffer, P. J., and Ham, J. (2003). FOXO transcription factors directly activate bim gene expression and promote apoptosis in sympathetic neurons. J Cell Biol 162, 613–622.

Giodini, A., Kallio, M. J., Wall, N. R., Gorbsky, G. J., Tognin, S., Marchisio, P. C., Symons, M., and Altieri, D. C. (2002). Regulation of microtubule stability and mitotic progression by survivin. Cancer Res 62, 2462–2467.

Gonda, T. J., Sheiness, D. K., and Bishop, J. M. (1982). Transcripts from the cellular homologs of retroviral oncogenes: distribution among chicken tissues. Mol Cell Biol 2, 617–624.

Gong, B. and Almasan, A. (2000). Genomic organization and transcriptional regulation of human Apo2/TRAIL gene. Biochem Biophys Res Commun 278, 747–752.

Gourdier, I., Del Rio, M., Crabbe, L., Candeil, L., Copois, V., Ychou, M., Auffray, C., Martineau, P., Mechti, N., Pommier, Y., and Pau, B. (2002). Drug specific resistance to oxaliplatin is associated with apoptosis defect in a cellular model of colon carcinoma. FEBS Lett 529, 232–236.

Gozuacik, D. and Kimchi, A. (2004). Autophagy as a cell death and tumor suppressor mechanism. Oncogene 23, 2891–2906.

Green, D. R. and Evan, G. I. (2002). A matter of life and death. Cancer Cell 1, 19–30.

Green, D. R. and Kroemer, G. (2004). The pathophysiology of mitochondrial cell death. Science 305, 626–629.

Gross, A., Yin, X. M., Wang, K., Wei, M. C., Jockel, J., Milliman, C., Erdjument-Bromage, H., Tempst, P., and Korsmeyer, S. J. (1999). Caspase cleaved BID targets mitochondria and is required for cytochrome c release, while BCL-XL prevents this release but not tumor necrosis factor-R1/Fas death. J Biol Chem 274, 1156–1163.

Gudas, J. M., Nguyen, H., Li, T., Sadzewicz, L., Robey, R., Wosikowski, K., and Cowan, K. H. (1996). Drug-resistant breast cancer cells frequently retain expression of a functional wild-type p53 protein. Carcinogenesis 17, 1417–1427.

Hanahan, D. and Weinberg, R. A. (2000). The hallmarks of cancer. Cell 100, 57–70.

Harada, T., Ogura, S., Yamazaki, K., Kinoshita, I., Itoh, T., Isobe, H., Yamashiro, K., Dosaka-Akita, H. and Nishimura, M. (2003). Predictive value of expression of P53, Bcl-2 and lung resistance-related protein for response to chemotherapy in non-small cell lung cancers. Cancer Sci 94, 394–399.

Harris, A. W., Pinkert, C. A., Crawford, M., Langdon, W. Y., Brinster, R. L., and Adams, J. M. (1988). The E mu-myc transgenic mouse. A model for high-incidence spontaneous lymphoma and leukemia of early B cells. J Exp Med 167, 353–371.

Heath, E. I., Gaskins, M., Pitot, H. C., Pili, R., Tan, W., Marschke, R., Liu, G., Hillman, D., Sarkar, F., Sheng, S., et al. (2005). A phase II trial of 17-allylamino-17- demethoxygeldanamycin in patients with hormone-refractory metastatic prostate cancer. Clin Prostate Cancer 4, 138–141.

Hennessy, B. T., Smith, D. L., Ram, P. T., Lu, Y., and Mills, G. B. (2005). Exploiting the PI3K/ AKT pathway for cancer drug discovery. Nat Rev Drug Discov 4, 988–1004.

Herr, I. and Debatin, K. M. (2001). Cellular stress response and apoptosis in cancer therapy. Blood 98, 2603–2614.

Hickman, J. A. (2002). Apoptosis and tumourigenesis. Curr Opin Genet Dev 12, 67–72.

Hill, M. M., Adrain, C., and Martin, S. J. (2003). Portrait of a killer: the mitochondrial apoptosome emerges from the shadows. Mol Interv 3, 19–26.

Hirsimaki, Y. and Hirsimaki, P. (1984). Vinblastine-induced autophagocytosis: the effect of disorganization of microfilaments by cytochalasin B. Exp Mol Pathol 40, 61–69.

Horvitz, H. R. (1999). Genetic control of programmed cell death in the nematode *Caenorhabditis elegans*. Cancer Res 59, 1701s–1706s.

Hu, M. C., Lee, D. F., Xia, W., Golfman, L. S., Ou-Yang, F., Yang, J. Y., Zou, Y., Bao, S., Hanada, N., Saso, H., et al. (2004). IkappaB kinase promotes tumorigenesis through inhibition of forkhead FOXO3a. Cell 117, 225–237.

Huang, Y., Lu, M., and Wu, H. (2004). Antagonizing XIAP-mediated caspase-3 inhibition. Achilles' heel of cancers? Cancer Cell 5, 1–2.

Hutchinson, J. N. and Muller, W. J. (2000). Transgenic mouse models of human breast cancer. Oncogene 19, 6130–6137.

Ilaria, R. L., Jr. and Van Etten, R. A. (1996). P210 and P190(BCR/ABL) induce the tyrosine phosphorylation and DNA binding activity of multiple specific STAT family members. J Biol Chem 271, 31704–31710.

Ingvarsson, S. (1990). The myc gene family proteins and their role in transformation and differentiation. Semin Cancer Biol 1, 359–369.

Ionov, Y., Yamamoto, H., Krajewski, S., Reed, J. C., and Perucho, M. (2000). Mutational inactivation of the proapoptotic gene BAX confers selective advantage during tumor clonal evolution. Proc Natl Acad Sci USA 97, 10872–10877.

Irwin, M. S., Kondo, K., Marin, M. C., Cheng, L. S., Hahn, W. C., and Kaelin, W. G., Jr. (2003). Chemosensitivity linked to p73 function. Cancer Cell 3, 403–410.

Jansen, B., Schlagbauer-Wadl, H., Eichler, H. G., Wolff, K., van Elsas, A., Schrier, P. I., and Pehamberger, H. (1997). Activated N-ras contributes to the chemoresistance of human melanoma in severe combined immunodeficiency (SCID) mice by blocking apoptosis. Cancer Res 57, 362–365.

Jansen, B., Wacheck, V., Heere-Ress, E., Schlagbauer-Wadl, H., Hoeller, C., Lucas, T., Hoermann, M., Hollenstein, U., Wolff, K., and Pehamberger, H. (2000). Chemosensitisation of malignant melanoma by BCL2 antisense therapy. Lancet 356, 1728–1733.

Johnstone, R. W., Ruefli, A. A., and Lowe, S. W. (2002). Apoptosis: a link between cancer genetics and chemotherapy. Cell 108, 153–164.

Kabarowski, J. H., Allen, P. B., and Wiedemann, L. M. (1994). A temperature sensitive p210 Bcr-Abl mutant defines the primary consequences of Bcr-Abl tyrosine kinase expression in growth factor dependent cells. EMBO J 13, 5887–5895.

Kagawa, S., Gu, J., Swisher, S. G., Ji, L., Roth, J. A., Lai, D., Stephens, L. C., and Fang, B. (2000). Antitumor effect of adenovirus-mediated Bax gene transfer on p53-sensitive and p53-resistant cancer lines. Cancer Res 60, 1157–1161.

Kaliberov, S. A., Buchsbaum, D. J., Gillespie, G. Y., Curiel, D. T., Arafat, W. O., Carpenter, M., and Stackhouse, M. A. (2002). Adenovirus-mediated transfer of BAX driven by the vascular endothelial growth factor promoter induces apoptosis in lung cancer cells. Mol Ther 6, 190–198.

Kamesaki, S., Kamesaki, H., Jorgensen, T. J., Tanizawa, A., Pommier, Y., and Cossman, J. (1993). bcl-2 protein inhibits etoposide-induced apoptosis through its effects on events subsequent to topoisomerase II-induced DNA strand breaks and their repair. Cancer Res 53, 4251–4256.

Kanzawa, T., Germano, I. M., Komata, T., Ito, H., Kondo, Y., and Kondo, S. (2004). Role of autophagy in temozolomide-induced cytotoxicity for malignant glioma cells. Cell Death Differ 11, 448–457.

Kasibhatla, S., Brunner, T., Genestier, L., Echeverri, F., Mahboubi, A., and Green, D. R. (1998). DNA damaging agents induce expression of Fas ligand and subsequent apoptosis in T lymphocytes via the activation of NF-kappa B and AP-1. Mol Cell 1, 543–551.

Kau, T. R., Schroeder, F., Ramaswamy, S., Wojciechowski, C. L., Zhao, J. J., Roberts, T. M., Clardy, J., Sellers, W. R., and Silver, P. A. (2003). A chemical genetic screen identifies inhibitors of regulated nuclear export of a Forkhead transcription factor in PTEN-deficient tumor cells. Cancer cell 4, 463–476.

Kaufmann, S. H. and Earnshaw, W. C. (2000). Induction of apoptosis by cancer chemotherapy. Exp Cell Res 256, 42–49.

Keeshan, K., Mills, K. I., Cotter, T. G., and McKenna, S. L. (2001). Elevated Bcr-Abl expression levels are sufficient for a haematopoietic cell line to acquire a drug-resistant phenotype. Leukemia 15, 1823–1833.

Kelly, W. K., O'Connor, O. A., Krug, L. M., Chiao, J. H., Heaney, M., Curley, T., MacGregore-Cortelli, B., Tong, W., Secrist, J. P., Schwartz, L., et al. (2005). Phase I study of an oral histone deacetylase inhibitor, suberoylanilide hydroxamic acid, in patients with advanced cancer. J Clin Oncol 23, 3923–3931 (epub 2005 May 3916).

Khosravi-Far, R., Campbell, S., Rossman, K. L., and Der, C. J. (1998). Increasing complexity of Ras signal transduction: involvement of Rho family proteins. Adv Cancer Res 72, 57–107.

Khosravi-Far, R. and Der, C. J. (1994). The Ras signal transduction pathway. Cancer Metastasis Rev 13, 67–89.

Kim, D., Dan, H. C., Park, S., Yang, L., Liu, Q., Kaneko, S., Ning, J., He, L., Yang, H., Sun, M., et al. (2005). AKT/PKB signaling mechanisms in cancer and chemoresistance. Front Biosci 10, 975–987 (print 2005 Jan 2001).

Kim, R., Emi, M., Tanabe, K., and Toge, T. (2004). Therapeutic potential of antisense Bcl-2 as a chemosensitizer for cancer therapy. Cancer 101, 2491–2502.

Kim, R., Emi, M., Tanabe, K., Uchida, Y., and Arihiro, K. (2006). The role of apoptotic or nonapoptotic cell death in determining cellular response to anticancer treatment. Eur J Surg Oncol 20, 20.

Kim, R., Tanabe, K., Emi, M., Uchida, Y., Inoue, H., and Toge, T. (2003). Inducing cancer cell death by targeting transcription factors. Anticancer Drugs 14, 3–11.

Kirkin, V., Joos, S., and Zornig, M. (2004). The role of Bcl-2 family members in tumorigenesis. Biochim Biophys Acta 1644, 229–249.

Klasa, R. J., Gillum, A. M., Klem, R. E., and Frankel, S. R. (2002). Oblimersen Bcl-2 antisense: facilitating apoptosis in anticancer treatment. Antisense Nucleic Acid Drug Dev 12, 193–213.

Knies-Bamforth, U. E., Fox, S. B., Poulsom, R., Evan, G. I., and Harris, A. L. (2004). c-Myc interacts with hypoxia to induce angiogenesis in vivo by a vascular endothelial growth factor-dependent mechanism. Cancer Res 64, 6563–6570.

Knuefermann, C., Lu, Y., Liu, B., Jin, W., Liang, K., Wu, L., Schmidt, M., Mills, G. B., Mendelsohn, J., and Fan, Z. (2003). HER2/PI-3K/Akt activation leads to a multidrug resistance in human breast adenocarcinoma cells. Oncogene 22, 3205–3212.

Kondo, Y., Kanzawa, T., Sawaya, R., and Kondo, S. (2005). The role of autophagy in cancer development and response to therapy. Nat Rev Cancer 5, 726–734.

Kostanova-Poliakova, D. and Sabova, L. (2005). Anti-apoptotic proteins-targets for chemosensitization of tumor cells and cancer treatment. Neoplasma 52, 441–449.

Koturbash, I., Rugo, R. E., Hendricks, C. A., Loree, J., Thibault, B., Kutanzi, K., Pogribny, I., Yanch, J. C., Engelward, B. P., and Kovalchuk, O. (2006). Irradiation induces DNA damage and modulates epigenetic effectors in distant bystander tissue in vivo. Oncogene 25, 4267–4275.

Kroemer, G. (2003). Mitochondrial control of apoptosis: an introduction. Biochem Biophys Res Commun 304, 433–435.

Kutzki, O., Park, H. S., Ernst, J. T., Orner, B. P., Yin, H., and Hamilton, A. D. (2002). Development of a potent Bcl-x(L) antagonist based on alpha-helix mimicry. J Am Chem Soc 124, 11838–11839.

Kuwana, T., Mackey, M. R., Perkins, G., Ellisman, M. H., Latterich, M., Schneiter, R., Green, D. R., and Newmeyer, D. D. (2002). Bid, Bax, and lipids cooperate to form supramolecular openings in the outer mitochondrial membrane. Cell 111, 331–342.

Lawen, A. (2003). Apoptosis-an introduction. Bioessays 25, 888–896.

le Coutre, P., Mologni, L., Cleris, L., Marchesi, E., Buchdunger, E., Giardini, R., Formelli, F., and Gambacorti-Passerini, C. (1999). In vivo eradication of human BCR/ABL-positive leukemia cells with an ABL kinase inhibitor. J Natl Cancer Inst 91, 163–168.

Letai, A., Sorcinelli, M. D., Beard, C., and Korsmeyer, S. J. (2004). Antiapoptotic BCL-2 is required for maintenance of a model leukemia. Cancer Cell 6, 241–249.

Levine, A. J., Finlay, C. A., and Hinds, P. W. (2004). P53 is a tumor suppressor gene. Cell 116, S67–S69, 61 p following S69.

Levy, E., Baroche, C., Barret, J. M., Alapetite, C., Salles, B., Averbeck, D., and Moustacchi, E. (1994). Activated ras oncogene and specifically acquired resistance to cisplatin in human mammary epithelial cells: induction of DNA cross-links and their repair. Carcinogenesis 15, 845–850.

Li, H. and Yuan, J. (1999). Deciphering the pathways of life and death. Curr Opin Cell Biol 11, 261–266.

Li, H., Zhu, H., Xu, C. J., and Yuan, J. (1998). Cleavage of BID by caspase 8 mediates the mitochondrial damage in the Fas pathway of apoptosis. Cell 94, 491–501.

Lima, R. T., Martins, L. M., Guimaraes, J. E., Sambade, C., and Vasconcelos, M. H. (2004). Specific downregulation of bcl-2 and xIAP by RNAi enhances the effects of chemotherapeutic agents in MCF-7 human breast cancer cells. Cancer Gene Ther 11, 309–316.

Lin, W. C., Lin, F. T., and Nevins, J. R. (2001). Selective induction of E2F1 in response to DNA damage, mediated by ATM-dependent phosphorylation. Genes Dev 15, 1833–1844.

Longley, D. B. and Johnston, P. G. (2005). Molecular mechanisms of drug resistance. J Pathol 205, 275–292.

Longley, D. B., Wilson, T. R., McEwan, M., Allen, W. L., McDermott, U., Galligan, L., and Johnston, P. G. (2006). c-FLIP inhibits chemotherapy-induced colorectal cancer cell death. Oncogene 25, 838–848.

Lowe, S. W., Bodis, S., McClatchey, A., Remington, L., Ruley, H. E., Fisher, D. E., Housman, D. E., and Jacks, T. (1994). p53 status and the efficacy of cancer therapy in vivo. Science 266, 807–810.

Lowe, S. W., Cepero, E., and Evan, G. (2004). Intrinsic tumour suppression. Nature 432, 307–315.

Lowe, S. W. and Lin, A. W. (2000). Apoptosis in cancer. Carcinogenesis 21, 485–495.

Luo, X., Budihardjo, I., Zou, H., Slaughter, C., and Wang, X. (1998). Bid, a Bcl2 interacting protein, mediates cytochrome c release from mitochondria in response to activation of cell surface death receptors. Cell 94, 481–490.

Luqmani, Y. A. (2005). Mechanisms of drug resistance in cancer chemotherapy. Med Princ Pract 14, 35–48.

Luu, Y., Bush, J., Cheung, K. J., Jr., and Li, G. (2002). The p53 stabilizing compound CP-31398 induces apoptosis by activating the intrinsic Bax/mitochondrial/caspase-9 pathway. Exp Cell Res 276, 214–222.

Marchenko, N. D., Zaika, A., and Moll, U. M. (2000). Death signal-induced localization of p53 protein to mitochondria. A potential role in apoptotic signaling. J Biol Chem 275, 16202–16212.

Martelli, A. M., Tabellini, G., Bortul, R., Tazzari, P. L., Cappellini, A., Billi, A. M., and Cocco, L. (2005). Involvement of the phosphoinositide 3-kinase/Akt signaling pathway in the resistance to therapeutic treatments of human leukemias. Histol Histopathol 20, 239–252.

McCormick, F. (2004). Cancer: survival pathways meet their end. Nature 428, 267–269.

McGahon, A., Bissonnette, R., Schmitt, M., Cotter, K. M., Green, D. R., and Cotter, T. G. (1994). Bcr-Abl maintains resistance of chronic myelogenous leukemia cells to apoptotic cell death. Blood 83, 1179–1187.

McManus, D. C., Lefebvre, C. A., Cherton-Horvat, G., St-Jean, M., Kandimalla, E. R., Agrawal, S., Morris, S. J., Durkin, J. P., and Lacasse, E. C. (2004). Loss of XIAP protein expression by RNAi and antisense approaches sensitizes cancer cells to functionally diverse chemotherapeutics. Oncogene 23, 8105–8117.

Meek, D. W. (2004). The p53 response to DNA damage. DNA Repair (Amst) 3, 1049–1056.

Mendelsohn, J. and Fan, Z. (1997). Epidermal growth factor receptor family and chemosensitization. J Natl Cancer Inst 89, 341–343.

Mesner, P. W., Jr., Budihardjo, II, and Kaufmann, S. H. (1997). Chemotherapy-induced apoptosis. Adv Pharmacol 41, 461–499.

Michalak, E., Villunger, A., Erlacher, M., and Strasser, A. (2005). Death squads enlisted by the tumour suppressor p53. Biochem Biophys Res Commun 331, 786–798.

Mihara, M., Erster, S., Zaika, A., Petrenko, O., Chittenden, T., Pancoska, P., and Moll, U. M. (2003). p53 has a direct apoptogenic role at the mitochondria. Mol Cell 11, 577–590.

Minucci, S. and Pelicci, P. G. (2006). Histone deacetylase inhibitors and the promise of epigenetic (and more) treatments for cancer. Nat Rev Cancer 6, 38–51.

Miyashita, T. and Reed, J. C. (1992). bcl-2 gene transfer increases relative resistance of S49.1 and WEHI7.2 lymphoid cells to cell death and DNA fragmentation induced by glucocorticoids and multiple chemotherapeutic drugs. Cancer Res 52, 5407–5411.

Modur, V., Nagarajan, R., Evers, B. M., and Milbrandt, J. (2002). FOXO proteins regulate tumor necrosis factor-related apoptosis inducing ligand expression. Implications for PTEN mutation in prostate cancer. J Biol Chem 277, 47928–47937.

Morgensztern, D. and McLeod, H. L. (2005). PI3K/Akt/mTOR pathway as a target for cancer therapy. Anticancer Drugs 16, 797–803.

Moroni, M. C., Hickman, E. S., Lazzerini Denchi, E., Caprara, G., Colli, E., Cecconi, F., Muller, H., and Helin, K. (2001). Apaf-1 is a transcriptional target for E2F and p53. Nat Cell Biol 3, 552–558.

Nagane, M., Levitzki, A., Gazit, A., Cavenee, W. K., and Huang, H. J. (1998). Drug resistance of human glioblastoma cells conferred by a tumor-specific mutant epidermal growth factor receptor through modulation of Bcl-XL and caspase-3-like proteases. Proc Natl Acad Sci USA95, 5724–5729.

Nahle, Z., Polakoff, J., Davuluri, R. V., McCurrach, M. E., Jacobson, M. D., Narita, M., Zhang, M. Q., Lazebnik, Y., Bar-Sagi, D., and Lowe, S. W. (2002). Direct coupling of the cell cycle and cell death machinery by E2F. Nat Cell Biol 4, 859–864.

Nakanishi, C. and Toi, M. (2005). Nuclear factor-kappaB inhibitors as sensitizers to anticancer drugs. Nat Rev Cancer 5, 297–309.

Nelson, D. A. and White, E. (2004). Exploiting different ways to die. Genes Dev 18, 1223–1226.

Newmeyer, D. D. and Ferguson-Miller, S. (2003). Mitochondria: releasing power for life and unleashing the machineries of death. Cell 112, 481–490.

Nguyen, D. M., Chen, G. A., Reddy, R., Tsai, W., Schrump, W. D., Cole, G., Jr., and Schrump, D. S. (2004). Potentiation of paclitaxel cytotoxicity in lung and esophageal cancer cells by pharmacologic inhibition of the phosphoinositide 3-kinase/protein kinase B (Akt)-mediated signaling pathway. J Thorac Cardiovasc Surg 127, 365–375.

Nicholson, K. M., Quinn, D. M., Kellett, G. L., and Warr, J. R. (2003). LY294002, an inhibitor of phosphatidylinositol-3-kinase, causes preferential induction of apoptosis in human multidrug resistant cells. Cancer Lett 190, 31–36.

Nieborowska-Skorska, M., Wasik, M. A., Slupianek, A., Salomoni, P., Kitamura, T., Calabretta, B., and Skorski, T. (1999). Signal transducer and activator of transcription (STAT)5 activation by BCR/ABL is dependent on intact Src homology (SH)3 and SH2 domains of BCR/ABL and is required for leukemogenesis. J Exp Med 189, 1229–1242.

Notarbartolo, M., Cervello, M., Dusonchet, L., Cusimano, A., and D'Alessandro, N. (2002). Resistance to diverse apoptotic triggers in multidrug resistant HL60 cells and its possible relationship to the expression of P-glycoprotein, Fas and of the novel anti-apoptosis factors IAP (inhibitory of apoptosis proteins). Cancer Lett 180, 91–101.

O'Gorman, D. M., McKenna, S. L., McGahon, A. J., Knox, K. A., and Cotter, T. G. (2000). Sensitisation of HL60 human leukaemic cells to cytotoxic drug-induced apoptosis by inhibition of PI3-kinase survival signals. Leukemia 14, 602–611.

Okada, M., Adachi, S., Imai, T., Watanabe, K., Toyokuni, S. Y., Ueno, M., Zervos, A. S., Kroemer, G., and Nakahata, T. (2004). A novel mechanism for imatinib mesylate-induced cell death of

Bcr-Abl-positive human leukemic cells: caspase-independent, necrosis-like programmed cell death mediated by serine protease activity. Blood 103, 2299–2307 (epub 2003 Nov 2226).

Oltersdorf, T., Elmore, S. W., Shoemaker, A. R., Armstrong, R. C., Augeri, D. J., Belli, B. A., Bruncko, M., Deckwerth, T. L., Dinges, J., Hajduk, P. J., et al. (2005). An inhibitor of Bcl-2 family proteins induces regression of solid tumours. Nature 435, 677–681 (epub 2005 May 2015).

Orian, A., van Steensel, B., Delrow, J., Bussemaker, H. J., Li, L., Sawado, T., Williams, E., Loo, L. W., Cowley, S. M., Yost, C., et al. (2003). Genomic binding by the Drosophila Myc, Max, Mad/Mnt transcription factor network. Genes Dev 17, 1101–1114.

Orlowski, R. Z. and Baldwin, A. S., Jr. (2002). NF-kappaB as a therapeutic target in cancer. Trends Mol Med 8, 385–389.

Osaki, M., Oshimura, M., and Ito, H. (2004). PI3K-Akt pathway: its functions and alterations in human cancer. Apoptosis 9, 667–676.

Oster, S. K., Ho, C. S., Soucie, E. L., and Penn, L. Z. (2002). The myc oncogene: MarvelouslY Complex. Adv Cancer Res 84, 81–154.

Ozoren, N. and El-Deiry, W. S. (2002). Defining characteristics of Types I and II apoptotic cells in response to TRAIL. Neoplasia 4, 551–557.

Ozoren, N. and El-Deiry, W. S. (2003). Cell surface death receptor signaling in normal and cancer cells. Semin Cancer Biol 13, 135–147.

Page, C., Lin, H. J., Jin, Y., Castle, V. P., Nunez, G., Huang, M., and Lin, J. (2000). Overexpression of Akt/AKT can modulate chemotherapy-induced apoptosis. Anticancer Res 20, 407–416.

Pegram, M. D., Finn, R. S., Arzoo, K., Beryt, M., Pietras, R. J., and Slamon, D. J. (1997). The effect of HER-2/neu overexpression on chemotherapeutic drug sensitivity in human breast and ovarian cancer cells. Oncogene 15, 537–547.

Pelengaris, S., Khan, M., and Evan, G. (2002). c-MYC: more than just a matter of life and death. Nat Rev Cancer 2, 764–776.

Perkins, C. L., Fang, G., Kim, C. N., and Bhalla, K. N. (2000). The role of Apaf-1, caspase-9, and bid proteins in etoposide- or paclitaxel-induced mitochondrial events during apoptosis. Cancer Res 60, 1645–1653.

Peter, M. E. and Krammer, P. H. (1998). Mechanisms of CD95 (APO-1/Fas)-mediated apoptosis. Curr Opin Immunol 10, 545–551.

Peter, M. E. and Krammer, P. H. (2003). The CD95(APO-1/Fas) DISC and beyond. Cell Death Differ 10, 26–35.

Pisha, E., Chai, H., Lee, I. S., Chagwedera, T. E., Farnsworth, N. R., Cordell, G. A., Beecher, C. W., Fong, H. H., Kinghorn, A. D., Brown, D. M., and et al. (1995). Discovery of betulinic acid as a selective inhibitor of human melanoma that functions by induction of apoptosis. Nat Med 1, 1046–1051.

Pommier, Y., Sordet, O., Antony, S., Hayward, R. L., and Kohn, K. W. (2004). Apoptosis defects and chemotherapy resistance: molecular interaction maps and networks. Oncogene 23, 2934–2949.

Pompetti, F., Rizzo, P., Simon, R. M., Freidlin, B., Mew, D. J., Pass, H. I., Picci, P., Levine, A. S., and Carbone, M. (1996). Oncogene alterations in primary, recurrent, and metastatic human bone tumors. J Cell Biochem 63, 37–50.

Proskuryakov, S. Y., Konoplyannikov, A. G., and Gabai, V. L. (2003). Necrosis: a specific form of programmed cell death? Exp Cell Res 283, 1–16.

Ravagnan, L., Roumier, T., and Kroemer, G. (2002). Mitochondria, the killer organelles and their weapons. J Cell Physiol 192, 131–137.

Ray, S., Lu, Y., Kaufmann, S. H., Gustafson, W. C., Karp, J. E., Boldogh, I., Fields, A. P., and Brasier, A. R. (2004). Genomic mechanisms of p210Bcr-Abl signaling: induction of heat shock protein 70 through the GATA response element confers resistance to paclitaxel-induced apoptosis. J Biol Chem 279, 35604–35615 (epub 32004 May 35621).

Reed, J. C. (1996). Mechanisms of Bcl-2 family protein function and dysfunction in health and disease. Behring Inst Mitt, 72–100.

Reed, J. C. (2000). Mechanisms of apoptosis. Am J Pathol 157, 1415–1430.

Reinke, V. and Lozano, G. (1997). Differential activation of p53 targets in cells treated with ultra-violet radiation that undergo both apoptosis and growth arrest. Radiat Res 148, 115–122.

Reuther, J. Y., Reuther, G. W., Cortez, D., Pendergast, A. M., and Baldwin, A. S., Jr. (1998). A requirement for NF-kappaB activation in Bcr-Abl-mediated transformation. Genes Dev 12, 968–981.

Rokudai, S., Fujita, N., Kitahara, O., Nakamura, Y., and Tsuruo, T. (2002). Involvement of FKHR-dependent TRADD expression in chemotherapeutic drug-induced apoptosis. Mol Cell Biol 22, 8695–8708.

Sakamuro, D., Eviner, V., Elliott, K. J., Showe, L., White, E., and Prendergast, G. C. (1995). c-Myc induces apoptosis in epithelial cells by both p53-dependent and p53-independent mechanisms. Oncogene 11, 2411–2418.

Salvesen, G. S. and Dixit, V. M. (1997). Caspases: intracellular signaling by proteolysis. Cell 91, 443–446.

Salvesen, G. S. and Dixit, V. M. (1999). Caspase activation: the induced-proximity model. Proc Natl Acad Sci USA 96, 10964–10967.

Salvesen, G. S. and Renatus, M. (2002). Apoptosome: the seven-spoked death machine. Dev Cell 2, 256–257.

Sawyers, C. L., McLaughlin, J., and Witte, O. N. (1995). Genetic requirement for Ras in the transformation of fibroblasts and hematopoietic cells by the Bcr-Abl oncogene. J Exp Med 181, 307–313.

Scaffidi, C., Fulda, S., Srinivasan, A., Friesen, C., Li, F., Tomaselli, K. J., Debatin, K. M., Krammer, P. H., and Peter, M. E. (1998). Two CD95 (APO-1/Fas) signaling pathways. EMBO J 17, 1675–1687.

Scaffidi, C., Kirchhoff, S., Krammer, P. H., and Peter, M. E. (1999). Apoptosis signaling in lymphocytes. Curr Opin Immunol 11, 277–285.

Schimmer, A. D., Dalili, S., Batey, R. A., and Riedl, S. J. (2006). Targeting XIAP for the treatment of malignancy. Cell Death Differ 13, 179–188.

Schmitt, C. A., McCurrach, M. E., de Stanchina, E., Wallace-Brodeur, R. R., and Lowe, S. W. (1999). INK4a/ARF mutations accelerate lymphomagenesis and promote chemoresistance by disabling p53. Genes Dev 13, 2670–2677.

Schuler, M. and Green, D. R. (2005). Transcription, apoptosis and p53: catch-22. Trends Genet 21, 182–187.

Schulze-Bergkamen, H., and Krammer, P. H. (2004). Apoptosis in cancer – implications for therapy. Semin Oncol 31, 90–119.

Schulze-Osthoff, K., Ferrari, D., Los, M., Wesselborg, S., and Peter, M. E. (1998). Apoptosis signaling by death receptors. Eur J Biochem 254, 439–459.

Seelan, R. S., Irwin, M., van der Stoop, P., Qian, C., Kaelin, W. G., Jr., and Liu, W. (2002). The human p73 promoter: characterization and identification of functional E2F binding sites. Neoplasia 4, 195–203.

Sheikh, M. S. and Huang, Y. (2004). Death receptors as targets of cancer therapeutics. Curr Cancer Drug Targets 4, 97–104.

Shi, Y. (2002). Apoptosome: the cellular engine for the activation of caspase-9. Structure (Camb) 10, 285–288.

Shuai, K., Halpern, J., ten Hoeve, J., Rao, X., and Sawyers, C. L. (1996). Constitutive activation of STAT5 by the Bcr-Abl oncogene in chronic myelogenous leukemia. Oncogene 13, 247–254.

Sillaber, C., Gesbert, F., Frank, D. A., Sattler, M., and Griffin, J. D. (2000). STAT5 activation contributes to growth and viability in Bcr/Abl-transformed cells. Blood 95, 2118–2125.

Skorski, T. (2002). BCR/ABL regulates response to DNA damage: the role in resistance to genotoxic treatment and in genomic instability. Oncogene 21, 8591–8604.

Skorski, T., Bellacosa, A., Nieborowska-Skorska, M., Majewski, M., Martinez, R., Choi, J. K., Trotta, R., Wlodarski, P., Perrotti, D., Chan, T. O., et al. (1997). Transformation of hematopoietic cells by BCR/ABL requires activation of a PI-3k/Akt-dependent pathway. EMBO J 16, 6151–6161.

Soengas, M. S., Capodieci, P., Polsky, D., Mora, J., Esteller, M., Opitz-Araya, X., McCombie, R., Herman, J. G., Gerald, W. L., Lazebnik, Y. A., et al. (2001). Inactivation of the apoptosis effector Apaf-1 in malignant melanoma. Nature 409, 207–211.

Song, L., Coppola, D., Livingston, S., Cress, D., and Haura, E. B. (2005). Mcl-1 regulates survival and sensitivity to diverse apoptotic stimuli in human non-small cell lung cancer cells. Cancer Biol Ther 4, 267–276 (epub 2005 Mar 2020).

Sorice, M., Circella, A., Cristea, I. M., Garofalo, T., Renzo, L. D., Alessandri, C., Valesini, G., and Esposti, M. D. (2004). Cardiolipin and its metabolites move from mitochondria to other cellular membranes during death receptor-mediated apoptosis. Cell Death Differ 11, 1133–1145.

Stahl, M., Dijkers, P. F., Kops, G. J., Lens, S. M., Coffer, P. J., Burgering, B. M., and Medema, R. H. (2002). The forkhead transcription factor FoxO regulates transcription of p27Kip1 and Bim in response to IL-2. J Immunol 168, 5024–5031.

Stegh, A. H. and Peter, M. E. (2001). Apoptosis and caspases. Cardiol Clin 19, 13–29.

Stiewe, T. and Putzer, B. M. (2000). Role of the p53-homologue p73 in E2F1-induced apoptosis. Nat Genet 26, 464–469.

Strasser, A., Harris, A. W., Bath, M. L., and Cory, S. (1990). Novel primitive lymphoid tumours induced in transgenic mice by cooperation between myc and bcl-2. Nature 348, 331–333.

Strasser, A., O'Connor, L., and Dixit, V. M. (2000). Apoptosis signaling. Annu Rev Biochem 69, 217–245.

Suhara, T., Kim, H. S., Kirshenbaum, L. A., and Walsh, K. (2002). Suppression of Akt signaling induces Fas ligand expression: involvement of caspase and Jun kinase activation in Akt-mediated Fas ligand regulation. Molecular and Cellular Biology 22, 680–691.

Sunters, A., Fernandez de Mattos, S., Stahl, M., Brosens, J. J., Zoumpoulidou, G., Saunders, C. A., Coffer, P. J., Medema, R. H., Coombes, R. C., and Lam, E. W. (2003). FoxO3a transcriptional regulation of Bim controls apoptosis in paclitaxel-treated breast cancer cell lines. J Biol Chem 278, 49795–49805 (epub 42003 Oct 49793).

Tai, Y. T., Lee, S., Niloff, E., Weisman, C., Strobel, T., and Cannistra, S. A. (1998). BAX protein expression and clinical outcome in epithelial ovarian cancer. J Clin Oncol 16, 2583–2590.

Teitz, T., Lahti, J. M., and Kidd, V. J. (2001). Aggressive childhood neuroblastomas do not express caspase-8: an important component of programmed cell death. J Mol Med 79, 428–436.

Teitz, T., Wei, T., Valentine, M. B., Vanin, E. F., Grenet, J., Valentine, V. A., Behm, F. G., Look, A. T., Lahti, J. M., and Kidd, V. J. (2000). Caspase 8 is deleted or silenced preferentially in childhood neuroblastomas with amplification of MYCN. Nat Med 6, 529–535.

Thompson, C. B. (1995). Apoptosis in the pathogenesis and treatment of disease. Science 267, 1456–1462.

Thorburn, A. (2004). Death receptor-induced cell killing. Cell Signal 16, 139–144.

Tomita, Y., Marchenko, N., Erster, S., Nemajerova, A., Dehner, A., Klein, C., Pan, H., Kessler, H., Pancoska, P., and Moll, U. M. (2006). WTp53 but not tumor-derived mutants bind to BCL2 via the DNA binding domain and induce mitochondrial permeabilization. J Biol Chem 26, 26.

Toretsky, J. A., Thakar, M., Eskenazi, A. E., and Frantz, C. N. (1999). Phosphoinositide 3-hydroxide kinase blockade enhances apoptosis in the Ewing's sarcoma family of tumors. Cancer Res 59, 5745–5750.

Tourneur, L., Delluc, S., Levy, V., Valensi, F., Radford-Weiss, I., Legrand, O., Vargaftig, J., Boix, C., Macintyre, E. A., Varet, B., et al. (2004). Absence or low expression of fas-associated protein with death domain in acute myeloid leukemia cells predicts resistance to chemotherapy and poor outcome. Cancer Res 64, 8101–8108.

Tzung, S. P., Kim, K. M., Basanez, G., Giedt, C. D., Simon, J., Zimmerberg, J., Zhang, K. Y., and Hockenbery, D. M. (2001). Antimycin A mimics a cell-death-inducing Bcl-2 homology domain 3. Nat Cell Biol 3, 183–191.

Underhill-Day, N., Pierce, A., Thompson, S. E., Xenaki, D., Whetton, A. D., and Owen-Lynch, P. J. (2006). Role of the C-terminal actin binding domain in BCR/ABL-mediated survival and drug resistance. Br J Haematol 132, 774–783.

Varticovski, L., Daley, G. Q., Jackson, P., Baltimore, D., and Cantley, L. C. (1991). Activation of phosphatidylinositol 3-kinase in cells expressing abl oncogene variants. Mol Cell Biol 11, 1107–1113.

Vassilev, L. T., Vu, B. T., Graves, B., Carvajal, D., Podlaski, F., Filipovic, Z., Kong, N., Kammlott, U., Lukacs, C., Klein, C., et al. (2004). In vivo activation of the p53 pathway by small-molecule antagonists of MDM2. Science 303, 844–848 (epub 2004 Jan 2002).

Vaux, D. L. (2002). Apoptosis timeline. Cell Death Differ 9, 349–354.

Vaux, D. L. and Korsmeyer, S. J. (1999). Cell death in development. Cell 96, 245–254.

Vogelstein, B., Lane, D., and Levine, A. J. (2000). Surfing the p53 network. Nature 408, 307–310.

Walczak, H., Miller, R. E., Ariail, K., Gliniak, B., Griffith, T. S., Kubin, M., Chin, W., Jones, J., Woodward, A., Le, T., et al. (1999). Tumoricidal activity of tumor necrosis factor-related apoptosis-inducing ligand in vivo. Nature medicine 5, 157–163.

Walensky, L. D., Kung, A. L., Escher, I., Malia, T. J., Barbuto, S., Wright, R. D., Wagner, G., Verdine, G. L., and Korsmeyer, S. J. (2004). Activation of apoptosis in vivo by a hydrocarbon-stapled BH3 helix. Science 305, 1466–1470.

Wallach, D. (1997). Apoptosis. Placing death under control. Nature 388, 123, 125–126.

Wallach, D., Varfolomeev, E. E., Malinin, N. L., Goltsev, Y. V., Kovalenko, A. V., and Boldin, M. P. (1999). Tumor necrosis factor receptor and Fas signaling mechanisms. Annu Rev Immunol 17, 331–367.

Walton, M. I., Whysong, D., O'Connor, P. M., Hockenbery, D., Korsmeyer, S. J., and Kohn, K. W. (1993). Constitutive expression of human Bcl-2 modulates nitrogen mustard and camptothecin induced apoptosis. Cancer Res 53, 1853–1861.

Wang, G. Q., Gastman, B. R., Wieckowski, E., Goldstein, L. A., Gambotto, A., Kim, T. H., Fang, B., Rabinovitz, A., Yin, X. M., and Rabinowich, H. (2001). A role for mitochondrial Bak in apoptotic response to anticancer drugs. J Biol Chem 276, 34307–34317 (epub 32001 July 34310).

Wang, J. L., Liu, D., Zhang, Z. J., Shan, S., Han, X., Srinivasula, S. M., Croce, C. M., Alnemri, E. S., and Huang, Z. (2000). Structure-based discovery of an organic compound that binds Bcl-2 protein and induces apoptosis of tumor cells. Proc Natl Acad Sci USA 97, 7124–7129.

Wang, J. Y. and Ki, S. W. (2001). Choosing between growth arrest and apoptosis through the retinoblastoma tumour suppressor protein, Abl and p73. Biochem Soc Trans 29, 666–673.

Wang, L. G., Liu, X. M., Kreis, W., and Budman, D. R. (1999). The effect of antimicrotubule agents on signal transduction pathways of apoptosis: a review. Cancer Chemother Pharmacol 44, 355–361.

Wang, W. and El-Deiry, W. S. (2004). Targeting FOXO kills two birds with one stone. Chem Biol 11, 16–18.

Waxman, D. J. and Schwartz, P. S. (2003). Harnessing apoptosis for improved anticancer gene therapy. Cancer Res 63, 8563–8572.

Wei, M. C., Zong, W. X., Cheng, E. H., Lindsten, T., Panoutsakopoulou, V., Ross, A. J., Roth, K. A., MacGregor, G. R., Thompson, C. B., and Korsmeyer, S. J. (2001a). Proapoptotic BAX and BAK: a requisite gateway to mitochondrial dysfunction and death. Science 292, 727–730.

Wei, M. C., Zong, W. X., Cheng, E. H., Lindsten, T., Panoutsakopoulou, V., Ross, A. J., Roth, K. A., MacGregor, G. R., Thompson, C. B., and Korsmeyer, S. J. (2001b). Proapoptotic BAX and BAK: a requisite gateway to mitochondrial dysfunction and death. Science 292, 727–730.

Weller, M. (1998). Predicting response to cancer chemotherapy: the role of p53. Cell Tissue Res 292, 435–445.

Wendel, H. G., De Stanchina, E., Fridman, J. S., Malina, A., Ray, S., Kogan, S., Cordon-Cardo, C., Pelletier, J., and Lowe, S. W. (2004). Survival signalling by Akt and eIF4E in oncogenesis and cancer therapy. Nature 428, 332–337.

Wennerberg, K., Rossman, K. L., and Der, C. J. (2005). The Ras superfamily at a glance. J Cell Sci 118, 843–846.

Whitesell, L. and Lindquist, S. L. (2005). HSP90 and the chaperoning of cancer. Nat Rev Cancer 5, 761–772.

Xu, H., Tai, J., Ye, H., Kang, C. B., and Yoon, H. S. (2006). The N-terminal domain of tumor suppressor p53 is involved in the molecular interaction with the anti-apoptotic protein Bcl-Xl. Biochem Biophys Res Commun 341, 938–944 (epub 2006 Jan 2023).

Yagita, H., Takeda, K., Hayakawa, Y., Smyth, M. J., and Okumura, K. (2004). TRAIL and its receptors as targets for cancer therapy. Cancer Sci 95, 777–783.

Yang, X., Chang, H. Y., and Baltimore, D. (1998). Autoproteolytic activation of pro-caspases by oligomerization. Mol Cell 1, 319–325.

Yin, H., Lee, G. I., Sedey, K. A., Kutzki, O., Park, H. S., Orner, B. P., Ernst, J. T., Wang, H. G., Sebti, S. M., and Hamilton, A. D. (2005). Terphenyl-Based Bak BH3 alpha-helical proteomimetics as low-molecular-weight antagonists of Bcl-xL. J Am Chem Soc 127, 10191–10196.

Yoo, C. B. and Jones, P. A. (2006). Epigenetic therapy of cancer: past, present and future. Nat Rev Drug Discov 5, 37–50.

Yu, J. and Zhang, L. (2005). The transcriptional targets of p53 in apoptosis control. Biochem Biophys Res Commun 331, 851–858.

Yuan, X. J. and Whang, Y. E. (2002). PTEN sensitizes prostate cancer cells to death receptor-mediated and drug-induced apoptosis through a FADD-dependent pathway. Oncogene 21, 319–327.

Zaffaroni, N., Pennati, M., Colella, G., Perego, P., Supino, R., Gatti, L., Pilotti, S., Zunino, F., and Daidone, M. G. (2002). Expression of the anti-apoptotic gene survivin correlates with taxol resistance in human ovarian cancer. Cell Mol Life Sci 59, 1406–1412.

Zamzami, N. and Kroemer, G. (2001). The mitochondrion in apoptosis: how Pandora's box opens. Nat Rev Mol Cell Biol 2, 67–71.

Zhang, L., Yu, J., Park, B. H., Kinzler, K. W., and Vogelstein, B. (2000). Role of BAX in the apoptotic response to anticancer agents. Science 290, 989–992.

Zhou, P., Qian, L., Kozopas, K. M., and Craig, R. W. (1997). Mcl-1, a Bcl-2 family member, delays the death of hematopoietic cells under a variety of apoptosis-inducing conditions. Blood 89, 630–643.

Zinkel, S. S., Ong, C. C., Ferguson, D. O., Iwasaki, H., Akashi, K., Bronson, R. T., Kutok, J. L., Alt, F. W., and Korsmeyer, S. J. (2003). Proapoptotic BID is required for myeloid homeostasis and tumor suppression. Genes Dev 17, 229–239.

Zong, W. X., Ditsworth, D., Bauer, D. E., Wang, Z. Q., and Thompson, C. B. (2004). Alkylating DNA damage stimulates a regulated form of necrotic cell death. Genes Dev 18, 1272–1282 (epub 2004 May 1214).

Zornig, M., Hueber, A., Baum, W., and Evan, G. (2001). Apoptosis regulators and their role in tumorigenesis. Biochim Biophys Acta 1551, F1–F37.

Chapter 5
Therapeutic Targeting of Death Pathways in Cancer

Mechanisms for Activating Cell Death in Cancer Cells

Ting-Ting Tan and Eileen White*

Abstract Defects in apoptosis that evolve during the course of cancer progression not only provide cancer cells with intrinsic survival advantage, but also provide inherent resistance to chemotherapeutic agents. Thus, modulation of apoptosis by targeting components of the apoptotic machinery and its regulators to restore apoptotic function is a rational approach for treating cancer. With our increasing knowledge of the mechanisms of apoptosis regulation and of how apoptosis is disabled in cancer cells, numerous novel approaches targeting apoptotic pathways can now be exploited for cancer therapy. While most of these therapies are still in preclinical development, some have shown considerable promise and progressed into the clinic. This chapter summarizes the current knowledge of the apoptotic pathways and provides a selective review on the development of drugs that target the apoptotic machinery.

Keywords apoptosis, chemotherapy, targeted therapy, BCL-2 family, death receptors, signal transduction inhibitors

1 Introduction

Inactivation of apoptosis is selected for in cancer, endowing cells with intrinsic survival advantage and the capacity to evade surveillance by the immune system. Furthermore, killing of cancer cells by currently used cytotoxic therapies, including chemotherapy, γ-irradiation, and immunotherapy largely depends on activation or

Ting-Ting Tan
Center for Advanced Biotechnology and Medicine and
Rutgers University, Piscataway, NJ 08854, USA

Eileen White
Center for Advanced Biotechnology and Medicine; Rutgers University, Piscataway,
NJ 08854, USA; Department of Molecular Biology and Biochemistry, 679 Hoes Lane,
Room 140 and Cancer Institue of New Jersey, New Brunswick, NJ 08901, USA

*To whom correspondence should be addressed: Tel.: 732-235-5329; fax: 732-235-5795;
e-mail: ewhite@cabm.rutgers.edu

R. Khosravi-Far and E. White (eds.), *Programmed Cell Death in Cancer Progression and Therapy*.
© Springer 2008

reactivation of the apoptosis program. Accordingly, failure to engage in apoptosis produces resistance to treatment. Defective apoptosis can allow genetically unstable and damaged cells to avoid elimination, further facilitating tumor progression (Nelson et al., 2004) and treatment failure (Johnstone et al., 2002). Advances in the understanding of the molecular mechanisms of apoptosis have laid the foundation for discovery of new drugs targeting various components of the apoptotic pathway to increase the effectiveness of cancer treatment. Several new approaches are being investigated that include gene therapy, small molecule peptide mimetics, antibodies, kinase inhibitors, and proteasome inhibitors to target specific apoptosis regulators. Knowing how apoptosis is regulated, identifying the key components that control the apoptotic response in cancer cells, and how common mutations found in human tumors alter apoptotic signaling, has provided a rational approach to cancer therapy.

2   Apoptosis Signaling Pathways

Apoptosis is a stringently regulated, evolutionarily conserved mechanism of cell death that is considered a critical regulatory process for development and for maintaining a homeostatic balance between cell survival and cell death. Disruption of apoptosis contributes to the pathogenesis of a wide variety of diseases. Too much cell death can contribute to degenerative disorders, whereas too little cell death leads to autoimmunity and cancer (Cory and Adams, 2002; Danial and Korsmeyer, 2004). Two alternative pathways can initiate apoptosis: one is mediated by death receptors on the cell surface, and is referred to as the "extrinsic pathway"; and the other is referred to as " intrinsic pathway" and involves the BCL-2 family proteins that regulate mitochondrial function. Ultimately, the two pathways converge on downstream effector cysteine aspartyl-specific proteases (caspases), activation of which leads to the biochemical and morphological changes that are characteristic of apoptosis (Shi, 2002). Caspase activation results in a collapse of cellular ultrastructure and function through internal proteolytic digestion, which is evident as dismantling of the cytoskeleton, metabolic dysfunction, and genomic fragmentation. In the end, the condensed cell corpse is engulfed by nearby cells in tissues and eliminated without inflammation (Wyllie, 1980).

2.1   The Intrinsic Pathway

The BCL-2 family members serve as key regulators of the intrinsic apoptotic pathway that signals through mitochondria (Fig. 5.1). About 20 BCL-2 family members in mammals fall into three interacting groups that share at least one of four relatively conserved BCL-2 homology domains (BH1–4). Multidomain antiapoptotic BCL-2 and its homologues (e.g., BCL-x_L BCL-w, BFL-1/A1, MCL-1, and adenoviral homolog E1B19K) act predominantly to inhibit apoptosis (Cuconati and White,

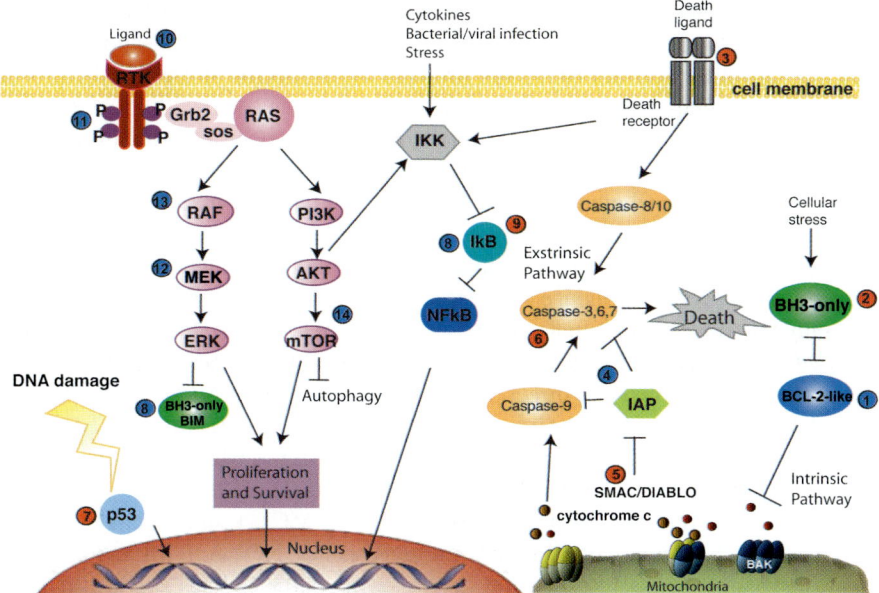

Fig. 5.1 Schematic representation of the major apoptotic pathway components and the acting points of agents that target the regulators of apoptosis. Apoptosis occur through two main pathways: the extrinsic and intrinsic pathways. Both pathways converge on activation of caspases that culminate in cell death. Extracellular signals via cytokines and growth factors are central to cell survival. Loss of p53 and hyperactivation of survival pathways are commonly found in cancer cells to deregulate cell cycle control and interfere with apoptotic signaling. Comprehensive knowledge of these pathways provides a variety of options for targeted therapy. Summarized here are major acting points of targeted agents indicated as numbered and colored circles. Red solid circles represent activation and blue solid circles represent inhibition of the target/pathway. The numbers stand for different classes of drugs: (1) represents antisense oligonucleotides or small-molecule inhibitors targeting antiapoptotic BCL-2-like proteins; (2) BH3 mimetics; (3) soluble death receptor ligands or agonistic antibodies against death receptors; (4) IAP inhibitors; (5) SMAC mimetics; (6) caspase activators; (7) p53 activators; (8) proteasome inhibitors; (9) IκB stabilizers; (10) antireceptor tyrosine kinase antibodies; (11) tyrosine kinase inhibitors; (12) MEK kinase inhibitors; (13) RAF kinase inhibitors; and (14) mTOR inhibitors

2002; Danial and Korsmeyer, 2004). The other two groups instead are proapoptotic. One of these proapoptotic groups comprises the multidomain proteins represented by BAX, BAK, and BOK that share BH1–3 with BCL-2. These three conserved regions in multidomain BCL-2 family members form a hydrophobic surface groove for binding of either a putative transmembrane helical domain at the carboxyl terminus, or a BH3 (Fesik, 2000; Suzuki et al., 2000). The other proapoptotic group is comprised of the BH3-only proteins (e.g., BAD, BID, BIM, HRK, PUMA, NOXA, and NBK/BIK) that are the most apical regulators of this intrinsic death signaling (Gelinas and White, 2005; Willis and Adams, 2005). BH3-only proteins typically initiate the apoptotic activity of the BCL-2 family in response to diverse cytotoxic stimuli. The BH3 is an amphipathic α-helix that serves as a binding motif for interaction with the hydrophobic groove on either multidomain antiapoptotic or proapoptotic BCL-2 proteins. Systematic study of the binding of BH3-only proteins to BCL-2 antiapoptotic proteins has shown that certain BH3-only proteins target specific subsets of the prosurvival proteins (Letai et al., 2002; Chen et al., 2005; Kuwana et al., 2005; Willis et al., 2005). BIM, PUMA, and tBID bind avidly to all five antiapoptotic proteins and demonstrate potent killing. In contrast, BAD and BMF bind preferentially to BCL-2, BCL-x_L, and BCL-w, whereas NOXA binds preferentially to MCL-1 and BFL-1/A1. Although BH3-only proteins with restricted targets can be less-potent inducers of apoptosis, BAD and NOXA with complementary affinity to antiapoptotic BCL-2 family members cooperate to induce substantial cell death (Chen et al., 2005). Although BH3-only proteins cannot initiate apoptosis in the absence of BAX and BAK (Cheng et al., 2001; Zong et al., 2001), it remains unresolved whether they activate BAX and BAK directly or indirectly. A direct binding model has been proposed for BAX activation where binding of BH3-only sensitizers (e.g., BAD or BIK) to BCL-2 displaces the normally sequestered BH3-only activators (e.g., tBID or BIM), releasing tBID or BIM to trigger BAX oligomerization (Letai et al., 2002). Evidence suggests a displacement model for the activation of BAK whereby MCL-1 and BCL-x_L sequester BAK, and that the binding of BH3-only proteins such as NOXA to MCL-1 displaces BAK as an apoptosis-activating step (Cuconati et al., 2003; Gelinas and White, 2005; Willis and Adams, 2005; Willis et al., 2005).

BH3-only proteins initiate apoptosis in response to a wide range of damage and stress, including DNA damage, deregulated growth, survival factor deficiency, hypoxia, anoikis, and Ca^{+2} overload (Adams, 2003; Cory and Adams, 2005; Willis and Adams, 2005). Although these diverse apoptotic stimuli activate different upstream components in the apoptotic signaling pathway, in most cells, these signals are transduced to and converge on mitochondria and cause permeabilization of outer mitochondrial membrane causing the release of cytochrome c and other apoptogenic proteins (e.g., SMAC/DIABLO). Cytochrome c release promotes the formation of the apoptosome, a large protein complex that contains cytochrome c, apoptotic protease-activating factor 1 (APAF1) and caspase-9 (Li et al., 1997; Zou et al., 1997). Apoptosome formation triggers activation of caspase-9, which further cleaves and activates the effector caspase-3, resulting in selective destruction of subcellular structures and organelles, and of the genome (Earnshaw et al., 1999).

2.2   The Extrinsic Pathway

Activation of the extrinsic apoptotic pathway is initiated by ligand-mediated activation of cell surface death receptors (DRs). It plays an important role in immune surveillance of transformed or virus-infected cells and in the removal of self-reactive lymphocytes. Death receptors form a subgroup of the tumor necrosis factor (TNF) receptor superfamily that includes TNF-R1, CD95 (also called APO-1 or FAS), DR3 (APO-2), DR4 (TNF-related apoptosis-inducing ligand receptor 1

[TRAIL R1]), DR5 (TRAIL R2), and DR6 (Zapata et al., 2001). Upon ligand binding, the death receptors interact via their intracellular motif called the death domain (DD) with the DD of adapter proteins such as FAS-associated death domain (FADD). These adapter proteins also contain a second protein interaction motif, the death effector domain (DED), that facilitates binding to a corresponding DED in the amino-terminal prodomains of initiator caspase-8 (or in some cases, its relative caspase-10) to form the death-inducing signaling complex (DISC) (Wallach et al., 1999). DISC formation activates caspase-8, which subsequently cleaves and activates caspase-3, resulting in further cleavage of cellular targets.

In many cells, however, DISC formation mediated caspase-3 activation is insufficient to complete the cell death program, and death receptor signaling must be amplified by engagement of the mitochondria-mediated cell death pathway through the caspase-8-mediated cleavage of the BH3-only protein BID (Fig. 5.1). Once cleaved, truncated BID translocates to mitochondria, where it can activate BAX and BAK and induce the release of cytochrome c and SMAC/DIABLO serving to amplify apoptosis signaling (Danial and Korsmeyer, 2004).

The intrinsic and extrinsic apoptotic pathways converge on downstream effector caspases that implement cell elimination. The caspase family forms the engine of apoptosis and is divided into two major groups (Fischer et al., 2003; Fuentes-Prior and Salvesen, 2004). The subset of caspases that cleave selected substrates to produce the typical alteration associated with apoptosis are known as executioner caspases, which in mammals are caspase-3, caspase-6, and caspase-7. Executioner caspases are activated by apical initiator caspases, including caspase-8, caspase-9, and caspase-10. Effector caspases are targets of suppression by an endogenous family of antiapoptotic proteins called inhibitor of apoptosis proteins (IAPs). The IAP family, characterized by one or more baculovirus IAP repeat (BIR) domains, includes X-linked IAP (XIAP), c-IAP1, c-IAP2, Survivin, Livin (ML-IAP), ILP2, and Apollon. Different BIR domains are responsible for suppression of specific caspases. Structural studies have revealed that the BIR3 of XIAP is responsible for binding and inhibition of caspase-9 and that a region adjacent to BIR2 is the major determinant for inhibition of caspase-3 and caspase-7. ML-IAP contains a single BIR and inhibits caspase-9 but not caspase-3 and caspase-7. BIRs are sometimes accompanied by really interesting new gene (RING) and ubiquitin-conjugating enzyme domains, which are associated with the ability to target them and other proteins for proteasome degradation. SMAC/DIABLO is an IAP antagonist that is released into the cytoplasm upon mitochondrial permeabilization. SMAC/DIABLO binds to IAPs in a manner similar to caspases, thereby promoting apoptosis by liberating caspases from IAPs (Fesik and Shi, 2001). The amino-terminal tetrapeptide motif of SMAC/DIABLO is responsible for binding to BIR domains of IAPs (Wu et al., 2000). Evidence suggests that apoptosis requires or is facilitated by coordinate inhibition of IAPs by SMAC/DIABLO and activation of the apoptosome by cytochrome c.

3   Targeting the Apoptotic Machinery

3.1   BCL-2 Family Proteins as Targets

Proteins of the BCL-2 family are crucial checkpoints of the intrinsic mitochondrial death pathway. Overexpression of antiapoptotic BCL-2 family proteins such as BCL-2, BCL-x_L, BFL-1/A1, or MCL-1 has been observed in various malignancies and can confound cancer treatment (Reed and Pellecchia, 2005). Reduction of expression levels of antiapoptotic BCL-2 family genes is considered to potentially contribute to the proapoptotic effects of some novel anticancer agents, such as retinoids, histone deacetylase inhibitors, and peroxisome-proliferator-activated receptor-γ (PPARγ)-modulating drugs (Reed and Pellecchia, 2005). Drug design targeting antiapoptotic BCL-2 family members has been focused on three strategies inducing mRNA degradation with antisense oligonucleotides, BH3-domain peptido-mimetics, and synthetic small molecule drugs interfering directly with BCL-2 family member protein function (Fig. 5.1).

Currently, oblimersen sodium (G3139, Genasense; Genta, Inc., Berkeley Heights, New Jersey) is the only nucleic acid-based inhibitor of BCL-2 and its antiapoptotic relatives to enter clinical trials. Oblimersen is a DNA-based synthetic 18-mer antisense oligonucleotide to BCL-2 and has been reported to induce RNaseH-mediated degradation of BCL-2 mRNA (Fig. 5.1). A completed phase 3 clinical trial of oblimersen for advanced melanoma in combination with dacarbazine demonstrated slowed disease progression, but failed to extend the survival time (Fischer and Schulze-Osthoff, 2005). Several phase 3 clinical trials of oblimersen in combination with conventional chemotherapy involving patients with other tumors are still being evaluated. Antisense directed against BCL-x_L also displays proapoptotic effects in cancer cells (Fennell et al., 2001). Bispecific BCL-2/BCL-x_L-suppressing antisense oligonucleotides under preclinical studies may optimize efficacy because simultaneous overexpression of multiple antiapoptotic members of the BCL-2 family may occur in malignant cells (Del Bufalo et al., 2003). An MCL-1 antisense compound has recently demonstrated efficacy in a sarcoma xenograft model (Thallinger et al., 2004). Efficient delivery of these agents to tumor cells remains as a potential limitation of this approach.

An alternative means to interfere with BCL-2 antiapoptotic proteins is direct inhibition by modified BH3 peptides or small molecules (Fig. 5.1). A hydrocarbon-stapled BID BH3 peptide engineered to be helical, protease resistant, and cell permeable, potently induces apoptosis in Jurkat T leukemia cells and slows the growth of a transplanted leukemia (Walensky et al., 2004). A BAD-BH3 peptide preferably kills BCL-x_L and BCL-2 overexpressing Jurkat cells, and a BAX BH3 peptide is slightly more effective in BCL-2 overexpressing cells, suggesting that the efficacy of BH3 peptides might depend on the affinity of a certain BH3 domain for a limited set of antiapoptotic BCL-2 proteins (Shangary and Johnson, 2002).

Small-molecule inhibitors of BCL-2 or related antiapoptotic relatives have recently been identified through high-throughput screening of chemical libraries for the ability to dock onto the BH3 pocket of antiapoptotic BCL-2 family proteins, negating their prosurvival activity. The most advanced among these is the natural product gossypol, and its semisynthetic analogs with less toxicity, which are undergoing late steps of preclinical and clinical testing (Qiu et al., 2002). Chelerythrine, identified as an inhibitor of BCL-x_L/BAK-BH3 interaction from a natural compound library, induces apoptosis effectively in BCL-2 or BCL-x_L overexpressing cells (Chan et al. 2003). Others include BH3I-1 and BH3I-2, identified by screening using a BH3 peptide displacement assay, and HA14–1 and antimycin analogs, identified by computational modeling (Reed and Pellecchia, 2005). Most promising is ABT737 (Abbott, Abbott Park, Illinois), a synthetic small molecule developed by NMR-guided, structure-based drug design (Oltersdorf et al., 2005) that exhibits very high affinity for the hydrophobic pocket of BCL-2, BCL-x_L, and BCL-w (Cory and Adams, 2005). ABT737 significantly sensitizes many tumors to cytotoxic agents and is effective as a single agent against certain lymphomas and solid tumors, inducing tumor regression in xenograft models. The BAD-like selectivity of ABT737 suggests that tumor resistance to ABT737 could result from high-level expression of MCL-1 or BFL-1/A1, and small molecules that target these proteins should synergize with ABT737 (Cory and Adams, 2005). A chapter by Moore and Letai will provide more detail on rationale designs of therapeutics that target BCL-2 family of proteins.

3.2   *Targeting Death Receptors*

Most chemotherapeutic agents and radiation therapy induce apoptosis in cancer cells primarily by engagement of the mitochondrial apoptosis machinery. Accordingly, chemorefractory tumor cells often evolve defects in their intrinsic apoptotic pathway. By directly activating the caspase cascade, death receptor-mediated apoptosis, in contrast, can bypass the mitochondria and thereby sensitize resistant tumor cells to conventional chemotherapeutic agents or ionizing radiation (Fig. 5.1).

Despite the selective antitumor activity, the proinflammatory actions of TNF preclude its systemic administration in cancer therapy. Nonetheless, because TNF destroys tumor-associated blood vessels by apoptosis and improves vascular permeability to cytotoxic drugs, local application of TNF has been exploited for cancer therapy. Low-dose TNF was shown to improve penetration of doxorubicin for the treatment of melanoma and lymphoma (Curnis et al., 2000). Isolated limb perfusion of high-dose TNF combined with chemotherapeutic drugs demonstrated significant synergistic effect in treatment of locally advanced melanomas and sarcomas (Eggermont and ten Hagen, 2001).

CD95L and TRAIL are expressed on cytolytic T cells, natural killer cells, and other immune cells and play an important role in eradication of virus-infected and transformed cells (Locksley et al., 2001). Unlike TNF, these death ligands do not induce concomitant NF-κB activation (Karin and Lin, 2002). Unfortunately, severe hepatotoxicity precluded the systemic administration of CD95 ligand (Ogasawara et al., 1993). Most promising is TRAIL and agonistic antibodies that bind TRAIL receptors, which selectively kill tumor cells in mouse xenograft models without harming normal tissues (Ashkenazi et al., 1999). Genentech (South San Francisco, California) and Amgen (Thousand Oaks, California) have initiated phase 1 clinical trials with soluble TRAIL. Human Genome Sciences, Inc. (Rockville, Maryland) has recently completed phase 1 clinical trials with an agonistic monoclonal antibody against TRAIL-R1 (HGS ETR1) and advanced to phase 2 clinical trials for the treatment of a variety of cancers, such as non-small-cell lung cancer (NSCLC), colorectal carcinoma, and non-Hodgkins lymphoma. So far, patients had little toxicity (Le, 2004; Georgakis et al., 2005; Pukac et al., 2005). Humphreys and Halpern will discuss targeting of TRAIL receptors for cancer therapy in Chapter 7.

3.3   Therapeutic Inactivation of IAPs

IAPs inhibit executioner caspases activated by extrinsic or intrinsic pathways. XIAP, cIAP1, cIAP2, ML-IAP, and Survivin are upregulated in many tumors including leukemias and neuroblastomas, and have been correlated with adverse prognosis (Salvesen and Duckett, 2002). Targeted therapy attacking XIAP is currently under preclinical and clinical investigation and includes XIAP antisense, XIAP antagonists that specifically target BIR2 domain of XIAP, and SMAC-peptide and nonpeptide mimetics.

Antisense molecules targeting XIAP have been shown to sensitize a variety of tumor cell lines to radiotherapy and chemotherapy. Second-generation oligonucleotides, comprising DNA/RNA hybrid backbones with improved pharmacokinetics and reduced toxicity are in phase 1 clinical trials in patients with solid tumors (Fischer and Schulze-Osthoff, 2005). XIAP antagonists that target BIR2 and displace caspase-3 were identified by an enzyme derepression assay where XIAP-mediated suppression of caspase-3 is overcome by chemical compounds (Wu et al., 2003; Schimmer et al., 2004). These compounds display proapoptotic effects in tumor cell lines through a BCL-2/BCL-x_L-independent pathway (Wang et al., 2004).

Proapoptotic SMAC/DIABLO is released from mitochondria during the apoptotic process and relieves inhibition of caspase-3, caspase-7, and caspase-9 by IAPs. The four N-terminal residues (AVPI) of SMAC/DIABLO recognize a surface groove on BIR3 of XIAP normally occupied by processed caspase-9, thereby dislodging caspase-9 from the XIAP-inhibitory complex (Shiozaki et al., 2003). A series of SMAC peptido-mimetics consisting of 4, 5, or 7 amino acids of the amino-terminus of SMAC/DIABLO fused to a carrier peptide for intracellular delivery (cell-permeable SMAC peptides), overcome resistance of cancer cells with high levels of XIAP expression to apoptosis, and enhance the activity of conventional anticancer drugs in vitro and in vivo (Fulda et al., 2002; Guo et al., 2002; Yang et al., 2003; Sun et al., 2005). A series of nonpeptidic small-molecule

XIAP antagonists are being developed to improve proteolytic stability, cell permeability, and pharmacokinetics (Li et al., 2004; Nikolovska-Coleska et al., 2004; Oost et al., 2004; Park et al., 2005). These potent mimetics represent a novel class of anticancer drugs particularly useful in combination chemotherapy.

Survivin is an IAP member that plays a major role in both cell division and apoptosis. Survivin is highly expressed in cancer cells and is implicated in tumor resistance to radiotherapy and chemotherapy (Altieri, 2003). In addition, Survivin deficiency results in abnormal spindle formation and mitotic catastrophe independent of p53 and BCL-2 (Okada and Mak, 2004). Several preclinical studies demonstrated that inhibition of Survivin by antisense oligonucleotides, ribozymes, small interfering RNAs, dominant negative mutants, and cyclin-dependent kinase inhibitors was able to promote spontaneous apoptosis in tumor cells and to enhance the efficacy of conventional treatments including chemotherapy, radiotherapy, and immunotherapy (Zaffaroni et al., 2005). The high level and specificity of Survivin expression in cancer cells make it an attractive target for anticancer drug discovery.

3.4   Caspase Activators

Selective activation of caspases might be a valuable strategy for cancer therapy. Several approaches to trigger caspase activation in tumor cells are presently being developed. Inducible caspases have been engineered by fusing them to chemical dimerization domains. After delivery of these chimeric, regulatable caspases by adenoviral gene transfer, they can be activated to trigger apoptosis in tumor cells by cell permeable dimerization drugs (Shariat et al., 2001). Tumor-specific delivery is also achieved by fusing caspases with antibodies against receptors that are overexpressed in human cancers. For instance, caspase-3 linked to a anti-HER2 antibody is internalized via endocytosis by HER2 overexpressing tumors (Xu et al., 2004). In addition, high-throughput drug screening has identified a series of small molecule caspase activators, which have been shown to induce apoptosis in multiple cancer cell lines including prostate, breast, colorectal, lung cancer (Jiang et al., 2003; Nguyen and Wells, 2003). How specificity to tumor cells will be achieved in this case is not yet clear.

3.5   Modulation of the p53 Tumor Suppressor

The p53 pathway is composed of a network of genes and their products that respond to stresses, which disturb the fidelity of DNA replication and cell division (Balint and Vousden, 2001). Loss of p53 leads to genomic instability, impaired cell cycle regulation, and inhibition of apoptosis. p53 mutation and thereby inactivation are found in more than 50% of human cancers, and lack of functional p53 may render tumor cells resistant to apoptosis induced by chemotherapy and radiotherapy. In Chapter 10, El-Diery describes the regulation of programmed cell death by p53.

Three main therapeutic strategies are currently in development that target the p53 pathway (Fig. 5.1). First is reconstitution of wild type p53 in cancer cells by introduction of exogenous p53 with viral vectors. The most commonly used viral p53 delivery mechanism is the use of an adenoviral vector carrying the wild-type p53 gene. Due to the low efficiency of gene delivery of gene therapy vectors and hepatotoxicity associated with systemic applications, current clinical trials evaluate the efficacy of Ad-p53 through intratumoral injection in advanced solid tumors. Ad-p53 gene therapy alone failed to demonstrate beneficial effects in patients and new trials in combination with chemotherapy or radiotherapy are being investigated (Khuri et al., 2000).

Second is reactivation of mutant p53 to the wild-type form to induce apoptosis. Several compounds have been identified by screening or rational design with the capability of restoring the transcriptional function of mutant p53 and thereby apoptosis. These p53 reactivators include PRIMA-1 (Bykov et al., 2002), CP-31398 (Foster et al., 1999), and CDB3 (Friedler et al., 2002). As p53-mediated apoptosis is induced by transcriptional upregulation of the BH3-only proteins PUMA and NOXA, and the majority of human tumors have mutant p53, this approach has enormous potential for success.

Third, is the interruption of the regulatory interaction between p53 and MDM2 to prevent p53 degradation by the E3 ligase activity of MDM2. Nutlins and RITA increase p53 levels by binding to the p53 pocket for MDM2 interaction thereby inhibiting tumor growth in mice (Issaeva et al., 2004; Vassilev et al., 2004). However, restoration of p53 function in tumors with a lower frequency of p53 mutations maybe counterproductive as certain tumors may adapt normal p53 activation to achieve cell cycle arrest and DNA damage repair inflicted by chemotherapy and/or radiotherapy (Scott et al., 2003; Stoklosa et al., 2004).

4   Targeting Survival Signaling Pathways

Another set of targeted therapies aims to inhibit survival signaling pathways that are regulated by cytokines, hormones, and growth factors. In cancer cells, key components of these pathways are altered by oncogene activation or loss of tumor suppressor gene function, resulting in deregulated cell proliferation, inhibition of apoptosis, and enhanced angiogenesis. Strategies targeting survival signaling include neutralization of ligands, inhibition of receptors, and inhibition of cytoplasmic secondary messengers.

4.1   The NF-κB Pathway

Growth factors, cytokines such as interleukin-1 and TNF, hormones and other signals activate NF-κB by phosphorylation of inhibitor of κB (IκB), which has been linked to enhancement of both survival and tumorigenesis (see Chapter 11). Substantial evidence indicates that NF-κB plays an important role in tumorigenesis. Tumor suppressor genes such as CYLD and ING4 have been shown to negatively regulate NF-κB (Brummelkamp et al., 2003; Garkavtsev, 2004), and NF-κB activation is implicated in the increased incidence of cancer associated with inflammatory diseases (Greten et al., 2004; Pikarsky et al., 2004). Constitutive activation of NF-κB observed in tumors contributes to chemoresistance, perhaps by blocking apoptosis through direct transcriptional induction of expression of antiapoptotic proteins such as BCL-x_L, BFL-1/A1, or IAP1/2 and XIAP and/or repression of proapoptotic p53 (Nakanishi and Toi, 2005). Several anticancer agents stimulate NF-κB, such as taxanes, vinca alkaloids, and topoisomerase inhibitors, which can potentially lead to chemoresistance (Nakanishi and Toi, 2005). Thus, the possibility of increasing the efficacy of anticancer drugs by inactivation of NF-κB makes this pathway an attractive chemotherapeutic target (Fig. 5.1).

BAY11-7082 and BAY11-7085 inhibit IκB phosphorylation and stabilize IκB, allowing it to sequester NF-κB in the cytoplasm in an inactivated state. In preclinical studies, these two drugs sensitized tumor cells to conventional chemotherapy agents. For example, the histone deacetylase inhibitor suberoylanilide hydroxamic acid (SAHA) increases the NF-κB transcriptional activity through the enhanced nuclear translocation of the p65 subunit, which may diminish its effectiveness as a cancer therapeutic. However, inhibition of NF-κB by BAY11-7085 coordinately administered with SAHA increases cell death in NSCLC cell lines (Rundall et al., 2004). BAY11-7085 also increases the efficacy of cisplatin and paclitaxel in an in vivo ovarian cancer model (Mabuchi et al., 2004a; Mabuchi et al., 2004b).

Inhibition of NF-κB activity is considered as one of the major mechanisms of proteasome inhibitors, a novel class of anticancer drugs. IκB is polyubiquitinated upon phosphorylation by the IκB kinase, IKK, which then targets it for degradation by the 26S proteasome. Proteasome inhibitors thereby induce IκB accumulation, which retains NF-κB in the cytoplasm and prevents transcriptional activation of target genes. The proteasome inhibitor bortezomib (Velcade, PS-341; Millennium, Cambridge, Massachusetts) (also see Chapter 12) exhibits antitumor activity against a wide range of malignancies either as a single agent or combined with conventional chemotherapeutic drugs, and has been approved by US Food and Drug Administration (FDA) for the treatment of relapsed or refractory multiple myeloma (Rajkumar et al., 2005).

4.2   Therapeutic Modulation of the BH3-Only Protein BIM

While proteasome inhibitors may work in some situations as NFκB inhibitors, other targets for the anticancer activity of proteasome inhibitors include cell cycle regulatory proteins, p53-mediated apoptosis, unfolded protein response pathway, intrinsic and extrinsic apoptosis pathway. For example, the H-ras/MAP kinase pathway suppresses apoptosis induced by the proapoptotic BH3-only protein BIM in response to taxanes (paclitaxel) by phosphorylating BIM and targeting BIM for degradation in proteasomes. The proteasome inhibitor bortezomib restores BIM induction and apoptosis, abrogating resistance to paclitaxel conferred by H-ras, promoting BIM-dependent tumor regression. This suggests the potential benefits of combinatorial chemotherapy of bortezomib and paclitaxel preferentially in tumors with MAP kinase activation (Fig. 5.1) (Tan et al., 2005). The newly developed orally bioactive proteasome inhibitor NPI-0052 induces apoptosis in multiple myeloma cells resistant to conventional chemotherapeutic drugs and bortezomib, with less toxicity as it is mechanistically distinct from bortezomib (Chauhan et al., 2005) (also see Chapter 12). More drugs of this class, whether the ultimate therapeutic target is NF-κB, BIM, or yet another protein, are likely to be entering the clinic in the near future.

4.2.1   Tyrosine Kinase Inhibitors

Tyrosine kinases (TKs), particularly receptor tyrosine kinases (RTKs), are key factors in the promotion of cancer cell survival, and as such represent an attractive therapeutic target for cancer therapy (Fig. 5.1). Deregulated TK activity can cause increased cell proliferation, reduced apoptosis, invasion and angiogenesis. Small-molecule inhibitors of TKs compete with the ATP-binding site of the catalytic domain of oncogenic TKs and thereby prevent their activation. This has been the basis for the success in the treatment of chronic myelogenous leukemia (CML), where development of the inhibitor of the oncogenic BCR-ABL TK fusion protein, imatinib mesylate (Gleevec, STI571; Novartis, Basel, Switzerland), has produced dramatic clinical responses (see also Chapter 4). Over 90% of CML patients carry the Philadelphia chromosome, a translocation between chromosomes 9 and 22 that generates the *bcr-abl* oncogene (Faderl et al., 1999). The constitutively activated BCR-ABL kinase leads to growth factor independence and apoptosis resistance by activation of RAS-MAP kinase and Janus activating kinases-signal transducers and activators of transcription (JAK-STAT) pathways (Yamauchi et al., 1998). BCR-ABL also activates the antiapoptotic PI3K/AKT pathway, increases BCL-2, and suppresses BIM expression (Kuribara et al., 2004; Essafi et al., 2005). Imatinib produces major hematologic and cytogenetic responses in 65–90% of CML patients after failed interferon-α therapy and in 80–90% of newly diagnosed and untreated patients (Kantarjian et al., 2002a; Kantarjian et al., 2002b). However, most patients experienced relapse after treatment discontinuance and a significant number of newly diagnosed patients start out resistant (Druker, 2004). The most common resistance mechanism involves mutations that affect the conformation of the BCR-ABL kinase domain and prevent binding to imatinib (Gorre et al., 2001). Second-generation kinase inhibitors retain activity against almost all imatinib-resistant mutants and are currently under early clinical evaluation (Shah et al., 2004).

Combination therapy of imatinib with other therapeutics is also being investigated to overcome resistance. Imatinib also targets the receptor c-KIT and platelet-derived growth factor receptor (PDGFR). Gastrointestinal stromal tumors (GISTs), where mutated c-KIT is implicated in the pathogenesis, also show significant responses to imatinib (Debiec-Rychter et al., 2004).

A member of the epidermal growth factor receptor (EGFR) family, Her-2/*neu* is overexpressed in 20–30% of malignant breast tumors. A recombinant humanized monoclonal antibody targeted to the extracellular domain of the Her-2/*neu* receptor, trastuzumab (Herceptin; Genentech, Inc., San Francisco, California) has demonstrated overall tumor response rates between 15% and 26% in the metastatic setting. In combination with chemotherapy, trastuzumab produces prolonged disease-free and overall survival when compared to standard chemotherapeutic treatment regimens. Trastuzumab induces induction of G1 arrest of cell cycle progression and apoptosis and is now part of the treatment of choice for Her-2-positive breast cancers (Emens, 2005).

The EGFR is overexpressed in a variety of tumors, including tumors of the breast, lung, ovaries, and kidney, and thus a rational target for cancer therapy (Jones et al., 2005). Presently, two classes of EGFR antagonists are in phase 2 and 3 trials: anti-EGFR monoclonal antibodies and TK inhibitors. Cetuximab (Erbitux; ImClone Systems, Inc., New York), the most established monoclonal antibody, is approved for use as a single agent or in combination with irinotecan in patients with metastatic colorectal cancer (Wong, 2005). Erlotinib (Tarceva; OSI, Long Island, New York), an orally available selective inhibitor of the EGFR (ErbB1) TK, has received FDA approval for the treatment of patients with locally advanced or metastatic NSCLC after failure of at least one prior chemotherapy regimen (Comis, 2005). Erlotinib is the only EGFR TK inhibitor that showed survival improvement in NSCLC patients in a randomized phase 3 clinical trial (Perez-Soler, 2004). When administered in combination with gemcitabine, erlotinib also significantly improved survival in patients with advanced or metastatic pancreatic cancer in a phase 3 trial (Thomas and Grandis, 2004). Gefitinib (Iressa; AstraZeneca, Wilmington, Delaware), is another EGFR TK inhibitor that is approved for refractory NSCLC. In the subset of patients with specific EFGR TK domain mutations, the response rate to gefitinib was high, suggesting that screening for these mutations in lung cancers to identify patients that respond is advisable (Lynch et al., 2004; Sordella et al., 2004).

4.2.2 RAS-MAP Kinase Pathway

The RAS-RAF-MEK-ERK pathway represents a common downstream pathway for several key RTKs such as EGFR, PDGFR, and VEGFR, which are frequently mutated or overexpressed in human malignancies and thus is a logical therapeutic target (Fig. 5.1). Constitutive activation of the MAP kinase pathway not only promotes tumor cell proliferation, but may also interfere with apoptosis. Activation of the MEK-ERK cascade upregulates antiapoptotic proteins BCL-2, BCL-X_L, and MCL-1 (Liu et al., 1999; Leu et al., 2000; Jost et al., 2001), and promotes survival by phosphorylating BCL-2 and blocking its degradation in proteasomes (Dimmeler et al., 1999) and by phosphorylating BIM and accelerating its proteasomal degradation (Ley et al., 2003; Luciano et al., 2003; Tan et al., 2005).

Two novel MEK inhibitors CI-1040 (PD 184352) and PD 0325901 are in clinical trials (Pfizer, Inc., New York). A phase 2 study testing the MEK inhibitor CI-1040 in NSCLC, breast, colorectal, and pancreatic cancers was performed with negative results (Rinehart et al., 2004). The second-generation agent PD 0325901 with better bioavailability and increased potency is currently in clinical trials. In addition, MEK inhibitors suppress the expression of several antiapoptotic players, thus lowering the apoptotic threshold and have shown striking synergistic effects with conventional chemotherapy. For example, MEK blockade sensitizes leukemic cells to classical cytotoxics including nucleoside analogs, microtubule-targeted drugs, and γ-irradiation (Milella et al., 2005).

Although no mutations in A-RAF or C-RAF have been found in human cancers, B-RAF is mutated and constitutively activated in 70% of melanomas and other cancer types (ovarian, thyroid, colon, lung) with a moderate to high frequency, suggesting its implication in cancer development (Wan et al., 2004). Sorafenib (Nexavar, BAY43-9006; Bayer, West Haven, Connecticut and Onyx, Emeryville, California) is one of the most promising agents of the class of RAF kinase inhibitors and has shown significant efficacy and minimal toxicity both as a single agent and in combination with standard chemotherapies in renal cell, hepatocellular, colorectal, ovarian, and breast cancers in phase 1 and 2 studies (Thompson and Lyons, 2005). Sorafenib can also indirectly inhibit several important TKs including VEGFR-2, VEGFR-3 Flt-3, and c-Kit that are upstream of RAF, which contributes to its antiproliferative, antiapoptotic, and antiangiogenic properties (Wilhelm et al., 2004).

4.2.3 PI3KAKT/mTOR Pathway

The PI3K/AKT/mTOR pathway regulates cell proliferation and cell survival and is commonly found aberrantly activated in a variety of tumors due to amplification of the PI3KC gene encoding for the p110α catalytic subunit of PI3K, gene amplification of AKT, and loss of PTEN tumor suppressor function (Morgensztern and McLeod, 2005). AKT promotes cell survival by inhibiting BAD, caspase-9 and FORKHEAD, and activating several antiapoptotic proteins including IKK (Downward, 2004). A downstream effector in the PI3K pathway is the protein kinase mTOR, which is inhibited by tuberous sclerosis complex (TSC1/2). Activation of AKT results in phosphorylation of TSC2 which disrupts TSC1/2 complex, leading to derepression of mTOR (Inoki et al., 2002). mTOR promotes cell proliferation by regulating translation initiation, mediated by activation of the 40S ribosomal protein p70S6 kinase (S6K1) and inactivation of 4E-binding protein (4E-BP1). The increase in the translation of a subset of mRNAs produces proteins that are required for G1/S phase cell cycle progression (Hay and Sonenberg, 2004). Rapamycin, a natural mTOR inhibitor, was not developed as an anticancer drug due to poor solubility and instability. Rapamycin analogs, including CCI-779 (Wyeth-Ayerst, Princeton, New Jersey), RAD001 (Novartis, Basel, Switzerland), and AP23573 (Ariad, Cambridge, Massachusetts), have improved pharmacokinetics and are currently under clinical evaluation for cancer treatment (Fig. 5.1). While RAD001 and AP23573 are in the early stage of phase 1 clinical trials, CCI-779 has completed phase 1 and 2 studies with good tolerance and impressive response rate in patients with renal cell carcinoma (RCC), breast, lung and neuroendocrine tumors, which has led to phase 3 studies in patients with RCC and breast cancer (Morgensztern and McLeod, 2005). Identification of biomarkers to predict tumor sensitivity and the synergy between CCI-779 and standard chemotherapy, hormone or growth factor inhibitors are also being investigated (Vignot et al., 2005).

5   Targeting Pathways for Alternate Forms of Cell Death in Cancer Therapy

Although apoptosis represents the predominant mechanism by which cancer cells are eliminated, other modes of cell death, such as necrosis, autophagy, and mitotic catastrophe are also considered as cell death response to cytotoxic therapies. What determines the form of cell death induced by a particular anticancer agent depends on the cell type, the genotype of the cell, the type of cellular damage that the drug induces, the dose of the agent used, as well as the microenvironment. Thus, a better understanding of these diverse modes of cell death in cancer therapy may lead to new approaches to overcome drug resistance.

Necrosis refers to cell death characterized by cell swelling and rupture in response to profound damage or a physical insult, that subsequently releases its intracellular components into the surrounding tissue. A major consequence of this is the activation of an inflammatory response and thereby immune surveillance. Chronic inflammation is thought to promote tumor formation and progression, which is the basis for current efforts to use nonsteroidal anti-inflammatory agents for chemoprevention (Balkwill et al., 2005). DNA-alkylating agents cause necrotic cell death, which is equally effective in cells with and without apoptotic defects, and is independent of p53 or BCL-2 family proteins (Zong et al., 2004). Interestingly, alkylating agents selectively target cells using aerobic glycolysis, as is characteristic of many cancer cells, but not normal cells that use mitochondrial substrates for oxidative phosphorylation (Zong et al., 2004). Induction of necrosis has also been reported with arsenic trioxide, which triggers a regulated form of caspase-independent necrotic cell death (Scholz et al., 2005). Thus, stimulation of necrotic cell death may be an alternative in cancer cells with a defective apoptotic response, but may be coupled to inhibition of inflammation.

Autophagy is also an ordered cellular process where cell compartmentalizes to form autophagic vacuoles in cytoplasm and digests itself (Klionsky and Emr,

2000). It is a bulk protein degradation system that is essential for normal cell activity and survival when nutrients are scarce. Recent studies have linked defective autophagy to tumor development. Loss of *beclin1*, the mammalian ortholog of the yeast autophagy gene *apg6* that is monoallelically deleted in many human tumors, correlates with reduced autophagy and promotes tumorigenesis in mice (Qu et al., 2003; Yue et al., 2003). It is likely that activation of the AKT and mTOR signaling pathway contributes to malignant transformation by simultaneous inhibition of autophagy and apoptosis. Autophagic cell death is reportedly activated in cancer cells in response to various chemotherapeutic drugs, such as paclitaxel, vinblastine, and rapamycin, as well as to irradiation (Kim, 2005), although the clinical significance of autophagy in cancer therapy is unclear. If autophagy functions to promote survival of cancer cells by enabling catabolism, then autophagy inhibitors may be therapeutically useful. Alternatively, if autophagic cell death is a significant mechanism of cancer cell elimination, then inhibition of mTOR and activation of autophagy may be therapeutically beneficial.

Finally, driving cells past mitotic checkpoints and into aberrant mitoses that lead to death by mitotic catastrophe has recently attracted interest as a means to kill tumor cells independently of a defective apoptotic response (Castedo et al., 2004). Inhibiting normal mitosis in tumor cells can result in death due to mitotic failure and many current antimicrotubule drugs already in use in the clinic may induce death this way, and others that directly target regulation of mitosis are in development.

6 Future Prospects

Defective apoptosis is essential in tumor development and renders cancer cells refractory to chemotherapeutic agents. The identification of genes and gene products that regulate apoptosis at different molecular levels, along with an increased knowledge about their mechanisms of action provides a variety of therapeutic options for rational drug design targeting apoptosis.

In targeted anticancer drug development, high-throughput screening of chemical libraries, along with modification by structural biology and combinatorial chemistry to generate potent drugs with highly specific targets and favorable pharmacology has replaced the previous random screening. However, the sheer number of targets requires the development of a rapid, efficient preclinical and clinical screening system to eliminate ineffective agents with the minimal cost. In the preclinical setting to provide the proof of concept in vivo, it requires judicious application of the most appropriately genetically defined animal models of human cancers based on the proposed target of the drugs. In clinical trials, careful patient selection based on genetic information of the tumors being treated and the therapeutic target of the drug being tested is considered critical. The hypothesis with this approach is that those tumors in which the targeted apoptotic pathway or survival pathway is critical will be more susceptible to the therapeutic agents. The successful application of TK inhibitor imatinib mesylate in CML patients with Philadelphia chromosome supports this hypothesis. Similarly, the recent identification of point mutations in the EGFR gene in tumors from patients responding to gefitinib and their absence in nonresponders provides a means for patient selection. These findings illustrate the importance of matching the therapy to the tumor as a form of personalized medicine.

Cancer cells may be more dependent on apoptosis suppression because of oncogene activation, deregulated cell cycle control, and environmental stress. Improved understanding of how cancer cells interfere with apoptotic pathways in contrast with normal cells is required for selectively killing cancer cells without affecting normal tissues. The conventional chemotherapy and radiotherapy will remain the mainstay in cancer treatment, however, specific apoptosis-targeted drugs will tip the balance in favor of death, thereby sensitizing tumor cells to lower doses of chemotherapy and reducing side effects. Thus, molecularly targeted drugs will be evaluated in select patient populations as a platform for combinatorial chemotherapy to achieve optimal synergistic effect.

Future targeted cancer therapy will be characterized by individualized treatment, matching the genetic lesions in tumors to the optimal agents. However, tumor cells represent a heterogeneous and constantly evolving population where multiple apoptosis resistance mechanisms may be involved or a single cell may have acquired mutations paralyzing more than one apoptotic pathway. Thus, for targeted therapy to work successfully, drugs that target common apoptotic pathways will be needed and alternatively, combinatorial chemotherapy will be more effective to achieve maximal efficacy. Because of the complex nature of cancer, analysis of clinical samples using genomic and proteomic arrays is necessary to study the impact of targeted drugs on apoptosis signaling molecules and to correlate the genotype of tumors with the therapeutic outcome of particular treatment regimens, thereby providing the basis for targeted, personalized cancer therapy.

Acknowledgments We thank Thomasina Sharkey for assistance with preparation of the manuscript. We also thank Drs. Degenhardt, Mathew, Karantza-Wadsworth, and Nieves-Neira for critical reading of the manuscript.

References

Adams, J. M. (2003). Ways of dying: multiple pathways to apoptosis. Genes Dev 17 2481–2495.
Altieri, D. C. (2003). Survivin, versatile modulation of cell division and apoptosis in cancer. Oncogene 22, 8581–8589.
Ashkenazi, A., Pai, R. C., Fong, S., Leung, S., Lawrence, D. A., Marsters, S. A., Blackie, C., Chang, L., McMurtrey, A. E., Hebert, A., DeForge, L., Koumenis, I. L., Lewis, D., Harris, L., Bussiere, J., Koeppen, H., Shahrokh, Z., and Schwall, R. H. (1999). Safety and antitumor activity of recombinant soluble Apo2 ligand. J Clin Invest 104, 155–162.
Balint, E. E. and Vousden, K. H. (2001). Activation and activities of the p53 tumour suppressor protein. Br J Cancer 85, 1813–1823.

Balkwill, F., Charles, K. A., and Mantovani, A. (2005). Smoldering and polarized inflammation in the initiation and promotion of malignant disease. Cancer Cell 7, 211–217.

Brummelkamp, T. R., Nijman, S. M. B., Dirac, A. M. G., and Bernards, R. (2003). Loss of the cylindromatosis tumor suppressor inhibits apoptosis by activating NF-κB. Nature 424, 797–801.

Bykov, V. J., Issaeva, N., Shilov, A., Hultcrantz, M., Pugacheva, E., Chumakov, P., Bergman, J., Wiman, K. G., and Selivanova, G. (2002). Restoration of the tumor suppressor function to mutant p53 by a low-molecular-weight compound. Nat Med 8, 282–288.

Castedo, M., Perfettini, J. L., Roumier, T., Andreau, K., Medema, R., and Kroemer, G. (2004). Cell death by mitotic catastrophe: a molecular definition. Oncogene 23, 2825–2837.

Chauhan, D., Catley, L., Li, G., Podar, K., Hideshima, T., Velankar, M., Mitsiades, C., Mitsiades, N., Yasui, H., Letai, A., Ovaa, H., Berkers, C., Nicholson, B., Chao, T. H., Neuteboom, S. T. C., Richardson, P., Palladino, M. A., and Anderson, K. C. (2005). A novel orally active proteasome inhibitor induces apoptosis in multiple myeloma cells with mechanisms distinct from Bortezomib. Cancer Cell 8, 407–354.

Chan, S. L., Lee, M. C., Tan, K. O., Yang, L. K., Lee, A. S., Flotow, H., Fu, N. Y., Butler, M. S., Soejarto, D. D., Buss, A. D., and Yu, V. C. (2003). Identification of chelerythrine as an inhibitor of Bcl-X$_L$ function. J Biol Chem 278, 20453–20456.

Chen, L., Willis, S. N., Wei, A., Smith, B. J., Fletcher, J. I., Hinds, M. G., Colman, P. M., Day, C. L., Adams, J. M., and Huang, D. C. S. (2005). Differential targeting of prosurvival Bcl-2 proteins by their BH3-only ligands allows complementary apoptotic function. Mol Cell 17, 393–403.

Cheng, E., Wei, M., Weiler, S., Flavell, R., Mak, T., Lindsten, T., and Korsmeyer, S. (2001). BCL-2, BCL-X$_L$ sequester BH3 domain-only molecules preventing BAX- and BAK-mediated mitochondrial apoptosis. Mol Cell 8, 705–711.

Comis, R. L. (2005). The current situation: Erlotinib (Tarceva) and Gefitinib (Iressa) in non-small cell lung cancer. Oncologist 10, 467–470.

Cory, S. and Adams, J. M. (2002). The Bcl2 family: regulators of the cellular life-or-death switch. Nat Rev Cancer 2, 647–656.

Cory, S. and Adams, J. M. (2005). Killing cancer cells by flipping the BCL-2/BAX switch. Cancer Cell 7, 5–6.

Cuconati, A., Mukherjee, C., Perez, D., and White, E. (2003). DNA damage response and MCL-1 destruction initiate apoptosis in adenovirus-infected cells. Genes Dev, 2922–2932.

Cuconati, A. and White, E. (2002). Viral homologues of Bcl-2: role of apoptosis in the regulation of virus infection. Genes Dev 16, 2465–2478.

Curnis, F., Sacchi, A., Borgna, L., Magni, F., Gasparri, A., and Corti, A. (2000). Enhancement of tumor necrosis factor alpha antitumor immunotherapeutic properties by targeted delivery to aminopeptidase N (CD13). Nat Biotechnol 18, 1185–1190.

Danial, N. N. and Korsmeyer, S. J. (2004). Cell death: critical control points. Cell 116, 205–219.

Debiec-Rychter, M., Dumez, H., Judson, I., Wasag, B., Verweij, J., Brown, M., Dimitrijevic, S., Sciot, R., Stul, M., Vranck, H., Scurr, M., Hagemeijer, A., van Oosterom, A. T., and Group (2004). Use of c-KIT/PDGFRA mutational analysis to predict the clinical response to imatinib in patients with advanced gastrointestinal stromal tumors entered on phase I and II studies of the EORTC Soft Tissue and Bone Sarcoma Group. Eur J Cancer 40, 689–695.

Del Bufalo, D., Trisciuoglio, D., Scarsella, M., Zangemeister-Wittke, U., and Zupi, G. (2003). Treatment of melanoma cells with a bcl-2/bcl-xL antisense oligonucleotide induces antiangiogenic activity. Oncogene 22, 8441–8447.

Dimmeler, S., Breitschopf, K., Haendeler, J., and Zeiher, A. M. (1999). Dephosphorylation targets Bcl-2 for ubiquitin-dependent degradation: a link between the apoptosome and the proteasome pathway. J Exp Med 189, 1815–1822.

Downward, J. (2004). PI3-kinase, Akt and cell survival. Semin Cell Dev Biol 15, 177–182.

Druker, B. J. (2004). Imatinib as a paradigm of targeted therapies. Adv Cancer Res 91, 1–30.

Earnshaw, W. C., Martins, L. M., and Kaufmann, S. H. (1999). Mammalian caspases: structure, activation, substrates, and functions during apoptosis. Annu Rev Biochem 68, 383–424.

Eggermont, A. M. and ten Hagen, T. L. (2001). Isolated limb perfusion for extremity soft-tissue sarcomas, in-transit metastases, and other unresectable tumors: credits, debits and future perspectives. Curr Oncol Rep 3, 359–367.

Emens, L. A. (2005). Trastuzumab: targeted therapy for the management of HER-2/neu-overexpressing metastatic breast cancer. Am J Ther 12, 243–253.

Essafi, A., Fernandez de Mattos, S., and Hassen, Y. A. M. (2005). Direct transcriptional regulation of Bim by FoxO3a mediates STI571-induced apoptosis in Bcr-Abl-expressing cells. Oncogene 24, 2317–2329.

Faderl, S., Talaz, M., Estrov, Z., O'Brien, S., Kurzrock, R., and Kantarjian, H. M. (1999). The biology of chronic myeloid leukemia. N Engl J Med 341, 164–172.

Fennell, D. A., Corbo, M. V., Dean, N. M., Monia, B. P., and Cotter, F. E. (2001). In vivo suppression of Bcl-XL expression facilitates chemotherapy-induced leukemia cell death in a SCID/NOD Hu model. Br J Haematol 112, 706–713.

Fesik, S. W. (2000). Insights into programmed cell death through structural biology. Cell 103, 273–282.

Fesik, S. W. and Shi, Y. (2001). Controlling the caspases. Science 294, 1477–1478.

Fischer, U. and Schulze-Osthoff, K. (2005). New approaches and therapeutics targeting apoptosis in disease. Pharmacological reviews 57, 187–215.

Fischer, U., Janicke, R. U., and Schulze-Osthoff, K. (2003). Many cuts to ruin: a comprehensive update of caspase substrates. Cell Death Differ 10, 76–100.

Foster, B. A., Coffey, H. A., Morin, M. J., and Rastinejad, F. (1999). Pharmacological rescue of mutant p53 conformation and function. Science 286, 2507–2510.

Friedler, A., Hansson, L. O., Veprintsev, D. B., Freund, S. M., Rippin, T. M., Nikolova, P. V., Proctor, M. R., Rudiger, S., and Fersht, A. R. (2002). A peptide that binds and stabilizes p53 core domain: chaperone strategy for rescue of oncogenic mutants. Proc Natl Acad Sci USA 99, 937–942.

Fuentes-Prior, P. and Salvesen, G. S. (2004). The protein structures that shape caspase activity, specificity, activation and inhibition. Biochemical J 384, 201–232.

Fulda, S., Wick, W., Weller, M., and Debatin, K. M. (2002). Smac agonists sensitize for Apo2L/TRAIL- or anticancer drug-induced apoptosis and induce regression of malignant glioma in vivo. Nat Med 8, 808–815.

Garkavtsev, I. (2004). The candidate tumor suppressor protein ING4 regulates brain tumor growth and angiogenesis. Nature 428, 328–332.

Gelinas, C. and White, E. (2005). BH3-only proteins in control: Specificity regulates MCL-1 and BAK-mediated apoptosis. Genes Dev 19, 1263–1268.

Georgakis, G. V., Li, Y., Humphreys, R., Andreeff, M., O'Brien, S., Younes, M., Carbone, A., Albert, V., and Younes, A. (2005). Activity of selective fully human agonistic antibodies to the TRAIL death receptors TRAIL-R1 and TRAIL-R2 in primary and cultured lymphoma cells: induction of apoptosis and enhancement of doxorubicin- and bortezomib = induced cell death. Br J Haematol 130, 501–510.

Gorre, M. E., Mohammed, M., Ellwood, K., Hsu, N., Paquette, R., Rao, P. N., and Sawyers, C. L. (2001). Clinical resistance to STI-571 cancer therapy caused by BCR-ABL gene mutation or amplification. Science (Wash DC) 293, 876–880.

Greten, F. R., Echmann, L., Greten, T. F., Park, J. M., Li, Z. W., Egan, L. J., Kagnoff, M. F., and Karin, M. (2004). IKKbeta links inflammation and tumorigenesis in a mouse model of colitis-associated cancer. Cell 118, 285–296.

Guo, F., Nimmanapalli, R., Paranawithana, S., Wittman, S., Griffin, D., Bali, P., O'Bryan, E., Fumero, C., Wang, H. G., and Bhalla, K. (2002). Ectopic overexpression of second mitochondria-derived activator of caspases (Smac/DIABLO) or cotreatment with N-terminus of Smac/DIABLO peptide potentiates epothilone B derivative-(BMS 247550) and Apo-2L/TRAIL-induced apoptosis. Blood 99, 3419–3426.

Hay, N. and Sonenberg, N. (2004). Upstream and downstream of mTOR. Genes Dev 18, 1926–1945.

Inoki, K., Li, Y., Zhu, T., Wu, J., and Guan, K. L. (2002). TSC2 is phosphorylated and inhibited by Akt and suppresses mTOR signaling. Nat Cell Biol 4, 648–657.

Issaeva, N., Bozko, P., Enge, M., Protopopova, M., Verhoef, L. G., Masucci, M., Pramanik, A., and Selivanova, G. (2004). Small molecule RITA binds to p53, blocks p53-HDM-2 interaction and activates p53 function in tumors. Nat Med 10, 1321–1328.

Jiang, X., Kim, H.-E., Shu, H., Zhao, Y., Zhang, H., Kofron, J., Donnelly, J., Burns, D., Ng, S.-c., Rosenberg, S., and Wang, X. (2003). Distinctive roles of PHAP proteins and prothymosin-α in a death regulatory pathway. Science 299, 223–226.

Johnstone, R. W., Ruefli, A., and Lowe, S. W. (2002). Apoptosis: A link between cancer genetics and chemotherapy. Cell 108, 153–164.

Jones, H. E., Gee, J. M., Taylor, K. M., Barrow, D., Williams, H. D., Rubini, M., and NIcholson, R. I. (2005). Development of strategies for the use of anti-growth factor treatments. Endocr Relat Cancer 12 (Suppl 1), S173–S182.

Jost, M., Huggett, T. M., Kari, C., Boise, L. H., and Rodeck, U. (2001). Epidermal growth factor receptor-dependent control of keratinocyte survival and Bcl-xL expression through a MEK-dependent pathway. J Biol Chem 276, 6320–6326.

Kantarjian, H., Sawyers, C. L., Hochhaus, A., Guilhot, F., Schiffer, C., Gambacorti-Passerini, C., Niederwieser, D., Resta, D. J., Capdeville, R., Zoellner, U., Talpaz, M., Druker, B., Goldman, J., O'Brien, S. G., Russell, N., and Fischer, T. (2002a). Hematologic and cytogenetic responses to imatinib mesylate in chronic myelogenous leukemia. N Engl J Med 346, 645–652.

Kantarjian, H. M., Talpaz, M., O'Brien, S., Smith, T. L., Giles, F. J., Faderl, S., Thomas, D. A., Garcia-Manero, G., Issa, J. P., Andreeff, M., Kornblau, S. M., Koller, C., and Beran, M. 2002b. Imatinib mesylate for Philadelphia chromosome-positive, chronic-phase myeloid leukemia after failure of interferon-alpha: follow-up results. Clin Cancer Res 8, 2177–2187.

Karin, M. and Lin, A. (2002). NF-kappaB at the crossroads of life and death. Nat Immunol 3, 221–227.

Khuri, F. R., Nemunaitis, J., Ganly, I., Arseneau, J., Tannock, I. F., Romel, L., Gore, M., Ironside, J., MacDougall, R. H., and Heise, C. (2000). A controlled trial of intratumoral ONYX-015, a selectively -replicating adenovirus, in combination with cisplatin and 5-fluorouracil in patients with recurrent head and neck cancer. Nat Med 6, 879–885.

Kim, R. (2005). Recent advances in understanding the cell death pathways activated by anticancer therapy. Cancer 103, 1551–1560.

Klionsky, D. J. and Emr, S. D. (2000). Autophagy as a regulated pathway of cellular degradation. Science 290, 1717–1721.

Kuribara, R., Honda, H., Matsui, H., Shinjyo, T., Inukai, T., Sugita, K., Nakazawa, S., Hirai, H., Ozawa, K., and Inaba, T. (2004). Roles of Bim in apoptosis of normal and Bcr-Abl-expressing hematopoietic progenitors. Mol Cell Biol 24, 6172–6183.

Kuwana, T., Bouchier-Hayes, L., Chipuk, J., Bonzon, C., Sullivan, B., Green, D., and Newmeyer, D. (2005). BH3 domains of BH3-only proteins differentially regulate Bax-mediated mitochondrial membrane permeabilization both directly and indirectly. Mol Cell 17, 525–535.

Le, L. (2004). Phase 1 study of a fully human monoclonal antibody to the tumor necrosis factor-related apoptosis-inducing ligand death receptor 4 (TRAIL-R1) in subjects with advanced solid malignancies or non-Hodgkin's lymphoma. In American Society of Clinical Oncology Annual Meeting, New Orleans.

Letai, A., Bassik, M. C., Walensky, L. D., Sorcinelli, M. D., Weiler, S., and Korsmeyer, S. J. (2002). Distinct BH3 domains either sensitize or activate mitochondrial apoptosis, serving as prototype cancer therapeutics. Cancer Cell 2, 183–192.

Leu, C. M., Chang, C., and Hu, C. (2000). Epidermal growth factor (EGF) suppresses staurosporine-induced apoptosis by inducing mcl-1 via the mitogen-activated protein kinase pathway. Oncogene 19, 1665–1675.

Ley, R., Balmanno, K., Hadfield, K., Weston, C., and Cook, S. J. (2003). Activation of the ERK1/2 signaling pathway promotes phosphorylation and proteasome-dependent degradation of the BH3-only protein, Bim. J Biol Chem 278, 18811–18816.

Li, L., Thomas, R. M., Suzuki, H., De Brabander, J. K., Wang, X., and Harran, P. G. (2004). A small molecule Smac mimic potentiates TRAIL- and TNF-alpha-mediated cell death. Science 305, 1471–1474.

Li, P., Nijihawan, D., Budihardjo, I., Srinivasula, S. M., Ahmad, M., Alnemri, E. S., and Wang, X. (1997). Cytochrome c and dATP-dependent formation of Apaf-1/caspase-9 complex initiates an apoptotic protease cascade. Cell 91, 479–489.

Liu, Y. Z., Boxer, L. M., and Latchman, D. S. (1999). Activation of the Bcl-2 promoter by nerve growth factor is mediated by the p42/p44 MAPK cascade. Nucleic Acids Res 27, 2086–2090.

Locksley, R. M., Killeen, N., and Lenardo, M. J. (2001). The TNF and TNF receptor superfamilies: integrating mammalian biology. Cell 104, 487–501.

Luciano, F., Jacquel, A., Colosetti, P., Herrant, M., Cagnol, S., Pages, G., and Auberger, P. (2003). Phosphorylation of Bim-EL by Erk 1/2 on serine 69 promotes its degradation via the proteasome pathway and regulates its proapoptotic function. Oncogene 22, 6785–6793.

Lynch, T. J., Bell, D. W., Sordella, R., Gurubhagavatula, S., Okimoto, R. A., Brannigan, B. W., Harris, P. L., Haserlat, S. M., Supko, J. G., Haluska, F. G., et al. (2004). Activating mutations in the epidermal growth factor receptor underlying responsiveness of non-small-cell lung cancer to gefitinib. N Engl J Med 350, 2129–2139.

Mabuchi, S., Ohmichi, M., Nishio, Y., Hayasaka, T., Kimura, A., Ohta, T., Kawagoe, J., Takahashi, K., Yada-Hashimoto, N., Seino-Noda, H., Sakata, M., Motoyama, T., Kurachi, H., Testa, J. R., Tasaka, K., and Murata, Y. (2004a). Inhibition of inhibitor of nuclear factor-kappaB phosphorylation increases the efficacy of paclitaxel in in vitro and in vivo ovarian cancer models. Clin Cancer Res 10, 7645–7654.

Mabuchi, S., Ohmichi, M., Nishio, Y., Hayasaka, T., Kimura, A., Ohta, T., Saito, M., Kawagoe, J., Takahashi, K., Yada-Hashimoto, N., Sakata, M., Motoyama, T., Kurachi, H., Tasaka, K., and Murata, Y. (2004b). Inhibition of NF-κB increases the efficacy of cisplatin in in vitro and in vivo ovarian cancer models. J Biol Chem 279, 23477–23485.

Milella, M., Precupanu, C. M., Gregorj, C., Ricciardi, M. R., Petrucci, M. T., Kornblau, S. M., Tafuri, A., and Andreeff, M. (2005). Beyond single pathway inhibition: MEK inhibitors as a platform for the development of pharmacological combinations with synergistic antileukemic effects. Curr Pharm Des 11, 2779–2795.

Morgensztern, D. and McLeod, H. (2005). PI3K/Akt/mTOR pathway as a target for cancer therapy. Anti-Canc Drugs 16, 797–803.

Nakanishi, C. and Toi, M. (2005). Nuclear factor-κB inhibitors as sensitizers to anticancer drugs. Nat Rev 5, 297309.

Nelson, D., Tan, T.-T., Rabson, A. B., Anderson, D., Degenhardt, K., and White, E. (2004). Hypoxia and defective apoptosis drive genomic instability and tumorigenesis. Genes Dev 18, 2095–2107.

Nguyen, J. T. and Wells, J. A. (2003). Direct activation of the apoptosis machinery as a mechanism to target cancer cells. Proc Natl Acad Sci USA 100, 7533–7538.

Nikolovska-Coleska, Z., Xu, L., Hu, Z., Tomita, Y., LI, P., Roller, P. P., Wang, R., Fang, X., Guo, R., and Zhang, M. (2004). Discovery of embelin as a cell-permeable, small-molecular weight inhibitor of XIAP through structure-based computational screening of a traditional herbal medicine three-dimensional structure database. J Med Chem 47, 2430–2440.

Ogasawara, J., Watanabe-Fukunaga, R., Adachi, M., Matsuzawa, A., Kasugai, T., Kitamura, Y., Itoh, N., Suda, T., and Nagata, S. (1993). Lethal effect of the anti-Fas antibody in mice. Nature 364, 806–809.

Okada, H. and Mak, T. W. (2004). Pathways of apoptotic and non-apoptotic death in tumor cells. Nature Rev Cancer 4, 592–603.

Oltersdorf, T., Elmore, S. W., Shoemaker, A. R., Armstrong, R. C., Augeri, D. J., Belli, B. A., Bruncko, M., Deckwerth, T. L., Dinges, J., Hajduk, P. J., Joseph, M. K., Kitada, S., Korsmeyer, S. J., Kunzer, A. R., Letai, A., Li, C., Mitten, M. J., Nettesheim, D. G., and Ng, S. E. A. (2005). An inhibitor of Bcl-2 family proteins induces regression of solid tumours. Nature 435, 677–681.

Oost, T. K., Sun, C., Armstrong, R. C., Al-Assaad, A.-S., Betz, S. F., Deckwerth, T. L., Ding, H., Elmore, S. W., Meadows, R. P., Olejniczak, E. T., Oltersdorf, T., Rosenberg, S. H., Shoemaker, A. R., Tomaselli, K. J., Zou, H., and Fesik, S. W. (2004). Discovery of potent antagonists of the antiapoptotic protein XIAP for the treatment of cancer. J Med Chem 47, 4417–4426.

Park, C.-M., Sun, C., Olejniczak, E. T., Wilson, A. E., Meadows, R. P., Betz, S. F., Elmore, S. W., and Fesik, S. W. (2005). Non-peptidic small molecule inhibitors of XIAP. Bioorg Med Chem Lett 15, 771–775.

Perez-Soler, R. (2004). The role of erlotinib (Tarceva, OSI 774) in the treatment of non-small cell lung cancer. Clin Cancer Res 10, 4238s–4240s.

Pikarsky, E., Porat, R. M., Stein, I., Abramovitch, R., Amit, S., Kasem, S., Gutkovich-Pyest, E., Urieli-Shoval, S., Galun, E., and Ben-Neriah, Y. (2004). NF-kappaB functions as a tumour promoter in inflammation-associated cancer. Nature 431, 461–466.

Pukac, L., Kanakaraj, P., Humphreys, R., Alderson, R., Bloom, M., Sung, C., Riccobene, T., Johnson, R., Fiscella, M., Mahoney, A., Carrell, J., Boyd, E., Yao, X. T., Zhang, L., Zhong, I., vonKerczek, A., Shepard, L., Vaughan, T., Edwards, B., Dobson, C., Salcedo, T., and Albert, V. (2005). HGS-ETR1, a fully human TRAIL-receptor 1 monoclonal antibody, induces cell death in multiple tumour types in vitro and in vivo. Br J Cancer 92, 1430–1441.

Qiu, J., Levin, L. R., Buck, J., and Reidenberg, M. M. (2002). Different pathways of cell killing by gossypol enantiomers. Exp Biol Med 227, 398–401.

Qu, X., Yu, J., Bhagat, G., Furuya, N., Hibshoosh, H., Troxel, A., Rosen, J., Eskelinen, E. L., Mizushima, N., Ohsumi, Y., Cattoretti, G., and Levine, B. (2003). Promotion of tumorigenesis by heterozygous disruption of the beclin 1 autophagy gene. J Clin Invest 112, 1809–1820.

Rajkumar, S. V., Richardson, P. G., Hideshima, T., and Anderson, K. C. (2005). Proteasome inhibition as a novel therapeutic target in human cancer. J Clin Oncol 23, 630–639.

Reed, J. C. and Pellecchia, M. (2005). Apoptosis-based therapies for hematologic malignancies. Blood 106, 408–418.

Rinehart, J., Adjei, A. A., and Lorusso, P. M. (2004). Multicenter phase 2 study of the oral MEK inhibitor, CI-1040, in patients with advanced non-small-cell lung, breast, colon, and pancreatic cancer. J Clin Oncol 22, 4456–4462.

Rundall, B. K., Denlinger, C. E., and Jones, D. R. (2004). Combined histone deacetylase and NF-κB inhibition sensitizes non-small cell lung cancer to cell death. Surgery 136, 416–425.

Salvesen, G. S. and Duckett, C. S. (2002). IAP proteins: blocking the road to death's door. Nat Rev Mol Cell Biol 3, 401–410.

Schimmer, A. D., Welsh, K., Pinilla, C., Wang, Z., Krajewska, M., Bonneau, M. J., I. M., P., Kitada, S., Scott, F. L., Bailly-Maitre, B., Glinsky, G., Scudiero, D., Sausville, E., Salvesen, G., Nefzi, A., Ostresh, J. M., Houghten, R. A., and Reed, J. C. (2004). Small-molecule antagonists of apoptosis suppressor XIAP exhibit broad antitumor activity. Cancer Cell 5, 25–35.

Scholz, C., Wieder, T., Starck, L., Essmann, F., Schultz-Osthoff, K., Dorken, B., and Daniel, P. T. (2005). Arsenic trioxide triggers a regulated form of caspase-independent necrotic cell death via the mitochondrial death pathway. Oncogene 24, 1904–1913.

Scott, S. L., Earle, J. D., and Gumerlock, P. H. (2003). Functional p53 increases prostate cancer cell survival after exposure to fractionated doses of ionizing radiation. Cancer Res 63, 7190–7196.

Shah, N. P., Tran, C., Lee, F. Y., Chen, P., Norris, D., and Sawyers, C. L. (2004). Overriding imatinib resistance with a novel ABL kinase inhibitor. Science (Wash DC) 305, 399–401.

Shangary, S. and Johnson, D. E. (2002). Peptides derived from BH3 domains of Bcl-2 family members: a comparative analysis of inhibition of Bcl-2, Bcl-x(L) and Bax oligomerization, induction of cytochrome c release and activation of cell death. Biochemistry 41, 9485–9495.

Shariat, S. F., Desai, S., Song, W., Khan, T., Zhao, J., Nguyen, C., Foster, B. A., Greenberg, N., Spencer, D. M., and Slawin, K. M. (2001). Adenovirus-mediated transfer of inducible caspases: a novel 'death switch' gene therapeutic approach of prostate cancer. Cancer Res 61, 2562–2571.

Shi, Y. (2002). Mechanisms of caspase activation and inhibition during apoptosis. Mol Cell 9, 459–470.

Shiozaki, E. N., Chai, J., Rigotti, D. J., Riedl, S. J., Li, P., Srinivasula, S. M., Alnemri, E. S., Frirman, R., and Shi, Y. (2003). Mol Cell 11, 519–527.

Sordella, R., Bell, D. W., Haber, D. A., and Settleman, J. (2004). Gefitinib-sensitizing EGFR mutations in lung cancer activate anti-apoptotic pathways. Science 305, 1163–1167.

Stoklosa, T., Slupianek, A., Datta, M., Nieborowska-Skorska, M., Nowicki, M. O., Koptyra, M., and Skorski, T. (2004). BCR/ABL recruits p53 tumor suppressor protein to induce drug resistance. Cell Cycle 3, 1463–1472.

Sun, H., Nikolovska-Coleska, Z., Chen, J., Yang, C. Y., Tomita, Y., Pan, H., Yoshioka, Y., Krajewski, K., Roller, P. P., and Wang, S. (2005). Structure-based design, synthesis and biochemical testing of novel and potent Smac peptido-mimetics. Bioorg Med Chem Lett 15, 793–797.

Suzuki, M., Youle, R. J., and Tjandra, N. (2000). Structure of Bax: coregulation of dimer formation and intracellular localization. Cell 103, 645–654.

Tan, T.-T., Degenhardt, K., Nelson, D. A., Beaudoin, B., Nieves-Neira, W., Bouillet, P., Villunger, A., Adams, J. M., and White, E. (2005). Key roles of BIM-driven apoptosis in epithelial tumors and rational chemotherapy. Cancer Cell 7, 227–238.

Thallinger, C., Wolschek, M. F., Maierhofer, H., Skvara, H., Pehamberger, H., and Monia, B. P. (2004). Mcl-1 is a novel therapeutic target for human sarcoma: synergistic inhibition of human sarcoma xenotransplants by a combination of mcl-1 antisense oligonucleotides with low-dose cyclophosphamide. Clin Cancer Res 10, 4185–4191.

Thomas, S. M. and Grandis, J. R. (2004). Pharmacokinetics and pharmacodynamic properties of EGFR inhibitors under clinical investigation. Cancer Treat Rev 30, 255–268.

Thompson, N. and Lyons, J. (2005). Recent progress in targeting the Raf/MEK/ERK pathway with inhibitors in cancer drug discovery. Curr Opin Pharmacol 5, 350–356.

Vassilev, L. T., Vu, B. T., Graves, B., Carvajal, D., Podlaski, F., Filipovic, Z., Kong, N., Kammlott, U., Lukacs, C., and Klein, C. (2004). In vivo activation of the p53-pathway by small-molecule antagonists of MDM2. Science (Wash DC) 303, 844–848.

Vignot, S., Faivre, S., Aguirre, D., and Raymond, E. (2005). mTOR-targeted therapy of cancer with rapamycin derivatives. Ann Oncol 16, 525–537.

Walensky, L. D., Kung, A. L., Escher, I., Malia, T. J., Barbuto, S., Wright, R. D., Wagner, G., Verdine, G. L., and Korsmeyer, S. J. (2004). Activation of apoptosis in vivo by a hydrocarbon-stapled BH3 helix. Science 305, 1466–1470.

Wallach, D., Varfolomeev, E. E., Malinin, N. L., Goltsev, Y. V., Kovalenko, A. V., and Boldin, M. P. (1999). Tumor necrosis factor receptor and Fas signaling mechanisms. Annu Rev Immunol 17, 331–367.

Wan, P. T., Garnett, M. J., Roe, S. M., Lee, S., Niculescu-Duvaz, D., Good, V. M., Jones, C. M., Marshall, C. J., Springer, C. J., Barford, D., and Marais, R. (2004). Mechanism of activation of the RAF-ERK signaling pathway by oncogenic mutations of B-RAF. Cell 116, 855–867.

Wang, Z., Cuddy, M., and Samuel, T., et al. (2004). Cellular, biochemical, and genetic analysis of mechanism of small-molecule IAP inhibitors. J Biol Chem 279, 48168–48176.

Wilhelm, S. M., Carter, C., Tang, L., Wilkie, D., McNabola, A., Rong, H., Chen, C., Zhang, X., Vincent, P., McHugh, M., Cao, Y., Shujath, J., Gawlak, S., Eveleigh, D., Rowlev, B., Liu, L., Adnane, L., Lynch, M., Auclair, D., Taylor, I., Gedrich, R., Voznesenskv, A., Riedl, B., Post, L. E., Bollag, G., and Trail, P. A. (2004). BAY43–9006 exhibits broad spectrum oral antitumor activity and targets the RAF/MEK/ERK pathway and receptor tyrosine kinases involved in tumor progression and angiogenesis. Cancer Res 64, 7099–7109.

Willis, S. N. and Adams, J. M. (2005). Life in the balance: how BH3-only proteins induce apoptosis. Curr Opin Cell Biol 17, 1–9.

Willis, S. N., Chen, L., Dewson, G., Wei, A., Naik, E., Fletcher, J. I., Adams, J. M., and Huang, D. C. (2005). Proapoptotic Bak is sequestered by Mcl-1 and Bcl-xL, but not Bcl-2, until displaced by BH3-only proteins. Genes Dev 19, 1294–1305.

Wong, S. F. (2005). Cetuximab: an epidermal growth factor receptor monoclonal antibody for the treatment of colorectal cancer. Clin Ther 6, 684–694.

Wu, G., Chai, J., Suber, T. L., Wu, J. W., Du, C., Wang, X., and Shi, Y. (2000). Structural basis of IAP recognition by Smac/DIABLO. Nature 408, 1008–1012.

Wu, T., Wagner, K. W., Bursulaya, B., Schultz, P. G., and Deveraux, Q. L. (2003). Development and characterization of nonpeptidic small molecule inhibitors of the XIAP/caspase-3 interaction. Chem Biol 10, 759–767.

Wyllie, A. H. (1980). Glucocorticoid-induced thymocyte apoptosis is associated with endogenous endonuclease activation. Nature (London) 284, 555–556.

Xu, Y. M., Wang, L. F., Jia, L. T., Qiu, X. C., Zhao, J., Yu, C. J., Zhang, R., Zhu, F., Wang, C. J., and Jin, B. Q. (2004). A caspase-6 and anti-human epidermal growth factor receptor-2 (HER2) antibody chimeric molecule suppresses the growth of HER2-overexpressing tumors. J Immunol 173, 61–67.

Yamauchi, T., Ueki, K., Tobe, K., Tamemoto, H., Sekine, N., Wada, M., Honjo, M., Takahashi, M., Takahashi, T., Hirai, H., Tsushima, T., Akanuma, Y., Fujita, T., Komuro, I., Yazaki, Y., and Kadowaki, T. (1998). Growth hormone-induced tyrosine phosphorylation of EGF receptor as an essential element leading to MAP kinase activation and gene expression. Endocr J 45 (Suppl), S27–S31.

Yang, L., Mashima, T., Sato, S., M., M., Sakamoto, H., Yamori, T., Oh-Hara, T., and Tsuruo, T. (2003). Predominant suppression of apoptosome by inhibitor of apoptosis protein in non-small cell lung cancer H460 cells: therapeutic effect of a novel polyarginine-conjugated Smac peptide. Cancer Res 63, 831–837.

Yue, Z., Jin, S., Yang, C., Levine, A. J., and Heintz, N. (2003). Beclin 1, an autophagy gene essential for early embryonic development, is a haploinsufficient tumor suppressor. Proc Natl Acad Sci USA 100, 15077–15082.

Zaffaroni, N., Pennati, M., and Daidone, M. G. (2005). Survivin as a target for new anticancer interventions. J Cell Mol Med 9, 360–372.

Zapata, J. M., Pawlowski, K., Haas, E., Ware, C. F., Godzik, A., and Reed, J. C. (2001). A diverse family of proteins containing tumor necrosis factor receptor-associated factor domains. J Biol Chem 276, 24242–24252.

Zong, W.-X., Lindsten, T., Ross, A. J., MacGregor, G. R., and Thompson, C. B. (2001). BH3-only proteins that bind pro-survival Bcl-2 family members fail to induce apoptosis in the absence of Bax and Bak. Genes Dev 15, 1481–1486.

Zong, W.-X., Ditsworth, D., Bauer, D. E., Wang, Z.-Q., and Thompson, C. B. (2004). Alkylating DNA damage stimulates a regulated form of necrotic cell death. Genes Dev 18, 1272–1282.

Zou, H., Henzel, W. J., Liu, X., Lutschg, A., and Wang, X. (1997). Apaf-1, a human protein homologous to C. elegans CED-4, participates in cytochrome c-dependent activation of caspase-3. Cell 90, 405–413.

Chapter 6
Overcoming Resistance to Apoptosis in Cancer Therapy

Peter Hersey*, Xu Dong Zhang, and Nizar Mhaidat

Keywords apoptosis, resistance, oncogene, melanoma

1 Introduction

A fundamental characteristic of cancer cells is suppression of apoptosis and increased cell survival.[1,2] These properties, when combined with deregulated cell proliferation, are the basic requirements for development of cancer. Increased deregulated cell proliferation by itself paradoxically may trigger cell death pathways which prevent outgrowth of the cancer cell unless the cell death pathways are inhibited.[3] Another consequence of the latter may be resistance to treatments that depend on induction of apoptosis in the cancer cell. These widely held concepts have given rise to intense study of the antiapoptotic mechanisms generated in different cancer cells that are driven by different oncogenic stimuli and how these mechanisms may operate against different therapies used against cancers. The mechanisms by which different therapies induce apoptosis are in turn poorly understood and answers to both questions are needed in development of effective treatment approaches. In the following sections, we review recent information about regulation of apoptosis, how oncogenes interact with apoptotic pathways, and some of the therapeutic opportunities that are developing as a consequence of this information. Emphasis is given to studies on melanoma as a model system in these developments.

2 Recent Concepts About Regulation of Apoptosis

Although apoptosis is traditionally described in terms of intrinsic and extrinsic pathways in most instances, apoptosis induced by oncogenes proceeds via the mitochondrial "intrinsic" pathway. Much is known about this pathway and in particular the proteins involved in regulation of the pathway.

P. Hersey, X.D. Zhang, and N. Mhaidat
Immunology and Oncology Unit, Newcastle Mater Hospital, Newcastle, NSW 2300, Australia

* To whom correspondence should be addressed: Immunology and Oncology Unit, Room 443, David Maddison Clinical Sciences Building, Cnr King & Watt Streets, Newcastle NSW 2300, Australia. Tel.: 61 2 49236828; fax: 61 2 49236184, e-mail: Peter.Hersey@newcastle.edu.au

R. Khosravi-Far and E. White (eds.), *Programmed Cell Death in Cancer Progression and Therapy*.
© Springer 2008

2.1 Bcl-2 Family Proteins in Regulation of Apoptosis

Apoptosis via the mitochondrial pathway is regulated by the Bcl-2 family of proteins which share at least one conserved Bcl-2 homology (BH) domain. The prosurvival Bcl-2 proteins share four such domains and act to protect intracellular membranes associated with mitochondria, nuclei, and endoplasmic reticulum. The proapoptotic Bax and Bak proteins have three BH domains and are located in the cytosol (Bax) and mitochondrial outer membrane (Bak). They are essential for apoptosis to proceed and mice lacking both genes have a number of developmental abnormalities.[1] Similarly, apoptosis of cancer cells induced by several chemotherapy agents is dependent on Bax.[4–6]

Once activated, Bax and Bak oligomerize and insert into the outer mitochondrial membrane and thereby cause the release of several factors from mitochondria that can trigger apoptosis. These include cytochrome-c, Smac/DIABLO, Omi, apoptosis-inducing factor (AIF), and endonuclease G. These factors are located in the membrane or intermembranous space between the outer and inner mitochondrial membranes. Two models have been proposed to explain the release of these proteins during apoptosis. In one model, an autonomous channel formed by Bax or Bak is formed and this allows the release of the factors from the intermembrane space.[2] Another model depends on specific interaction of Bax or Bcl-2 with components of the permeability transition pore (PTP), which exists at sites of contact between outer and inner mitochondrial membranes. This results in opening of the PTP, swelling of the mitochondrial matrix, and rupture of the outer mitochondrial membrane.[7]

2.1.1 Bcl-2 Sensor Proteins

The discovery of a third group of Bcl-2 proteins which share a single BH3 domain has had a major influence on concepts regarding initiation of apoptosis.[8,9] They are regarded as sensors of damage to cells and different members respond to a diverse array of damaging agents by activating the Bax/Bak proteins to damage mitochondria. Two of the members, Bid and Bim, may be able to directly cause changes in Bax and Bak, which result in their oligomerization and insertion into mitochondria.[9] The other members, such as Bad, Noxa, and P53-upregulated modulator of apoptosis (PUMA), appear to function by binding to and neutralizing the antiapoptotic proteins. In addition, they may displace other BH3 proteins such as Bid, Bim, and p53, which have the ability to activate Bax and Bak.[10]

Bid appears to mediate apoptosis induced by tumor necrosis factor (TNF) family ligands and by granzyme B from cytotoxic T lymphocyte (CTL). Bid is cleaved by caspase-8 at Asp59 into tBid or by granzyme B at Asp75 into active (gtBid) form.[11] tBid is able to cause oligomerization of cytosolic Bax or Bak associated with mitochondria which facilitates binding of the Bax/Bak oligomers to the outer mitochondrial membrane and release of aptogenic proteins as referred to above.

Several sensor proteins appear to be located in the cytoskeleton of cells. BimEL, BimL, and BimS are the main splice variants of Bim. BimEL is a relatively weak inducer of apoptosis and is degraded in proteasomes after phosphorylation by ERK1/2 on Serine 65. BimL and BimS appear to be the main inducers of apoptosis.[12,13] BimL (and BimEL) is associated with microtubules by attachment to the Dynein motor complex and is released from this by agents such as the vinca alkaloids and taxols. Once released into the cytosol, Bim binds to prosurvival Bcl-2 proteins (Bcl-2, Bcl-XL, and Mcl-1) and may also bind to and activate Bax similar to that proposed for interaction of tBid and Bax. In some hematopoietic cell types Bim is located predominantly with Bcl-2 antiapoptotic proteins on mitochondria. Bak may be activated simply by releasing it from the antiapoptotic Mcl-1 and Bcl-XL proteins (not Bcl-2) due to competitive binding of BH3 (Bim) proteins to antiapoptotic proteins.[9] Bmf is associated with the Actin Myosin V motor complex[14] and is released by anoikis (cell detachment) and appears to have more restricted binding to the antiapoptotic proteins Bcl-2 and Bcl-XL.[15]

Agents which damage DNA and upregulate p53 result in p53-dependent upregulation of several BH3 proteins, Bad, Noxa, and PUMA (see also Chapter 3). Noxa appears to bind predominantly to the antiapoptotic protein Mcl-1 and competitively inhibits binding of Bak to Mcl-1. This results in release of Bak, allowing it to oligomerize and bind to the outer mitochondrial membrane. PUMA is also believed to mediate its effects by binding to the antiapoptotic proteins Bcl-XL (and Bcl-2) and thereby cause the release of proapoptotic proteins bound to them. One of the proteins so released may be p53 itself, which may be able to induce mitochondrial permeabilization directly[16] or by binding to Bak on the outer mitochondrial membrane and thereby induce apoptosis.[17] This nontranscriptional role of p53 is dependent on it being transported into the cytosol[18] Noxa has a more restricted specificity to the antiapoptotic proteins Mcl-1 and A1.[19] In addition to transcriptional regulation by p53, Noxa may be increased by inhibition of proteasome degradation[20] and by a gamma-secretase tripeptide inhibitor.[21]

Another transcription target of p53 is a relatively little studied protein called PIDD, which is believed to combine with an adaptor protein, RAIDD, and form a complex with caspase-2 called the PIDDOSOME.[22] Caspase 2 is an initiator caspase that appears to act upstream of mitochondria. Substrates may include Bid[23] and PKCδ, as well as proteins in the Golgi apoptosis complex and cytoskeleton. It may have direct effects on mitochondria and cause release of aptogenic proteins.[24] Caspase 2 may also be activated by casein kinase 2 and sensitize cells to TNF-related apoptosis-inducing ligand (TRAIL) by processing of caspase 8.[25]

2.2 Inhibitor of Apoptosis Proteins

Apoptosis is also regulated by another family of proteins referred to as inhibitor of apoptosis proteins (IAPs).[26–28] These include IAP 1 and 2, XIAP, ML-IAP, and Survivin. In general, they bind to caspases and prevent their activation (caspase-9) or inhibit their effector function (caspases 3 and 7). As discussed elsewhere, they also have other roles as E3 ligases and in ubiquination of proteins for degradation by proteasomes.[29] Binding of IAPs to caspases is competitively inhibited by Smac/DIABLO and OMI released from mitochondria and this allows effector caspases to induce apoptosis. This mechanism was shown to be the principal pathway in TRAIL-induced apoptosis of melanoma.[30]

3   Oncogenes and Apoptosis

3.1   Drivers of Cell Proliferation

Several transcription factors appear to be key players in cell proliferation, such as E2F, which is under the control of the Retinoblastoma protein (Rb), and c-Myc, which targets a number of proteins involved in cell division. Transition from G1 to S phase is regulated mainly by cyclin D, CDK4/6 complexes which phosphorylate Rb proteins and thereby activate E2F1–3 transcription of proteins involved in cell division. c-Myc is believed to play an essential role in this process by increasing cyclin D1/CK4/6 levels and suppression of CDK inhibitors such as p27.[31] Regulatory control of the G1/S transition is believed to vary widely between different tissues and different cancers. In melanoma the Ras and P1(3)K pathways appear to be key drivers of cell division in response to a number of mitogenic factors acting on Tyrosine kinase and G protein-coupled receptors. Activating mutations of BRAF are relatively common in melanoma[32,33] and naevi,[34] and have focused attention on this particular pathway. There is some evidence that loss of inhibitors of the cell cycle such as P16 may differentiate melanoma from naevi.[35] There may also be subsets of melanoma that are particularly dependent on constitutive activation of this pathway, e.g., melanoma in skin without signs of chronic sun damage were more likely (81%) to have activating mutations of N-RAS or BRAF than melanoma in chronic sun-damaged skin. The latter had increases in gene copies for Cyclin D1 (CCND1) and CDK4[36] but whether this is caused by elevated c-Myc levels, as discussed above, is not known. Immunohistological studies on nodular melanoma also found correlations with nuclear staining for Rb, Cyclin D1 and high mitotic rate measured by Ki-67 staining.[37]

3.2   Oncogene Pathways and Apoptosis

The transcription factors E2F and c-Myc target a number of genes involved in initiation of apoptosis. Most important of these is the transcription factor p53, which in turn targets genes for proteins such as PUMA, Noxa, PIDD, Bid, Bax, and Apaf-1 (see also Chapter 10). As discussed above, p53 itself may have a direct nontranscriptional role in inducing apoptosis.[9] Oncogenes may also upregulate p53 via the alternate reading frame (ARF) of the CDKN2A gene and thereby increase p53 levels by inhibition of HDM2, which ubiquinates and degrades p53. c-Myc may also be more important in induction of ARF than E2F and in induction of Bim, which mediates p53-independent apoptosis.

Given the evidence that oncogenes may also drive apoptosis, the survival of cancer cells implies that cancer cells that have been selected by outgrowth of apoptosis-resistant cells. Rb is inactivated (and E2F thereby activated) in melanoma by high cyclin D/CDK4 levels associated with extracellular stimuli or mutated signal pathway intermediates such as BRAF. This is complemented in some cells by mutated CDKN2a genes and low p16 protein levels which normally would inhibit the CDKs. In normal cells the increased levels of E2F would induce proteins associated with apoptosis, but it is speculated that Rb may selectively dissociate from E2F promoter regions involved in cell cycle regulation, but may not dissociate from those inducing apoptosis.[38]

Mutations in CDKN2a are present in approximately 20% of patients with familial melanoma, but are uncommon in sporadic melanoma which account for over 90% of melanoma cases.[39] Mutations in the p53 gene is also uncommon in melanoma compared to some other cancers,[40,41] but were reported to be higher in melanoma from sun-exposed sites.[42] Protein levels of p53 appear to be elevated in 18–40% of melanoma.[40,43,44] The reasons for the elevated p53 levels are not clear. p53 in some melanoma appeared functionally inactive and could not induce cell cycle arrest.[45] This question is of much interest in view of reports of splice variants which may act like dominant negatives to inhibit the function of wild type p53.[46]

These studies on oncogenic pathways do not adequately explain a number of changes in cancer cells, e.g., in melanoma, contrary to expectations, the antiapoptotic protein Bcl-2 was reduced in progressive forms of the disease, whereas Bcl-XL and Mcl-1 increased in thick primary melanoma and in metastases.[47] The basis for elevation of Mcl-1 in melanoma is not clear. Activation of signal transducer and transcription activator 3 (Stat 3) by Src kinases in melanoma cells was reported to upregulate both Mcl-1 and Bcl-XL[48] and Stat 3 was regarded as a critical transcriptional activator of Mcl-1, Bcl-XL, and survivin.[49] Activation of Akt was also held responsible for upregulation of Mcl-1 levels in Cholangiocarcinoma cells.[50] Akt is frequently activated in melanoma and may therefore in part be responsible for elevation of Mcl-1. Mcl-1 levels were downregulated by the multikinase BRAF inhibitor BAY 43-9006 (Sorafenib), but this was apparently due to increased proteasome degradation.[51] Mcl-1 was reported to be ubiquinated by a specific Mcl-1 ubiquitin ligase E3 (Mule),[52] which also targets p53.[19,53]

The decrease in Bcl-2 expression in metastatic melanoma is also hard to explain. The microphthalmia-associated transcription factor (MITF) appears to be a key factor in its regulation. MITF in turn is regulated through the receptor c-kit and is believed to be responsible for differentiation and survival of melanocytes.[54] C-kit is downregulated in melanoma cells[55] and this may play some role via decreased activation of MITF in the decreased levels of Bcl-2. Another transcription factor regulating Bcl-2 and c-kit is activator protein 2 (AP-2).[56] This was shown to be lost in progression of melanoma and loss of AP-2 was associated with short overall and relapse-free survival.[57] AP-2 proteins were reported to bind with p53 to p53 target genes such as p21 and so act as a tumor suppressor.[58] It is not clear why AP-2 is lost in melanoma but AP-2 appears essential for development of neural crest lineages.[56]

4   Signal Pathways Involved in Resistance of Cancer Cells to Cell Death

The above-mentioned studies suggest that activation of signal pathways may be all important in driving both cell division and resistance to apoptosis. Some of the principal pathways are described as follows.

4.1   The ERK1/2 Kinase Pathway in Inhibition of Apoptosis

The RAS, RAF, MEK ERK1/2 pathway has received particular attention in melanoma.[59] In previous studies on melanoma cell lines, we found that activation of this pathway was a common cause of resistance to apoptosis.[60] Similar results were found in studies on other cancers.[61] Activation of MAPK (ERK1/2) was detected more frequently in primary melanoma than in naevi, and activation of ERK was higher in thick melanoma and subcutaneous metastases.[62] Introduction of activated MAPK kinase into melanocytes resulted in tumorigenesis in nude mice.[63]

As discussed earlier, a high proportion of melanoma has activating mutations (such as the V600E) in BRAF downstream of Ras.[32] A smaller proportion has activating mutations in Ras that were exclusively seen in melanoma without BRAF mutations.[33] These findings suggested this pathway may be responsible for induction of melanoma, but this idea was tempered by the finding that benign naevi also frequently had BRAF mutations.[34] Further insights into growth arrest of naevi was the finding that expression of the p16 protein was high in naevi and may account for growth arrest of naevus cells despite activation of the ERK1/2 pathway. p16 expression was not uniform and other senescence-inducing factors were thought to be involved.[35] One study suggested that melanoma with BRAF mutations were more sensitive to MEK inhibitors[64] but this was not the finding in studies by Zhang et al.[60] Clinical responses to the BRAF inhibitor BAY 43-9006 (Sorafenib) also did not correlate with BRAF mutation in the melanoma.[65]

Apart from activating mutations of BRAF and NRAS, the RAS, RAF, and ERK pathways are activated by a number of external factors such as $\beta3$ integrin/adhesion interactions[66,67] and autocrine growth factors acting through receptor tyrosine kinases such as c-kit, IL-6, insulin growth factor, basic fibroblast growth factor (bFGF), hepatocyte growth factor.[59] Factors acting on G protein-coupled receptors such as MSH also activate adenylate cyclase and thereby RAS.[59]

Several target proteins in the apoptosis pathway are phosphorylated by the ERK1/2 kinases. BimEL is phosphorylated directly by ERK1/2 on Serine69 and possibly two other sites.[68] This promotes proteasomal degradation of BimEL and may prevent interactions with Bax. In both cases the effect is to limit apoptosis mediated by BimEL. It is not clear whether BimL is phosphorylated by ERK1/2.[68]

The ERK1/2 pathway has also been implicated in transcriptional upregulation of Mcl-1 by the transcription factor ElK-1[69] and of Bcl-2 and Bcl-XL.[61] Bcl-2 is known to be regulated by the microphthalmic transcription factor (MITF).[54] MITF, however, may be suppressed by activation of ERK perhaps due to degradation of the protein.[70] ERK phosphorylates and stabilizes c-Myc, which in turn induces cyclin D1 and cell proliferation.[2]

4.2   Inhibition of Apoptosis by Akt Signaling

An equally important cell survival signal pathway appears to be the Akt/PKB pathway. This is initiated by tyrosine kinase and G protein-coupled receptor activation of phosphoinositide-3-kinase PI(3)K, which in turn phosphorylates phosphatidylinositol biphosphate (PIP2) to PIP3. This causes translocation of PIP3 to the cell membrane and phosphorylation of Akt by phosphoinositide dependent kinase-1 (PDK-1) on threonine 308 and on Serine 473 in the hydrophobic tail by the rictor–m TOR complex.[71,72] Akt consists of three family members; Akt, Akt2, and Akt3. The latter appears to be preferentially upregulated in melanoma.[71]

Akt is constitutively activated in many melanoma cells[73] and is able to suppress apoptosis via a number of mechanisms. These include phosphorylation of forkhead transcription factors, which regulate several proapoptotic proteins such as Bim and Fas ligand. The phosphorylated forkhead proteins are trapped in the cytosol and cannot enter the nucleus. Akt also phosphorylates and inactivates several proapoptotic proteins such as Bad and caspase 9.[71,74.] Importantly, it activates I Kappa B Kinase (IKK) and thereby activates the transcription factor NF-κB, leading to transcription of several antiapoptotic proteins such as Bcl-XL, AI, and XIAP.[75]

The main factors involved in upregulation of Akt in melanoma are not clearly defined, but may result from growth factor stimulation of surface receptors as proposed for the RAS, RAF, MEK, and ERK pathway. Activating mutations of proteins in this pathway have not been described in melanoma, but have been in colon carcinoma (PDK1, ATK2, and PAK4).[76] Another possibility is that the downregulatory mechanisms in this pathway are abnormal, e.g., there has been much interest in PTEN status in melanoma as this phosphatase inactivates PIP3. Abnormalities in PTEN appear, however, to be a low-frequency event[77] and would not account for activation of this pathway in the majority of melanoma.

4.3 The Protein Kinase C Pathway in Apoptosis

Protein kinase C (PKC) is a family of phospholipid-dependent serine/threonine kinases comprising at least 11 isoforms that play fundamental roles in signal transduction pathways that regulate cellular proliferation, differentiation, and apoptosis.[78,79] Activation of PKC by phorbol esters (PMA) has been shown to have variable effects on apoptosis.[80–86] In particular, activation of PKCδ seemed proapoptotic,[80–82] whereas activation of PKCε and PKCα was antiapoptotic.[83–86] Activation of PKC has been reported to abrogate Fas-induced apoptosis through inhibition of death-inducing signaling complex formation by blocking Fas-associated death domain (FADD) recruitment and thus caspase-8 activation.[87–89] A similar mechanism has also been implicated in protection of HeLa cells from TRAIL-induced apoptosis.[90] Moreover, inhibition of TRAIL-induced apoptosis by PKC activation was suggested to occur at the level of proteolytic cleavage of caspase-8 or downstream of caspase-8-mediated Bid cleavage.[89,91]

The expression levels of PKCε may play an important role in determining sensitivity of melanoma to apoptosis induced by TRAIL.[92] This was supported by studies using an adenovirus vector expression systems to express PKCε in the PKCε-deficient melanoma cells, which reversed the potentiating effect of PMA on TRAIL-induced apoptosis whereas expression of a dominant-negative PKCε in PKCε-expressing Mel-RM cells reversed the protective effect of PMA on TRAIL-induced apoptosis. In contrast, PKCδ in melanoma cells increased TRAIL-induced apoptosis. Hence, activation of PKC by TRAIL may provide positive or negative regulation of sensitivity of cells to TRAIL-induced apoptosis depending on the levels of these two PKC isoforms. Activation of PKC was found to regulate TRAIL-induced apoptosis of melanoma by modulating Bax activation and did not cause significant changes in the expression levels of TRAIL death receptors, alterations in activation of caspase-8, or cleavage of Bid. The protective effect of PKCε was found in part to be associated with activation of ERK1/2 induced by TRAIL[60] as inhibition of ERK1/2 by the MAPK kinase-specific inhibitor partially reversed the protective effect on melanoma cells. In addition, activation of ERK1/2 was downstream of PKC as inhibition of PKC blocked TRAIL-induced activation of ERK1/2. These results suggest that measurement of the relative amounts of PKC isoforms may help define melanoma that is sensitive or resistant to treatment.

4.4 The Jun NH2 Kinase Pathway

The Jun NH2 kinase (JNK) pathway may be activated by TNF or TRAIL receptor-associated factors (TRAFs) or by environmental and genotoxic stresses such as ultraviolet (UV) or gamma radiation. The JNK proteins are coded for by three separate genes and these give rise to approximately ten different splice variants. All JNK are able to phosphorylate c-Jun and thereby upregulate activator protein-1 (AP-1)-dependent genes. One of the genes so regulated is Bim.[68] In addition, JNK has important posttranscriptional effects on Bim, which include phosphorylation of the Bim motif binding to the Dynein motor complex of microtubules and thereby release of Bim into the cytosol.[93] In addition, JNK may phosphorylate Bcl-2 and Bcl-XL[94] and inhibit their ability to bind to the BH3-only sensor proteins such as Bim, PUMA, and human protein harakiri (HRK). It may also phosphorylate 14-3-3 proteins in the cytosol and promote translocation of Bax to mitochondria.[95] These proapoptotic effects of JNK may be inhibited by the Akt pathway.[96]

Activation of JNK by TRAIL was reported to occur predominantly via TRAIL-R2 rather than –R1 death receptors[97] and to involve FADD and caspase activation.[98] Subsequent studies showed that activation of JNK was dependent on formation of a secondary complex of FADD, TRAF2, RIP1, and IKK.[99] TRAIL may therefore mediate some of its apoptotic effects via the JNK Bim pathway. This secondary complex is also responsible for activation of NF-κB which exerts antiapoptotic effects by upregulation of antiapoptotic proteins Bcl-2, Bcl-XL, and A1, as well as the IAP proteins, some of which bind to and inhibit TRAF2.[100]

Inhibitors of JNK have attracted much attention, particularly in treatment of neurological diseases.[101] SP600125 is a direct inhibitor of JNK and has been used to treat arthritis in animal models. CEP-1347 acts to inhibit MAP kinases upstream of JNK. Peptide inhibitors that inhibit substrate-binding sites or regulatory regions have also been studied with some success.[101]

5   Therapeutic Opportunities

The rational development of treatments against cancer would ideally be based on the known oncogenic pathways involved and the resistance mechanisms which prevent oncogene-induced apoptosis. In practice, several factors act against implementing such an idealized approach. Principal among these is the heterogeneity of most solid cancers so that treatments focused on any particular pathway or against particular targets may only be effective against 10–20% of patients with particular cancers. The second limitation is the state of ignorance surrounding particular mechanisms involved in resistance to cell death.

5.1   Understanding how Commonly used Agents Kill Cancer Cells and Resistance Mechanisms Against them

In the case of melanoma with apparently normal p53 pathways, it should be a simple matter of using DNA-damaging agents such as Cisplatin or Doxorubicin to activate p53 and thereby the apoptotic pathway to cell death discussed earlier. In practice, melanoma shows low response rates to Doxorubicin and Cisplatin. In the case of Cisplatin, cell death, when it occurs, may be more related to necrosis induced by activation of poly (ADP-ribose) polymerase (PARP) in DNA repair and consumption of ATP[102] as described by others.[103] In studies on melanoma cell lines several proapoptotic BH3 proteins (PUMA, Noxa, and Bim) appeared constitutively upregulated, but there was no evidence that this adversely affected the cell lines. The mechanism involved in resistance to apoptosis of these cells remains uncertain. Inhibition of ERK1/2 and Akt pathways increased apoptosis in the lines but other factors were clearly involved.

p53-independent initiators such as BimEL were also detectable in melanoma. Agents targeting microtubules such as the Taxols and Vinca alkaloids are postulated to release BimEL and possibly BimL from the microtubules. Nevertheless, studies on Doxetaxel showed a wide variation in susceptibility to Doxetaxel-induced apoptosis. ERK1/2 inhibitors potentiated apoptosis induced by Doxetaxel strongly in some melanoma cell lines and this was in proportion to activation of ERK1/2 by Doxetaxel. In contrast, there was a good correlation between activation of JNK and induction of apoptosis. Hence, in the case of Doxetaxel the relative activation of these two pathways appears to largely determine the overall degree of apoptosis[103a]. These results provide further support for the use of inhibitors of the RAF/MEK/ERK pathway in combination with Taxols. Similar results were seen with Vincristine, but the mechanism of induction of apoptosis differed from that of Doxetaxel[144]. Whether results from such studies can be utilized to define responsive tumor subgroups remains unknown. It is encouraging however to think that further insights into the mechanisms of induction and resistance to commonly used agents may help to define responsive tumor subsets.

Apart from more intelligent use of existing agents, studies over the past few years have generated a number of new agents designed to overcome resistance to apoptosis. These are summarized as follows.

5.2 Therapeutic Approaches Targeting Signal Pathways

5.2.1 The RAS, RAF, MEK, and ERK Pathway

Several inhibitors of this pathway have been produced, such as the Onyx/Bayer 43-9006 agent (Sorafenib)[104] and the Pfizer compound CI-1040.[105] In phase II studies with Sorafenib as a single agent there was only one response in 34 patients, but when given in combination with Carboplatin and Paclitaxel there were 20 partial responses and 26 with stable disease in 54 patients. Response rates in 23 previously untreated patients was 48%.[65] These results are now being tested in a randomized trial in previously untreated patients (ECOG trial 2603) and in previously treated patients (Onyx/Bayer [117118 protocol]).

RAS is upstream of RAF and requires a farnesyl group to be attached for membrane anchorage. It may therefore be possible to inhibit the pathway with inhibitors of farnesyl transferase. These have shown antitumor activity in preclinical studies[106,107] and sensitized human melanoma cells to Cisplatin[108] but further evaluation is needed[108a]. Recent studies suggest that the MEK inhibitor, VO126, induced by upregulation of Bim and PUMA and down regulation of Mcl-1 (Wang et al).

5.2.2 Inhibitors of the Akt Pathway

Relatively few studies have been carried out with inhibitors of this pathway. PX-866 is a specific inhibitor of PI3K which was shown to have single agent activity and to enhance chemotherapy and radiation in preclinical studies.[109] Heat shock protein 90 (HSP90) is a chaperone for a number of signal proteins, including Akt and RAF. A geldanamycin derivative (17AAG) was shown to deplete Akt and cyclin D1 in melanoma lines.[110,111] Phase I studies have been conducted in patients with advanced malignancies and phase II studies on melanoma patients in the Memorial Sloan Kettering Institute are in progress. A more soluble preparation, referred to as KOS-953, is about to enter clinical trials (Kosan Biosciences, Inc.). A nonpeptide small-molecule compound API-59-OME was shown to inhibit Akt activity in ovarian carcinoma lines, but not a wide range of other kinases. Studies were in vitro.[112]

CCI-779, a rapamycin analogue, was tested in 33 patients with melanoma. Only one partial response was seen[113] but studies in combination with apoptosis-inducing agents may be needed. Rapamycin was found to inhibit activation of NF-κB by Doxorubicin but the mechanism of action appeared independent of P13K.[114] Specific inhibitors of NF-κB activation do not appear to have been clinically evaluated, but proteasome inhibitors such as PS-341/Bortezomib have been thought to act by inhibiting activation of NF-κB and account for its effects in potentiating chemotherapy[115] and radiotherapy[116] (also see Chapter 12). Nevertheless, proteasome inhibitors affect a wide range of apoptosis regulators. One study in fact found no effect on NF-κB activity, but instead apoptosis appeared to be due to upregulation of Noxa.[20] A number of agents inhibit NF-κB activation in vitro, such as Curcumin,[117] but are yet to be tested in vivo.

5.2.3 Protein Kinase C Inhibitors

PKC as a target for anticancer drugs has been recognized for some time. Bryostatin is an activator of PKC that has been evaluated in phase II trials in melanoma. No responses were seen when used as a single agent.[118,119] Aprinocarsen is an antisense reagent against the PKCα isoform that was tested in patients with ovarian carcinoma. But it had no activity as a single agent.[120] Some of the difficulty in evaluating such agents is the diverse functions of different PKC isoforms and cross talk with other signal pathways.[121] As noted earlier, TRAIL appears to activate the ERK1/2 pathway via PKC activation. Similarly, PKC may activate the JNK pathway in the presence of the receptor for casein kinase 1 (CK1).[122] These two pathways may have opposing effects on apoptosis and illustrate the potential difficulty in targeting PKC in treatment.

5.3   *Histone Deacetylase Inhibitors*

Histone deacetylase (HDAC) inhibitors may directly induce apoptosis of cancer cells, e.g., by activation of Bim or by a number of other mechanisms, as reviewed elsewhere.[123,124] Some drugs in this class, however, appear to have relatively weak direct cytotoxic effects, but may synergize with other agents such as TRAIL to markedly enhance apoptosis.[123,102] These drugs are discussed further in a Chapter 13.

5.4   *Activating the Extrinsic Pathway*

The agents discussed earlier are also applicable to attempts to treat melanoma by agents such as TRAIL or Fas Ligand. These pathways have several additional obstacles that may need to be overcome. Principal among these is the low or absent death receptor expression on many melanoma, particularly on fresh isolates.[125] The main death receptor for TRAIL, TRAIL-R2 (DR5), was shown to be transcriptionally regulated by p53 and non-p53-dependent mechanisms as reviewed elsewhere[126] and also discussed in Chapter 10. In melanoma mRNA for the death receptors appeared at normal levels and nontranscriptional events appeared more important in regulation.[127] It was shown in TRAIL-resistant colon carcinoma that TRAIL-R1 appeared located in the Golgi and treatment with tunicamycin resulted in upregulation of TRAIL-R1.[128] Similarly, tunicamycin was shown to upregulate TRAIL-R2 in prostate carcinoma cells.[129] These findings have been reproduced in cultured melanoma cells[143]. Further studies are needed to investigate their clinical applicability.

5.5   *Agents Targeting Antiapoptotic Proteins*

Arguably, some of the most exciting new agents are those being developed against the antiapoptotic Bcl-2 proteins (also see Chapter 8) and IAPs. A list of these is given in Table 6.1. Evaluation of most of these is at an early stage and only one at this stage has gone through to phase III clinical trials. This was the antisense molecule against Bcl-2. This particular trial did not reach its primary end point of an effect on overall survival when all patients were included but did so when only patients with normal lactic dehydrogenase (LDH) levels were included in the analysis. The trial has been criticized on several grounds, but clear benefit was seen in some patients. It is hoped that experience gained from this trial will be used to plan future trials. In particular, antisense agents against Mcl-1 would appear an attractive target in melanoma.

Table 6.1 Agents against anti-apoptotic proteins

Target/action	Drug	Study reference
Anti-apoptotic Bcl-2 proteins		
BH3 mimics	ABT-737 (Abbot)	130
	GX015–070, BL-193	131
	Gossypol	132
Bcl-2 antisense	Oblimersen (Genta)	133, 134
Bcl-XL antisense		135–137
Mcl-1 antisense		135, 138
Inhibitor of apoptosis proteins		
Smac/DIABLO mimics	IDN-13389 (Idun Pharmaceuticals)	139–142

6   Conclusion

The widely held view that the oncogenic process involves deregulated cell division as well as resistance to apoptosis has been useful in focusing attention on how cancer cells evade cell death induced by the many therapeutic agents available to treat cancers. Part of the evolution of this concept is the realization that the selection pressures acting against cancer cells generate a variety of defects in the cell death pathways. These selection pressures include apoptotic pathways generated by oncogenes, neighboring cells, or the immune system. One striking conclusion is that information about these resistance mechanisms is still very limited even in particular cancer types such as melanoma. In the case of killing by the immune system through TRAIL, individual cell lines can be identified with a variety of defects in the apoptotic pathway such as absence of caspase-8, loss of Bid, or the death receptors. Downregulation of death receptors seems a more general cause of resistance to TRAIL that may be a worthwhile target in therapy.

Inactivating mutations in the p53 pathway are well known and common in many cancer types, but we suspect this particular pathway may also be inactivated by other as yet poorly characterized mechanisms. Whether it will become the focus of new therapies is uncertain. Activating mutations in signal pathway intermediaries appears common, as discussed earlier, and perhaps provides the best therapeutic options with agents targeting such pathways. Experience with BAY-43-9006 (Sorafenib, Nexator) however indicates that it has a number of unwanted toxicities such as skin rashes, diarrhea, hypertension, and hand-foot syndrome. Much remains to be learnt about the basis of these toxicities and whether other agents may have different toxicity profiles.

Therapeutic agents also become part of the selective process acting to generate resistant cancer cells. It is well known that cancer, which recurs after treatment with chemotherapy often have increased growth rates and metastatic potential. The taxols also appear to activate antiapoptotic pathways such as the ERK1/2 MAP kinases. It is therefore quite possible that such agents will select cancer cells where this pathway is dominant over proapoptotic mechanisms. Such insights should translate quickly into new protocols and provide optimism that agents for control of cancer may already be at hand provided we know how to use them and which cancers to use them against.

References

1. Cory, S. and Adams, J. M. (2002). The Bcl2 family: regulators of the cellular life-or-death switch. Nat Rev Cancer 2(9), 647–656.
2. Green, D. R. and Evan, G. I. (2002). A matter of life and death. Cancer Cell 1(1), 19–30.
3. Lowe, S. W., Cepero, E., and Evans, G. (2004). Intrinsic tumour suppression. Nature 432, 307–315.
4. Zhang, L., Yu, J., Park, B. H., Kinzler, K. W., and Vogelstein, B. (2000). Role of BAX in the apoptotic response to anticancer agents. Science 290(5493), 989–992.
5. Bellosillo, B., Villamor, N., Lopez-Guillermo, A., Marce, S., Bosch, F., Campo, E., Montserrat, E., and Colomer, D. (2002). Spontaneous and drug-induced apoptosis is mediated by conformational changes of Bax and Bak in B-cell chronic lymphocytic leukemia. Blood 100(5), 1810–1816.
6. Deng, Y., Lin, Y., and Wu, X. (2002). TRAIL-induced apoptosis requires Bax-dependent mitochondrial release of Smac/DIABLO. Genes Dev 16, 33–45.
7. Zamzami, N. and Kroemer, G. (2001). The mitochondrion in apoptosis: how Pandora's box opens. Mol Cell Biol 2, 67–71.
8. Puthalakath, H. and Strasser, A. (2002). Keeping killers on a tight leash: transcriptional and post-translational control of the pro-apoptotic activity of BH3-only proteins. Cell Death Differ 9(5), 505–512.
9. Willis, S. N. and Adams, J. M. (2005). Life in the balance: how BH3-only proteins induce apoptosis. Curr Opin Cell Biol 17, 617–625.
10. Letai, A., Bassik, M. C., Walensky, L. D., Sorcinelli, M. D., Weiler, S. and Korsmeyer, S. J. (2002). Distinct BH3 domains either sensitize or activate mitochondrial apoptosis, serving as prototype cancer therapeutics. Cancer Cell 2(3), 183–192.
11. Trapani, J. A. and Sutton, V. R. (2003). Granzyme B: pro-apoptotic, antiviral and antitumor functions. Curr Opin Immunol 15(5), 533–543.
12. Ley, R., Hadfield, K., Howes, E., and Cook, S. J. (2005). Identification of a DEF-type docking domain for extracellular signal-regulated kinases 1/2 that directs phosphorylation and turnover of the BH3-only protein BimEL. J Biol Chem 280(18), 17657–17663.
13. Ley, R., Ewings, K. E., Hadfield, K., Howes, E., Balmanno, K., and Cook, S. J. (2004). Extracellular signal-regulated kinases 1/2 are serum-stimulated "Bim(EL) kinases" that bind to the BH3-only protein Bim(EL) causing its phosphorylation and turnover. J Biol Chem 279(10), 8837–8847.
14. Puthalakath, H., Villunger, A., O'Reilly, L. A., Beaumont, J. G., Coultas, L., Cheney, R. E., Huang, D. C. and Strasser, A. (2001). Bmf: a proapoptotic BH3-only protein regulated by interaction with the myosin V actin motor complex, activated by anoikis, Science 293(5536), 1829–1832.
15. Collins, N. L., Reginato, M. J., Paulus, J. K., Sgroi, D. C., Labaer, J., and Brugge, J. S. (2005). G1/S cell cycle arrest provides anoikis resistance through Erk-mediated Bim suppression. Mol Cell Biol 25(12), 5282–5291.
16. Chipuk, J. E., Bouchier-Hayes, L., Kuwana, T., Newmeyer, D. D., and Green, D. R. (2005). PUMA couples the nuclear and cytoplasmic proapoptotic function of p53. Science 309(5741), 1732–1735.

17. Leu, J. I.-J., Dumont, P., Hafey, M., Murphy, M. E. and George, D. L. (2004). Mitochondrial p53 activates Bak and causes disruption of Bak-Mcl1 complex. Nat Cell Biol 6, 443–450.
18. Moll, U. M., Wolff, S., Speidel, D., and Deppert, W. (2005). Transcription-independent pro-apoptotic function of p53. Curr Opin Cell Biol 17(6), 631–636.
19. Chen, L., Willis, S. N., Wei, A., Smith, B. J., Fletcher, J. I., Hinds, M. G., Colman, P. M., Day, C. L., Adams, J. M., and Huang, D. C. (2005). Differential targeting of prosurvival Bcl-2 proteins by their BH3-only ligands allows complementary apoptotic function. Mol Cell 17(3), 393–403.
20. Fernandez, Y., Verhagen, M., Miller, T. P., Rush, J. L., Steiner, P., Opipari, A. W., Jr., Lowe, S. W. and Soengas, M. S. (2005). Differential regulation of noxa in normal melanocytes and melanoma cells by proteasome inhibition: therapeutic implications. Cancer Res 65(14), 6294–6304.
21. Qin, J. Z., Stennett, L., Bacon, P., Bodner, B., Hendrix, M. J., Seftor, R. E., Seftor, E. A., Margaryan, N. V., Pollock, P. M., Curtis, A., Trent, J. M., Bennett, F., Miele, L., and Nickoloff, B. J. (2004). p53-independent NOXA induction overcomes apoptotic resistance of malignant melanomas. Mol Cancer Ther 3(8), 895–902.
22. Tinel, A. and Tschopp, J. (2004). The PIDDosome: a protein complex implicated in activation of caspase-2 in response to genotoxic stress. Science 304(5672), 843–846.
23. Gao, Z., Shao, Y., and Jiang, X. (2005). Essential roles of the Bcl-2 family of proteins in caspase-2-induced apoptosis. J Biol Chem 280(46), 38271–38275.
24. Zhivotovsky, B. and Orrenius, S. (2005). Caspase-2 function in response to DNA damage. Biochem Biophys Res Commun 331, 859–867.
25. Shin, S., Lee, Y., Kim, W., Ko, H., Choi, H., and Kim, K. (2005). Caspase-2 primes cancer cells for TRAIL-mediated apoptosis by processing procaspase-8. EMBO J 24, 3532–3542.
26. Deveraux, Q. L., Roy, N., Stennicke, H. R., Van Arsdale, T., Zhou, Q., Srinivasula, S. M., Alnemri, E. S., Salvesen, G. S., and Reed, J. C. (1998). IAPs block apoptotic events induced by caspase-8 and cytochrome c by direct inhibition of distinct caspases. EMBO J 17(8), 2215–2223.
27. Verhagen, A. M., Coulson, E. J. and Vaux, D. L. (2001). Inhibitor of apoptosis proteins and their relatives: IAPs and other BIRPs. Genome Biol 2(7), 3009.1–3009.10.
28. Schimmer, A. D. (2004). Inhibitor of apoptosis proteins: translating basic knowledge into clinical practice. Cancer Res 64(20), 7183–7190.
29. Vaux, D. L. and Silke, J. (2005). IAPs, RINGs and ubiquitylation. Nat Rev Mol Cell Biol 6(4), 287–297.
30. Zhang, X. D., Zhang, X. Y., Gray, C. P., Nguyen, T., and Hersey, P. (2001). Tumor necrosis factor-related apoptosis-inducing ligand-induced apoptosis of human melanoma is regulated by Smac/DIABLO release from mitochondria. Cancer Res 61(19), 7339–7348.
31. Massague, J. (2004). G1 cell-cycle control and cancer. Nature 432, 298–306.
32. Davies, H., Bignell, G. R., Cox, C., Stephens, P., Edkins, S., Clegg, S., Teague, J., Woffendin, H., Garnett, M. J., Bottomley, W., Davis, N., Dicks, E., Ewing, R., Floyd, Y., Gray, K., Hall, S., Hawes, R., Hughes, J., Kosmidou, V., Menzies, A., Mould, C., Parker, A., Stevens, C., Watt, S., Hooper, S., Wilson, R., Jayatilake, H., Gusterson, B. A., Cooper, C., Shipley, J., Hargrave, D., Pritchard-Jones, K., Maitland, N., Chenevix-Trench, G., Riggins, G. J., Bigner, D. D., Palmieri, G., Cossu, A., Flanagan, A., Nicholson, A., Ho, J. W. C., Leung, S. Y., Yen, S. T., Weber, B. L., Seigler, H. F., Darrow, T. L., Paterson, H., Marais, R., Marshall, C. J., Wooster, R., Stratton, M. R., and Futreal, P. A. (2002). Mutations of the BRAF gene in human cancer. Nature 417, 949–954.
33. Pavey, S., Johansson, P., Packer, L., Taylor, J., Stark, M., Pollock, P. M., Walker, G. J., Boyle, G. M., Harper, U., Cozzi, S. J., Hansen, K., Yudt, L., Schmidt, C., Hersey, P., Ellem, K. A. O., O'Rourke, M. G. E., Parsons, P. G., Meltzer, P., Ringner, M., and Hayward, N. K. (2004). Microarray expression profiling in melanoma reveals a BRAF mutation signature. Oncogene 23, 4060–4067.
34. Pollock, P. M., Harper, U. L., Hansen, K. S., Yudt, L. M., Stark, M., Robbins, C. M., Moses, T. Y., Hostetter, G., Wagner, U., Kakareka, J., Salem, G., Pohida, T., Heenan, P., Duray, P.,

Kallioniemi, O., Hayward, N. K., Trent, J. M., and Meltzer, P. S. (2003). High frequency of BRAF mutations in nevi. Nat Genet 33(1), 19–20.

35. Michaloglou, C., Vredeveld, L. C., Soengas, M. S., Denoyelle, C., Kuilman, T., van der Horst, C. M., Majoor, D. M., Shay, J. W., Mooi, W. J., and Peeper, D. S. (2005). BRAFE600-associated senescence-like cell cycle arrest of human naevi. Nature 436(7051), 720–724.

36. Curtin, J. A., Fridlyand, J., Kageshita, T., Patel, H. N., Busam, K. J., Kutzner, H., Cho, K. H., Aiba, S., Brocker, E. B., LeBoit, P. E., Pinkel, D., and Bastian, B. C. (2005). Distinct sets of genetic alterations in melanoma. N Engl J Med 353(20), 2104–2107.

37. Bachmann, I. M., Straume, O. and Akslen, L. A. (2004). Altered expression of cell cycle regulators Cyclin D1, p14, p16, CDK4 and Rb in nodular melanomas. Int J Oncol 25(6), 1559–1565.

38. Halaban, R. (2005). Rb/E2F: a two-edged sword in the melanocytic system. Cancer Metastasis Rev 24(2), 339–356.

39. Thompson, J. F., Scolyer, R. A., and Kefford, R. F. (2005). Cutaneous melanoma. Lancet 365, 687–701.

40. Albino, A. P., Vidal, M. J., McNutt, N. S., Shea, C. R., Prieto, V. G., Nanus, D. M., Palmer, J. M., and Hayward, N. K. (1994). Mutation and expression of the p53 gene in human malignant melanoma. Melanoma Res 4, 35–45.

41. Vogelstein, B. and Kinzler, K. W. (2004). Cancer genes and the pathways they control. Nat Med 10(8), 789–799.

42. Zerp, S. F., van Elsas, A., Peltenburg, L. T. C., and Schrier, P. I. (1999). p53 mutations in human cutaneous melanoma correlate with sun exposure but are not always involved in melanomagenesis. Brit J Cancer 79, 921–926.

43. Whiteman, D. C., Parsons, P. G., and Green, A. C. (1998). p53 expression and risk factors for cutaneous melanoma: a case-control study. Int J Cancer 77(6), 843–848.

44. Purdue, M. P., From, L., Kahn, H. J., Armstrong, B. K., Kricker, A., Gallagher, R. P., McLaughlin, J. R., Klar, N. S. and Marrett, L. D. (2005). Etiologic factors associated with p53 immunostaining in cutaneous malignant melanoma. Int J Cancer 117, 486–493.

45. Satyamoorthy, K., Chehab, N. H., Waterman, M. J. F., Lien, M. C., el-Deiry, W. S., Herlyn, M., and Halazonetis, T. D. (2000). Aberrant regulation and function of Wild-type p53 in radioresistant melanoma cells. Cell Growth Differ 11, 467–474.

46. Ghosh, A., Stewart, D., and Matlashewski, G. (2004). Regulation of human p53 activity and cell localization by alternative splicing, Mol Cell Biol 24(18), 7987–7997.

47. Zhuang, L. (2006) (in preparation).

48. Niu, G., Bowman, T., Huang, M., Shivers, S., Reintgen, D., Daud, A., Chang, A., Kraker, A., Jove, R., and Yu, H. (2002). Roles of activated Src and Stat3 signaling in melanoma tumor cell growth. Oncogene 21(46), 7001–7010.

49. Jing, N. and Tweardy, D. J. (2005). Targeting Stat3 in cancer therapy. Anticancer Drugs 16(6), 601–607.

50. Kobayashi, S., Werneburg, N. W., Bronk, S. F., Kaufmann, S. H., and Gores, G. J. (2005). Interleukin-6 contributes to Mcl-1 up-regulation and TRAIL resistance via an Akt-signaling pathway in cholangiocarcinoma cells. Gastroenterology 128(7), 2054–2065.

51. Yu, C., Bruzek, L. M., Meng, X. W., Gores, G. J., Carter, C. A., Kaufmann, S. H. and Adjei, A. A. (2005). The role of Mcl-1 downregulation in the proapoptotic activity of the multikinase inhibitor BAY 43–9006. Oncogene 24(46), 6861–6869.

52. Zhong, Q., Gao, W., Du, F., and Wang, X. (2005). Mule/ARF-BP1, a BH3-only E3 ubiquitin ligase catalyzes the polyubiquitination of Mcl-1 and regulates apoptosis. Cell 121(7), 1085–1095.

53. Shmueli A. and Oren, M. (2005). Life, death, and ubiquitin: taming the mule. Cell 121(7), 963–965.

54. McGill, G. G., Horstmann, M., Widlund, H. R., Du, J., Motyckova, G., Nishimura, E. K., Lin, Y. L., Ramaswamy, S., Avery, W., Ding, H. F., Jordan, S. A., Jackson, I. J., Korsmeyer, S. J., Golub, T. R., and Fisher, D. E. (2002). Bcl2 regulation by the melanocyte master regulator Mitf modulates lineage survival and melanoma cell viability. Cell 109, 707–718.

55. Janku, F., Novotny, J., Julis, I., Julisova, I., Pecen, L., Tomancova, V., Kocmanova, G., Krasna, L., Krajsova, I., Stork, J., and Petruzelka, L. (2005). KIT receptor is expressed in more than 50% of early-stage malignant melanoma: a retrospective stud of 261 patients. Melanoma Res 15(4), 251–256.

56. Leslie, M. C. and Bar-Eli, M. (2005). Regulation of gene expression in melanoma: new approaches for treatment. J Cell Biochem 94(1), 25–38.

57. Karjalainen, J. M., Kellokoski, J. K., Eskelinen, M. J., Alhava, E. M., and Kosma, V. M. (1998). Downregulation of transcription factor AP-2 predicts poor survival in stage I cutaneous malignant melanoma. J Clin Oncol 16(11), 3584–3591.

58. McPherson, L. A., Loktev, A. V., and Weigel, R. J. (2002). Tumor suppressor activity of AP2alpha mediated through a direct interaction with p53. J Biol Chem 277(47), 45028–45033.

59. Smalley, K. S. M. (2003). A pivotal role for ERK in the oncogenic behaviour of malignant melanoma? Int J Cancer 104, 527–532.

60. Zhang, X. D., Borrow, J. M., Zhang, X. Y., Nguyen, T., and Hersey, P. (2003). Activation of ERK1/2 protects melanoma cells from TRAIL-induced apoptosis by inhibiting Smac/DIABLO release from mitochondria. Oncogene 222, 869–2881.

61. Boucher, M.-J., Morisset, J., Vachon, P. H., Reed, J. C., Laine, J., and Rivard, N. (2000). MEK/ERK signaling pathway regulates the expression of Bcl-2, Bcl-X$_L$ and Mcl-1 and promotes survival of human pancreatic cancer cells. J Cell Biol 79, 355–369.

62. Zhuang, L., Lee, C. S., Scolyer, R. A., McCarthy, S. W., Palmer, A. A., Zhang, X. D., Thompson, J. F., Bron, L. P. and Hersey, P. (2005). Activation of the extracellular-signal related kinase (ERK) pathway in human melanoma. J Clin Pathol 58(11), 1163–1169.

63. Govindarajan, B., Bai, X., Cohen, C., Zhong, H., Kilroy, S., Louis, G., Moses, M., and Arbiser, J. L. (2003). Malignant transformation of melanocytes to melanoma by constitutive activation of mitogen-activated protein kinase kinase (MAPKK) signaling. J Biol Chem 278(11), 9790–9795.

64. Solit, D. B., Garraway, L. A., Pratilas, C. A., Sawai, A., Getz, G., Basso, A., Ye, Q., Lobo, M., She, Y., Osman, I., Golub, T. R., Sebolt-Leopold, J., Sellers, W. R. and Rosen, N. (2006). BRAF mutation predicts sensitivity to MEK inhibition. Nature 439, 358–362.

65. Flaherty (2005). 6th World Congress on Melanoma, Vancouver, British Columbia.

66. Woods, D., Cherwinski, H., Venetsanakos, E., Bhat, A., Gysin, S., Humbert, M., Bray, P. F., Saylor, V. L., and McMahon, M. (2001). Induction of β3-integrin gene expression by sustained activation of the Ras-regulated Raf-MEK-extracellular signal-regulated kinase signaling pathway. Mol Cell Biol 21(9), 3192–3205.

67. Aplin, A. E., Stewart, S. A., Assoian, R. K., and Juliano, R. L. (2001). Integrin-mediated adhesion regulates ERK nuclear translocation and phosphorylation of Elk-1. J Cell Biol 153(2), 273–282.

68. Ley, R., Ewings, K. E., Hadfield, K., and Cook, S. J. (2005). Regulatory phosphorylation of Bim: sorting out the ERK from the JNK. Cell Death Differ 12, 1008–1014.

69. Townsend, K. J., Zhou, P., Qian, L., Bieszczad, C. K., Lowrey, C. H., Yen, A., and Craig, R. W. (1999). Regulation of MCL1 through a serum response factor/Elk-1-mediated mechanism links expression of a viability-promoting member of the BCL2 family to the induction of hematopoietic cell differentiation. J Biol Chem 274(3), 1801–1813.

70. Wellbrock, C. and Marais, R. (2005). Elevated expression of MITF counteracts B-RAF-stimulated melanocyte and melanoma cell proliferation. J Cell Biol 170(5), 703–708.

71. Robertson, G. P. (2005). Functional and therapeutic significance of Akt deregulation in malignant melanoma. Cancer Metastasis Rev 24(2), 273–285.

72. Sarbassov, D. D., Guertin, D. A., Ali, S. M., and Sabatini, D. M. (2005). Phosphorylation and regulation of Akt/PKB by the Rictor-mTOR complex. Science 307, 1098–1101.

73. Dai, D. L., Martinka, M., and Li, G. (2005). Prognostic significance of activated Akt expression in melanoma: a clinicopathologic study of 292 cases. J Clin Oncol 23(7), 1473–1482.

74. Brazil, D. P. and Hemmings, B. A. (2001). Ten years of protein kinase B signalling: a hard Akt to follow. Trends Biochem Sci 26(11), 657–664.

75. Amiri, K. I. and Richmond, A. (2005). Role of nuclear factor-kB in melanoma. Cancer Metastasis Rev 24, 301–313.
76. Parsons, D. W., Wang, T. L., Samuels, Y., Bardelli, A., Cummins, J. M., DeLong, L., Silliman, N., Ptak, J., Szabo, S., Willson, J. K. V., Markowitz, S., Kinzler, K. W., Vogelstein, B., Lengauer, C., and Velculescu, V. E. (2005). Colorectal cancer: mutations in a signalling pathway. Nature 436(7052), 792.
77. Pollock, P. M., Walker, G. J., Glendening, J. M., Que Noy, T., Bloch, N. C., Fountain, J. W., and Hayward, N. K. (2002). PTEN inactivation is rare in melanoma tumours but occurs frequently in melanoma cell lines. Melanoma Res 12(6), 565–575.
78. Nishizuka, Y. (1992). Intracellular signaling by hydrolysis of phospholipids and activation of protein kinase C. Science 258, 607–614.
79. Newton, A. C. (1995). Protein kinase C: structure, function, and regulation. J Biol Chem 270, 28495–28498.
80. Majumder, P. K., Pandey, P., Sun, X., Cheng, K., Datta, R., Saxena, S., Kharbanda, S., and Kufe, D. (2000). Mitochondrial translocation of protein kinase Cδ in phorbol ester-induced cytochrome c release and apoptosis. J Biol Chem 275(29), 21793–21796.
81. Li, L., Lorenzo, P. S., Bogi, K., Blumberg, P. M. and Yuspa, S. H. (1999). Protein kinase targets mitochondria, alters mitochondrial membrane potential, and induces apoptosis in normal and neoplastic keratinocytes when overexpressed by an adenoviral vector Cδ. Mol Cell Biol 19, 8547–8558.
82. Fujii, T., Garcia-Bermejo, M. L., Bernabo, J. L., Caamano, J., Ohba, M., Kuroki, T., Li, I., Yuspa, S. H., and Kazanietz, M. G. (2000). Involvement of protein kinase Cδ (PKCδ) in phorbol ester-induced apoptosis in LNCaP prostate cancer cells. Lack of proteolytic cleavage of PKCδ. J Biol Chem 275, 7574–7582.
83. Shinohara, H., Kayagaki, N., Yagita, H., Oyaizu, N., Ohba, M., Kuroki, T., and Ikawa, Y. (2001). A protective role of PKCa against TNF-related apoptosis-inducing ligand (TRAIL)-induced apoptosis in glioma cells. Biochem Biophys Res Commun 284, 1162–1167.
84. Whelan, R. D. and Parker, P. J. (1998). Loss of protein kinase C function induces an apoptotic response. Oncogene 16, 1939–1944.
85. Matassa, A. A., Kalkofen, R. L., Carpenter, L., Biden, T. J., and Reyland, M. E. (2003). Inhibition of PKCα induces a PKCδ-dependent apoptotic program in salivary epithelial cells. Cell Death Differ 10, 269–277.
86. Basu, A., Lu, D., Sun, B., Moor, A. N., Akkaraju, G. R., and Huang, J. (2002). Proteolytic activation of protein kinase Cε by caspase-mediated processing and transduction of antiapoptotic signals. J Biol Chem 277, 41850–41856.
87. Gomez-Angelats, M. and Cidlowski, J. A. (2001). Protein kinase C regulates FADD recruitment and death-inducing signaling complex formation in Fas/CD95-induced apoptosis. J Biol Chem 276, 44944–44952.
88. Sarker, M., Ruiz-Ruiz, C., and Lopez-Rivas, A. (2001). Activation of protein kinase C inhibits TRAIL-induced caspases activation mitochondrial events and apoptosis in a human leukemic T cell line. Cell Death Differ 8(2), 172–181.
89. Meng, W. H., Heldebrant, M. P., and Kaufmann, S. H. (2002). Phorbol 12-myristate 13-acetate inhibits death receptor-mediated apoptosis in Jurkat cells by disrupting recruitment of Fas-associated polypeptide with death domain. J Biol Chem 277, 3776–3783.
90. Harper, N., Hughes, M. A., Farrow, S. N., Cohen, G. M., and MacFarlane, M. (2003). Protein kinase C modulates tumor necrosis factor-related apoptosis-inducing ligand-induced apoptosis by targeting the apical events of death receptor signaling. J Biol Chem 278, 44338–44347.
91. Sarker, M., Ruiz-Ruiz, C., Robledo, G., and Lopez-Rivas, A. (2002). Stimulation of the mitogen-activated protein kinase pathway antagonizes TRAIL-induced apoptosis downstream of BID cleavage in human breast cancer MCF-7 cells. Oncogene 21(27), 4323–4327.
92. Gillespie, S., Zhang, X. D., and Hersey, P. (2005). Variable expression of protein kinase Cε in human melanoma cells regulates sensitivity to TRAIL-induced apoptosis. Mol Cancer Ther 4, 668–676.

93. Lei, K. and Davis, R. J. (2003). JNK phosphorylation of Bim-related members of the Bcl2 family induces Bax-dependent apoptosis. Proc Natl Acad Sci USA 100(5), 2432–2437.

94. Brichese, L., Cazettes, G., and Valette, A. (2004). JNK is associated with Bcl-2 and PP1 in mitochondria: Paclitaxel induces its activation and its association with the phosphorylated form of Bcl-2. Cell Cycle 3(10), 1312–1319.

95. Tsuruta, F., Sunayama, J., Mori, Y., Hattori, S., Shimizu, S., Tsujimoto, Y., Yoshioka, K., Masuyama, N., and Gotoh, Y. (2004). JNK promotes Bax translocation to mitochondria through phosphorylation of 14–3–3 proteins. EMBO J 23(8), 1889–1899.

96. Molton, S. A., Todd, D. E., and Cook, S. J. (2003). Selective activation of the c-Jun-N-terminal kinase (JNK) pathway fails to elicit Bax activation or apoptosis unless the phosphoinositide 3'-kinase (PI3K) pathway is inhibited. Oncogene 22(30), 4690–4701.

97. Muhlenbeck, F., Schneider, P., Bodmer, J. L., Schwenzer, R., Hauser, A., Schubert, G., Scheurich, P., Moosmayer, D., Tschopp, J., and Wajant, H. (2000). The tumor necrosis factor-related apoptosis-inducing ligand receptors TRAIL-R1 and TRAIL-R2 have distinct cross-linking requirements for initiation of apoptosis and are non-redundant in JNK activation. J Biol Chem 275(41), 32208–32213.

98. Herr, I., Wilhelm, D., Meyer, E., Jeremias, I., Angel, P., and Debatin, K. M. (1999). JNK/SAPK activity contributes to TRAIL-induced apoptosis. Cell Death Differ 6(2), 130–135.

99. Varfolomeev, E., Maecker, H., Sharp, D., Lawrence, D., Renz, M., Vucic, D. and Ashkenazi, A. (2005). Molecular determinants of kinase pathway activation by apo2 ligand/tumor necrosis factor related apoptosis-inducing ligand. J Biol Chem 280, 40599–40608.

100. Bharti, A. C. and Aggarwal, B. B. (2004). Ranking the role of RANK ligand in apoptosis. Apoptosis 9, 677–690.

101. Bogoyevitch, M. A., Boehm, I., Oakley, A., Ketterman, A. J., and Barr, R. K. (2004). Targeting the JNK MAPK cascade for inhibition: basic science and therapeutic potential. Biochim Biophys Acta 1697, 89–101.

102. Zhang, X. D., Wu, J. J., Gillespie, S., Borrow, J. and Hersey, P. (2006). Human melanoma cells selected for resistance to apoptosis by prolonged exposure to TRAIL are more vulnerable to necrotic cell death induced by Cisplatin. Clin Cancer Res 12, 1355–1364.

103. Zong, W. X., Ditsworth, D., Bauer, D. E., Wang, Z. Q., and Thompson, C. B. (2004). Alkylating DNA damage stimulates a regulated form of necrotic cell death. Genes Dev 18(11), 1272–1282.

103a. Mhaidat, N. M., Zhang, X. D., Jiang, C. C., and Hersey, P. (2007). Docetaxel-induced apoptosis of human melanoma is mediated by activation of c-Jun NH2-terminal kinase and inhibited by the mitogen-activated protein kinase extracellular signal-regulated kinase 1/2 pathway. Clin cancer Res 13, 1308–1314.

104. Strumberg, D., Richly, H., Hilger, R. A., Schleucher, N., Korfee, S., Tewes, M., Faghih, M., Brendel, E., Voliotis, D., Haase, C. G., Schwartz, B., Awada, A., Voigtmann, R., Scheulen, M. E., and Seeber, S. (2005). Phase I clinical and pharmacokinetic study of the Novel Raf kinase and vascular endothelial growth factor receptor inhibitor BAY 43–9006 in patients with advanced refractory solid tumors. J Clin Oncol 23(5), 965–972.

105. LoRusso, P. M., Adjei, A. A., Varterasian, M., Gadgeel, S., Reid, J., Mitchell, D. Y., Hanson, L., DeLuca, P., Bruzek, L., Piens, J., Asbury, P., VanBecelaere, K., Herrera, R., Sebolt-Leopold, J., and Meyer, M. B. (2005). Phase I and pharmacodynamic study of the oral MEK inhibitor CI-1040 in patients with advanced malignancies. J Clin Oncol 23(23), 5281–5293.

106. End, D. W., Smets, G., Todd, A. V., Applegate, T. L., Fuery, C. J., Angibaud, P., Venet, M., Sanz, G., Poignet, H., Skrzat, S., Devine, A., Wouters, W., and Bowden, C. (2001). Characterization of the antitumor effects of the selective farnesyl protein transferase inhibitor R115777 in vivo and in vitro. Cancer Res 61(1), 131–137.

107. Prevost, G. P., Pradines, A., Brezak, M. C., Lonchampt, M. O., Viossat, I., Ader, I., Toulas, C., Kasprzyk, P., Gordon, T., Favre, G. and Morgan, B. (2001). Inhibition of human tumor cell growth in vivo by an orally bioavailable inhibitor of human farnesyltransferase., BIM-46228. Int J Cancer 91(5), 718–722.

108. Smalley, K. S. and Eisen, T. G. (2003). Farnesyl transferase inhibitor SCH66336 is cyto-static pro-apoptotic and enhances chemosensitivity to cisplatin in melanoma cells. Int J Cancer 105(2), 165–175.

108a. Wang, Y.F., Jiang, C. C., Kiejda, K. A., Gillespie, S., Zhang, X. D., and Hersey, P. (2007). Apoptosis Induction in Human Melanoma Cells by Inhibition of MEK is Caspase-Independent and Mediated by the Bcl-2 Family Members PUMA, Bim, and Mcl-1. Clin Cancer Res.

109. Ihle, N. T., Williams, R., Chow, S., Chew, W., Berggren, M. I., Paine-Murrieta, G., Minion, D. J., Halter, R. J., Wipf, P., Abraham, R., Kirkpatrick, L. and Powis, G. (2004). Molecular pharmacology and antitumor activity of PX-866, a novel inhibitor of phosphoinositide-3-kinase signaling. Mol Cancer Ther 3(7), 763–772.

110. Smith, V., Sausville, E. A., Camalier, R. F., Fiebig, H. H., and Burger, A. M. (2005). Comparison of 17-dimethylaminoethylamino-17-demethoxy-geldanamycin (17DMAG) and 17-allylamino-17-demethoxygeldanamycin (17AAG) in vitro: effects on Hsp90 and client proteins in melanoma models. Cancer Chemother Pharmacol 56(2), 126–137.

111. Dou, F., Yuan, L. D., and Zhu, J. J. (2005). Heat shock protein 90 indirectly regulates ERK activity by affecting Raf protein metabolism. Acta Biochim Biophys Sin 37(7), 501–505.

112. Tang, H. J., Jin, X., Wang, S., Yang, D., Cao, Y., Chen, J., Gossett, D. R. and Lin, J. (2005). A small molecule compound inhibits AKT pathway in ovarian cancer cell lines. Gynecol. Oncol. Oct 3 (epub ahead of print).

113. Margolin, K., Longmate, J., Baratta, T., Synold, T., Christensen, S., Weber, J., Gajewski, T., Quirt, I., and Doroshow, J. H. (2005). CCI-779 in metastatic melanoma: a phase II trial of the California Cancer Consortium. Cancer 104(5), 1045–1048.

114. Romano, M. F., Avellino, R., Petrella, A., Bisogni, R., Romano, S., and Venuta, S. (2004). Rapamycin inhibits doxorubicin-induced NF-kappaB/Rel nuclear activity and enhances the apoptosis of melanoma cells, Eur J Cancer 40(18), 2829–2836.

115. Amiri, K. I., Horton, L. W., LaFleur, B. J., Sosman, J. A., and Richmond, A. (2004). Augmenting chemosensitivity of malignant melanoma tumors via proteasome inhibition: implication for bortezomib (VELCADE, PS-341) as a therapeutic agent for malignant melanoma. Cancer Res 64(14), 4912–4918.

116. Munshi, A., Kurland, J. F., Nishikawa, T., Chiao, P. J., Andreeff, M., and Meyn, R. E. (2004). Inhibition of constitutively activated nuclear factor-kappaB radiosensitizes human melanoma cells. Mol Cancer Ther 3(8), 985–992.

117. Siwak, D. R., Shishodia, S., Aggarwal, B. B., and Kurzrock, R. (2005). Curcumin-induced antiproliferative and proapoptotic effects in melanoma cells are associated with suppression of IkappaB kinase and nuclear factor kappaB activity and are independent of the B-Raf/mitogen-activated/extracellular signal-regulated protein kinase pathway and the Akt pathway. Cancer 104(4), 879–890.

118. Tozer, R. G., Burdette-Radoux, S., Berlanger, K., Davis, M. L., Lohmann, R. C., Rusthoven, J. R., Wainman, N., Zee, B., Seymour, L. and National Cancer Institute of Canada Clinical Trials Group. (2002). A randomized phase II study of two schedules of bryostatin-1 (NSC339555) in patients with advanced malignant melanoma: a National Cancer Institute of Canada Clinical Trials Group study. Invest New Drugs 20(4), 407–412.

119. Bedikian, A. Y., Plager, C., Stewart, J. R., O'Brian, C. A., Herdman, S. K., Ross, M., Papadopoulos, N., Eton, O., Ellerhorst, J., and Smith, T. (2001). Phase II evaluation of bryostatin-1 in metastatic melanoma. Melanoma Res 11(2), 183–188.

120. Advani, R., Peethambaram, P., Lum, B. L., Fisher, G. A., Hartmann, L., Long, H. J., Halsey, J., Holmlund, J. T., Dorr, A. and Sikic, B. I. (2004). A phase II trial of aprinocarsen, an antisense oligonucleotide inhibitor of protein kinase C alpha, administered as a 21-day infusion to patients with advanced ovarian carcinoma. Cancer 100(2), 321–326.

121. Swannie, H. C. and Kaye, S. B. (2002). Protein kinase C inhibitors. Curr Oncol Rep 4(1), 37–46.

122. Lopez-Bergami, P., Habelhah, H., Bhoumik, A., Zhang, W., Wang, L. H., and Ronai, Z. (2005). RACK1 mediates activation of JNK by protein kinase C. Mol Cell 19(3), 309–320.
123. Zhang, X. D., Gillespie, S. K., Borrow, J. M., and Hersey, P. (2004). The histone deacetylase inhibitor suberic bishydroxamate regulates the expression of multiple apoptotic mediators and induces mitochondria-dependent apoptosis of melanoma cells. Mol Cancer Ther 3(4), 425–435.
124. Dokmanovic, M. and Marks, P. A. (2005). Prospects: histone deacetylase inhibitors. J Cell Biochem 96(2), 293–304.
125. Nguyen, T., Zhang, X. D. and Hersey, P. (2001). Relative resistance of fresh isolates of melanoma to tumor necrosis factor-related apoptosis-inducing ligand (TRAIL) induced apoptosis. Clin Cancer Res 7 (Suppl 3), 966s–973s.
126. Hersey, P., Zhang, S. Y. and Zhang, X. D. (2005). Regulation of trail receptor expression in human melanoma. In: Cancer Drug Discovery and Development: Death Receptors in Cancer Therapy, ed. el-Deiry W. S. Humana Press, Totowa, NJ, pp. 175–187.
127. Zhang, X. Y., Zhang, X. D., Borrow, J. M., Nguyen, T. and Hersey, P. (2004). Translational control of tumor necrosis factor-related apoptosis-inducing ligand (TRAIL) death receptor expression in melanoma cells. J Biol Chem 279, 10606–10614.
128. Jin, Z., McDonald, E. R., III, Dicker, D. T. and el-Deiry, W. S. (2004). Deficient tumor necrosis factor-related apoptosis-inducing ligand (TRAIL) death receptor transport to the cell surface in human colon cancer cells selected for resistance to TRAIL-induced apoptosis. J Biol Chem 279(34), 35829–35239.
129. Shiraishi, T., Yoshida, T., Nakata, S., Horinaka, M., Wakada, M., Mizutani, Y., Miki, T., and Sakai, T. (2005). Tunicamycin enhances tumor necrosis factor-related apoptosis-inducing ligand-induced apoptosis in human prostate cancer cells. Cancer Res 65(14), 6364–6370.
130. Oltersdorf, T., Elmore, S. W., Shoemaker, A. R., Armstrong, R. C., Augeri, D. J., Belli, B. A., Bruncko, M., Deckwerth, T. L., Dinges, J., Hajduk, P. J., Joseph, M. K., Kitada, S., Korsmeyer, S. J., Kunzer, A. R., Letai, A., Li, C., Mitten, M. J., Nettesheim, D. G., Ng, S., Nimmer, P. M., O'Connor, J. M., Oleksijew, A., Petros, A. M., Reed, J. C., Shen, W., Tahir, S. K., Thompson, C. B., Tomaselli, K. J., Wang, B., Wendt, M. D., Zhang, H., Fesik, S. W., and Rosenberg, S. H. (2005). An inhibitor of Bcl-2 family proteins induces regression of solid tumours. Nature 435(7042), 677–681.
131. Karl, E., Warner, K., Zeitlin, B., Kaneko, T., Wurtzel, L., Jin, T., Chang, J., Wang, S., Wang, C. Y., Strieter, R. M., Nunez, G., Polverini, P. J., and Nor, J. E. (2005). Bcl-2 acts in a proangiogenic signaling pathway through nuclear factor-kappaB and CXC chemokines. Cancer Res 65(12), 5063–5069.
132. Oliver, C. L., Miranda, M. B., Shangary, S., Land, S., Wang, S., and Johnson, D. E. (2005). (-)-Gossypol acts directly on the mitochondria to overcome Bcl-2- and Bcl-X(L)-mediated apoptosis resistance. Mol Cancer Ther 4(1), 23–31.
133. Jansen, B., Wacheck, V., Heere-Ress, E., Schlagbauer-Wadl, H., Hoeller, C., Lucas, T., Hoermann, M., Hollenstein, U., Wolff, K., and Pehamberger, H. (2000). Chemosensitisation of malignant melanoma by BCL2 antisense therapy. Lancet 356(9243), 1728–1733.
134. Bedikian, A. Y., Millward, M., Pehamberger, H., Conry, R., Gore, M., Trefzer, U., Pavlick, A. C., DeConti, R., Hersh, E. M., Hersey, P., et al. (2006). Bcl-2 antisense (oblimersen sodium) plus dacarbazine in patients with advanced melanoma: the Oblimersen Melanoma Study Group. J Clin Oncol 24, 4738–4745.
135. Zangemeister-Wittke, U., Leech, S. H., Olie, R. A., Simoes-Wust, A. P., Gautschi, O., Luedke, G. H., Natt, F., Haner, R., Martin, P., Hall, J., Nalin, C. M., and Stahel, R. A. (2000). A novel bispecific antisense oligonucleotide inhibiting both bcl-2 and bcl-xL expression efficiently induces apoptosis in tumor cells. Clin Cancer Res 6(6), 2547–2555.
136. Heere-Ress, E., Thallinger, C., Lucas, T., Schlagbauer-Wadl, H., Wacheck, V., Monia, B. P., Wolff, K., Pehamberger, H., and Jansen, B. (2002). Bcl-X(L) is a chemoresistance factor in human melanoma cells that can be inhibited by antisense therapy. Int J Cancer 99(1), 29–34.

137. Wacheck, V., Losert, D., Gunsberg, P., Vornlocher, H. P., Hadwiger, P., Geick, A., Pehamberger, H., Muller, M., and Jansen, B. (2003). Small interfering RNA targeting bcl-2 sensitizes malignant melanoma. Oligonucleotides 13(5), 393–400.

138. Thallinger, C., Wolschek, M. F., Maierhofer, H., Skvara, H., Pehamberger, H., Monia, B. P., Jansen, B., Wacheck, V., and Selzer, E. (2004). Mcl-1 is a novel therapeutic target for human sarcoma: synergistic inhibition of human sarcoma xenotransplants by a combination of mcl-1 antisense oligonucleotides with low-dose cyclophosphamide. Clin Cancer Res 10(12 Pt 1), 4185–4191.

139. Schimmer, A. D., Welsh, K., Pinilla, C., Wang, Z., Krajewska, M., Bonneau, M. J., Pedersen, I. M., Kitada, S., Scott, E. L., Bailly-Maitre, B., Glinsky, G., Scudiero, D., Sausville, E., Salvesen, G., Nefzi, A., Ostresh, J. M., Houghten, R. A., and Reed, J. C. (2004). Small-molecule antagonists of apoptosis suppressor XIAP exhibit broad antitumor activity. Cancer Cell 5(1), 25–35.

140. Fulda, S., Wick, W., Weller, M., and Debatink, K. M. (2002). Smac agonists sensitize for Apo2L/TRAIL- or anticancer drug-induced apoptosis and induce regression of malignant glioma in vivo. Nat Med 8(8), 808–815.

141. Arnt, C. R., Chiorean, M. V., Heldebrant, M. P., Gores, G. J., and Kaufmann, S. H. (2002). Synthetic Smac/DIABLO peptides enhance the effects of chemotherapeutic agents by binding XIAP and cIAP1 in situ. J Biol Chem 277(46), 44236–44243.

142. Guo, F., Nimmanapalli, R., Paranawithana, S., Wittman, S., Griffin, D., Bali, P., O'Bryan, E., Fumero, C., Wang, H. G., and Bhalla, K. (2002). Ectopic overexpression of second mitochondria-derived activator of caspases (Smac/DIABLO) or cotreatment with N-terminus of Smac/DIABLO peptide potentiates epothilone B derivative-(BMS 247550) and Apo-2L/TRAIL-induced apoptosis. Blood 99(9), 3419–3426.

143. Jiang, C. C., Chen, L. H., Gillespie, S., Kiejda, K. A., Mhaidat, N., Wang, Y. F., Thorne, R., Zhang, X. D., and Hersey, P. (2007). Tunicamycin sensitizes human melanoma cells to tumor necrosis factor-related apoptosis-inducing ligand-induced apoptosis by up-regulation of TRAIL-R2 via the unfolded protein response. Cancer research 67, 5880–5888.

144. Zhu et al. Activation of JNK is a mediator of Vincristine-induced apoptosis in melanoma cells, under editorial review.

Chapter 7
Trail Receptors: Targets for Cancer Therapy

Robin C. Humphreys* and Wendy Halpern

Abstract A human tumor cell's ability to avoid the normal regulatory mechanisms of cell growth, division, and death are the hallmarks of transformation and cancer. Numerous novel therapeutic agents currently in preclinical or clinical evaluation aim to revive the normal regulation or evade these regulatory defects and induce growth arrest and cell death. One of the cell death pathways that has garnered significant interest, as a potential target for therapeutic intervention, is the programmed cell death pathway regulated by the tumor necrosis factor-related apoptosis-inducing ligand receptors (TRAIL-RS). Receptor agonist molecules including forms of the native ligand and monoclonal antibodies are being developed and tested as therapeutics in the treatment of human cancer.

Keywords apoptosis, monoclonal antibody, agonist, TRAIL, TRAIL receptor

1   Introduction

This review will focus on the tumor necrosis factor-related apoptosis-inducing ligand receptor (TRAIL-R) signaling pathway and the therapeutic agents currently in development that activate this cell death pathway as a treatment for cancer. The TRAIL receptors are an attractive therapeutic target because of their relatively restricted expression on tumor cells, their capacity, when activated, to induce cell death in a spectrum of human tumor cells and their ability to act in concert with various chemotherapeutic agents to promote tumor cell death. TRAIL agonists, including various forms of the ligand and agonist antibodies, have demonstrated significant antitumor activity in preclinical studies across a spectrum of different

Robin C. Humphreys
Human Genome Sciences, Oncology Research Department, Rockville, MD 20850, USA

Wendy Halpern
Human Oncology Research Genome Sciences, Pathology, Pharmacokinetics and Toxicology, Rockville, MD 20850, USA

* To whom correspondence should be addressed: e-mail: robin_humphreys@hgsi.com

R. Khosravi-Far and E. White (eds.), *Programmed Cell Death in Cancer Progression and Therapy*.
© Springer 2008

human tumors. Recently, some of these agents have begun evaluation in the clinical setting. In addition, emerging molecular therapies are being developed to act on specific key regulatory molecules in the TRAIL-R apoptosis pathway to complement the action of TRAIL-R agonists. The combination of receptor agonist activation and attenuation of the anti-apoptotic threshold with targeted molecular therapy holds promise as a rational approach to cancer treatment.

2   Trail Receptor Signaling

The ability to induce programmed cell death is mediated in all eukaryotic cells through distinct signaling pathways that are responsive to both external and internal inputs. A spectrum of sources can induce programmed cell death, including secreted and membrane-bound proteins, DNA damage, radiation, Ca^{+2} stress, viral and oncogenic transformation, and serum and growth factor starvation.

The active induction of cell death is effected through a family of related cell surface proteins and their cognate ligands. One of these death-inducing ligands, TRAIL, can instigate cell death through a cell surface receptor-mediated catalytic activation of a series of cysteine proteases, leading to cleavage of key cellular structural and signaling components. Binding to either of the competent TRAIL-R's by ligand or antibody agonists can activate this protease cascade through two distinct but intersecting pathways; an extrinsic receptor-mediated pathway and an intrinsic pathway associated with the mitochondria. TRAIL binds to five cognate cell surface receptors, but only two of these receptors TRAIL-R1 and TRAIL-R2, are death receptors that have the ability to transmit a complete death signal. TRAIL-R1 (DR4, TNFSFR10a) and TRAIL-R2 (DR5, TNFSF10b) are members of the TNF receptor superfamily (TNFRSF). Only these two receptors possess the capability to competently transmit a TRAIL death signal. The other members of this family capable of binding to the ligand, TRAIL; DcR1, DcR2, and osteoprotegerin, lack a required cytoplasmic signaling domain, known as a death domain (DD) (Table 7.1).

TRAIL-Rs exist as a functional homotrimeric subunit. Members of the TNFSFR can form and function as heterotrimers. The TRAIL-Rs have been identified in a heteromeric structure in cells transfected with TRAIL-R1 and TRAIL-R2 expression constructs. It is unclear whether this is a physiologically relevant formation as this heterotrimer has not been isolated in immunoprecipitation experiments from nontransfected cells. (Kischkel et al., 2000; Schneider et al., 1997). Although initial reports suggested that ligand is required for receptor trimerization, studies of

Table 7.1 TRAIL receptors

Receptor	TNFSF	Other names	Death domain
TRAIL-R1	10A	DR4, Apo2	Complete
TRAIL-R2	10B	DR5, TRICK, KILLER	Complete
TRAIL-R3	10C	Decoy receptor 1	None
TRAIL-R4	10D	Decoy receptor 1	Truncated
Osteoprotegerin	11B	OPG, OCIF, TR1	None

TNFR1 and Fas have demonstrated the presence of a preligand association domain (PLAD) that is required for ligand-independent trimerization. Interestingly, the PLAD domains interactions are very specific and only permit homotrimeric formations (Chan et al., 2000). This data suggests that TRAIL-R1 and TRAIL-R2 only form homotrimers. However, a recent report suggested that the TRAIL-binding decoy receptor, DcR1, may regulate TRAIL-R2 activity by forming a heterocomplex through the PLAD (Clancy et al., 2005). A common structural feature present in all TNFSF receptors is a series of extracellular cysteine-rich domains (CRD). The number of these domains can vary between different TNFSFRs from 1 to 6. Each CRD domain is defined by six highly conserved cysteines that form three intrachain disulfide bridges. TRAIL-R1 and TRAIL-R2 possess three such CRD repeats that contain seven intrachain disulfide bridges (Hymowitz et al., 1999; Locksley et al., 2001; Marsters et al., 1992; Mongkolsapaya et al., 1999). The TRAIL- Rs also possess a structural feature that is unique to death-inducing receptors in the TNFSFR. Each of the receptors in this class possesses a short (65–80 aa) cytoplasmic protein–protein domain that is required for interaction with a key adaptor protein that is required for transmission of the death signal. Consequently, this structure is known as the DD. Seven members of the TNFSFR, including TRAIL-R1 and TRAIL-R2, possess DD. (Igney and Krammer, 2002)

3 Trail Receptor Expression

Two of the most intriguing and attractive features of the TRAIL-Rs are that TRAIL-R1 and TRAIL-R2 are proapoptotic and that these two receptors are expressed on many types of tumor cells. These features make the proapoptotic TRAIL-Rs an extremely appealing target for the generation of therapeutic agents.

Surface expression of TRAIL receptors has been reported for both normal (Atkins et al., 2002; Dorr et al., 2002; Jo et al., 2000; Leverkus et al., 2000b; Mundt et al., 2003) and tumor cells (Arts et al., 2004; Ashkenazi et al., 1999; Bouralexis et al., 2004; Clodi et al., 2000; Cuello et al., 2001; Frank et al., 1999; Frese et al., 2002; Ibrahim et al., 2001; Mitsiades et al., 2000; Odoux et al., 2002; Shin et al., 2001; Song et al., 2003a; van Geelen et al., 2003; Vignati et al., 2002). Weak but detectable TRAIL-R1 and TRAIL-R2 expression has been identified by flow cytometry on the surface of a limited number of normal (diploid) cell types, including hepatocytes, keratinocytes, astrocytes, and osteoblasts (Atkins et al., 2002; Dorr et al., 2002; Jo et al., 2000; Leverkus et al., 2000a; Mundt et al., 2003). However, a broad spectrum of tumor cell types has been identified with variable levels of TRAIL-R1 and /or TRAIL-R2 including some examples of relatively high expression. Cells isolated from primary tumors of the lung (Odoux et al., 2002), blood (Cappellini et al., 2005; Clodi et al., 2000), skin (Song et al., 2003a), bone (Bouralexis et al., 2004), and the brain (Ciusani et al., 2005) have detectable cell surface expression of TRAIL-R1 and TRAIL-R2 by flow cytometry. Likewise, human tumor cell lines derived from carcinomas of the colon (van Geelen et al.,

2003), breast (Ashkenazi et al., 1999), ovary (Cuello et al., 2001; Vignati et al., 2002), thyroid (Mitsiades et al., 2000), lung (Frese et al., 2002), pancreas (Ibrahim et al., 2001), and liver (Griffith et al., 1998), as well as from melanomas (Song et al., 2003a), sarcomas (Bouralexis et al., 2003), and tumors of the brain (Song et al., 2003b), have variable and high-level FACS-detectable TRAIL-R1 and TRAIL-R2. In many tumor cell lines where resistance to TRAIL-R agonism was observed, the relevance of cell surface expression of the TRAIL-Rs was complicated by the fact that receptor levels did not have a role in regulating response. A clear relationship between receptor expression level and potential for activation of apoptosis through proapoptotic TRAIL-Rs has not been established. However, evaluation of receptor expression in a tissue context is desirable in understanding more about the TRAIL-Rs as targets of systemic therapies.

Antibody reagents specific for linear peptides of the C-terminal, intracellular portion of TRAIL-R1 and TRAIL-R2 have been utilized in studies of TRAIL-R distribution in tissues (Arts et al., 2004; Koornstra et al., 2003; Reesink-Peters et al., 2005; Spierings et al., 2003; Spierings et al., 2004). Arts et al. demonstrated that most ovarian tumors expressed one or both proapoptotic TRAIL receptors, and that TRAIL-R2 expression was increased after chemotherapy in paired samples collected pre-therapy and post-therapy. Likewise, Koornstra et al. highlighted that expression of these death receptors was increased in colon tumors vs normal colon, and that both TRAIL-R1 and TRAIL-R2 were detected on all adenomas and carcinomas evaluated. In parallel, Spierings et al. (2003) evaluated a large panel of stage III non-small-cell lung (NSCL) tumors ($n = 87$) and related the staining to available clinical outcome data. In this study, TRAIL-R1 was identified on essentially all specimens (99%), with staining often strongest at the basal cell layers in tumors with squamous differentiation. TRAIL-R2 was also identified on the majority of the specimens (82%); interestingly, TRAIL-R2 expression was correlated with increased risk of death (odds ratio 5.76). A second study by Spierings et al. (2004) evaluated distribution of TRAIL and TRAIL-RS on normal tissues from humans and chimpanzees. In this study, as with the tumor panels, there was fairly widespread labeling of tissues evaluated for both TRAIL-R1 and TRAIL-R2, but staining patterns were similar across the two species. Finally, a recent study by Reesink-Peters et al. evaluated the distribution of TRAIL-Rs and markers of proliferation and apoptosis in cervical neoplasia. TRAIL-R1 and TRAIL-R2 were each identified in >80% of the specimens evaluated, with slightly more staining for TRAIL-R2; however, there was no correlation of TRAIL-Rs to either proliferation or ongoing apoptosis in these specimens.

Interestingly, in several of the studies listed above, staining was often restricted to the cytoplasmic compartment; therefore, it is unclear whether this distribution is relevant to the potential activity of therapeutics that target the extracellular portion of the receptor. It should also be noted that the peptides used for immunization to produce these polyclonal antibodies include considerable homology between the published TRAIL-R1 and TRAIL-R2 sequences. Although these antibody reagents perform well for specific recognition of the linear peptide in a western blot, it may be difficult to demonstrate highly specific staining in an immunohistochemical assay format where the receptor protein has not been denatured and stabilized as a linear peptide target.

Others have reported tissue distribution of the proapoptotic TRAIL-Rs using monoclonal antibody reagents raised against the extracellular domains of these TRAIL-Rs, including use on formalin-fixed tissues (Daniels et al., 2005), frozen sections (Strater et al., 2002a), or a fluorescence-based method of quantitative tissue staining (McCarthy et al., 2005). Daniels et al. reported widespread staining of TRAIL-R1 and TRAIL-R2 in both tumor and normal tissues, with tumors staining more intensely than the adjacent normal tissue, but noted that the staining was often patchy in breast carcinomas, and that there was much less staining than expected on lymphoid tumors. Strater reported widespread TRAIL-R1 and TRAIL-R2 staining in tumors of the colon, but reported also that there was a positive correlation between TRAIL-R1 expression and survival. In contrast, McCarthy et al. identified a strong negative correlation between TRAIL-R2 expression and survival in breast cancer, with TRAIL-R2 expression associated with increased node-positive tumors. The TRAIL-R2 specific monoclonal antibody described in these studies can also be used for flow cytometry applications for determination of surface receptor levels, but is not currently recommended by the manufacturer for immunohistochemical studies in tissue specimens.

In evaluating the distribution of potential targets for agonist TRAIL receptor antibodies, it was considered critical to focus efforts specifically on detection of the extracellular portion of these receptors in order to understand the distribution of the part of the receptor recognized by TRAIL-R agonists. Antibodies have been developed to TRAIL-R1 and TRAIL-R2, respectively and the specificity of these antibodies has been tested by western blotting, flow cytometry, and immunohistochemistry utilizing fixed and embedded cell pellets and xenografts. These antibodies have been utilized for development of sensitive and specific immunohistochemical tests for TRAIL-R1 and TRAIL-R2 in formalin-fixed tissue specimens as described (Roach et al., 2004). To evaluate TRAIL-R distribution using these tests, approximately 270 tumor and normal tissue specimens have been evaluated. A summary of the expanded results is presented in Figs. 7.1–7.3.

After screening several proprietary and commercially available antibodies selected for specificity to TRAIL-R1 or TRAIL-R2, we concluded that, for both TRAIL-R1 and TRAIL-R2, rabbit polyclonal antibodies represented the best option for developing these tests (Roach et al., 2004). These antibodies performed well on formalin-fixed paraffin-embedded tissues and had minimal background staining. Importantly, staining of sectioned cell pellets or xenografts of cell lines was consistent with receptor expression levels identified by other methods such as flow cytometry, using different TRAIL-R antibodies, and TaqMan to quantitate RNA levels ((Roach et al., 2004) and hierarchical genetic search [HGS] data not shown).

The mean staining scores were determined for the 10 tumor types for which there were at least 10 evaluable specimens, and are presented in Figs. 7.1 and 7.2. Like the other studies reported, we identified stronger staining for both TRAIL-R1 and TRAIL-R2 in tumor specimens overall than in normal tissues. In addition, for most tumor types, although granular to diffuse cytoplasmic staining was noted,

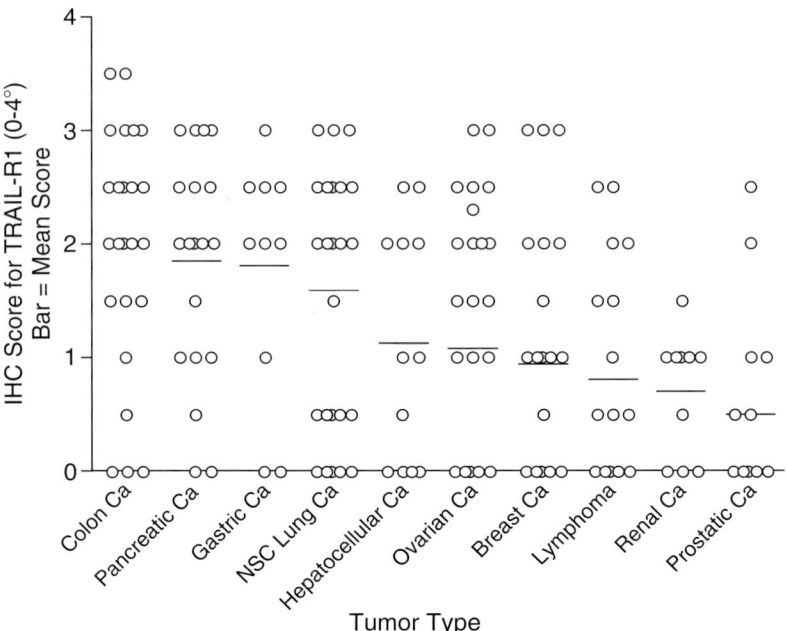

Fig. 7.1 General tumor survey for TRAIL-R1: Tumor types with at least 10 samples. Immunohistochemical score for TRAIL-R1 staining, using a subjective scale of 0–4+, is indicated for 10 general tumor types for which at least 10 specimens were evaluated. Circles indicate individual specimen scores, and the mean score is indicated by a bar for each tumor type. Abbreviations in tumor type: carcinoma (Ca), non-small-cell (NSC)

distinct membrane staining was also observed and was often the stronger pattern. We identified general patterns of relatively high staining in colorectal carcinomas and relatively little staining in most lymphomas, as has been reported by others (Daniels et al., 2005; Koornstra et al., 2003). However, we identified less staining overall, in both tumor and normal tissues, than in most other reports in the literature. Instead of >80% staining for each receptor in tumor tissues, an average of 43% of tumors scored at least a 2+ for TRAIL-R1 ($N = 235$), and an average of 64% of tumors scored at least a 2+ for TRAIL-R2 ($N = 227$). These scores were based upon both distribution and intensity of receptor staining (Fig. 7.3). It was noted that there was often heterogeneous staining of tumor cells within individual specimens, as well as occasionally between specimens from a single individual. There was more staining identified for TRAIL-R2 than for TRAIL-R1 in both normal and tumor tissues using this method. This was in agreement with other nonclinical data suggesting more widespread distribution of TRAIL-R2 than TRAIL-R1 on several human tumor cell lines.

Importantly, the potential clinical relevance of TRAIL receptor expression has yet to be determined. Agonist monoclonal antibodies specific for TRAIL receptors can cause decreases in cell line viability even in cell lines with very low surface TRAIL receptor as determined by flow cytometry (HGS data not shown). It is possible

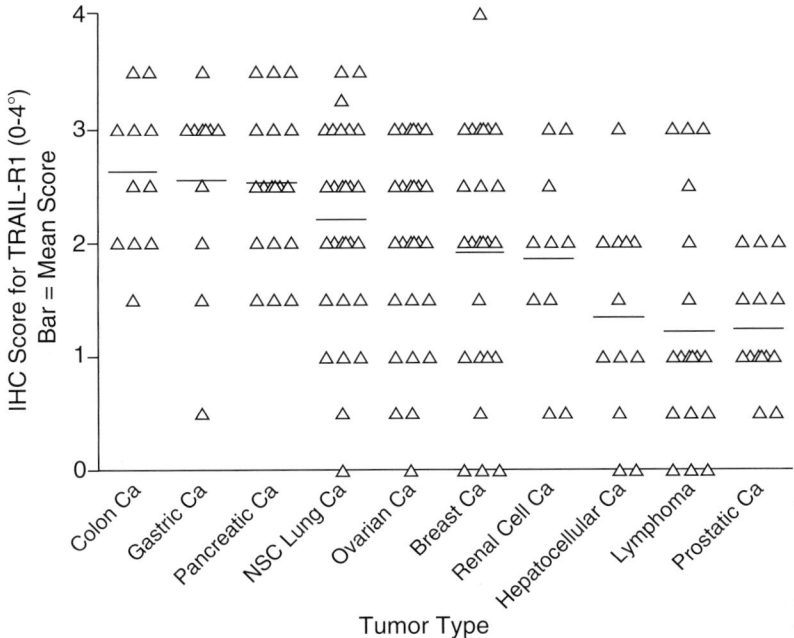

Fig. 7.2 General tumor survey for TRAIL-R2: Tumor types with at least 10 samples. Immunohistochemical score for TRAIL-R2 staining, using a subjective scale of 0–4+, is indicated for 10 general tumor types for which at least 10 specimens were evaluated. Triangles indicate individual specimen scores, and the mean score is indicated by a bar for each tumor type. Abbreviations in tumor type: carcinoma (Ca), non-small-cell (NSC)

that immunostaining methodology on formalin-fixed tissues will not be sensitive enough to detect functionally meaningful levels of the extracellular portion of the receptor, which would be targeted by TRAIL-R agonists. However, this possibility cannot be fully evaluated in the absence of clinical outcome data. Therefore, the immunohistochemical staining of tumor tissue specimens from subjects enrolled in the ongoing clinical trial programs represents a critical next step in the development of these tests. In addition, further studies of the downstream signaling pathways will likely contribute to our understanding of the appropriate application of TRAIL-R agonist therapies.

4 The Extrinsic Pathway and the Death-Inducing Signaling Complex

There are two essential pathways for TRAIL-mediated cell death (see also Chapter 3): one "extrinsic," which is dependent on the signal from the ligand and activation of a TRAIL-R protein complex, which leads directly to cell death; the other,

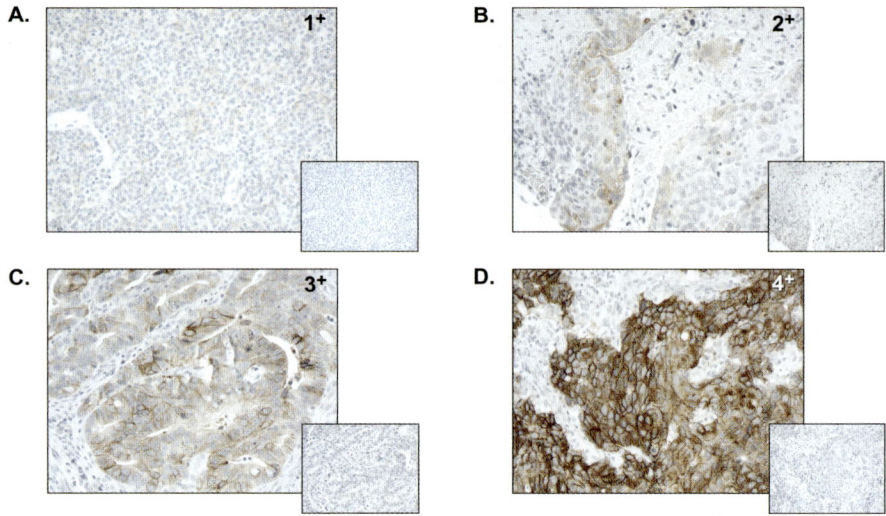

Fig. 7.3 Examples of IHC scoring scale for TRAIL-receptors. Panels A–D indicate examples of the 0–4+ scale used to evaluate TRAIL-R1 and TRAIL-R2 staining, and illustrate some of the typical patterns observed. Panel A, illustrating 1+ staining, has weak, but widespread, staining (non-Hodgkin's lymphoma); panel B, illustrating 2+ staining, has focally stronger staining of the tumor population (cervical carcinoma); panel C, illustrating 3+ staining, has widespread staining with variability in intensity (gastric carcinoma); and panel D, illustrating 4+ staining, considered exceptional, has uniformly strong staining of the tumor cell population, highlighting membrane areas and excluding nuclei (colon carcinoma). The smaller image includes the same field stained with a non-specific IgG as a control (scored 0). All photomicrographs were taken using a 20× objective

"intrinsic," which is activated by TRAIL binding to the TRAIL-Rs, but initiation of cell death is mediated through the mitochondria.

The extrinsic pathway of TRAIL-R cell death mediated through the formation of a ligand–receptor complex. Each DD on a TRAIL-R molecule interacts with a similar DD on a cytoplasmic adaptor protein, Fas-associating protein with a death domain (FADD). FADD acts as a bridge between the ligand–receptor complex and the receptor proximal caspase-8, through the death effector domain (DED). Transfection of dominant negative forms of FADD, or wild-type TRAIL-R1 or TRAIL-R2 into cells lacking FADD, blocks apoptosis demonstrating that FADD is a critical component of TRAIL-R signaling. Recently, it has been shown that the C-terminal tails of TRAIL-R1 and TRAIL-R2 are required for efficient FADD binding, caspase cleavage, and TRAIL-dependent apoptosis. (Ashkenazi, 2002; Bodmer et al., 2000a; Kuang et al., 2000; Luschen et al., 2000; Muhlenbeck et al., 1998; Thomas et al., 2004a; Yeh et al., 1998). The multiprotein complex of ligand, death receptor, adaptor, and protease is known as the death-inducing signaling complex (DISC). This signaling structure is unique amongst cell surface receptor signaling pathways for its threefold symmetry. The formation of this complex is a critical regulatory event in the process of apoptosis. Inactive caspase-8 molecules are recruited into the DISC by FADD and are cleaved into active proteases through an unknown mechanism. It has been suggested in the "induced proximity" model that inactive initiator caspases brought into close proximity during DISC formation, promotes mutual cleavage and activation. (Boatright et al., 2003; Boatright and Salvesen, 2003; Muzio et al., 1998; Salvesen and Dixit, 1999). Autocleavage and activation of the receptor-associated caspase-8 leads to its release from the DISC and formation of heterodimeric active subunits. The initiator caspase is now able to target the "effector caspases" 3, 6, and 7. These terminal caspases, once activated, cleave key structural and signaling components of the cell and begin the physical destruction of apoptosis. This relatively short signaling cascade emphasizes the potential for rapid induction of cell death. Various apoptosis assays have demonstrated cellular and molecular changes associated with apoptosis appearing within 30 min after TRAIL-R engagement (Houghton, 1999; Walczak and Sprick, 2001). Importantly, the TRAIL-R pathway can activate cell death independently of p53, a primary target for apoptosis regulation by tumor cells (Galligan et al., 2005; Igney and Krammer, 2002; Wang and El-Deiry, 2003). Interestingly, p53 regulates expression of TRAIL-R2, suggesting p53 can increase sensitivity to TRAIL-R agonists in response to other apoptotic stimuli (Sheikh and Fornace, 2000).

5 The Intrinsic Pathway

The bridge from the DISC to the intrinsic pathway is formed through an intervening catalytic event. One of the cytoplasmic targets for the TRAIL-R-activated initiator caspases is the cytoplasmic protein Bid. Bid is a member of the Bcl-2 family of proteins responsible for regulating the mitochondrial pathway of apoptosis. In addition, the intrinsic pathway is also activated through several molecular monitors of cellular health such as p53 and AKT. In response to apoptotic stimuli from various metabolic and structural insults, including DNA damage, serum starvation and radiation, there is a loss of mitochondrial membrane integrity that precipitates the activation of another initiator caspase, caspase-9, and subsequently the effector caspases. Here, at the mitochondria the two pathways of apoptosis intersect emphasizing the importance of the regulation of this intersection.

The cleavage of Bid by caspase-8 creates a truncated form of Bid (tBid) that can translocate to the mitochondrial membrane (Srivastava, 2001). Bid is thought to form a heteromeric complex with other apoptosis-promoting molecules, Bax and Bak (Luo et al., 1998; Wei et al., 2000). Bax is liberated from its complex with the antiapoptotic protein Bcl-2 in response to apoptotic stimuli. This translocation of Bax or Bid to the mitochondrial membrane disrupts membrane integrity and induces release of cytochrome c and the formation of a protein complex known as the apoptosome (Adams and Cory, 2002). The apoptosome is comprised of cytochrome c, pro-caspase-9 and a scaffolding protein, apoptotic protease activating factor (APAF1). APAF1 forms a heptamer after binding cytochrome c and recruits several molecules of pro-caspase-9 through reciprocal caspase recruitment domains

(CARD) present in APAF1 and caspase-9 (Pan et al., 1998). This recruitment and oligomerization leads to caspase-9 activation and suggests again a role for the induced proximity model of caspase activation. Deletion of APAF1 demonstrates its necessary role in caspase-9 activation (Yoshida et al., 1998). Active caspase-9 can now cleave and activate the executioner caspase-3, caspase-6, and caspase-7.

The intrinsic pathway can be activated independent of the TRAIL-R pathway through other signals such as those transmitted by p53. One of the dominant mechanisms of chemotherapeutic resistance in cancer cells is the gene deletion or acquisition of inactivating mutations in TP53. Conversely, the ability of the TRAIL-R pathway to bypass the loss or inactivation of p53, via Bid cleavage, and still induce apoptosis through the mitochondria is one of the distinct advantages of targeting the TRAIL-Rs. Therefore, there are two pathways, extrinsic and intrinsic, for activation and execution of the TRAIL-R-mediated signals that lead to cell death.

6   Regulation of Death Signaling

Not surprisingly, given the activation of the death signal and its resulting dire consequences for the cell, the apoptotic pathway is highly regulated at several key points. Importantly, tumor cells have exploited these normal regulatory check points through acquired or induced modifications to attenuate the activity of caspases, alter the formation or composition of the DISC, or alter the interaction of intrinsic apoptosis regulatory proteins Bcl-2 and Bax, or their family members (Igney and Krammer, 2002).

FLICE-like inhibitory protein (FLIP) is a dominant negative form of caspase-8 that competes with caspase-8 for binding in the DISC. FLIP plays an important role in regulating sensitivity to TRAIL signaling (Griffith et al., 1998). Chemotherapy or FLIP siRNA can modify FLIP levels in tumors and promote TRAIL-induced apoptosis (Chawla-Sarkar et al., 2004; Galligan et al., 2005; Kang et al., 2005; Song et al., 2003a; Xiao et al., 2005). Interestingly, in support of the data that FLIP receptor complexes exist prior to ligand binding, a peptide sequence at the COOH terminus of FLIP (L) and TRAIL-R2 interact preventing FADD binding to TRAIL-R2. Upon ligand binding, FLIP is dislodged and a competent DISC is formed (Jin et al., 2004). The intimate interaction of FLIP with the receptor makes it an attractive target for pharmacologic intervention (Roth and Reed, 2004).

All of the apoptotic caspases described are regulated not only by a requirement for death receptor- or mitochondrial-mediated cleavage, but also by endogenous inhibitory proteins as well. These caspase-inhibitory proteins contain a protein interaction domain that classifies them as inhibitor of apoptosis proteins (IAPs). Their baculovirus IAP repeat (BIR) domains are zinc-binding folds that play a role in forming binding grooves for the active caspase. Once bound within the groove, caspase-9 cannot self-activate. Several members of this family are overexpressed in tumors (Igney and Krammer, 2002). Interestingly, the protein SMAC/DIABLO

Table 7.2 Tumor modifications of the extrinsic and intrinsic pathways

Location	Target	Modification and consequence	References
Upstream of mitochondria	AKT	AKT constitutive activity promotes Bad phosphorylation	Bortul et al. (2003); Cenni et al. (2004); Chen et al. (2001); Whang et al. (2004)
Upstream of mitochondria	PTEN	Loss of PTEN yields an inability to dephosphorylate AKT	Deocampo et al. (2003); Nesterov et al. (2001)
Upstream of mitochondria	Bcl-2	Overexpression, blocks apoptosis	Nencioni et al. (2005)
Upstream of mitochondria	Bcl-XL	Overexpression, blocks apoptosis	Dole et al. (1995); Foreman et al. (1996); Nagane et al. (1998)
Upstream of mitochondria	Mcl-1	Overexpressed in AML, blocks apoptosis	Kaufmann et al. (1998); Taniai et al. (2004); Yu et al. (2005)
Downstream of mitochondria	Survivin	Overexpressed in neuroblastoma blocks apoptosis	Adida et al. (2000); Kim et al. (2005); Wang et al. (2005); Yamaguchi et al. (2005b)
Downstream of mitochondria	cIAP2	Gene rearranged in MALT	Dierlamm et al. (1999)
Downstream of mitochondria	ML-IAP	Overexpressed in melanoma	Vucic et al. (2000)
Downstream of mitochondria	APAF1	Loss of APAF1 Blocks Caspase 9	Soengas et al. (2001); Soengas et al. (2006)
Downstream of mitochondria	XAF	Binds to XIAP	Leaman et al. (2002)
Receptor complex	Caspase-8	Methylation of gene represses expression, blocks cytoplasmic apoptosis signal	Ashley et al. (2005); Poulaki et al. (2005); van Noesel et al. (2003); Zuzak et al. (2002)
Receptor complex	TRAIL-R1, TRAIL-R2	Point mutations and genetic deletion	Fisher et al. (2001); Kuraoka et al. (2005); McDonald et al. (2001); Ozoren et al. (2000); Pai et al. (1998); Wolf et al. (2006)
Receptor complex	Decoy receptors, DcR1, DcR2	Occasional elevated expression	Meng et al. (2000)
Upstream of mitochondria	Bax	Inactivating mutation prevents apoptosis	Rampino et al. (1997); Zhang et al. (2000); Ionov et al. (2000)
Upstream of mitochondria	c-Myc	Represses FLIP expression	Ricci et al. (2004)
Downstream of mitochondria	SMAC/ Diablo	Reduced release of SMAC/Diablo	Zhang and Fang (2005)

is released from the mitochondria and antagonizes the binding of caspase-9 to the IAP family member, X-linked IAP (XIAP), thereby promoting apoptosis (Ng and Bonavida, 2002). A summary of the known modifications of proteins involved in regulating apoptosis found in tumor cells is described in Table 7.2.

7   Agonists of the Trail-R Apoptotic Pathway

7.1   *Tumor Necrosis Factor-Related Apoptosis-Inducing Ligand*

Members of the tumor necrosis factor (TNF) superfamily have demonstrated the ability to induce apoptosis in virally and oncogenically transformed cells, human tumor cell lines and activated lymphocytes, NK, and monocytes (TRAIL, TNFα, TNFβ, and FASL). The ability to induce cell death, 14 other TNF ligands possess a diverse array of immunomodulatory and growth-stimulatory capabilities, including stimulation and proliferation of B-cells (BLyS) and T-cells (CD40L, LIGHT, OX40L, and 4–1-BBL) and regulation of bone metabolism (RANKL) (reviewed in Locksley et al. (2001) and Ashkenazi (2002)). TRAIL is a type II membrane-bound protein which exists as a self-assembling homotrimeric molecule that possesses apoptotic activity in a membrane bound or soluble form. The membrane form can be cleaved from the cell surface by an extracellular cysteine protease (Lawrence et al., 2001; Mariani and Krammer, 1998). TRAIL exists as a trimer in solution and requires elemental $Zn+2$ and a cysteine residue to coordinate and properly organize the trimeric structure. (Hymowitz et al., 2000) TRAIL that is generated in the absence of zinc permits the formation of cysteine disulfide bonds that result in an asymmetric molecule, which is less stable and insoluble in solution (Ashkenazi, 2002; Lawrence et al., 2001). Crystal studies of the ligand bound to TRAIL-R2 have revealed that the inverted pyramid-shaped trimeric ligand binds in the pocket between three receptor molecules (Hymowitz et al., 2000; Mongkolsapaya et al., 1999).

Cell surface expression of the ligand TRAIL has been observed on a variety of immune cells including IL-15- or IL-2-activated NK cells, virally infected T-cells, interferon gamma-activated monocytes, and dendritic cells, as well as CD4+ and CD3+ T-cells. TRAIL can confer tumoricidal activity to monocytes and NK cells and plays a role in immune surveillance against tumor development (Kayagaki et al., 1999a, b; Mariani and Krammer, 1998; Nieda et al., 2001; Takeda et al., 2002). Recently, a "window of TRAIL sensitivity" was observed in CD34 erythroid progenitor cells that is promoted initially by the expression of TRAIL-Rs and then inhibited by intercellular expression of Bcl-2 (Mirandola et al., 2006a). TRAIL has also been detected on the surface of colonic epithelium (Strater et al., 2002b). The soluble and membrane-bound form of TRAIL-induced apoptosis in a wide variety of human tumor cells both in vitro and in vivo without affecting the viability of normal cells.

Several forms of recombinant TRAIL have been generated to evaluate the ligand in preclinical studies. Histidine-tagged (Pitti et al., 1996), leucine zipper (Walczak et al., 1999), Flag-tagged (Bodmer et al., 2000b; Schneider and Tschopp, 2000), and $Zn+2$-stabilized versions (Ashkenazi and Dixit, 1999; Kelley et al., 2001) have all been generated and tested for activity against tumor and normal cells in preclinical studies. These different forms of the ligand have displayed a spectrum of antitumor activity in human cell lines in vitro, in xenograft models and primary tissues transplanted into nude mice. TRAIL, either alone or in combination with chemotherapeutic agents, has demonstrated apoptosis activity in tumor cell lines derived from a broad array of human tumors including colon, brain, uterus, ovary, liver, breast, prostate, kidney, liver, lung, thyroid, and blood (Asakuma et al., 2003; Ashkenazi et al., 1999; Bouralexis et al., 2003, 2004; Chen et al., 2003; El-Zawahry et al., 2005; Jazirehi et al., 2001; Jeon et al., 2003; Keane et al., 1999; Kelly et al., 2002; LeBlanc and Ashkenazi, 2003; Miao et al., 2003; Mitsiades et al., 2001a; Muhlethaler-Mottet et al., 2004; Nagane et al., 2001; Naka et al., 2002; Ohtsuka et al., 2003; Pitti et al., 1996; Secchiero et al., 2002; Singh et al., 2003; Srivastava, 2001). TRAIL can overcome chemoresistance or radioresistance when administered in combination with chemotherapy in adriamycin-resistant myeloma, radio-resistant lymphoma, and taxane- and platinum-insensitive breast and osteosarcoma cell lines (Belka et al., 2001; Clayer et al., 2001; Cuello et al., 2001; Evdokiou et al., 2002; Frese et al., 2002; Jazirehi et al., 2001; Johnston et al., 2003; Keane et al., 1999; Liu et al., 2001; Mitsiades et al., 2001b; Nagane et al., 2000, 2001; Voelkel-Johnson, 2003).

While the epitope-tagged forms of the ligand assisted the isolation and purification of the recombinant protein, and in many instances enhanced the activity of TRAIL, they also enhanced the toxicity on normal cells. HIS-tagged, leucine-zipper or Flag-tagged antibody cross-linked forms of TRAIL-induced apoptosis in normal hepatocytes in vitro(Jo et al., 2000; Lawrence et al., 2001). Conflicting results were obtained when no apoptosis was observed with soluble TRAIL administered to normal primary cells from the lung, bone, liver, endothelium, breast, brain, and kidney (Ashkenazi et al., 1999). Safety studies of Zn2+-stabilized TRAIL administered in short-term treatment of mouse, monkey, and chimpanzees showed no detectable toxicities (Lawrence et al., 2001). Additional studies with soluble TRAIL were performed in chimeric mice whose livers were reconstituted with human hepatocytes. Repeated injection of soluble nontagged form of TRAIL did not generate any hepatotoxicity (Hao et al., 2004). These conflicting results suggested that nonphysiologically or inappropriately aggregated forms of TRAIL can be toxic. Whereas a soluble, correctly organized Zn2+-stabilized TRAIL was not toxic. It is important to note that there is a role for native TRAIL in response to inflammation or infection. Acute bacterial or viral infection of the liver or pancreas or in mouse models of hepatitis or pancreatitis TRAIL can induce apoptosis (Mundt et al., 2003; Hasel et al., 2003). Membrane-bound TRAIL has been shown to induce liver damage in adenoviral-transfected hepatocytes in vivo (Ichikawa et al., 2001). These types of responses coincide with the predicted role for TRAIL in mediating an immune surveillance response to acute bacterial- or viral-induced infection or inflammation.

The substantial preclinical antitumor data observed with the ligand implied that TRAIL-R agonism could potentially yield significant clinical antitumor activity. In fact, a recombinant form of the TRAIL ligand is currently is phase 1 clinical development. Nonetheless, this optimism should be tempered with the knowledge that certain versions of the TRAIL ligand, albeit in nonphysiological forms, did induce severe cytotoxicity of normal cells. Therefore, clinical development should be prudently conducted with awareness toward potential indicators of toxicity.

7.2   Antibodies

A spectrum of mouse and human monoclonal and polyclonal antibodies has demonstrated the ability to agonize the TRAIL-Rs and induce death in tumor cells. They have proven to be valuable tools to explore mechanism of action, define chemotherapeutic combinations agents that enhance apoptosis, and describe functional differences between the TRAIL-R1 and TRAIL-R2 pathways. Importantly, human monoclonal antibodies selected for high-affinity binding and maximal agonism have been advanced into clinical development as therapeutic cancer agents.

Experiments using antibodies, which target the TRAIL-Rs, revealed that only TRAIL-R1 and TRAIL-R2 were capable of inducing apoptosis and not the decoy receptors TRAIL-R3 (DcR1) and TRAIL-R4 (DcR2) (Griffith et al., 1999). TRAIL-R2 specific antibodies when cross-linked, generated a distinct activation of NF-κB, apoptosis, and Jun NH2 kinase (JNK) activation compared to the NF-κB activation and apoptosis induced by cross-linking of TRAIL-R1 antibodies (Muhlenbeck et al., 2000). Mouse monoclonal antibodies against TRAIL-R1 were potent agonists in vivo but minimally active in vitro. In vitro activity was enhanced by secondary cross-linking antibodies, presumably through multimerization of receptor complexes. These antibodies, however, were very active against human xenografts when administered in vivo (Chuntharapai et al., 2001; Griffith et al., 1998). This result suggested that the mouse contributed a cross-linking function possibly through Fc receptors on immune cells. However, the use of agonist antibodies of immunoglobin isotypes that preferentially bind to Fc receptors or have the ability to fix complement were not significantly more active in vivo suggesting that this is not the mechanism that enhances these antibodies in vivo.

Receptor-specific antibodies selected for high-affinity binding and their TRAIL-R agonism have been identified and generated from phage display libraries, hybridomas, and transgenic mice containing human immunoglobin genes (Dobson et al., 2002; Ichikawa et al., 2001; Motoki et al., 2005; Pukac et al., 2005).

TRA-8, an agonist mouse monoclonal antibody to the human TRAIL-R2, was generated by immunization of mice with the extracellular domain of human TRAIL-R2 fused to the Fc portion of human IgG$_1$. TRA-8 bound to TRAIL-R2 specifically induced apoptosis in human T-cell leukemia, B-cell lymphoma, and glioma lines, and enhanced antitumor activity in combination with chemotherapeutic agents including Adriamycin (doxorubicin hydrochloride) and cisplatin. In TRA-8-resistant glioma lines sensitivity was restored after overexpression of Bax mediated by adenoviral transfer. Importantly, TRA-8 was also tested against hepatocytes in vitro and did not display any evidence of apoptosis (Choi et al., 2002; Ichikawa et al., 2001; Kaliberov et al., 2004; Ohtsuka et al., 2003).

Human Genome Sciences in collaboration with Cambridge Antibody Technology generated a series of fully human monoclonal antibodies, which target TRAIL-R1 or TRAIL-R2. The most active of these candidates were selected for evaluation in preclinical studies and are now advancing through clinical development. HGS-ETR1 (mapatumumab) an antibody specifically targeting TRAIL-R1, demonstrated potent in vitro apoptotic activity against human tumor cell lines derived from colon, lung, pancreas, ovary, uterus, renal, and hematologic malignancies. This in vitro activity was achieved in the absence of cross-linking agents. HGS-ETR1 enhanced the cytotoxicity of chemotherapeutic agents (camptothecin, cisplatin, carboplatin, or 5-fluorouracil) even in tumor cell lines that were not sensitive to HGS-ETR1 alone. In preestablished colon, NSCL, and renal xenografts, HGS-ETR1 treatment resulted in rapid tumor regression or repression of tumor growth. Addition of chemotherapeutic agents like topotecan, 5-fluorouracil, and irinotecan in colon xenograft models enhanced antitumor efficacy and in some models a synergistic antitumor activity was observed (Pukac et al., 2005).

Phase 1 trials of HGS-ETR1 have been conducted in advanced solid tumor patients and have demonstrated the safety and tolerability of single agent HGS-ETR1 up to 20 mg/kg. Single agent phase 2 studies were conducted in colorectal cancer, non-small-cell lung carcinoma (NSCLC), and non-Hodgkin's lymphoma (NHL). While stable disease was the best response observed in the two solid tumor studies, objective responses, including one complete response, were observed in the NHL study. Further, phase 1b studies have demonstrated that HGS-ETR1 can be safely administered in combination with standard doses of chemotherapy agents, such as carboplatin and paclitaxel. Additional phase 2 studies are planned to assess the activity of HGS-ETR1 in combination with chemotherapy.

HGS-ETR2 (lexatumumab), a fully human antibody identified via screening of phage display libraries for high-affinity, single-chain antibodies to TRAIL-R2, has been evaluated in similar human tumor cell lines for apoptotic activity. HGS-ETR2 produced potent apoptotic activity in a spectrum of human tumor cell lines including NSCL, colon, renal, and ovarian in the absence of cross-linking agents. (Alderson et al., 2003; Humphreys et al., 2003; Johnson et al., 2003, 2004). HGS-ETR2 has demonstrated the ability to enhance the activity of chemotherapeutic agents from various classes including taxanes and platinums (Georgakis et al., 2003; Humphreys et al., 2003; Johnson et al., 2004; Zeng et al., 2006). HGS-ETR2 induced cell death in two human RCC cell lines and nine human primary RCC cell cultures. This in vitro effect was enhanced with addition of a cross-linking antibody. In a renal xenograft model using primary renal carcinoma tumor cells HGS-ETR2 was able to induce tumor regression (Zeng et al., 2006). HGS-ETR1 and HGS-ETR2 were effective in cell lines from multiple myeloma, acute lymphoblastic leukemia (ALL), NHL, and chronic myelogenous leukemia and in primary hematological tumor cells from NHL, chronic lymphocytic leukemia, and multiple myeloma patients (Georgakis et al., 2003; Johnson et al., 2003). Phase 1 trials of HGS-ETR2 have been conducted in advanced solid tumor patients. This agent has demonstrated that it can be safely and repetitively administered up to 10 mg/kg. The results of the phase 1 studies support the additional study of HGS-ETR2 in phase 2 trials to evaluate its potential for use in the treatment of cancer.

Another TRAIL-R2 mAb (HGS-TR2J, KMTR2) was identified in collaboration between Human Genome Sciences and Kirin Brewery, Inc. This agonist antibody, derived from transchromosomal mice expressing human Ig locus, showed in vitro and in vivo activity against human tumor cell lines. Importantly, HGS-TR2J generated significant apoptotic activity without cross-linking and was active in many human tumor cell lines. It was also shown that ligation of HGS-TR2J to cell surface receptors induced clustering of TRAIL-R2 (Motoki et al., 2005). HGS-TR2J is currently in phase 1 clinical development.

7.3 *Agonist Signaling*

Receptor oligomerization is potentially a key event in TRAIL-R signaling. In vitro and in vivo experiments have shown that cross-linking TRAIL-R agonists, including various forms of the recombinant ligand and antibodies, altered antitumor activity. Antibodies, because of their bivalent binding, have the potential to oligomerize receptor molecules, which could lead to activation of TRAIL-R signaling, DISC formation and cell death. The recombinant ligand, generated in several forms that permitted cross-linking or aggregation, demonstrated potent antiapoptotic activity. Additionally, chemotherapeutic treatment has been able to induce TRAIL-R1 and TRAIL-R2 receptor aggregation and enhance apoptosis (Bergeron et al., 2004; Delmas et al., 2004). Experiments have shown that cross-linking an agonist, including the ligand and antibodies, can improve apoptosis in vitro. Even in those experiments where cross-linking was required for activity in vitro, agonists were readily effective in vivo without cross-linking. In addition, some agonists can achieve maximal apoptosis activity without any enhancement from in vitro cross-linking. This conflicting data suggests several possible mechanisms of killing by TRAIL-R agonists.

Conceivably, both cross-linking-dependent and cross-linking-independent mechanisms may exist for TRAIL-R agonists. Where cross-linking is involved in vivo this function may be provided by the host through immune cells that can cross-link IgG molecules, i.e., Fc receptors. Alternatively, in the absence of cross-linking a TRAIL-R agonist could bind to the trimerized receptor and induce a conformational change similar to the alteration that is theorized to occur with the native ligand. Conformational change in the receptor could expose relevant binding domains on FADD and induce the apoptotic cascade. In fact, the ability to expose different protein-binding domains of FADD has been observed with TRAIL-R agonist antibodies (Thomas et al., 2004b). The ability to cross-link cell surface receptors with antibodies induces capping and increases agonistic activity that has been shown in other signaling systems including those within the TNFSFR. (Cremesti et al., 2001; Liu et al., 2003; Ludwig et al., 2003; Miller et al., 2003). While the precise nature of the interaction between TRAIL-R agonists and the formation of the DISC remains to be determined, their ability to activate this pathway and induce tumor cell death has been proven in preclinical studies and is being validated in the clinical setting.

8 Agents Targeting the Apoptosis Pathway

The availability of human tumor cell lines that are refractory to TRAIL-R agonism has allowed exploration of potential mechanisms of resistance (Igney and Krammer, 2002; Wang and El-Deiry, 2003). Both extrinsic and intrinsic regulatory proteins have been blamed for this resistance, including FLIP (Griffith et al., 1998; Kim et al., 2000; Leverkus et al., 2000a), XIAP, survivin (Kim et al., 2004) Bcl-2 (Fulda et al., 2002a), and Bax (Deng et al., 2002; He et al., 2003; Kandasamy et al., 2003; LeBlanc et al., 2002). Genetic alterations have been identified in TRAIL-R1 and TRAIL-R2 in NSCLC, colon cancer, head and neck cancer, and lymphoma. Some of these modifications induced a loss of apoptotic signaling. Unfortunately, their role in TRAIL resistance in the clinic has not been validated. (Arai et al., 1998; Fisher et al., 2001; Jeng and Hsu, 2002; Lee et al., 1999; Ozoren et al., 2000; Pai et al., 1998; Wolf et al., 2006; Wu et al., 2000). Changes in the level of cell surface receptor expression, caspase-8/FLIP ratio and loss of caspase-8 have all been discovered as mechanisms of resistance to TRAIL-R agonism (Poulaki et al., 2005; Van Geelen et al., 2004; Wachter et al., 2004).

Consequently, many strategies have been evaluated for their ability to enhance sensitivity or maximize responsiveness to TRAIL-R agonism. Early obvious strategies involved combining standard, approved chemotherapeutic agents with TRAIL-R agonists. Chemotherapy agents or radiation improved response in breast, colorectal, and NSCLC cell lines that displayed resistance to TRAIL-R agonism (Adams and Cory, 2002; Ganten et al., 2005; Kondo et al., 2006; Wendt et al., 2005; Zhang et al., 2005). The use of chemotherapy agents modified levels of specific molecules including TRAIL-R1, TRAIL-R2, FLIP, XIAP, or the proapoptotic protein Bad and restored TRAIL-R responsiveness (Fesik, 2005; Galligan et al., 2005; Mirandola et al., 2006b; Xiao et al., 2005; Yamaguchi et al., 2005a). Other strategies have targeted specific molecules known to regulate the pathway at important catalytic or survival signaling steps. For example, many new compounds have targeted the ubiquitous, antiapoptotic protein Bcl-2, or related family members, through antisense or small molecules (Chawla-Sarkar et al., 2004; Sinicrope et al., 2004; Zhu et al., 2005a) (Table 7.3) (also see Chapter 8). Oblimersen sodium (Bcl-2 antisense) as a single agent or in combination with chemotherapy has shown some clinical activity. (Marcucci et al., 2005; O'Brien et al., 2005; Tolcher et al., 2005). Many strategies are focused on the elimination or reduction of inhibitors that block activation of the initiator caspase-8 and caspase-9, namely XIAP, survivin, and FLIP. Small-molecule and antisense techniques have yielded promising results in preclinical models. FLIP, survivin, and XIAP inhibitors in combination with TRAIL-R agonists have significantly enhanced apoptosis across

Table 7.3 Apoptosis therapeutics in development

Compound	Type	Target	Institute/company	Status
HGS-ETR1	Human agonist mAb	TRAIL-R1	Human Genome Sciences	Ph2
HGS-ETR2	Human agonist mAb	TRAIL-R2	Human Genome Sciences	Ph1
HGS-TR2J	Human agonist mAb	TRAIL-R2	Human Genome Sciences	Ph1
TRA-8	Agonist mAb TRA-8	TRAIL-R2	Sankyo	Preclinical
APO2L/TRAIL-PRO1762	Recombinant TRAIL ligand	TRAIL-R1 TRAIL-R2	Amgen/Genentech	Ph1
Genasense (oblimersen sodium)	Antisense	Bcl-2	Genta	Ph2/3
GX15–070	Small molecule	Bcl-2	GeminX	Ph1
AT101	Small molecule	Bcl-2	Ascenta	Ph1/2
ApoGossypol	Small molecule	Bcl-2	Burnham Institute/NCI	Preclinical
EGCG	Small molecule	Bcl-2	Mayo Clinic	Preclinical
ABT-737	Small molecule	Bcl-2	Abbott/Idun	Preclinical
HA14–1	Small molecule	Bcl-2	Raylight	Preclinical
CDDO	Triterpenoid	FLIP	Reata Discovery/ Dartmouth	Preclinical
ISIS 2181308	Antisense	Survivin	Isis/Lilly	Ph1
AG35156	Antisense	XIAP	Aegera	Ph1
Not defined	SMAC mimetic peptide	XIAP	Joyant Pharmaceuticals	Preclinical
Not defined	SMAC mimetic peptide	XIAP	Tetralogics	Preclinical

various cancer cell lines. (Amantana et al., 2004; Chawla-Sarkar et al., 2004; McManus et al., 2004; Ou et al., 2005; Wang et al., 2005; Yamaguchi et al., 2005a, b). There are other compounds that mimic the action of the mitochondrially released XIAP inhibitor, SMAC/DIABLO (Bockbrader et al., 2005; Fulda et al., 2002b; Li et al., 2004; Pei et al., 2004; Roa et al., 2003). There are examples of single-agent activity in tumor cell lines and xenografts for many of these targeted therapies. More importantly, where they have been evaluated, the apoptosis activity of these agents shows a dramatic enhancement in combination with TRAIL-R agonists. These data demonstrate that the use of TRAIL-R agonists and compounds that lower hurdles for active apoptosis signaling may be potent therapeutic agents and importantly active in TRAIL insensitive cells.

Another avenue that has generated encouraging results has come from the use of agents with less direct action on TRAIL-R signaling. The proteosome inhibitor, bortezomib, has broad-ranging effects on receptor expression, upregulation of proapoptotic proteins such as Bik and Bim, and TRAIL production (also see Chapter 12). Bortezomib has also shown activity in combination with the agonist antibodies HGS-ETR1 and HGS-ETR2 in hematological cell lines and primary cells from NHL and CLL patients. (Georgakis et al., 2005; Lashinger et al., 2005; Matta and Chaudhary, 2005; Nencioni et al., 2005; Nikrad et al., 2005; Papageorgiou et al., 2004; Sayers and Murphy, 2006; Zhang et al., 2004; Zhu et al., 2005b). Histone deacetylase (HDAC) inhibitors have demonstrated significant antitumor activity in combination with TRAIL-R agonists. Effects with HDAC inhibitors include changes in TRAIL-R2 expression, decreasing levels of Bcl-2 and FLIP, and increasing the proapoptotic protein Bik. Some early HDAC inhibitors are now progressing through clinical trial development and show early signs of activity (Ganten et al., 2005; Guo et al., 2004; Kelly and Marks, 2005; Kelly et al., 2005; Marks et al., 2004; Yoshida et al., 2005; Zhu et al., 2005b) (also see Chapter 13).

The use of these apoptosis-promoting compounds as single agents or in combination with standard chemotherapy has, in those agents being advanced into clinical development, shown signs of biological activity. These strategies directly targeting the apoptosis pathway are exploiting the potential that they will confer greater effectiveness to chemotherapy. Alternatively, the elimination or obstruction of antiapoptotic molecules may lower the threshold for induction of apoptosis when used in combination with a TRAIL-R agonist. This combination strategy of TRAIL-R agonists and proapoptotic-targeted therapy has the potential to significantly enhance antitumor activity and eliminate the need for nonspecific chemotherapeutic agents that elicit toxic side effects. While a broad range of exciting preclinical data has verified the activity of this amalgamation, a combinatorial apoptotic strategy needs to be validated in a clinical setting.

9 Conclusion

Targeting the TRAIL-R pathway with therapeutic agents provides an opportunity to induce apoptosis selectively in tumor cells. In preclinical studies the use of TRAIL-R agonists like recombinant TRAIL ligand or monoclonal antibodies have demonstrated significant, potent antitumor activity and have enhanced chemotherapeutic agent activity in a spectrum of human tumor cell lines and xenografts. Several human monoclonal antibodies and a recombinant TRAIL ligand have advanced through preclinical evaluation and are now in clinical development. Hopefully, other novel agents that target the apoptotic pathway will enter and advance successfully through the clinical arena, strengthening, and diversifying the armamentarium against the tumor cell.

Acknowledgements The author would like to thank Jessica Farnsworth for critical review of the manuscript.

References

Adams, J. M. and Cory, S. (2002). Apoptosomes: engines for caspase activation. Curr Opin Cell Biol 14, 715–720.

Adida, C., Haioun, C., Gaulard, P., Lepage, E., Morel, P., Briere, J., Dombret, H., Reyes, F., Diebold, J., Gisselbrecht, C., et al. (2000). Prognostic significance of survivin expression in diffuse large B-cell lymphomas. Blood 96, 1921–1925.

Alderson, R. F., Birse, C. E., Connolly, K., Humphreys, R. C., Choi, G. H., Fox, N. L., Gilles Gallant, Ina Han,, Ron Johnson, et al. (2003). TRAIL-R2 mAb, a human agonistic monoclonal antibody to tumor necrosis factor-related apoptosis inducing ligand receptor 2, induces apoptosis in human tumor cells. Proceedings of 94th AACR Annual Meeting 44, 963.

Amantana, A., London, C. A., Iversen, P. L., and Devi, G. R. (2004). X-linked inhibitor of apoptosis protein inhibition induces apoptosis and enhances chemotherapy sensitivity in human prostate cancer cells. Mol Cancer Ther 3, 699–707.

Arai, T., Akiyama, Y., Okabe, S., Saito, K., Iwai, T., and Yuasa, Y. (1998). Genomic organization and mutation analyses of the DR5/TRAIL receptor 2 gene in colorectal carcinomas. Cancer Lett 133, 197–204.

Arts, H. J., de Jong, S., Hollema, H., ten Hoor, K., van der Zee, A. G., and de Vries, E. G. (2004). Chemotherapy induces death receptor 5 in epithelial ovarian carcinoma. Gynecol Oncol 92, 794–800.

Asakuma, J., Sumitomo, M., Asano, T., and Hayakawa, M. (2003). Selective Akt inactivation and tumor necrosis actor-related apoptosis-inducing ligand sensitization of renal cancer cells by low concentrations of paclitaxel. Cancer Res 63, 1365–1370.

Ashkenazi, A. (2002). Targeting death and decoy receptors of the tumour-necrosis factor superfamily. Nat Rev Cancer 2, 420–430.

Ashkenazi, A. and Dixit, V. M. (1999). Apoptosis control by death and decoy receptors. Curr Opin Cell Biol 11, 255–260.

Ashkenazi, A., Pai, R. C., Fong, S., Leung, S., Lawrence, D. A., Marsters, S. A., Blackie, C., Chang, L., McMurtrey, A. E., Hebert, A., et al. (1999). Safety and antitumor activity of recombinant soluble Apo2 ligand. J Clin Invest 104, 155–162.

Ashley, D. M., Riffkin, C. D., Muscat, A. M., Knight, M. J., Kaye, A. H., Novak, U., and Hawkins, C. J. (2005). Caspase-8 is absent or low in many ex vivo gliomas. Cancer 104, 1487–1496.

Atkins, G. J., Bouralexis, S., Evdokiou, A., Hay, S., Labrinidis, A., Zannettino, A. C., Haynes, D. R., and Findlay, D. M. (2002). Human osteoblasts are resistant to Apo2L/TRAIL-mediated apoptosis. Bone 31, 448–456.

Belka, C., Schmid, B., Marini, P., Durand, E., Rudner, J., Faltin, H., Bamberg, M., Schulze-Osthoff, K., and Budach, W. (2001). Sensitization of resistant lymphoma cells to irradiation-induced apoptosis by the death ligand TRAIL. Oncogene 20, 2190–2196.

Bergeron, S., Beauchemin, M., and Bertrand, R. (2004). Camptothecin- and etoposide-induced apoptosis in human leukemia cells is independent of cell death receptor-3 and -4 aggregation but accelerates tumor necrosis factor-related apoptosis-inducing ligand-mediated cell death. Mol Cancer Ther 3, 1659–1669.

Boatright, K. M. and Salvesen, G. S. (2003). Mechanisms of caspase activation. Curr Opin Cell Biol 15, 725–731.

Boatright, K. M., Renatus, M., Scott, F. L., Sperandio, S., Shin, H., Pedersen, I. M., Ricci, J. E., Edris, W. A., Sutherlin, D. P., Green, D. R., and Salvesen, G. S. (2003). A unified model for apical caspase activation. Mol Cell 11, 529–541.

Bockbrader, K. M., Tan, M., and Sun, Y. (2005). A small molecule Smac-mimic compound induces apoptosis and sensitizes TRAIL- and etoposide-induced apoptosis in breast cancer cells. Oncogene 24, 7381–7388.

Bodmer, J. L., Holler, N., Reynard, S., Vinciguerra, P., Schneider, P., Juo, P., Blenis, J., and Tschopp, J. (2000a). TRAIL receptor-2 signals apoptosis through FADD and caspase-8. Nat Cell Biol 2, 241–243.

Bodmer, J. L., Meier, P., Tschopp, J., and Schneider, P. (2000b). Cysteine 230 is essential for the structure and activity of the cytotoxic ligand TRAIL. J Biol Chem 275, 20632–20637.

Bortul, R., Tazzari, P. L., Cappellini, A., Tabellini, G., Billi, A. M., Bareggi, R., Manzoli, L., Cocco, L., and Martelli, A. M. (2003). Constitutively active Akt1 protects HL60 leukemia cells from TRAIL-induced apoptosis through a mechanism involving NF-kappaB activation and cFLIP(L) up-regulation. Leukemia 17, 379–389.

Bouralexis, S., Findlay, D. M., Atkins, G. J., Labrinidis, A., Hay, S., and Evdokiou, A. (2003). Progressive resistance of BTK-143 osteosarcoma cells to Apo2L/TRAIL-induced apoptosis is mediated by acquisition of DcR2/TRAIL-R4 expression: resensitisation with chemotherapy. Br J Cancer 89, 206–214.

Bouralexis, S., Clayer, M., Atkins, G. J., Labrinidis, A., Hay, S., Graves, S., Findlay, D. M., and Evdokiou, A. (2004). Sensitivity of fresh isolates of soft tissue sarcoma, osteosarcoma and giant cell tumour cells to Apo2L/TRAIL and doxorubicin. Int J Oncol 24, 1263–1270.

Cappellini, A., Mantovani, I., Tazzari, P. L., Grafone, T., Martinelli, G., Cocco, L., and Martelli, A. M. (2005). Application of flow cytometry to molecular medicine: detection of tumor necrosis factor-related apoptosis-inducing ligand receptors in acute myeloid leukaemia blasts. Int J Mol Med 16, 1041–1048.

Cenni, V., Maraldi, N. M., Ruggeri, A., Secchiero, P., Del Coco, R., De Pol, A., Cocco, L., and Marmiroli, S. (2004). Sensitization of multidrug resistant human ostesosarcoma cells to Apo2 Ligand/TRAIL-induced apoptosis by inhibition of the Akt/PKB kinase. Int J Oncol 25, 1599–1608.

Chan, F. K., Chun, H. J., Zheng, L., Siegel, R. M., Bui, K. L., and Lenardo, M. J. (2000). A domain in TNF receptors that mediates ligand-independent receptor assembly and signaling. Science 288, 2351–2354.

Chawla-Sarkar, M., Bae, S. I., Reu, F. J., Jacobs, B. S., Lindner, D. J., and Borden, E. C. (2004). Downregulation of Bcl-2, FLIP or IAPs (XIAP and survivin) by siRNAs sensitizes resistant melanoma cells to Apo2L/TRAIL-induced apoptosis. Cell Death Differ 11, 915–923.

Chen, X., Thakkar, H., Tyan, F., Gim, S., Robinson, H., Lee, C., Pandey, S. K., Nwokorie, C., Onwudiwe, N., and Srivastava, R. K. (2001). Constitutively active Akt is an important regulator of TRAIL sensitivity in prostate cancer. Oncogene 20, 6073–6083.

Chen, X. P., He, S. Q., Wang, H. P., Zhao, Y. Z., and Zhang, W. G. (2003). Expression of TNF-related apoptosis-inducing Ligand receptors and antitumor tumor effects of TNF-related apoptosis-inducing Ligand in human hepatocellular carcinoma. World J Gastroenterol 9, 2433–2440.

Choi, C., Kutsch, O., Park, J., Zhou, T., Seol, D. W., and Benveniste, E. N. (2002). Tumor necrosis factor-related apoptosis-inducing ligand induces caspase-dependent interleukin-8 expression and apoptosis in human astroglioma cells. Mol Cell Biol 22, 724–736.

Chuntharapai, A., Dodge, K., Grimmer, K., Schroeder, K., Marsters, S. A., Koeppen, H., Ashkenazi, A., and Kim, K. J. (2001). Isotype-dependent inhibition of tumor growth in vivo by monoclonal antibodies to death receptor 4. J Immunol 166, 4891–4898.

Ciusani, E., Croci, D., Gelati, M., Calatozzolo, C., Sciacca, F., Fumagalli, L., Balzarotti, M., Fariselli, L., Boiardi, A., and Salmaggi, A. (2005). In vitro effects of topotecan and ionizing radiation on TRAIL/Apo2L-mediated apoptosis in malignant glioma. J Neurooncol 71, 19–25.

Clancy, L., Mruk, K., Archer, K., Woelfel, M., Mongkolsapaya, J., Screaton, G., Lenardo, M. J., and Chan, F. K. (2005). Preligand assembly domain-mediated ligand-independent association between TRAIL receptor 4 (TR4) and TR2 regulates TRAIL-induced apoptosis. Proc Natl Acad Sci USA 102, 18099–18104.

Clayer, M., Bouralexis, S., Evdokiou, A., Hay, S., Atkins, G. J., and Findlay, D. M. (2001). Enhanced apoptosis of soft tissue sarcoma cells with chemotherapy: a potential new approach using TRAIL. J Orthop Surg (Hong Kong) 9, 19–22.

Clodi, K., Wimmer, D., Li, Y., Goodwin, R., Jaeger, U., Mann, G., Gadner, H., and Younes, A. (2000). Expression of tumour necrosis factor (TNF)-related apoptosis-inducing ligand (TRAIL) receptors and sensitivity to TRAIL-induced apoptosis in primary B-cell acute lymphoblastic leukaemia cells. Br J Haematol 111, 580–586.

Cremesti, A., Paris, F., Grassme, H., Holler, N., Tschopp, J., Fuks, Z., Gulbins, E., and Kolesnick, R. (2001). Ceramide enables fas to cap and kill. J Biol Chem 276, 23954–23961.

Cuello, M., Ettenberg, S. A., Nau, M. M., and Lipkowitz, S. (2001). Synergistic induction of apoptosis by the combination of trail and chemotherapy in chemoresistant ovarian cancer cells. Gynecol Oncol 81, 380–390.

Daniels, R. A., Turley, H., Kimberley, F. C., Liu, X. S., Mongkolsapaya, J., Ch'En, P., Xu, X. N., Jin, B. Q., Pezzella, F., and Screaton, G. R. (2005). Expression of TRAIL and TRAIL receptors in normal and malignant tissues. Cell Res 15, 430–438.

Delmas, D., Rebe, C., Micheau, O., Athias, A., Gambert, P., Grazide, S., Laurent, G., Latruffe, N., and Solary, E. (2004). Redistribution of CD95, DR4 and DR5 in rafts accounts for the synergistic toxicity of resveratrol and death receptor ligands in colon carcinoma cells. Oncogene 23, 8979–8986.

Deng, Y., Lin, Y., and Wu, X. (2002). TRAIL-induced apoptosis requires Bax-dependent mito-chondrial release of Smac/DIABLO. Genes Dev 16, 33–45.

Deocampo, N. D., Huang, H., and Tindall, D. J. (2003). The role of PTEN in the progression and survival of prostate cancer. Minerva Endocrinol 28, 145–153.

Dierlamm, J., Baens, M., Wlodarska, I., Stefanova-Ouzounova, M., Hernandez, J. M., Hossfeld, D. K., De Wolf-Peeters, C., Hagemeijer, A., Van den Berghe, H., and Marynen, P. (1999). The apoptosis inhibitor gene API2 and a novel 18q gene, MLT, are recurrently rearranged in the t(11;18)(q21;q21) associated with mucosa-associated lymphoid tissue lymphomas. Blood 93, 3601–3609.

Dobson, C., Edwards, B., Main, S., Minter, R., and Williams, L. (2002). Generation of human therapeutic anti-TRAIL-R1 agonistic antibodies by phage display. American Association for Cancer Research 93rd Annual Meeting, Abstract 2869.

Dole, M. G., Jasty, R., Cooper, M. J., Thompson, C. B., Nunez, G., and Castle, V. P. (1995). Bcl-xL is expressed in neuroblastoma cells and modulates chemotherapy-induced apoptosis. Cancer Res 55, 2576–2582.

Dorr, J., Bechmann, I., Waiczies, S., Aktas, O., Walczak, H., Krammer, P. H., Nitsch, R., and Zipp, F. (2002). Lack of tumor necrosis factor-related apoptosis-inducing ligand but presence of its receptors in the human brain. J Neurosci 22, RC209.

El-Zawahry, A., McKillop, J., and Voelkel-Johnson, C. (2005). Doxorubicin increases the effec-tiveness of Apo2L/TRAIL for tumor growth inhibition of prostate cancer xenografts. BMC Cancer 5, 2.

Evdokiou, A., Bouralexis, S., Atkins, G. J., Chai, F., Hay, S., Clayer, M., and Findlay, D. M. (2002). Chemotherapeutic agents sensitize osteogenic sarcoma cells, but not normal human bone cells, to Apo2L/TRAIL-induced apoptosis. Int J Cancer 99, 491–504.

Fesik, S. W. (2005). Promoting apoptosis as a strategy for cancer drug discovery. Nat Rev Cancer 5, 876–885.

Fisher, M. J., Virmani, A. K., Wu, L., Aplenc, R., Harper, J. C., Powell, S. M., Rebbeck, T. R., Sidransky, D., Gazdar, A. F., and El-Deiry, W. S. (2001). Nucleotide substitution in the ecto-domain of trail receptor DR4 is associated with lung cancer and head and neck cancer. Clin Cancer Res 7, 1688–1697.

Foreman, K. E., Wrone-Smith, T., Boise, L. H., Thompson, C. B., Polverini, P. J., Simonian, P. L., Nunez, G., and Nickoloff, B. J. (1996). Kaposi's sarcoma tumor cells preferentially express Bcl-xL. Am J Pathol 149, 795–803.

Frank, S., Kohler, U., Schackert, G., and Schackert, H. K. (1999). Expression of TRAIL and its receptors in human brain tumors. Biochem Biophys Res Commun 257, 454–459.

Frese, S., Brunner, T., Gugger, M., Uduehi, A., and Schmid, R. A. (2002). Enhancement of Apo2L/TRAIL (tumor necrosis factor-related apoptosis-inducing ligand)-induced apoptosis in non-small cell lung cancer cell lines by chemotherapeutic agents without correlation to the expression level of cellular protease caspase-8 inhibitory protein. J Thorac Cardiovasc Surg 123, 168–174.

Fulda, S., Meyer, E., and Debatin, K. M. (2002a). Inhibition of TRAIL-induced apoptosis by Bcl-2 overexpression. Oncogene 21, 2283–2294.

Fulda, S., Wick, W., Weller, M., and Debatin, K. M. (2002b). Smac agonists sensitize for Apo2L/TRAIL- or anticancer drug-induced apoptosis and induce regression of malignant glioma in vivo. Nat Med 8, 808–815.

Galligan, L., Longley, D. B., McEwan, M., Wilson, T. R., McLaughlin, K., and Johnston, P. G. (2005). Chemotherapy and TRAIL-mediated colon cancer cell death: the roles of p53, TRAIL receptors, and c-FLIP. Mol Cancer Ther 4, 2026–2036.

Ganten, T. M., Koschny, R., Haas, T. L., Sykora, J., Li-Weber, M., Herzer, K., and Walczak, H. (2005). Proteasome inhibition sensitizes hepatocellular carcinoma cells, but not human hepatocytes, to TRAIL. Hepatology 42, 588–597.

Georgakis, G. V., Li, Y., Humphreys, R., Andreeff, M., O'Brien, S., Younes, M., Carbone, A., Albert, V., and Younes, A. (2005). Activity of selective fully human agonistic antibodies to the TRAIL death receptors TRAIL-R1 and TRAIL-R2 in primary and cultured lymphoma cells: induction of apoptosis and enhancement of doxorubicin- and bortezomib-induced cell death. Br J Haematol 130, 501–510.

Georgakis, G. V., Li, Y., Humphreys, R., Johnson, R., Andreeff, M., Fiscella, M., Pukac, L., O'Brien, S., Albert, V., and Younes, A. (2003). Activity of selective agonistic monoclonal antibodies to TRAIL death receptors R1 and R2 in primary and cultured tumor cells of hematological origin. Blood 102, 228.

Gillotte, D., Poortman, C., Zhang, L., Huang, X., Fiscella, M., Humphreys, R., and Johnson, R. L., (2004). Human agonistic anti-TRAIL receptor antibodies, HGS-ETR1 and HGS-ETR2, induce apoptosis in ovarian tumor lines and their activity is enhanced by taxol and carboplatin. Proceedings of the AACR 73: 3579.

Griffith, T. S., Chin, W. A., Jackson, G. C., Lynch, D. H., and Kubin, M. Z. (1998). Intracellular regulation of TRAIL-induced apoptosis in human melanoma cells. J Immunol 161, 2833–2840.

Griffith, T. S., Wiley, S. R., Kubin, M. Z., Sedger, L. M., Maliszewski, C. R., and Fanger, N. A. (1999). Monocyte-mediated tumoricidal activity via the tumor necrosis factor-related cytokine, TRAIL. J Exp Med 189, 1343–1354.

Guo, F., Sigua, C., Tao, J., Bali, P., George, P., Li, Y., Wittmann, S., Moscinski, L., Atadja, P., and Bhalla, K. (2004). Cotreatment with histone deacetylase inhibitor LAQ824 enhances Apo-2L/tumor necrosis factor-related apoptosis inducing ligand-induced death inducing signaling complex activity and apoptosis of human acute leukemia cells. Cancer Res 64, 2580–2589.

Hao, C., Song, J. H., Hsi, B., Lewis, J., Song, D. K., Petruk, K. C., Tyrrell, D. L., and Kneteman, N. M. (2004). TRAIL inhibits tumor growth but is nontoxic to human hepatocytes in chimeric mice. Cancer Res 64, 8502–8506.

Hasel, C., Durr, S., Rau, B., Strater, J., Schmid, R. M., Walczak, H., Bachem, M. G., and Moller, P. (2003). In chronic pancreatitis, widespread emergence of TRAIL receptors in epithelia coincides with neoexpression of TRAIL by pancreatic stellate cells of early fibrotic areas. Lab Invest 83, 825–836.

He, Q., Montalbano, J., Corcoran, C., Jin, W., Huang, Y., and Sheikh, M. S. (2003). Effect of Bax deficiency on death receptor 5 and mitochondrial pathways during endoplasmic reticulum calcium pool depletion-induced apoptosis. Oncogene 22, 2674–2679.

Houghton, J. A. (1999). Apoptosis and drug response. Curr Opin Oncol 11, 475–481.

Humphreys, R. C., Alderson, R. F., Bayever, E., Connolly, K., Choi, G. H., Fox, N. L., Gallant, G., Grzegorzewski, K. J., Roschke, V., Salcedo, T. W., et al. (2003). TRAIL R2-mAb, a human agonistic monoclonal antibody to tumor necrosis factor-related apoptosis inducing ligand receptor 2, affects tumor growth and induces apoptosis in human tumor xenograft models in vivo. 94th AACR Annual Meeting 44, 642.

Hymowitz, S. G., Christinger, H. W., Fuh, G., Ultsch, M., O'Connell, M., Kelley, R. F., Ashkenazi, A., and de Vos, A. M. (1999). Triggering cell death: the crystal structure of Apo2L/TRAIL in a complex with death receptor 5. Mol Cell 4, 563–571.

Hymowitz, S. G., O'Connell, M. P., Ultsch, M. H., Hurst, A., Totpal, K., Ashkenazi, A., de Vos, A. M., and Kelley, R. F. (2000). A unique zinc-binding site revealed by a high-resolution X-ray structure of homotrimeric Apo2L/TRAIL. Biochemistry 39, 633–640.

Ibrahim, S. M., Ringel, J., Schmidt, C., Ringel, B., Muller, P., Koczan, D., Thiesen, H. J., and Lohr, M. (2001). Pancreatic adenocarcinoma cell lines show variable susceptibility to TRAIL-mediated cell death. Pancreas 23, 72–79.

Ichikawa, K., Liu, W., Zhao, L., Wang, Z., Liu, D., Ohtsuka, T., Zhang, H., Mountz, J. D., Koopman, W. J., Kimberly, R. P., and Zhou, T. (2001). Tumoricidal activity of a novel anti-human DR5 monoclonal antibody without hepatocyte cytotoxicity. Nat Med 7, 954–960.

Igney, F. H. and Krammer, P. H. (2002). Death and anti-death: tumour resistance to apoptosis. Nat Rev Cancer 2, 277–288.

Ionov, Y., Yamamoto, H., Krajewski, S., Reed, J. C., and Perucho, M. (2000). Mutational inactivation of the proapoptotic gene BAX confers selective advantage during tumor clonal evolution. Proc Natl Acad Sci USA 97, 10872–10877.

Jazirehi, A. R., Ng, C. P., Gan, X. H., Schiller, G., and Bonavida, B. (2001). Adriamycin sensitizes the adriamycin-resistant 8226/Dox40 human multiple myeloma cells to Apo2L/tumor necrosis factor-related apoptosis-inducing ligand-mediated (TRAIL) apoptosis. Clin Cancer Res 7, 3874–3883.

Jeng, Y. M. and Hsu, H. C. (2002). Mutation of the DR5/TRAIL receptor 2 gene is infrequent in hepatocellular carcinoma. Cancer Lett 181, 205–208.

Jeon, K. I., Rih, J. K., Kim, H. J., Lee, Y. J., Cho, C. H., Goldberg, I. D., Rosen, E. M., and Bae, I. (2003). Pretreatment of indole-3-carbinol augments TRAIL-induced apoptosis in a prostate cancer cell line, LNCaP. FEBS Lett 544, 246–251.

Jin, T.-G., Kurakin, A., Benhaga, N., Abe, K., Mohseni, M., Sandra, F., Song, K., Kay, B. K., and Khosravi-Far, R. (2004). FADD-independent recruitment of c-FLIPL to death receptor 5. J Biol Chem 279, 55594–55601.

Jo, M., Kim, T. H., Seol, D. W., Esplen, J. E., Dorko, K., Billiar, T. R., and Strom, S. C. (2000). Apoptosis induced in normal human hepatocytes by tumor necrosis factor-related apoptosis-inducing ligand. Nat Med 6, 564–567.

Johnson, R. L., Huang, X., Fiscella, M., Cole, C., Pukac, L., Von Kerczek, A., Humphreys, R., Grzegorzewski, K. J., Gallant, G., and Albert, V. (2003). Human agonistic anti-TRAIL receptor antibodies, HGS-ETR1 and HGS-ETR2, induce apoptosis in diverse hematological tumor lines. Blood 102 (Abstract 3316).

Johnston, B., Kabore, A. F., Strutinsky, J., Hu, X., Paul, J. T., Kropp, D. M., Kuschak, B., Begleiter, A., and Gibson, S. B. (2003). Role of the TRAIL/APO2-L death receptors in chlorambucil- and fludarabine-induced apoptosis in chronic lymphocytic leukemia. Oncogene 22, 8356–8369.

Kaliberov, S., Stackhouse, M. A., Kaliberova, L., Zhou, T., and Buchsbaum, D. J. (2004). Enhanced apoptosis following treatment with TRA-8 anti-human DR5 monoclonal antibody and overexpression of exogenous Bax in human glioma cells. Gene Ther 11, 658–667.

Kandasamy, K., Srinivasula, S. M., Alnemri, E. S., Thompson, C. B., Korsmeyer, S. J., Bryant, J. L., and Srivastava, R. K. (2003). Involvement of proapoptotic molecules Bax and Bak in tumor necrosis factor-related apoptosis-inducing ligand (TRAIL)-induced mitochondrial disruption and apoptosis: differential regulation of cytochrome c and Smac/DIABLO release. Cancer Res 63, 1712–1721.

Kang, J., Bu, J., Hao, Y., and Chen, F. (2005). Subtoxic concentration of doxorubicin enhances TRAIL-induced apoptosis in human prostate cancer cell line LNCaP. Prostate Cancer Prostatic Dis 8, 274–279.

Kaufmann, S. H., Karp, J. E., Svingen, P. A., Krajewski, S., Burke, P. J., Gore, S. D., and Reed, J. C. (1998). Elevated expression of the apoptotic regulator Mcl-1 at the time of leukemic relapse. Blood 91, 991–1000.

Kayagaki, N., Yamaguchi, N., Nakayama, M., Kawasaki, A., Akiba, H., Okumura, K., and Yagita, H. (1999a). Involvement of TNF-related apoptosis-inducing ligand in human CD4 + T cell-mediated cytotoxicity. J Immunol 162, 2639–2647.

Kayagaki, N., Yamaguchi, N., Nakayama, M., Takeda, K., Akiba, H., Tsutsui, H., Okamura, H., Nakanishi, K., Okumura, K., and Yagita, H. (1999b). Expression and function of TNF-related apoptosis-inducing ligand on murine activated NK cells. J Immunol 163, 1906–1913.

Keane, M. M., Ettenberg, S. A., Nau, M. M., Russell, E. K., and Lipkowitz, S. (1999). Chemotherapy augments TRAIL-induced apoptosis in breast cell lines. Cancer Res 59, 734–741.

Kelley, S. K., Harris, L. A., Xie, D., Deforge, L., Totpal, K., Bussiere, J., and Fox, J. A. (2001). Preclinical studies to predict the disposition of Apo2L/tumor necrosis factor-related apoptosis-inducing ligand in humans: characterization of in vivo efficacy, pharmacokinetics, and safety. J Pharmacol Exp Ther 299, 31–38.

Kelly, M. M., Hoel, B. D., and Voelkel-Johnson, C. (2002). Doxorubicin pretreatment sensitizes prostate cancer cell lines to TRAIL induced apoptosis which correlates with the loss of c-FLIP expression. Cancer Biol Ther 1, 520–527.

Kelly, W. K. and Marks, P. A. (2005). Drug insight: histone deacetylase inhibitors – development of the new targeted anticancer agent suberoylanilide hydroxamic acid. Nat Clin Pract Oncol 2, 150–157.

Kelly, W. K., O'Connor, O. A., Krug, L. M., Chiao, J. H., Heaney, M., Curley, T., MacGregore-Cortelli, B., Tong, W., Secrist, J. P., Schwartz, L., et al. (2005). Phase I study of an oral histone deacetylase inhibitor, suberoylanilide hydroxamic acid, in patients with advanced cancer. J Clin Oncol 23, 3923–3931.

Kim, E. H., Kim, S. U., Shin, D. Y., and Choi, K. S. (2004). Roscovitine sensitizes glioma cells to TRAIL-mediated apoptosis by downregulation of survivin and XIAP. Oncogene 23, 446–456.

Kim, E. H., Kim, S. U., and Choi, K. S. (2005). Rottlerin sensitizes glioma cells to TRAIL-induced apoptosis by inhibition of Cdc2 and the subsequent downregulation of survivin and XIAP. Oncogene 24, 838–849.

Kim, K., Fisher, M. J., Xu, S. Q., and el-Deiry, W. S. (2000). Molecular determinants of response to TRAIL in killing of normal and cancer cells. Clin Cancer Res 6, 335–346.

Kischkel, F. C., Lawrence, D. A., Chuntharapai, A., Schow, P., Kim, K. J., and Ashkenazi, A. (2000). Apo2L/TRAIL-dependent recruitment of endogenous FADD and caspase-8 to death receptors 4 and 5. Immunity 12, 611–620.

Kondo, K., Yamasaki, S., Sugie, T., Teratani, N., Kan, T., Imamura, M., and Shimada, Y. (2006). Cisplatin-dependent upregulation of death receptors 4 and 5 augments induction of apoptosis by TNF-related apoptosis-inducing ligand against esophageal squamous cell carcinoma. Int J Cancer 118, 230–242.

Koornstra, J. J., Kleibeuker, J. H., van Geelen, C. M., Rijcken, F. E., Hollema, H., de Vries, E. G., and de Jong, S. (2003). Expression of TRAIL (TNF-related apoptosis-inducing ligand) and its receptors in normal colonic mucosa, adenomas, and carcinomas. J Pathol 200, 327–335.

Kuang, A. A., Diehl, G. E., Zhang, J., and Winoto, A. (2000). FADD is required for DR4- and DR5-mediated apoptosis: lack of trail-induced apoptosis in FADD-deficient mouse embryonic fibroblasts. J Biol Chem 275, 25065–25068.

Kuraoka, K., Matsumura, S., Sanada, Y., Nakachi, K., Imai, K., Eguchi, H., Matsusaki, K., Oue, N., Nakayama, H., and Yasui, W. (2005). A single nucleotide polymorphism in the extracellular domain of TRAIL receptor DR4 at nucleotide 626 in gastric cancer patients in Japan. Oncol Rep 14, 465–470.

Lashinger, L. M., Zhu, K., Williams, S. A., Shrader, M., Dinney, C. P., and McConkey, D. J. (2005). Bortezomib abolishes tumor necrosis factor-related apoptosis-inducing ligand resistance via a p21-dependent mechanism in human bladder and prostate cancer cells. Cancer Res 65, 4902–4908.

Lawrence, D., Shahrokh, Z., Marsters, S., Achilles, K., Shih, D., Mounho, B., Hillan, K., Totpal, K., DeForge, L., Schow, P., et al. (2001). Differential hepatocyte toxicity of recombinant Apo2L/TRAIL versions. Nat Med 7, 383–385.

Leaman, D. W., Chawla-Sarkar, M., Vyas, K., Reheman, M., Tamai, K., Toji, S., and Borden, E. C. (2002). Identification of X-linked inhibitor of apoptosis-associated factor-1 as an interferon-stimulated gene that augments TRAIL Apo2L-induced apoptosis. J Biol Chem 277, 28504–28511.

LeBlanc, H., Lawrence, D., Varfolomeev, E., Totpal, K., Morlan, J., Schow, P., Fong, S., Schwall, R., Sinicropi, D., and Ashkenazi, A. (2002). Tumor-cell resistance to death receptor–induced apoptosis through mutational inactivation of the proapoptotic Bcl-2 homolog Bax. Nat Med 8, 274–281.

LeBlanc, H. N. and Ashkenazi, A. (2003). Apo2L/TRAIL and its death and decoy receptors. Cell Death Differ 10, 66–75.

Lee, S. H., Shin, M. S., Kim, H. S., Lee, H. K., Park, W. S., Kim, S. Y., Lee, J. H., Han, S. Y., Park, J. Y., Oh, R. R., et al. (1999). Alterations of the DR5/TRAIL receptor 2 gene in non-small cell lung cancers. Cancer Res 59, 5683–5686.

Leverkus, M., Neumann, M., Mengling, T., Rauch, C. T., Brocker, E. B., Krammer, P. H., and Walczak, H. (2000a). Regulation of tumor necrosis factor-related apoptosis-inducing ligand sensitivity in primary and transformed human keratinocytes. Cancer Res 60, 553–559.

Leverkus, M., Walczak, H., McLellan, A., Fries, H. W., Terbeck, G., Brocker, E. B., and Kampgen, E. (2000b). Maturation of dendritic cells leads to up-regulation of cellular FLICE-inhibitory protein and concomitant down-regulation of death ligand-mediated apoptosis. Blood 96, 2628–2631.

Li, L., Thomas, R. M., Suzuki, H., De Brabander, J. K., Wang, X., and Harran, P. G. (2004). A small molecule Smac mimic potentiates TRAIL- and TNFalpha-mediated cell death. Science 305, 1471–1474.

Liu, W., Bodle, E., Chen, J. Y., Gao, M., Rosen, G. D., and Broaddus, V. C. (2001). Tumor necrosis factor-related apoptosis-inducing ligand and chemotherapy cooperate to induce apoptosis in mesothelioma cell lines. Am J Respir Cell Mol Biol 25, 111–118.

Liu, X. S., Zhu, Y., Han, W. N., Li, Y. N., Chen, L. H., Jia, W., Song, C. J., Liu, F., Yang, K., Li, Q., and Jin, B. Q. (2003). Preparation and characterization of a set of monoclonal antibodies to TRAIL and TRAIL receptors DR4, DR5, DcR1, and DcR2. Hybrid Hybridomics 22, 121–125.

Locksley, R. M., Killeen, N., and Lenardo, M. J. (2001). The TNF and TNF receptor super-families: integrating mammalian biology. Cell 104, 487–501.

Ludwig, D. L., Pereira, D. S., Zhu, Z., Hicklin, D. J., and Bohlen, P. (2003). Monoclonal antibody therapeutics and apoptosis. Oncogene 22, 9097–9106.

Luo, X., Budihardjo, I., Zou, H., Slaughter, C., and Wang, X. (1998). Bid, a Bcl2 interacting protein, mediates cytochrome c release from mitochondria in response to activation of cell surface death receptors. Cell 94, 481–490.

Luschen, S., Ussat, S., Scherer, G., Kabelitz, D., and Adam-Klages, S. (2000). Sensitization to death receptor cytotoxicity by inhibition of fas-associated death domain protein (FADD)/caspase signaling. Requirement of cell cycle progression. J Biol Chem 275, 24670–24678.

Marcucci, G., Stock, W., Dai, G., Klisovic, R. B., Liu, S., Klisovic, M. I., Blum, W., Kefauver, C., Sher, D. A., Green, M., et al. (2005). Phase I study of oblimersen sodium, an antisense to Bcl-2, in untreated older patients with acute myeloid leukemia: pharmacokinetics, pharmacodynamics, and clinical activity. J Clin Oncol 23, 3404–3411.

Mariani, S. M. and Krammer, P. H. (1998). Surface expression of TRAIL/Apo-2 ligand in activated mouse T and B cells. Eur J Immunol 28, 1492–1498.

Marks, P. A., Richon, V. M., Kelly, W. K., Chiao, J. H., and Miller, T. (2004). Histone deacetylase inhibitors: development as cancer therapy. Novartis Found Symp 259, 269–281; discussion 281–268.

Marsters, S. A., Frutkin, A. D., Simpson, N. J., Fendly, B. M., and Ashkenazi, A. (1992). Identification of cysteine-rich domains of the type 1 tumor necrosis factor receptor involved in ligand binding. J Biol Chem 267, 5747–5750.

Matta, H. and Chaudhary, P. M. (2005). The proteasome inhibitor bortezomib (PS-341) inhibits growth and induces apoptosis in primary effusion lymphoma cells. Cancer Biol Ther 4, 77–82.

McCarthy, M. M., Sznol, M., DiVito, K. A., Camp, R. L., Rimm, D. L., and Kluger, H. M. (2005). Evaluating the expression and prognostic value of TRAIL-R1 and TRAIL-R2 in breast cancer. Clin Cancer Res 11, 5188–5194.

McDonald, E. R., III, Chui, P. C., Martelli, P. F., Dicker, D. T., and El-Deiry, W. S. (2001). Death domain mutagenesis of KILLER/DR5 reveals residues critical for apoptotic signaling. J Biol Chem 276, 14939–14945.

McManus, D. C., Lefebvre, C. A., Cherton-Horvat, G., St-Jean, M., Kandimalla, E. R., Agrawal, S., Morris, S. J., Durkin, J. P., and Lacasse, E. C. (2004). Loss of XIAP protein expression by RNAi and antisense approaches sensitizes cancer cells to functionally diverse chemotherapeutics. Oncogene 23, 8105–8117.

Meng, R. D., McDonald, E. R., III, Sheikh, M. S., Fornace, A. J., Jr., and El-Deiry, W. S. (2000). The TRAIL decoy receptor TRUNDD (DcR2, TRAIL-R4) is induced by adenovirus-p53 overexpression and can delay TRAIL-, p53-, and KILLER/DR5-dependent colon cancer apoptosis. Mol Ther 1, 130–144.

Miao, L., Yi, P., Wang, Y., and Wu, M. (2003). Etoposide upregulates Bax-enhancing tumour necrosis factor-related apoptosis inducing ligand-mediated apoptosis in the human hepatocellular carcinoma cell line QGY-7703. Eur J Biochem 270, 2721–2731.

Miller, K., Meng, G., Liu, J., Hurst, A., Hsei, V., Wong, W. L., Ekert, R., Lawrence, D., Sherwood, S., DeForge, L., et al. (2003). Design, construction, and in vitro analyses of multivalent antibodies. J Immunol 170, 4854–4861.

Mirandola, P., Gobbi, G., Ponti, C., Sponzilli, I., Cocco, L., and Vitale, M. (2006a). PKC{epsilon} controls the protection against TRAIL in erythroid progenitors. Blood 107, 508–513.

Mirandola, P., Sponzilli, I., Gobbi, G., Marmiroli, S., Rinaldi, L., Binazzi, R., Piccari, G. G., Ramazzotti, G., Gaboardi, G. C., Cocco, L., and Vitale, M. (2006b). Anticancer agents sensitize osteosarcoma cells to TNF-related apoptosis-inducing ligand downmodulating IAP family proteins. Int J Oncol 28, 127–133.

Mitsiades, N., Poulaki, V., Tseleni-Balafouta, S., Koutras, D. A., and Stamenkovic, I. (2000). Thyroid carcinoma cells are resistant to FAS-mediated apoptosis but sensitive tumor necrosis factor-related apoptosis-inducing ligand. Cancer Res 60, 4122–4129.

Mitsiades, C. S., Treon, S. P., Mitsiades, N., Shima, Y., Richardson, P., Schlossman, R., Hideshima, T., and Anderson, K. C. (2001a). TRAIL/Apo2L ligand selectively induces apoptosis and overcomes drug resistance in multiple myeloma: therapeutic applications. Blood 98, 795–804.

Mitsiades, N., Mitsiades, C. S., Poulaki, V., Anderson, K. C., and Treon, S. P. (2001b). Concepts in the use of TRAIL/Apo2L: an emerging biotherapy for myeloma and other neoplasias. Expert Opin Investig Drugs 10, 1521–1530.

Mongkolsapaya, J., Grimes, J. M., Chen, N., Xu, X. N., Stuart, D. I., Jones, E. Y., and Screaton, G. R. (1999). Structure of the TRAIL-DR5 complex reveals mechanisms conferring specificity in apoptotic initiation. Nat Struct Biol 6, 1048–1053.

Motoki, K., Mori, E., Matsumoto, A., Thomas, M., Tomura, T., Humphreys, R., Albert, V., Muto, M., Yoshida, H., Aoki, M., et al. (2005). Enhanced apoptosis and tumor regression induced by a direct agonist antibody to tumor necrosis factor-related apoptosis-inducing ligand receptor 2. Clin Cancer Res 11, 3126–3135.

Muhlenbeck, F., Haas, E., Schwenzer, R., Schubert, G., Grell, M., Smith, C., Scheurich, P., and Wajant, H. (1998). TRAIL/Apo2L activates c-Jun NH2-terminal kinase (JNK) via caspase-dependent and caspase-independent pathways. J Biol Chem 273, 33091–33098.

Muhlenbeck, F., Schneider, P., Bodmer, J. L., Schwenzer, R., Hauser, A., Schubert, G., Scheurich, P., Moosmayer, D., Tschopp, J., and Wajant, H. (2000). The tumor necrosis factor-related apoptosis-inducing ligand receptors TRAIL-R1 and TRAIL-R2 have distinct cross-linking requirements for initiation of apoptosis and are non-redundant in JNK activation. J Biol Chem 275, 32208–32213.

Muhlethaler-Mottet, A., Bourloud, K. B., Auderset, K., Joseph, J. M., and Gross, N. (2004). Drug-mediated sensitization to TRAIL-induced apoptosis in caspase-8-complemented neuroblastoma cells proceeds via activation of intrinsic and extrinsic pathways and caspase-dependent cleavage of XIAP, Bcl-x(L) and RIP. Oncogene 23(32), 5415–5425.

Mundt, B., Kuhnel, F., Zender, L., Paul, Y., Tillmann, H., Trautwein, C., Manns, M. P., and Kubicka, S. (2003). Involvement of TRAIL and its receptors in viral hepatitis. FASEB J 17, 94–96.

Muzio, M., Stockwell, B. R., Stennicke, H. R., Salvesen, G. S., and Dixit, V. M. (1998). An induced proximity model for caspase-8 activation. J Biol Chem 273, 2926–2930.

Nagane, M., Levitzki, A., Gazit, A., Cavenee, W. K., and Huang, H. J. (1998). Drug resistance of human glioblastoma cells conferred by a tumor-specific mutant epidermal growth factor receptor through modulation of Bcl-XL and caspase-3-like proteases. Proc Natl Acad Sci USA 95, 5724–5729.

Nagane, M., Pan, G., Weddle, J. J., Dixit, V. M., Cavenee, W. K., and Huang, H. J. (2000). Increased death receptor 5 expression by chemotherapeutic agents in human gliomas causes synergistic cytotoxicity with tumor necrosis factor-related apoptosis-inducing ligand in vitro and in vivo. Cancer Res 60, 847–853.

Nagane, M., Huang, H. J., and Cavenee, W. K. (2001). The potential of TRAIL for cancer chemotherapy. Apoptosis 6, 191–197.

Naka, T., Sugamura, K., Hylander, B. L., Widmer, M. B., Rustum, Y. M., and Repasky, E. A. (2002). Effects of tumor necrosis factor-related apoptosis-inducing ligand alone and in combination with chemotherapeutic agents on patients' colon tumors grown in SCID mice. Cancer Res 62, 5800–5806.

Nencioni, A., Wille, L., Dal Bello, G., Boy, D., Cirmena, G., Wesselborg, S., Belka, C., Brossart, P., Patrone, F., and Ballestrero, A. (2005). Cooperative cytotoxicity of proteasome inhibitors and tumor necrosis factor-related apoptosis-inducing ligand in chemoresistant Bcl-2-overexpressing cells. Clin Cancer Res 11, 4259–4265.

Nesterov, A., Lu, X., Johnson, M., Miller, G. J., Ivashchenko, Y., and Kraft, A. S. (2001). Elevated AKT activity protects the prostate cancer cell line LNCaP from TRAIL-induced apoptosis. J Biol Chem 276, 10767–10774.

Ng, C. P. and Bonavida, B. (2002). X-linked inhibitor of apoptosis (XIAP) blocks Apo2 ligand/tumor necrosis factor-related apoptosis-inducing ligand-mediated apoptosis of prostate cancer cells in the presence of mitochondrial activation: sensitization by overexpression of second mitochondria-derived activator of caspase/direct IAP-binding protein with low pI (Smac/DIABLO). Mol Cancer Ther 1, 1051–1058.

Nieda, M., Nicol, A., Koezuka, Y., Kikuchi, A., Lapteva, N., Tanaka, Y., Tokunaga, K., Suzuki, K., Kayagaki, N., Yagita, H., et al. (2001). TRAIL expression by activated human CD4(+)V alpha 24NKT cells induces in vitro and in vivo apoptosis of human acute myeloid leukemia cells. Blood 97, 2067–2074.

Nikrad, M., Johnson, T., Puthalalath, H., Coultas, L., Adams, J., and Kraft, A. S. (2005). The proteasome inhibitor bortezomib sensitizes cells to killing by death receptor ligand TRAIL via BH3-only proteins Bik and Bim. Mol Cancer Ther 4, 443–449.

O'Brien, S. M., Cunningham, C. C., Golenkov, A. K., Turkina, A. G., Novick, S. C., and Rai, K. R. (2005). Phase I to II multicenter study of oblimersen sodium, a Bcl-2 antisense oligonucleotide, in patients with advanced chronic lymphocytic leukemia. J Clin Oncol 23, 7697–7702.

Odoux, C., Albers, A., Amoscato, A. A., Lotze, M. T., and Wong, M. K. (2002). TRAIL, FasL and a blocking anti-DR5 antibody augment paclitaxel-induced apoptosis in human non-small-cell lung cancer. Int J Cancer 97, 458–465.

Ohtsuka, T., Buchsbaum, D., Oliver, P., Makhija, S., Kimberly, R., and Zhou, T. (2003). Synergistic induction of tumor cell apoptosis by death receptor antibody and chemotherapy agent through JNK/p38 and mitochondrial death pathway. Oncogene 22, 2034–2044.

Ou, D., Wang, X., Metzger, D. L., James, R. F., Pozzilli, P., Plesner, A., Korneluk, R. G., Verchere, C. B., and Tingle, A. J. (2005). Synergistic inhibition of tumor necrosis factor-related apoptosis-inducing ligand-induced apoptosis in human pancreatic beta cells by Bcl-2 and X-linked inhibitor of apoptosis. Hum Immunol 66, 274–284.

Ozoren, N., Fisher, M. J., Kim, K., Liu, C. X., Genin, A., Shifman, Y., Dicker, D. T., Spinner, N. B., Lisitsyn, N. A., and El-Deiry, W. S. (2000). Homozygous deletion of the death receptor DR4 gene in a nasopharyngeal cancer cell line is associated with TRAIL resistance. Int J Oncol 16, 917–925.

Pai, S. I., Wu, G. S., Ozoren, N., Wu, L., Jen, J., Sidransky, D., and El-Deiry, W. S. (1998). Rare loss-of-function mutation of a death receptor gene in head and neck cancer. Cancer Res 58, 3513–3518.

Pan, G., O'Rourke, K., and Dixit, V. M. (1998). Caspase-9, Bcl-XL, and Apaf-1 form a ternary complex. J Biol Chem 273, 5841–5845.

Papageorgiou, A., Lashinger, L., Millikan, R., Grossman, H. B., Benedict, W., Dinney, C. P., and McConkey, D. J. (2004). Role of tumor necrosis factor-related apoptosis-inducing ligand in interferon-induced apoptosis in human bladder cancer cells. Cancer Res 64, 8973–8979.

Pei, Z., Chu, L., Zou, W., Zhang, Z., Qiu, S., Qi, R., Gu, J., Qian, C., and Liu, X. (2004). An onco-lytic adenoviral vector of Smac increases antitumor activity of TRAIL against HCC in human cells and in mice. Hepatology 39, 1371–1381.

Pitti, R. M., Marsters, S. A., Ruppert, S., Donahue, C. J., Moore, A., and Ashkenazi, A. (1996). Induction of apoptosis by Apo-2 ligand, a new member of the tumor necrosis factor cytokine family. J Biol Chem 271, 12687–12690.

Poulaki, V., Mitsiades, C. S., McMullan, C., Fanourakis, G., Negri, J., Goudopoulou, A., Halikias, I. X., Voutsinas, G., Tseleni-Balafouta, S., Miller, J. W., and Mitsiades, N. (2005). Human retinoblastoma cells are resistant to apoptosis induced by death receptors: role of caspase-8 gene silencing. Invest Ophthalmol Vis Sci 46, 358–366.

Pukac, L., Kanakaraj, P., Humphreys, R., Alderson, R., Bloom, M., Sung, C., Riccobene, T., Johnson, R., Fiscella, M., Mahoney, A., et al. (2005). HGS-ETR1, a fully human TRAIL-receptor 1 monoclonal antibody, induces cell death in multiple tumour types in vitro and in vivo. Br J Cancer 92, 1430–1441.

Rampino, N., Yamamoto, H., Ionov, Y., Li, Y., Sawai, H., Reed, J. C., and Perucho, M. (1997). Somatic frameshift mutations in the BAX gene in colon cancers of the microsatellite mutator phenotype. Science 275, 967–969.

Reesink-Peters, N., Hougardy, B. M., van den Heuvel, F. A., Ten Hoor, K. A., Hollema, H., Boezen, H. M., de Vries, E. G., de Jong, S., and van der Zee, A. G. (2005). Death receptors and ligands in cervical carcinogenesis: an immunohistochemical study. Gynecol Oncol 96, 705–713.

Ricci, M. S., Jin, Z., Dews, M., Yu, D., Thomas-Tikhonenko, A., Dicker, D. T., and El-Deiry, W. S. (2004). Direct repression of FLIP expression by c-myc is a major determinant of TRAIL sensitivity. Mol Cell Biol 24, 8541–8555.

Roa, W. H., Chen, H., Fulton, D., Gulavita, S., Shaw, A., Th'ng, J., Farr-Jones, M., Moore, R., and Petruk, K. (2003). X-linked inhibitor regulating TRAIL-induced apoptosis in chemoresist-ant human primary glioblastoma cells. Clin Invest Med 26, 231–242.

Roach, C., Sharifi, A., Askaa, J., Welcher, R., Chenoweth, D., Lincoln, C., Sosnovtseva, S., Zhao, Q., Johnson, R., Carrell, J., et al. (2004). Development of sensitive and specific immunohisto-chemical assays for pro-apoptotic TRAIL receptors. Paper presented at 95th AACR Meeting, Orlando, FL.

Roth, W. and Reed, J. C. (2004). FLIP protein and TRAIL-induced apoptosis. Vitam Horm 67, 189–206.

Salvesen, G. S. and Dixit, V. M. (1999). Caspase activation: the induced-proximity model. Proc Natl Acad Sci USA 96, 10964–10967.

Sayers, T. J. and Murphy, W. J. (2006). Combining proteasome inhibition with TNF-related apoptosis-inducing ligand (Apo2L/TRAIL) for cancer therapy. Cancer Immunol Immunother 55, 76–84.

Schneider, P. and Tschopp, J. (2000). Apoptosis induced by death receptors. Pharm Acta Helv 74, 281–286.

Schneider, P., Thome, M., Burns, K., Bodmer, J. L., Hofmann, K., Kataoka, T., Holler, N., and Tschopp, J. (1997). TRAIL receptors 1 (DR4) and 2 (DR5) signal FADD-dependent apoptosis and activate NF-kappaB. Immunity 7, 831–836.

Secchiero, P., Gonelli, A., Mirandola, P., Melloni, E., Zamai, L., Celeghini, C., Milani, D., and Zauli, G. (2002). Tumor necrosis factor-related apoptosis-inducing ligand induces monocytic maturation of leukemic and normal myeloid precursors through a caspase-dependent pathway. Blood 100, 2421–2429.

Sheikh, M. S. and Fornace, A. J., Jr. (2000). Death and decoy receptors and p53-mediated apop-tosis. Leukemia 14, 1509–1513.

Shin, E. C., Ahn, J. M., Kim, C. H., Choi, Y., Ahn, Y. S., Kim, H., Kim, S. J., and Park, J. H. (2001). IFN-gamma induces cell death in human hepatoma cells through a TRAIL/death receptor-mediated apoptotic pathway. Int J Cancer 93, 262–268.

Singh, T. R., Shankar, S., Chen, X., Asim, M., and Srivastava, R. K. (2003). Synergistic interac-tions of chemotherapeutic drugs and tumor necrosis factor-related apoptosis-inducing ligand/

Apo-2 ligand on apoptosis and on regression of breast carcinoma in vivo. Cancer Res 63, 5390–5400.

Sinicrope, F. A., Penington, R. C., and Tang, X. M. (2004). Tumor necrosis factor-related apoptosis-inducing ligand-induced apoptosis is inhibited by Bcl-2 but restored by the small molecule Bcl-2 inhibitor, HA 14–1, in human colon cancer cells. Clin Cancer Res 10, 8284–8292.

Soengas, M. S., Capodieci, P., Polsky, D., Mora, J., Esteller, M., Opitz-Araya, X., McCombie, R., Herman, J. G., Gerald, W. L., Lazebnik, Y. A., et al. (2001). Inactivation of the apoptosis effector Apaf-1 in malignant melanoma. Nature 409, 207–211.

Soengas, M. S., Gerald, W. L., Cordon-Cardo, C., Lazebnik, Y., and Lowe, S. W. (2006). Apaf-1 expression in malignant melanoma. Cell Death Differ 13, 352–353.

Song, J. H., Song, D. K., Herlyn, M., Petruk, K. C., and Hao, C. (2003a). Cisplatin down-regulation of cellular Fas-associated death domain-like interleukin-1beta-converting enzyme-like inhibitory proteins to restore tumor necrosis factor-related apoptosis-inducing ligand-induced apoptosis in human melanoma cells. Clin Cancer Res 9, 4255–4266.

Song, J. H., Song, D. K., Pyrzynska, B., Petruk, K. C., Van Meir, E. G., and Hao, C. (2003b). TRAIL triggers apoptosis in human malignant glioma cells through extrinsic and intrinsic pathways. Brain Pathol 13, 539–553.

Spierings, D. C., de Vries, E. G., Timens, W., Groen, H. J., Boezen, H. M., and de Jong, S. (2003). Expression of TRAIL and TRAIL death receptors in stage III non-small cell lung cancer tumors. Clin Cancer Res 9, 3397–3405.

Spierings, D. C., de Vries, E. G., Vellenga, E., van den Heuvel, F. A., Koornstra, J. J., Wesseling, J., Hollema, H., and de Jong, S. (2004). Tissue distribution of the death ligand TRAIL and its receptors. J Histochem Cytochem 52, 821–831.

Srivastava, R.K. (2001). TRAIL/Apo-2L: mechanisms and clinical applications in cancer. Neoplasia 3, 535–546.

Strater, J., Hinz, U., Walczak, H., Mechtersheimer, G., Koretz, K., Herfarth, C., Moller, P., and Lehnert, T. (2002a). Expression of TRAIL and TRAIL receptors in colon carcinoma: TRAIL-R1 is an independent prognostic parameter. Clin Cancer Res 8, 3734–3740.

Strater, J., Walczak, H., Pukrop, T., Von Muller, L., Hasel, C., Kornmann, M., Mertens, T., and Moller, P. (2002b). TRAIL and its receptors in the colonic epithelium: a putative role in the defense of viral infections. Gastroenterology 122, 659–666.

Takeda, K., Smyth, M. J., Cretney, E., Hayakawa, Y., Kayagaki, N., Yagita, H., and Okumura, K. (2002). Critical role for tumor necrosis factor-related apoptosis-inducing ligand in immune surveillance against tumor development. J Exp Med 195, 161–169.

Taniai, M., Grambihler, A., Higuchi, H., Werneburg, N., Bronk, S. F., Farrugia, D. J., Kaufmann, S. H., and Gores, G. J. (2004). Mcl-1 mediates tumor necrosis factor-related apoptosis-inducing ligand resistance in human cholangiocarcinoma cells. Cancer Res 64, 3517–3524.

Thomas, L. R., Henson, A., Reed, J. C., Salsbury, F. R., and Thorburn, A. (2004a). Direct binding of Fas-associated death domain (FADD) to the tumor necrosis factor-related apoptosis-inducing ligand receptor DR5 is regulated by the death effector domain of FADD. J Biol Chem 279, 32780–32785.

Thomas, L. R., Johnson, R. L., Reed, J. C., and Thorburn, A. (2004b). The C-terminal tails of tumor necrosis factor-related apoptosis-inducing ligand (TRAIL) and Fas receptors have opposing functions in Fas-associated death domain (FADD) recruitment and can regulate agonist-specific mechanisms of receptor activation. J Biol Chem 279, 52479–52486.

Tolcher, A. W., Chi, K., Kuhn, J., Gleave, M., Patnaik, A., Takimoto, C., Schwartz, G., Thompson, I., Berg, K., D'Aloisio, S., et al. (2005). A phase II, pharmacokinetic, and biological correlative study of oblimersen sodium and docetaxel in patients with hormone-refractory prostate cancer. Clin Cancer Res 11, 3854–3861.

van Geelen, C. M., de Vries, E. G., Le, T. K., van Weeghel, R. P., and de Jong, S. (2003). Differential modulation of the TRAIL receptors and the CD95 receptor in colon carcinoma cell lines. Br J Cancer 89, 363–373.

Van Geelen, C. M., de Vries, E. G., and de Jong, S. (2004). Lessons from TRAIL-resistance mechanisms in colorectal cancer cells: paving the road to patient-tailored therapy. Drug Resist Updat 7, 345–358.

van Noesel, M. M., van Bezouw, S., Voute, P. A., Herman, J. G., Pieters, R., and Versteeg, R. (2003). Clustering of hypermethylated genes in neuroblastoma. Genes Chromosomes Cancer 38, 226–233.

Vignati, S., Codegoni, A., Polato, F., and Broggini, M. (2002). Trail activity in human ovarian cancer cells: potentiation of the action of cytotoxic drugs. Eur J Cancer 38, 177–183.

Voelkel-Johnson, C. (2003). An antibody against DR4 (TRAIL-R1) in combination with doxorubicin selectively kills malignant but not normal prostate cells. Cancer Biol Ther 2, 283–290.

Vucic, D., Stennicke, H. R., Pisabarro, M. T., Salvesen, G. S., and Dixit, V. M. (2000). ML-IAP, a novel inhibitor of apoptosis that is preferentially expressed in human melanomas. Curr Biol 10, 1359–1366.

Wachter, T., Sprick, M., Hausmann, D., Kerstan, A., McPherson, K., Stassi, G., Brocker, E. B., Walczak, H., and Leverkus, M. (2004). cFLIPL inhibits tumor necrosis factor-related apoptosis-inducing ligand-mediated NF-kappaB activation at the death-inducing signaling complex in human keratinocytes. J Biol Chem 279, 52824–52834.

Walczak, H. and Sprick, M. R. (2001). Biochemistry and function of the DISC. Trends Biochem Sci 26, 452–453.

Walczak, H., Miller, R. E., Ariail, K., Gliniak, B., Griffith, T. S., Kubin, M., Chin, W., Jones, J., Woodward, A., Le, T., et al. (1999). Tumoricidal activity of tumor necrosis factor-related apoptosis-inducing ligand in vivo. Nat Med 5, 157–163.

Wang, S. and El-Deiry, W. S. (2003). TRAIL and apoptosis induction by TNF-family death receptors. Oncogene 22, 8628–8633.

Wang, Z., Sampath, J., Fukuda, S., and Pelus, L. M. (2005). Disruption of the inhibitor of apoptosis protein survivin sensitizes Bcr-abl-positive cells to STI571-induced apoptosis. Cancer Res 65, 8224–8232.

Wei, M. C., Lindsten, T., Mootha, V. K., Weiler, S., Gross, A., Ashiya, M., Thompson, C. B., and Korsmeyer, S. J. (2000). tBID, a membrane-targeted death ligand, oligomerizes BAK to release cytochrome c. Genes Dev 14, 2060–2071.

Wendt, J., von Haefen, C., Hemmati, P., Belka, C., Dorken, B., and Daniel, P. T. (2005). TRAIL sensitizes for ionizing irradiation-induced apoptosis through an entirely Bax-dependent mitochondrial cell death pathway. Oncogene 24, 4052–4064.

Whang, Y. E., Yuan, X. J., Liu, Y., Majumder, S., and Lewis, T. D. (2004). Regulation of sensitivity to TRAIL by the PTEN tumor suppressor. Vitam Horm 67, 409–426.

Wolf, S., Mertens, D., Pscherer, A., Schroeter, P., Winkler, D., Grone, H. J., Hofele, C., Hemminki, K., Kumar, R., Steineck, G., et al. (2006). Ala228 variant of trail receptor 1 affecting the ligand binding site is associated with chronic lymphocytic leukemia, mantle cell lymphoma, prostate cancer, head and neck squamous cell carcinoma and bladder cancer. Int J Cancer 118, 1831–1835.

Wu, W. G., Soria, J. C., Wang, L., Kemp, B. L., and Mao, L. (2000). TRAIL-R2 is not correlated with p53 status and is rarely mutated in non-small cell lung cancer. Anticancer Res 20, 4525–4529.

Xiao, C., Yang, B. F., Song, J. H., Schulman, H., Li, L., and Hao, C. (2005). Inhibition of CaMKII-mediated c-FLIP expression sensitizes malignant melanoma cells to TRAIL-induced apoptosis. Exp Cell Res 304, 244–255.

Yamaguchi, K., Uzzo, R. G., Pimkina, J., Makhov, P., Golovine, K., Crispen, P., and Kolenko, V. M. (2005a). Methylseleninic acid sensitizes prostate cancer cells to TRAIL-mediated apoptosis. Oncogene 24, 5868–5877.

Yamaguchi, Y., Shiraki, K., Fuke, H., Inoue, T., Miyashita, K., Yamanaka, Y., Saitou, Y., Sugimoto, K., and Nakano, T. (2005b). Targeting of X-linked inhibitor of apoptosis protein or survivin by short interfering RNAs sensitize hepatoma cells to TNF-related apoptosis-inducing ligand- and chemotherapeutic agent-induced cell death. Oncol Rep 14, 1311–1316.

Yeh, W. C., Pompa, J. L., McCurrach, M. E., Shu, H. B., Elia, A. J., Shahinian, A., Ng, M., Wakeham, A., Khoo, W., Mitchell, K., et al. (1998). FADD: essential for embryo development and signaling from some, but not all, inducers of apoptosis. Science 279, 1954–1958.

Yoshida, H., Kong, Y. Y., Yoshida, R., Elia, A. J., Hakem, A., Hakem, R., Penninger, J. M., and Mak, T. W. (1998). Apaf1 is required for mitochondrial pathways of apoptosis and brain development. Cell 94, 739–750.

Yoshida, T., Shiraishi, T., Nakata, S., Horinaka, M., Wakada, M., Mizutani, Y., Miki, T., and Sakai, T. (2005). Proteasome inhibitor MG132 induces death receptor 5 through CCAAT/enhancer-binding protein homologous protein. Cancer Res 65, 5662–5667.

Yu, C., Bruzek, L. M., Meng, X. W., Gores, G. J., Carter, C. A., Kaufmann, S. H., and Adjei, A. A. (2005). The role of Mcl-1 downregulation in the proapoptotic activity of the multikinase inhibitor BAY 43–9006. Oncogene 24, 6861–6869.

Zeng, Y., Wu, X. X., Fiscella, M., Shimada, O., Humphreys, R., Albert, V., and Kakehi, Y. (2006). Monoclonal antibody to tumor necrosis factor-related apoptosis-inducing ligand receptor 2 (TRAIL-R2) induces apoptosis in primary renal cell carcinoma cells in vitro and inhibits tumor growth in vivo. Int J Oncol 28, 421–430.

Zhang, H. G., Wang, J., Yang, X., Hsu, H. C., and Mountz, J. D. (2004). Regulation of apoptosis proteins in cancer cells by ubiquitin. Oncogene 23, 2009–2015.

Zhang, L. and Fang, B. (2005). Mechanisms of resistance to TRAIL-induced apoptosis in cancer. Cancer Gene Ther 12, 228–237.

Zhang, L., Yu, J., Park, B. H., Kinzler, K. W., and Vogelstein, B. (2000). Role of BAX in the apoptotic response to anticancer agents. Science 290, 989–992.

Zhang, X., Cheung, R. M., Komaki, R., Fang, B., and Chang, J. Y. (2005). Radiotherapy sensitization by tumor-specific TRAIL gene targeting improves survival of mice bearing human non-small cell lung cancer. Clin Cancer Res 11, 6657–6668.

Zhu, H., Guo, W., Zhang, L., Davis, J. J., Wu, S., Teraishi, F., Cao, X., Smythe, W. R., and Fang, B. (2005a). Enhancing TRAIL-induced apoptosis by Bcl-X(L) siRNA. Cancer Biol Ther 4, 393–397.

Zhu, H., Guo, W., Zhang, L., Wu, S., Teraishi, F., Davis, J. J., Dong, F., and Fang, B. (2005b). Proteasome inhibitors-mediated TRAIL resensitization and Bik Accumulation. Cancer Biol Ther 4, 781–786.

Zuzak, T. J., Steinhoff, D. F., Sutton, L. N., Phillips, P. C., Eggert, A., and Grotzer, M. A. (2002). Loss of caspase-8 mRNA expression is common in childhood primitive neuroectodermal brain tumour/medulloblastoma. Eur J Cancer 38, 83–91.

Chapter 8
Rational Design of Therapeutics Targeting the BCL-2 Family

Are Some Cancer Cells Primed for Death but Waiting for a Final Push?

Victoria Del Gaizo Moore and Anthony Letai*

Abstract A mechanism for circumventing apoptosis prevalent in many cancer cells is the overexpression of antiapoptotic BCL-2 family members. Upregulated expression of BCL-2 may be required to permit ongoing death signaling without a cellular response. Therefore, antagonizing BCL-2 function may cause death in many cancer cells. The selection for expression of BCL-2 or other antiapoptotic proteins during oncogenesis may derive from these proteins' ability to bind and sequester proapoptotic BH3-only proteins. This situation may be advantageous from a therapeutic viewpoint because cancer cells may be distinguished from normal cells by being primed with death signals. There are several strategies currently under investigation that may lead to improved treatment of many cancers by taking advantage of these differences.

Keywords apoptosis, BCL-2, BH3, therapeutics, peptide

1 The BCL-2 Family of Proteins

The BCL-2 family of proteins plays a critical role in controlling death via the intrinsic, or mitochondrial, programmed cell death pathway. BCL-2, the namesake of the family, was identified at the breakpoint of the t(14;18) translocation common to follicular lymphoma (1–3). More than 85% of follicular lymphomas contain a chromosomal translocation involving the fusion of the *bcl-2* gene at 18q21 to the immunoglobulin heavy chain locus on 14q32 (4). This translocation places the *BCL-2* gene under the control of the immunoglobulin heavy chain elements. Thus, overexpression of BCL-2 protein is driven in B-cells possessing the t(14;18). BCL-2 was credentialed as an oncogene when it was shown that overexpression was linked to the

Victoria Del Gaizo Moore and Anthony Letai
Medical Oncology, Dana-Farber Cancer Institute, 44 Binney Street, Dana 530B,
Boston, MA 02115, USA

*To whom correspondence should be addressed: e-mail: anthony_letai@dfci.harvard.edu

R. Khosravi-Far and E. White (eds.), *Programmed Cell Death in Cancer Progression and Therapy*.
© Springer 2008

induction of lymphoma in mice (5, 6). Until the discovery of BCL-2, only oncogenes that increased cell proliferation, like *myc, ras*, and *src* had been described. BCL-2's discovery and characterization opened a new class of oncogenes: inhibitors of cell death. The last 20 years have seen the discovery of a family of proteins related to BCL-2 by structural homology and by participation in control over the mitochondrial apoptotic pathway.

BCL-2 proteins largely interact at the mitochondria, the nexus of events that irreversibly commit a cell to programmed cell death via the intrinsic pathway. Some BCL-2 proteins are localized to the mitochondria even during normal cellular conditions while many have other subcellular locations. For example, BAK resides as a monomer at the mitochondrial outer membrane as well as the endoplamic reticulum (7). Prior to activation, BAX exists as a monomer, either in the cytosol or loosely attached to the mitochondrial outer membrane. When activated, however, BAX undergoes alkali-stable insertion into the mitochondrial membrane. BCL-2 itself is found not only at mitochondria, but also at the endoplasmic reticulum where it is implicated in calcium homeostasis (8). Many BCL-2 family members have identified roles outside of control of apoptosis, and it is likely that BCL-2 family members are important in other aspects of cellular homeostasis. The extra-apoptotic functions of BCL-2 family members remain an area of active investigation (9–11).

BCL-2 family members can be divided into three broad groups: antiapoptotic, multidomain proapoptotic, and BH3-only proapoptotic proteins (Fig. 8.1). Antiapoptotic proteins include BCL-2, MCL-1, BCL-X_L, BCL-w, and BFL-1, all of which have the ability to oppose cell death. These antiapoptotic proteins possess sequence homology in four alpha-helical BCL-2 homology or BH regions. Multidomain proapoptotic proteins, including BAX and BAK, promote the progression of cell death and share homology in the BH1–3 regions. BH3-only proapoptotic proteins also promote cell death but, as their name implies, have only a BH3 domain in common. The BH3 domains contain an amphipathic α-helix that is necessary for the proapoptotic function of BH3-only proteins. However, this pro-death function requires interaction with multidomain BAX or BAK (12–14).

Upon cellular stress such as oncogene activation, uncontrolled proliferation, DNA damage, or growth factor withdrawal, BH3-only proteins become functionally upregulated via transcriptional or posttranslational means (15, 16). Proapoptotic BH3-only proteins may be further categorized as "activators" or "sensitizers" (17)

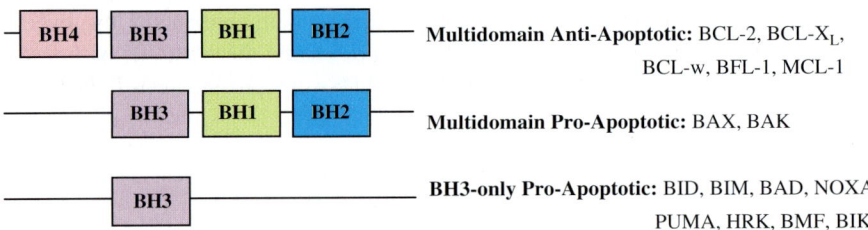

Fig. 8.1 Three classes of the BCL-2 family of proteins. BH3 domains are coded by color

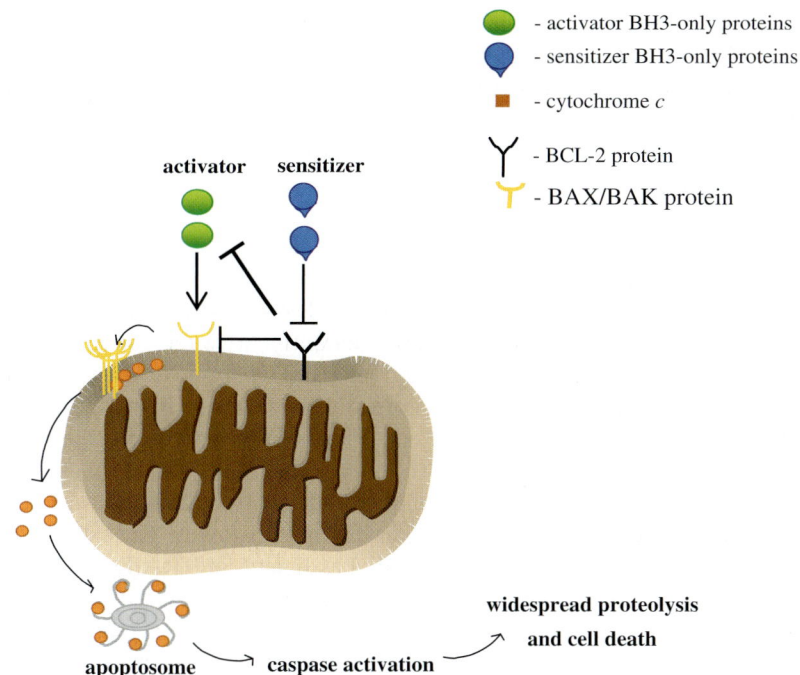

- activator BH3-only proteins
- sensitizer BH3-only proteins
- cytochrome *c*
- BCL-2 protein
- BAX/BAK protein

activator   sensitizer

apoptosome → caspase activation

widespread proteolysis
and cell death

Fig. 8.2 BCL-2 family "activators" vs "sensitizer." BH3 domain-only activators, such as BID or BIM, interact with BAX or BAK to induce their activation, leading to MOMP, caspase activation, and apoptosis. BCL-2 may also bind and sequester BID or BIM, preventing activation of BAX or BAK. Sensitizers binding to BCL-2 may either block activators from binding or displace them from BCL-2

(Fig. 8.2). "Activator" BH3-only proteins, such as BID or BIM, interact with BAX or BAK, inducing an allosteric change. Subsequently, activated BAX or BAK can oligomerize. Oligomerized BAX or BAK, perhaps in complex with other proteins, induce mitochondria outer membrane permeablization (MOMP) (14, 18–22). Permeablization allows certain mitochondrial factors such as cytochrome *c*, Smac/Diablo, and AIF, to be released into the cytosol (23–28). Once in the cytosol, cytochrome *c* forms a holoenzyme complex with caspase-9 and APAF-1, called the apoptosome, which cleaves procaspase-3, into an active protease (29). Widespread proteolysis ensues, leading to cellular dysfunction and death. Consequently, MOMP can be considered the step at which commitment to cell death occcurs. Notably, there are recent studies that suggest that a key proapoptotic function of p53 is mediated by its ability to act as an activator (30–33).

While antiapoptotic proteins like BCL-2 and MCL-1 have been shown to directly interact with multidomain BAX and BAK, their interaction with BH3-only proteins may be more important to their antiapoptotic function (13, 34). The BH1–3 domains of BCL-2 form a hydrophobic cleft where the BH3 domain of multidomain and BH3-only proteins can bind. BCL-2 binding of BID or BIM causes sequestration of these activator proteins, thereby preventing interaction and activation of BAX and BAK and thereby preventing MOMP (13, 17). Not all BH3-only proteins, however, are able to activate BAX or BAK. BH3-only proteins that do not activate BAX or BAK, including BAD, BIK, BMF, NOXA, and PUMA, we classify as "sensitizers" (17, 35). In contrast to activators that can activate BAX and BAK, these BH3 domains exert their proapoptotic function by binding to antiapoptotic BCL-2 proteins. In so doing, they compete with the binding of activators, either preventing activator binding, or displacing activators from BCL-2. In the presence of sensitizers, displaced activator BH3-only proteins are freed from antiapoptotic proteins to activate BAX and BAK and induce MOMP (17, 35, 36). While antiapoptotic proteins apparently share the common function of inhibiting apoptosis by sequestering activator BH3-only proteins, their binding pockets are nonetheless distinct. This is most clearly shown by the fact that each antiapoptotic protein has a distinct pattern of interaction with the range of sensitizer BH3 domains (35–37).

In addition to the intrinsic or mitochondrial pathway, apoptosis also can be initiated through the death receptor-mediated, or extrinsic, pathway. The extrinsic pathway is triggered when ligands, such as TNF, Fas ligand, or TRAIL, are bound by cell surface death receptors that cause changes in the intracellular domains of these receptors, resulting in assembly of a so-called death-inducing signaling complex (DISC) reviewed in (16). Activation of the initiator caspase-8 activation results, leading to activation of downstream effector caspases. In some systems, linkage to the intrinsic apoptotic pathway is accomplished by caspase-8 cleavage of the activator BH3-only protein BID, which can then trigger BAX or BAK oligomerization and MOMP (38, 39). Even though initiation of the intrinsic and extrinsic pathways is different, both converge at the activation of downstream effector caspase-3 and caspase-7.

2 The Link Between BCL-2 and Cancer

While elevated BCL-2 levels as a result of the t(14;18) translocation involving the *BCL-2* gene occurs in 80–90% of follicular non-Hodgkins lymphomas, aberrant expression of antiapoptotic expression has been implicated in many other cancers (4, 40, 41). 20–55% of diffuse large cell lymphomas have elevated BCL-2, either due to t(14;18) translocations, gene amplification, or other mechanisms, which may correlate with decreased patient survival (42–44). Many other cancers exhibit high levels of BCL-2 protein in the absence of a t(14;18); the mechanism of upregulated BCL-2 remains obscure in most of these instances. Examples include 70% of breast cancer (45, 46), 30–60% prostate cancer (47), and 90% of colorectal cancer cases (41, 48, 49). Chronic lymphoid leukemia (CLL) is largely considered a disease of failed apoptosis (50–52), but usually not due to t(14;18) (53). Nonetheless, the majority of CLL cells express high levels of BCL-2 (54). Recently, a more common chromosomal aberration, deletion, or translocation of 13q14.3, was implicated in elevated BCL-2 in CLL (55). Changes affecting region 13q14.3 downregulated two microRNAs (miRNA) *mir-15A* and *mir-16-1*, and occurred in >50% of all CLL cases. miRNAs are a class of genes involved in tumorigenesis that produce short, single-stranded RNAs that bind to specific mRNA sequences and either prevent the translation of the mRNA or hasten degradation of the mRNA, thereby lowering the levels of the corresponding protein (56, 57). Expression of *mir-15A* and *mir-16-1* inversely correlates to BCL-2 expression in CLL samples and both negatively regulate BCL-2 levels (58, 59).

Expression of other antiapoptotic proteins has been detected in many cancers, including BFL-1 in diffuse large-cell lymphoma (60), MCL-1 in myeloma (61), and BCL-X$_L$ in lung adenocarcinoma (62). Both BCL-2 and MCL-1 have been implicated as important contributors to melanoma development and maintenance (63–65). The oncogenic Epstein-Barr virus (EBV) and human herpes virus-8 (HHV-8; also known as Kaposi sarcoma herpes virus) encode BCL-2 homologs that oppose cell death from multiple stimuli, analogous to BCL-2 (66, 67). EBV has been implicated in the causation of HIV-related lymphoma, Burkitt lymphoma, nasopharyngeal carcinoma, and posttransplantation lymphomas, and HHV-8 in the causation of Kaposi sarcoma, Castleman disease, and body cavity lymphomas. The evolutionary selection for BCL-2 homologs in these viruses suggests that blocking the intrinsic pathway to programmed cell death is important in viral infection, and perhaps also for oncogenesis.

Multiple myeloma (MM) cells have been shown to express BCL-2, BCL-X$_L$, and MCL-1. Clinical and in vitro data suggest important roles for these proteins in MM cell survival as well as clinical resistance to therapy (68, 69). Despite the lack of chromosomal translocations, protein expression of each of these antiapoptotic proteins has been observed in clinical isolates (68–70). Antisense oligonucleotides (ASO) have been used with MM cells to determine if BCL-2, BCL-X$_L$, or MCL-1 expression is critical for the survival of these cancer cells, with mixed results (61, 71).

It has been hypothesized that oncogenesis requires an apoptotic defect (72, 73). One apparent strategy for apoptotic escape exploited by certain cancer cells is the overexpression of antiapoptotic BCL-2 family members. These proteins can bind and sequester activator BH3-only death signals likely initiated by cancer phenotypes including genomic instability, oncogene activation, and inappropriate cell contact. Therefore, in cancer cells that adopt such a strategy it seems likely that much of the antiapoptotic proteins will be "primed" with activator BH3-only proteins (Fig. 8.3). Primed cells are rendered exquisitely sensitive to mimetics of the sensitizer BH3 domains, which function as selective antagonists of BCL-2 and other antiapoptotic proteins (17, 35, 74) (Fig. 8.4). Certain, though probably not all, normal tissues may lack this priming, as they do not violate the rules of normal cellular behavior that provoke death signals in many cancer cells. Thus, the possibility exists of targeted intervention to exploit the therapeutic window between "primed" cancer cells and "unprimed" normal cells by antagonizing BCL-2 family antiapoptotic protein function.

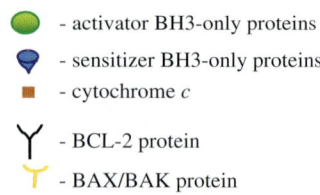

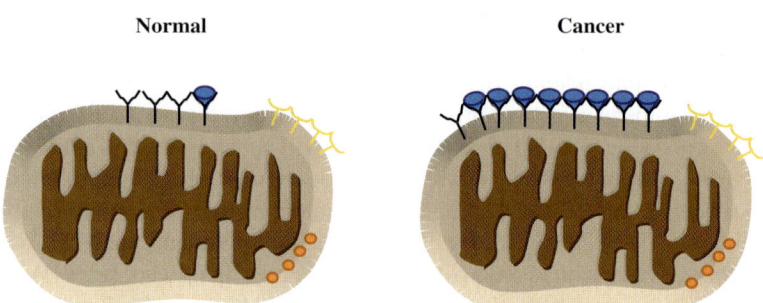

Fig. 8.3 Idealized cartoon representation of a normal mitochondrion compared to a cancer mitochondrion. Though they express more BCL-2 than the normal mitochondrion, the cancer mitochondrion has less antiapoptotic reserve due to significant priming by activator BH3-only proteins

3 Therapeutic Strategies Targeting Antiapoptotic BCL-2 Family Members

Efforts have begun to target the expression of antiapoptotic BCL-2 family members. One strategy is to downregulate antiapoptotic genes by ASO. An 18-mer phosphorothioated oligonucleotide directed against the first six codons of the human BCL-2 open reading frame, called Oblimersen or Genasense, was introduced by Genta, Inc. and has advanced through clinical trials (75, 76). Side effects have been tolerable, generally limited to thrombocytopenia, fatigue, back pain, weight loss, and dehydration (77). However, efficacy has been difficult to demonstrate. For example, treatment of metastatic melanoma with dacarbazine and oblimersen in a randomized phase III study showed no significant benefit in overall survival compared with dacarbazine alone (274 vs 238 days, $P = 0.18$). Even though significant benefit in progression free survival was observed (74 vs 49 days, $P = 0.0003$), overall survival was the primary end point, thus an FDA panel declared that clinical benefit was not demonstrated. In a phase III trial of myeloma, oblimersen plus high-dose dexamethasone was compared with dexamethasone alone; this trial also failed to meet its primary end point, time to disease progression. Furthermore, response to oblimersen in another myeloma trial did

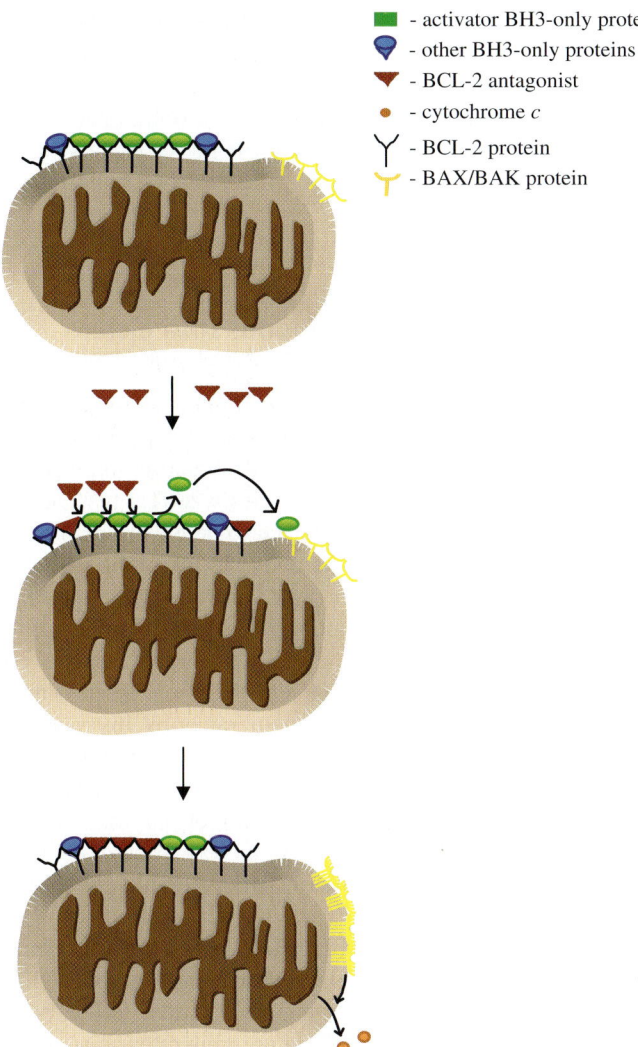

Fig. 8.4 Model of BCL-2 antagonist inducing death in a "primed" cancer mitochondrion. Cancer mitochondria have activator proteins like BIM sequestered by BCL-2 on the outer membrane. Upon addition of a BCL-2 antagonist, BIM is displaced and BCL-2 becomes occupied by the antagonist. Freed BIM then interacts with BAX or BAK, causing oligomerization and leading to cytochrome c release, MOMP, and apoptosis

not correlate with reduced BCL-2 protein levels, which provokes the question of whether oblimersen has significant off-target activity. In general, ASO has been a somewhat disappointing strategy for targeting BCL-2. The cellular effects of the lowering of BCL-2 levels by antisense oligonucelotides may not only provoke undesirable coregulation of other BCL-2 family members but decreasing the mRNA is likely very different from functional antagonism of the protein (78). Furthermore, BCL-2 protein levels tend to be in the 10–50% range, which is unlikely to have a widespread cellular effect. Finally, oblimersen contains 2 CpG dinucleotides which may well produce many off-target effects on the immune system (78). While some of these off-target effects may be beneficial, others may well limit its maximum tolerated dose.

4 Delivery of Therapeutic Compounds into Cells

Delivery of drugs and therapeutic compounds is limited by the ability to penetrate the cell membrane. Compounds cross membranes either by passive processes or by mechanisms involving active participation of membrane components. In general, water, small hydrophilic molecules, and molecules <200 Da (79) passively diffuse through membranes. Therefore, most drugs need to be either small and water soluble, or polar enough for absorption into the body yet lipophilic enough to promote passage through the nonpolar lipid bilayer (80). This narrow range of physical characteristics limits the success of many compounds. Additionally, the degree of ionization of the compound, the circulation to the site of absorption, and its concentration can affect a compound's ability to reach its site of action. Even if a compound is able to circumvent passive passage across membranes by interacting with membrane receptors, there are still stringent criteria that must be met. No matter how a drug enters cell, once inside its effects can be terminated by metabolism or excretion. An additional difficulty is that the compound not only has to cross into cells rapidly and efficiently, but it then needs to make its way through the cellular milieu, which is full of proteases and other proteins, and eventually travel to the desired subcellular location to be effective.

While peptides based on sensitizer BH3-domains have been demonstrated to function as selective inhibitors of antiapoptotic proteins, unmodified BH3 domain peptides are cell impermeant (81). One strategy to augment cell entry is use of protein transduction domains (PTD) (82). PTDs are generally small (~10–20 amino acids) peptide sequences enriched for positively charged amino acids that rapidly and efficiently cross cell membranes. When fused to larger molecules, they have been shown to transport into cells a wide variety of cargo along with such large proteins (83), liposomes (84), and even metallic beads (85). The transduction process is not receptor mediated and is temperature independent, making it unlikely that endocytosis or transporter mechanisms are involved (86–88); however, the exact mechanism is not known.

To facilitate cell internalization, BH3 peptides have been linked with PTD such as a poly-D-arginine or Antennapedia internalization sequence tags (17, 26, 89, 90). N-terminal poly-D- arginine octomer (r8) linkage to BH3 peptides from BAD or BID have been shown to kill a human leukemia cell line that expresses BCL-2, while r8BIDBH3 double point mutant did not. Furthermore, r8BADBH3 peptide caused no apoptosis on its own but when added with the r8BIDBH3 peptide increased apoptosis, suggesting that the moiety did indeed facilitate internalization and that an intact BH3 domain was necessary for killing (17). In a separate study, a 27-amino acid peptide derived from the BH3 domain of BAD was linked to decanoic acid (26).

Decanoic acid allows cell permeablization by a different mechanism than PTDs, which may involve activation of phospholipase C which causes intracellular stores of calcium to be released followed by contraction of calmodulin-dependent actin filaments (91). The BAD-decanoic acid compound, called, cpm-1285, but not a peptide bearing a point mutation at a residue necessary for BH3 function, induced apoptosis in a BCL-2-expressing human myeloid leukemia line, HL-60. Furthermore, immunodeficient mice injected with HL-60 cells survived longer when treated with cpm-1285. However, these studies do not conclusively demonstrate the mechanism of action of the peptide derivatives, and the cytotoxic effects could be independent of direct interaction with BCL-2 family members. Such off-target toxicity was demonstrated by Schimmer and coworkers where linking the BH3 domain of BAD to the Antennapedia internalization sequence had considerable off-target toxicity (89). Their compound was toxic to a wide variety of cells, including yeast, wherein BCL-2 family members have yet to be identified. Others have demonstrated that BH3 peptides derived from BAX and BCL-2 linked to an Antennapedia internalization sequence induce MOMP and apoptosis, but overexpression of either BCL-2 or BCL-X$_L$ did not rescue the cells from apoptosis (92). All of these effects may be due to a nonspecific membrane disruption rather than to interaction with the BCL-2 family pathway. For example, the Antennapedia internalization sequence is mainly a positively charged amphipathic α-helix that could interact and disrupt the negatively charged mitochondrial membranes independent of BCL-2 family protein interaction, in a manner similar to certain natural antibiotics (93–95). Therefore, nonspecific killing due to intrinsic biophysical properties of these internalization moieties make interpretation of cell killing by linked, some tagged BH3 peptides difficult. Further pharmaceutical development of such molecules would require considerable attention to reducing this toxicity.

The α-helix of BH3 peptides is vital for their function, but in aqueous solution the α-helical conformation can be less than 25%. Attempts have therefore been made to improve peptide function by stabilizing α-helicity. Small improvements have been gained by grafting a BAK BH3 domain to a helix-stabilizing miniprotein (96) or synthesizing BH3 peptide analogs with covalent molecular bridges, which improved affinity for BCL-2 or stabilized the α-helical conformation (97). Perhaps the most striking example of the potential of α-helix stabilization was provided by a BID BH3 peptide stabilized by an all-hydrocarbon "staple" (81). Not only did this modification enhance α-helicity, but it also increased affinity for BCL-2, cell entry, protease resistance, as well as leukemia cell line toxicity in vitro and in vivo. Mice bearing leukemia cell line xenografts demonstrated statistically significant survival improvement after 6 days and normal tissues appeared unaffected as measured by histological analysis. Since the molecule was modeled after a BID BH3 domain previously shown to be an activator (17), the compound was able to directly induce cytochrome c release in a BAK-dependent fashion in vitro. Even though the compound did not behave as a selective BCL-2 antagonist but rather an activator, it was still able to exploit an apparent therapeutic window between the tumor xenograft and the normal tissues. It remains to be seen whether an analogous sensitizer BH3-based compounds would provide an even greater therapeutic window.

5   Cell-Permeant Small Molecules

Cell-permeant small molecules that bind to antiapoptotic BCL-2 family members have been identified through structure-based computer screening. One molecule isolated was able to displace the BAK BH3 peptide from BCL-2 with an IC_{50} of 1–14 μM. Since the Kd for the BAK BH3 peptide is approximately 200 nM, it is reasonable to surmise that the Kd for binding of these molecules to BCL-2 may be significantly higher. Another molecule identified was toxic to four cell lines tested at concentrations of 10–20 μM and toxicity correlated with BCL-2 expression levels (98, 99). Screens of chemical libraries have also been used. Out of 16,320 screened, Degterev et al. identified two molecules that disrupt a BCL-X_L/BAK BH3 complex, both which had toxicity in the 10–90 μM range in a leukemia cell line (100). A screen of a library of natural products allowed the isolation of Tetrocarcin A, which is derived from *Actinomyces*, identified for its ability to counteract BCL-2 protection of anti-Fas/cycloheximide-treated HeLa cells at concentrations in the micromolar range (101). Antimycin A, an antimicrobial agent with antitumor properties in experimental systems and a known inhibitor of electron transport at mitochondrial respiratory chain complex III, was identified from a screen for inhibitors of mitochondrial respiration in mammalian cells (102). Further characterization demonstrated that antimycin A interacts with BCL-2 and BCL-X_L, and that increasing cellular levels of BCL-X_L correlated with increasing toxicity. Nuclear magnetic resonance (NMR) spectroscopy used to investigate natural products found certain polyphenols from green tea extracts were able to bind to BCL-X_L (103). In addition, these compounds displaced a BH3 domain from BCL-X_L and BCL-2 in the submicromolar range. Another screen of a small library of natural products identified two molecules, purpurogallin and gossypol, both of which resemble human BAD and inhibit binding of a BH3 domain to BCL-X_L (104). While chemical modification of purpurogallin did not lower the IC_{50} of peptide displacement of the parental compound, a racemic mixture of the (+) and (−) isomers of gossypol displaced the BH3 peptide with an IC_{50} of 0.5 μM. Molecular modeling suggested that removal of two aldehyde groups from gossypol might reduce steric hindrance in binding the hydrophobic pocket of BCL-X_L, however this modification actually decreased the binding to BCL-2 family members (105).

Small molecules that enter cells and bind the hydrophobic pocket of BCL-2 analogously to sensitizer BH3 peptides are currently in clinical development. The biotechnology company Gemin X has isolated a compound (GX01) that has been reported to bind BCL-2 and BCL-X_L and displace BH3 domains from their binding pockets (106). GX01 was identified from a high-throughput screen of chemical libraries and is in phase I clinical trials in both chronic lymphocytic leukemia (at the University of California, San Diego [UCSD]) and solid tumors (at Georgetown University). Ascenta Therapeutics has an orally administered gossypol derivative in an ongoing phase I cancer trial.

Using a strategy of combining high-throughput screening with interactive modulation of chemical structure based on NMR, Abbott Laboratories has developed compounds reported to displace BH3 domains from BCL-2, BCL-X$_L$, and BCL-w with an IC$_{50}$ of not more than 1 nM (74). One lead molecule, ABT-737, is a BAD-like sensitizer that can antagonize BCL-2 protection but cannot directly cause activation of BAX/BAK. ABT-737 was reported to have significant activity in primary CLL cells and mouse xenograft models of lung cancer and lymphoma. When injected into mice ABT-737 was well tolerated with minimal side effects in noncancerous tissues except for a reduction in platelets and lymphocytes. Furthermore, ABT-737 enhanced the cytotoxicity of paclitaxel against a cancerous cell line where single-agent activity was not achieved. Other preclinical studies have shown that the toxicity of ABT-737 is due to selective antagonism of BCL-2 in cells that require BCL-2 for survival (35). Given its high affinity for BCL-2, the data that support its function via its designed mechanism, and its effectiveness across several different cancer types in vitro, ABT-737 seems to be a promising lead compound, although clinical trials are yet to begin.

6 Conclusions

Our current understanding of the mechanisms by which BCL-2 family members control commitment to cell death gives good theoretical backing to strategies aimed at manipulating this system for clinical benefit. Certain cancers in which antiapoptotic BCL-2 is overexpressed and activator BH3-only proteins are upregulated may be "primed" for death, needing only a modest, targeted biochemical nudge for final execution of apoptosis. Small molecules designed to antagonize BCL-2 and related antiapoptotic proteins appear to be useful tools to generate this targeted signal. As the binding clefts among proteins like BCL-2, MCL-1, and BFL-1 are demonstrably distinct, it may be possible to design molecules which selectively antagonize individual proteins. Whether such "narrow spectrum" antagonists will be better cancer therapeutics than "broad spectrum" antagonists that might target the entire antiapoptotic group remains to be seen. Experimental evidence suggests that the state of protein–protein interactions among BCL-2 family members within cancer cells is different from those within normal cells. Therefore, even if an antagonizing compound entered all cells, induction of apoptosis might selectively be triggered within cancer cells. The promise of these molecules as anticancer therapeutics will soon be tested as clinical trials of compounds targeting BCL-2 are currently underway. It is exciting to witness the emergence of a potentially new class of anticancer drugs, those specifically designed to unleash the latent apoptotic potential within cancer cells.

References

1. Tsujimoto, Y., Gorham, J., Cossman, J., Jaffe, E., and Croce, C. M. The t(14;18) (1985). Chromosome translocations involved in B-cell neoplasms result from mistakes in VDJ joining. Science 229, 1390–1393.

2. Cleary, M. L. and Sklar, J. (1985). Nucleotide sequence of a t(14;18) chromosomal breakpoint in follicular lymphoma and demonstration of a breakpoint-cluster region near a transcriptionally active locus on chromosome 18. Proc Natl Acad Sci USA 82, 7439–7443.

3. Bakhshi, A., Jensen, J. P., Goldman, P., Wright, J. J., McBride, O. W., Epstein, A. L., and Korsmeyer, S. J. (1985). Cloning the chromosomal breakpoint of t(14;18) human lymphomas: clustering around JH on chromosome 14 and near a transcriptional unit on 18. Cell 41, 899–906.

4. Packham, G. (1998). Mutation of BCL-2 family proteins in cancer. Apoptosis 3, 75–82.

5. McDonnell, T. J. and Korsmeyer, S. J. (1991). Progression from lymphoid hyperplasia to high-grade malignant lymphoma in mice transgenic for the t(14; 18). Nature 349, 254–256.

6. Strasser, A., Harris, A. W., Bath, M. L., and Cory, S. (1990). Novel primitive lymphoid tumours induced in transgenic mice by cooperation between myc and bcl-2. Nature 348, 331–333.

7. Scorrano, L., Oakes, S. A., Opferman, J. T., Cheng, E. H., Sorcinelli, M. D., Pozzan, T., and Korsmeyer, S. J. (2003). BAX and BAK regulation of endoplasmic reticulum Ca2 +: a control point for apoptosis. Science 300, 135–139.

8. Bassik, M. C., Scorrano, L., Oakes, S. A., Pozzan, T., and Korsmeyer, S. J. (2004). Phosphorylation of BCL-2 regulates ER Ca2 + homeostasis and apoptosis. EMBO J 23, 1207–1216.

9. Kamer, I., Sarig, R., Zaltsman, Y., Niv, H., Oberkovitz, G., Regev, L., Haimovich, G., Lerenthal, Y., Marcellus, R. C., and Gross, A. (2005). Proapoptotic BID is an ATM effector in the DNA-damage response. Cell 122, 593–603.

10. Zinkel, S. S., Hurov, K. E., Ong, C., Abtahi, F. M., Gross, A., and Korsmeyer, S. J. (2005). A role for proapoptotic BID in the DNA-damage response. Cell 122, 579–591.

11. Danial, N. N., Gramm, C. F., Scorrano, L., Zhang, C. Y., Krauss, S., Ranger, A. M., Datta, S. R., Greenberg, M. E., Licklider, L. J., Lowell, B. B., Gygi, S. P., and Korsmeyer, S. J. (2003). BAD and glucokinase reside in a mitochondrial complex that integrates glycolysis and apoptosis. Nature 424, 952–956.

12. Zong, W. X., Lindsten, T., Ross, A. J., MacGregor, G. R., and Thompson, C. B. (2001). BH3-only proteins that bind pro-survival Bcl-2 family members fail to induce apoptosis in the absence of Bax and Bak. Genes Dev 15, 1481–1486.

13. Cheng, E. H., Wei, M. C., Weiler, S., Flavell, R. A., Mak, T. W., Lindsten, T., and Korsmeyer, S. J. (2001). BCL-2, BCL-X(L) sequester BH3 domain-only molecules preventing BAX- and BAK-mediated mitochondrial apoptosis. Mol Cell 8, 705–711.

14. Wei, M. C., Zong, W. X., Cheng, E. H., Lindsten, T., Panoutsakopoulou, V., Ross, A. J., Roth, K. A., MacGregor, G. R., Thompson, C. B., and Korsmeyer, S. J. (2001). Proapoptotic BAX and BAK: a requisite gateway to mitochondrial dysfunction and death. Science 292, 727–730.

15. Puthalakath, H. and Strasser, A. (2002). Keeping killers on a tight leash: transcriptional and post-translational control of the pro-apoptotic activity of BH3-only proteins. Cell Death Differ 9, 505–512.

16. Danial, N. N. and Korsmeyer, S. J. (2004). Cell death: critical control points. Cell 116, 205–219.

17. Letai, A., Bassik, M. C., Walensky, L. D., Sorcinelli, M. D., Weiler, S., and Korsmeyer, S. J. (2002). Distinct BH3 domains either sensitize or activate mitochondrial apoptosis, serving as prototype cancer therapeutics. Cancer Cell 2, 183–192.

18. Lindsten, T., Ross, A. J., King, A., Zong, W.-X., Rathmell, J. C., Shiels, H. A., Ulrich, E., Waymire, K. G., Mahar, P., Frauwirth, K., Chen, Y., Wei, M., Eng, V. M., Adelman, D. M.,

Simon, M. C., Ma, A., Golden, J. A., Evan, G., Korsmeyer, S. J., MacGregor, G. R., and Thompson, C. B. (2000). The combined functions of proapoptotic Bcl-2 family members Bak and Bax are essential for normal development of multiple tissues. Mol Cell 6, 1389–1399.

19. Wei, M. C., Lindsten, T., Mootha, V. K., Weiler, S., Gross, A., Ashiya, M., Thompson, C. B., and Korsmeyer, S. J. (2000). tBID, a membrane-targeted death ligand, oligomerizes BAK to release cytochrome c. Genes Dev 14, 2060–2071.

20. Suzuki, M., Youle, R. J., and Tjandra, N. (2000). Structure of Bax: coregulation of dimer formation and intracellular localization. Cell 103, 645–654.

21. Griffiths, G. J., Dubrez, L., Morgan, C. P., Jones, N. A., Whitehouse, J., Corfe, B. M., Dive, C., and Hickman, J. A. (1999). Cell damage-induced conformational changes in the pro-apoptotic protein Bak in vivo precede the onset of apoptosis. J Cell Biol 144, 903–914.

22. Desagher, S., Osen-Sand, A., Nichols, A., Eskes, R., Montessuit, S., Lauper, S., Maundrell, K., Antonsson, B., and Martinou, J. C. (1999). Bid-induced conformational change of Bax is responsible for mitochondrial cytochrome c release during apoptosis. J Cell Biol 144, 891–901.

23. Du, C., Fang, M., Li, Y., Li, L., and Wang, X. (2000). Smac, a mitochondrial protein that promotes cytochrome c-dependent caspase activation by eliminating IAP inhibition. Cell 102, 33–42.

24. Li, L. Y., Luo, X., and Wang, X. (2001). Endonuclease G is an apoptotic DNase when released from mitochondria. Nature 412, 95–99.

25. Li, P., Nijhawan, D., Budihardjo, I., Srinivasula, S. M., Ahmad, M., Alnemri, E. S., and Wang, X. (1997). Cytochrome c and dATP-dependent formation of Apaf-1/caspase-9 complex initiates an apoptotic protease cascade. Cell 91, 479–489.

26. Wang, J. L., Zhang, Z. J., Choksi, S., Shan, S., Lu, Z., Croce, C. M., Alnemri, E. S., Korngold, R., and Huang, Z. (2000). Cell permeable Bcl-2 binding peptides: a chemical approach to apoptosis induction in tumor cells. Cancer Res 60, 1498–1502.

27. Susin, S. A., Lorenzo, H. K., Zamzami, N., Marzo, I., Snow, B. E., Brothers, G. M., Mangion, J., Jacotot, E., Costantini, P., Loeffler, M., Larochette, N., Goodlett, D. R., Aebersold, R., Siderovski, D. P., Penninger, J. M., and Kroemer, G. (1999). Molecular characterization of mitochondrial apoptosis-inducing factor. Nature 397, 441–446.

28. Verhagen, A. M., Ekert, P. G., Pakusch, M., Silke, J., Connolly, L. M., Reid, G. E., Moritz, R. L., Simpson, R. J., and Vaux, D. L. (2000). Identification of DIABLO, a mammalian protein that promotes apoptosis by binding to and antagonizing IAP proteins. Cell 102, 43–53.

29. Wang, X. (2001). The expanding role of mitochondria in apoptosis. Genes Dev 15, 2922–2933.

30. Chipuk, J. E., Kuwana, T., Bouchier-Hayes, L., Droin, N. M., Newmeyer, D. D., Schuler, M., and Green, D. R. (2004). Direct activation of Bax by p53 mediates mitochondrial membrane permeabilization and apoptosis. Science 303, 1010–1014.

31. Moll, U. M., Wolff, S., Speidel, D., and Deppert, W. (2005). Transcription-independent pro-apoptotic functions of p53. Curr Opin Cell Biol 17, 631–636.

32. Murphy, M. E., Leu, J. I., and George, D. L. (2004). p53 moves to mitochondria: a turn on the path to apoptosis. Cell Cycle 3, 836–839.

33. Leu, J. I., Dumont, P., Hafey, M., Murphy, M. E., and George, D. L. (2004). Mitochondrial p53 activates Bak and causes disruption of a Bak-Mcl1 complex. Nat Cell Biol 6, 443–450.

34. Cheng, E. H., Levine, B., Boise, L. H., Thompson, C. B., and Hardwick, J. M. (1996). Bax-independent inhibition of apoptosis by Bcl-XL. Nature 379, 554–556.

35. Certo, M., Moore, V. D. G., Nishino, M., Wei, G., Korsmeyer, S., Armstrong, S. A., and Letai, A. (2006). Mitochondria primed by death signals determine cellular addiction to antiapoptotic BCL-2 family members. Cancer Cell 9(5), 351–365.

36. Kuwana, T., Bouchier-Hayes, L., Chipuk, J. E., Bonzon, C., Sullivan, B. A., Green, D. R., and Newmeyer, D. D. (2005). BH3 Domains of BH3-only proteins differentially regulate Bax-mediated mitochondrial membrane permeabilization both directly and indirectly. Mol Cell 17, 525–535.

37. Chen, L., Willis, S. N., Wei, A., Smith, B. J., Fletcher, J. I., Hinds, M. G., Colman, P. M., Day, C. L., Adams, J. M., and Huang, D. C. (2005). Differential targeting of prosurvival Bcl-2 proteins by their BH3-only ligands allows complementary apoptotic function. Mol Cell 17, 393–403.

38. Li, H., Zhu, H., Xu, C. J., and Yuan, J. (1998). Cleavage of BID by caspase-8 mediates the mitochondrial damage in the Fas pathway of apoptosis. Cell 94, 491–501.

39. Gross, A., McDonnell, J. M., and Korsmeyer, S. J. (1999). BCL-2 family members and the mitochondria in apoptosis. Genes Dev 13, 1899–1911.

40. Kitada, S., Pedersen, I. M., Schimmer, A. D., and Reed, J. C. (2002). Dysregulation of apoptosis genes in hematopoietic malignancies. Oncogene 21, 3459–3474.

41. Wang, S., Yang, D., and Lippman, M. E. (2003). Targeting Bcl-2 and Bcl-XL with nonpeptidic small-molecule antagonists. Semin Oncol 30, 133–142.

42. Mounier, N., Briere, J., Gisselbrecht, C., Emile, J. F., Lederlin, P., Sebban, C., Berger, F., Bosly, A., Morel, P., Tilly, H., Bouabdallah, R., Reyes, F., Gaulard, P., and Coiffier, B. (2003). Rituximab plus CHOP (R-CHOP) overcomes bcl-2-associated resistance to chemotherapy in elderly patients with diffuse large B-cell lymphoma (DLBCL). Blood 101, 4279–4284.

43. Gascoyne, R. D., Adomat, S. A., Krajewski, S., Krajewska, M., Horsman, D. E., Tolcher, A. W., O'Reilly, S. E., Hoskins, P., Coldman, A. J., Reed, J. C., and Connors, J. M. (1997). Prognostic significance of Bcl-2 protein expression and Bcl-2 gene rearrangement in diffuse aggressive non-Hodgkin's lymphoma. Blood 90, 244–251.

44. Hill, M. E., MacLennan, K. A., Cunningham, D. C., Vaughan Hudson, B., Burke, M., Clarke, P., Di Stefano, F., Anderson, L., Vaughan Hudson, G., Mason, D., Selby, P., and Linch, D. C. (1996). Prognostic significance of BCL-2 expression and bcl-2 major breakpoint region rearrangement in diffuse large cell non-Hodgkin's lymphoma: a British National Lymphoma Investigation Study. Blood 88, 1046–1051.

45. Joensuu, H., Pylkkanen, L., and Toikkanen, S. (1994). Bcl-2 protein expression and long-term survival in breast cancer. Am J Pathol 145, 1191–1198.

46. Leek, R. D., Kaklamanis, L., Pezzella, F., Gatter, K. C., and Harris, A. L. (1994). bcl-2 in normal human breast and carcinoma, association with oestrogen receptor-positive, epidermal growth factor receptor-negative tumours and in situ cancer. Br J Cancer 69, 135–139.

47. McDonnell, T. J., Troncoso, P., Brisbay, S. M., Logothetis, C., Chung, L. W., Hsieh, J. T., Tu, S. M., and Campbell, M. L. (1992). Expression of the protooncogene bcl-2 in the prostate and its association with emergence of androgen-independent prostate cancer. Cancer Res 52, 6940–6944.

48. Sinicrope, F. A., Hart, J., Michelassi, F., and Lee, J. J. (1995). Prognostic value of bcl-2 oncoprotein expression in stage II colon carcinoma. Clin Cancer Res 1, 1103–1110.

49. Buolamwini, J. K. (1999). Novel anticancer drug discovery. Curr Opin Chem Biol 3, 500–509.

50. Caligaris-Cappio, F. and Hamblin, T. J. (1999). B-cell chronic lymphocytic leukemia: a bird of a different feather. J Clin Oncol 17, 399–408.

51. Schimmer, A. D., Munk-Pedersen, I., Minden, M. D., and Reed, J. C. (2003). Bcl-2 and apoptosis in chronic lymphocytic leukemia. Curr Treat Options Oncol 4, 211–218.

52. Jewell, A. P. (2002). Role of apoptosis in the pathogenesis of B-cell chronic lymphocytic leukaemia. Br J Biomed Sci 59, 235–238.

53. Schena, M., Gottardi, D., Ghia, P., Larsson, L. G., Carlsson, M., Nilsson, K., and Caligaris-Cappio, F. (1993). The role of Bcl-2 in the pathogenesis of B chronic lymphocytic leukemia. Leuk Lymphoma 11, 173–179.

54. Hanada, M., Delia, D., Aiello, A., Stadtmauer, E., and Reed, J. (1993). C. bcl-2 gene hypomethylation and high-level expression in B-cell chronic lymphocytic leukemia. Blood 82, 1820–1828.

55. Calin, G. A., Dumitru, C. D., Shimizu, M., Bichi, R., Zupo, S., Noch, E., Aldler, H., Rattan, S., Keating, M., Rai, K., Rassenti, L., Kipps, T., Negrini, M., Bullrich, F., and Croce, C. M. (2002). Frequent deletions and down-regulation of micro-RNA genes miR15 and miR16 at 13q14 in chronic lymphocytic leukemia. Proc Natl Acad Sci USA 99, 15524–15529.

56. Bartel, D. P. (2004). MicroRNAs: genomics, biogenesis, mechanism, and function. Cell 116, 281–297.
57. Lim, L. P., Lau, N. C., Garrett-Engele, P., Grimson, A., Schelter, J. M., Castle, J., Bartel, D. P., Linsley, P. S., and Johnson, J. M. (2005). Microarray analysis shows that some microRNAs downregulate large numbers of target mRNAs. Nature 433, 769–773.
58. Calin, G. A., Ferracin, M., Cimmino, A., Di Leva, G., Shimizu, M., Wojcik, S. E., Iorio, M. V., Visone, R., Sever, N. I., Fabbri, M., Iuliano, R., Palumbo, T., Pichiorri, F., Roldo, C., Garzon, R., Sevignani, C., Rassenti, L., Alder, H., Volinia, S., Liu, C. G., Kipps, T. J., Negrini, M., and Croce, C. M. A (2005). MicroRNA signature associated with prognosis and progression in chronic lymphocytic leukemia. N Engl J Med 353, 1793–1801.
59. Cimmino, A., Calin, G. A., Fabbri, M., Iorio, M. V., Ferracin, M., Shimizu, M., Wojcik, S. E., Aqeilan, R. I., Zupo, S., Dono, M., Rassenti, L., Alder, H., Volinia, S., Liu, C. G., Kipps, T. J., Negrini, M., and Croce, C. M. (2005). miR-15 and miR-16 induce apoptosis by targeting BCL2. Proc Natl Acad Sci USA 102, 13944–13949.
60. Shipp, M. A., Ross, K. N., Tamayo, P., Weng, A. P., Kutok, J. L., Aguiar, R. C., Gaasenbeek, M., Angelo, M., Reich, M., Pinkus, G. S., Ray, T. S., Koval, M. A., Last, K. W., Norton, A., Lister, T. A., Mesirov, J., Neuberg, D. S., Lander, E. S., Aster, J. C., and Golub, T. R. (2002). Diffuse large B-cell lymphoma outcome prediction by gene-expression profiling and supervised machine learning. Nat Med 8, 68–74.
61. Derenne, S., Monia, B., Dean, N. M., Taylor, J. K., Rapp, M. J., Harousseau, J. L., Bataille, R., and Amiot, M. (2002). Antisense strategy shows that Mcl-1 rather than Bcl-2 or Bcl-x(L) is an essential survival protein of human myeloma cells. Blood 100, 194–199.
62. Berrieman, H. K., Smith, L., O'Kane, S. L., Campbell, A., Lind, M. J., and Cawkwell, L. (2005). The expression of Bcl-2 family proteins differs between nonsmall cell lung carcinoma subtypes. Cancer 103, 1415–1419.
63. Garraway, L. A., Widlund, H. R., Rubin, M. A., Getz, G., Berger, A. J., Ramaswamy, S., Beroukhim, R., Milner, D. A., Granter, S. R., Du, J., Lee, C., Wagner, S. N., Li, C., Golub, T. R., Rimm, D. L., Meyerson, M. L., Fisher, D. E., and Sellers, W. R. (2005). Integrative genomic analyses identify MITF as a lineage survival oncogene amplified in malignant melanoma. Nature 436, 117–122.
64. McGill, G. G., Horstmann, M., Widlund, H. R., Du, J., Motyckova, G., Nishimura, E. K., Lin, Y. L., Ramaswamy, S., Avery, W., Ding, H. F., Jordan, S. A., Jackson, I. J., Korsmeyer, S. J., Golub, T. R., and Fisher, D. E. (2002). Bcl2 regulation by the melanocyte master regulator Mitf modulates lineage survival and melanoma cell viability. Cell 109, 707–718.
65. Thallinger, C., Wolschek, M. F., Wacheck, V., Maierhofer, H., Gunsberg, P., Polterauer, P., Pehamberger, H., Monia, B. P., Selzer, E., Wolff, K., and Jansen, B. (2003). Mcl-1 antisense therapy chemosensitizes human melanoma in a SCID mouse xenotransplantation model. J Invest Dermatol 120, 1081–1086.
66. Henderson, S., Huen, D., Rowe, M., Dawson, C., Johnson, G., and Rickinson, A. (1993). Epstein-Barr virus-coded BHRF1 protein, a viral homologue of Bcl-2, protects human B cells from programmed cell death. Proc Natl Acad Sci USA 90, 8479–8483.
67. Cheng, E. H., Nicholas, J., Bellows, D. S., Hayward, G. S., Guo, H. G., Reitz, M. S., and Hardwick, J. M. (1997). A Bcl-2 homolog encoded by Kaposi sarcoma-associated virus, human herpesvirus 8, inhibits apoptosis but does not heterodimerize with Bax or Bak. Proc Natl Acad Sci USA 94, 690–694.
68. Gauthier, E. R., Piche, L., Lemieux, G., and Lemieux, R. (1996). Role of bcl-X(L) in the control of apoptosis in murine myeloma cells. Cancer Res 56, 1451–1456.
69. Miguel-Garcia, A., Orero, T., Matutes, E., Carbonell, F., Miguel-Sosa, A., Linares, M., Tarin, F., Herrera, M., Garcia-Talavera, J., and Carbonell-Ramon, F. (1998). bcl-2 expression in plasma cells from neoplastic gammopathies and reactive plasmacytosis: a comparative study. Haematologica 83, 298–304.
70. Jourdan, M., Veyrune, J. L., Vos, J. D., Redal, N., Couderc, G., and Klein, B. (2003). A major role for Mcl-1 antiapoptotic protein in the IL-6-induced survival of human myeloma cells. Oncogene 22, 2950–2959.

71. Zhang, B., Gojo, I., and Fenton, R. G. (2002). Myeloid cell factor-1 is a critical survival factor for multiple myeloma. Blood 99, 1885–1893.

72. Green, D. R. and Evan, G. I. (2002). A matter of life and death. Cancer Cell 1, 19–30.

73. Hahn, W. C. and Weinberg, R. A. (2002). Modelling the molecular circuitry of cancer. Nat Rev Cancer 2, 331–341.

74. Oltersdorf, T., Elmore, S. W., Shoemaker, A. R., Armstrong, R. C., Augeri, D. J., Belli, B. A., Bruncko, M., Deckwerth, T. L., Dinges, J., Hajduk, P. J., Joseph, M. K., Kitada, S., Korsmeyer, S. J., Kunzer, A. R., Letai, A., Li, C., Mitten, M. J., Nettesheim, D. G., Ng, S., Nimmer, P. M., O'Connor, J. M., Oleksijew, A., Petros, A. M., Reed, J. C., Shen, W., Tahir, S. K., Thompson, C. B., Tomaselli, K. J., Wang, B., Wendt, M. D., Zhang, H., Fesik, S. W., and Rosenberg, S. H. (2005). An inhibitor of Bcl-2 family proteins induces regression of solid tumours. Nature 435, 677–681.

75. Jansen, B., Schlagbauer-Wadl, H., Brown, B. D., Bryan, R. N., van Elsas, A., Muller, M., Wolff, K., Eichler, H. G., and Pehamberger, H. (1998). bcl-2 antisense therapy chemosensitizes human melanoma in SCID mice. Nat Med 4, 232–234.

76. O'Brien, S. M., Cunningham, C. C., Golenkov, A. K., Turkina, A. G., Novick, S. C., and Rai, K. R. (2005). Phase I to II multicenter study of oblimersen sodium, a Bcl-2 antisense oligonucleotide, in patients with advanced chronic lymphocytic leukemia. J Clin Oncol 23, 7697–7702.

77. Waters, J. S., Webb, A., Cunningham, D., Clarke, P. A., Raynaud, F., di Stefano, F., and Cotter, F. E. (2000). Phase I clinical and pharmacokinetic study of bcl-2 antisense oligonucleotide therapy in patients with non-Hodgkin's lymphoma. J Clin Oncol 18, 1812–1823.

78. Konopleva, M., Tari, A. M., Estrov, Z., Harris, D., Xie, Z., Zhao, S., Lopez-Berestein, G., and Andreeff, M. (2000). Liposomal Bcl-2 antisense oligonucleotides enhance proliferation, sensitize acute myeloid leukemia to cytosine-arabinoside, and induce apoptosis independent of other antiapoptotic proteins. Blood 95, 3929–3938.

79. Levin, V. A. (1980). Relationship of octanol/water partition coefficient and molecular weight to rat brain capillary permeability. J Med Chem 23, 682–684.

80. Begley, D. J. (1996). The blood-brain barrier: principles for targeting peptides and drugs to the central nervous system. J Pharm Pharmacol 48, 136–146.

81. Walensky, L. D., Kung, A. L., Escher, I., Malia, T. J., Barbuto, S., Wright, R. D., Wagner, G., Verdine, G. L., and Korsmeyer, S. J. (2004). Activation of apoptosis in vivo by a hydrocarbon-stapled BH3 helix. Science 305, 1466–1470.

82. Service, R. F. (2000). Biochemistry. Protein arrays step out of DNA's shadow. Science 289, 1673.

83. Schwarze, S. R., Ho, A., Vocero-Akbani, A., and Dowdy, S. F. (1999). In vivo protein transduction: delivery of a biologically active protein into the mouse. Science 285, 1569–1572.

84. Levchenko, T. S., Rammohan, R., Volodina, N., and Torchilin, V. P. (2003). Tat peptide-mediated intracellular delivery of liposomes. Methods Enzymol 372, 339–349.

85. Lewin, M., Carlesso, N., Tung, C. H., Tang, X. W., Cory, D., Scadden, D. T., and Weissleder, R. (2000). Tat peptide-derivatized magnetic nanoparticles allow in vivo tracking and recovery of progenitor cells. Nat Biotechnol 18, 410–414.

86. Derossi, D., Calvet, S., Trembleau, A., Brunissen, A., Chassaing, G., and Prochiantz, A. (1996). Cell internalization of the third helix of the Antennapedia homeodomain is receptor-independent. J Biol Chem 271, 18188–18193.

87. Mi, Z., Mai, J., Lu, X., and Robbins, P. D. (2000). Characterization of a class of cationic peptides able to facilitate efficient protein transduction in vitro and in vivo. Mol Ther 2, 339–347.

88. Mann, D. A. and Frankel, A. D. (1991). Endocytosis and targeting of exogenous HIV-1 Tat protein. EMBO J 10, 1733–1739.

89. Schimmer, A. D., Hedley, D. W., Chow, S., Pham, N. A., Chakrabartty, A., Bouchard, D., Mak, T. W., Trus, M. R., and Minden, M. D. (2001). The BH3 domain of BAD fused to the Antennapedia peptide induces apoptosis via its alpha helical structure and independent of Bcl-2. Cell Death Differ 8, 725–733.

90. Goldsmith, K. C., Liu, X., Dam, V., Morgan, B. T., Shabbout, M., Cnaan, A., Letai, A., Korsmeyer, S. J., and Hogarty, M. D. (2006). BH3 peptidomimetics potently activate apoptosis and demonstrate single agent efficacy in neuroblastoma. Oncogene 25, 4525–33.

91. Tomita, M., Hayashi, M., and Awazu, S. (1995). Absorption-enhancing mechanism of sodium caprate and decanoylcarnitine in Caco-2 cells. J Pharmacol Exp Ther 272, 739–743.

92. Vieira, H. L., Haouzi, D., Hamel, C. E., Jacotot, E., Belzacq, A., Brenner, C., and Kroemer, G. (2000). Permeabilization of the mitochondrial inner membrane during apoptosis: impact of the adenine nucleotide translocator. Cell Death Differ 7, 1146–1154.

93. Westerhoff, H. V., Juretic, D., Hendler, R. W., and Zasloff, M. (1989). Magainins and the disruption of membrane-linked free-energy transduction. Proc Natl Acad Sci USA 86, 6597–6601.

94. Ellerby, H. M., Arap, W., Ellerby, L. M., Kain, R., Andrusiak, R., Rio, G. D., Krajewski, S., Lombardo, C. R., Rao, R., Ruoslahti, E., Bredesen, D. E., and Pasqualini, R. (1999). Anticancer activity of targeted pro-apoptotic peptides. Nat Med 5, 1032–1038.

95. Matsuzaki, K. (2001). Why and how are peptide-lipid interactions utilized for self defence? Biochem Soc Trans 29, 598–601.

96. Chin, J. W. and Schepartz, A. (2001). Design and evolution of a miniature Bcl-2 binding protein. Angew. Chem Int Ed 40, 3806–3809.

97. Yang, B., Liu, D., and Huang, Z. (2004). Synthesis and helical structure of lactam bridged BH3 peptides derived from pro-apoptotic Bcl-2 family proteins. Bioorg Med Chem Lett 14, 1403–1406.

98. Wang, J. L., Liu, D., Zhang, Z. J., Shan, S., Han, X., Srinivasula, S. M., Croce, C. M., Alnemri, E. S., and Huang, Z. (2000). Structure-based discovery of an organic compound that binds Bcl-2 protein and induces apoptosis of tumor cells. Proc Natl Acad Sci USA 97, 7124–7129.

99. Enyedy, I. J., Ling, Y., Nacro, K., Tomita, Y., Wu, X., Cao, Y., Guo, R., Li, B., Zhu, X., Huang, Y., Long, Y. Q., Roller, P. P., Yang, D., and Wang, S. (2001). Discovery of small-molecule inhibitors of Bcl-2 through structure-based computer screening. J Med Chem 44, 4313–4324.

100. Degterev, A., Lugovskoy, A., Cardone, M., Mulley, B., Wagner, G., Mitchison, T., and Yuan, J. (2001). Identification of small-molecule inhibitors of interaction between the BH3 domain and Bcl-xL. Nat Cell Biol 3, 173–182.

101. Nakashima, T., Miura, M., and Hara, M. (2000). Tetrocarcin A inhibits mitochondrial functions of Bcl-2 and suppresses its anti-apoptotic activity. Cancer Res 60, 1229–1235.

102. Tzung, S. P., Kim, K. M., Basanez, G., Giedt, C. D., Simon, J., Zimmerberg, J., Zhang, K. Y., and Hockenbery, D. M. (2001). Antimycin A mimics a cell-death-inducing Bcl-2 homology domain 3. Nat Cell Biol 3, 183–191.

103. Leone, M., Zhai, D., Sareth, S., Kitada, S., Reed, J. C., and Pellecchia, M. (2003). Cancer prevention by tea polyphenols is linked to their direct inhibition of antiapoptotic Bcl-2-family proteins. Cancer Res 63, 8118–8121.

104. Kitada, S., Leone, M., Sareth, S., Zhai, D., Reed, J. C., and Pellecchia, M. (2003). Discovery, characterization, and structure-activity relationships studies of proapoptotic polyphenols targeting B-cell lymphocyte/leukemia-2 proteins. J Med Chem 46, 4259–4264.

105. Becattini, B., Kitada, S., Leone, M., Monosov, E., Chandler, S., Zhai, D., Kipps, T. J., Reed, J. C., and Pellecchia, M. (2004). Rational design and real time, in-cell detection of the proapoptotic activity of a novel compound targeting Bcl-X(L). Chem Biol 11, 389–395.

106. Shore, G. C. and Viallet, J. (2005). Modulating the bcl-2 family of apoptosis suppressors for potential therapeutic benefit in cancer. Hematology (Am Soc Hematol Educ Program) 226–230.

Chapter 9
Autophagy and Tumor Suppression

Recent Advances in Understanding the Link between Autophagic Cell Death Pathways and Tumor Development

Shani Bialik and Adi Kimchi*

Abstract Autophagy is a process by which the cell recycles its components through self-consumption of cellular organelles and bulk cytoplasm. In times of stress, it serves to generate much needed nutrients. When overactivated, however, the orderly destruction of organelles can lead to cell death. At times, autophagic cell death is used as an alternative to apoptosis to eliminate unwanted, damaged, or transformed cells. Consistent with this, tumorigenesis is associated with a down-regulation in autophagy, and genes that mediate the execution of the process have been shown to be tumor suppressors. At the same time, basal autophagy has been harnessed by some tumor cells as a survival mechanism to protect against ischemia and signals that induce apoptosis. Thus, the relationship between autophagy and tumor development is complex. Here, we discuss the basic machinery of mammalian autophagy and its regulators, with specific emphasis on those genes that have been linked to cancer. Research supporting the divergent nature of autophagy in both tumor suppression and tumor progression is presented. We conclude with a survey of recent approaches to treating cancer with strategies that modulate autophagy.

Keywords autophagy, programmed cell death, tumor suppressor, DAPk, Beclin 1, mTOR

1 Introduction

It is now an accepted dogma that cancer can develop from the imbalance of cells which results from disruptions in cell death. This realization gave impetus to analyze the molecular, cellular, and genetic mechanisms of programmed cell death, in particular apoptosis. However, apoptosis is not the only means by which a cell can die in a programmed, regulated manner. Different cell death morphologies were long

Shani Bialik and Adi Kimchi
Department of Molecular Genetics, Weizmann Institute of Science, Rehovot, Israel 76100

* To whom correspondence should be addressed: e-mail: adi.kimchi@weizmann.ac.il

R. Khosravi-Far and E. White (eds.), *Programmed Cell Death in Cancer Progression and Therapy*.
© Springer 2008

observed in tissue (e.g., Schweichel and Merker, 1973), however, it is only within the past few years that these death processes were more precisely classified and their molecular aspects deciphered. Of these alternate death pathways, autophagic cell death, also referred to as type 2 cell death, has recently been characterized in more detail (see, e.g., Gozuacik and Kimchi, 2004). Autophagic cell death results from the self-consumption of cellular organelles from within by means of the basic cellular autophagy machinery. This involves the de novo formation of double membrane- or multimembrane-enclosed vesicles called autophagosomes that elongate and surround portions of cytosol, including organelles such as mitochondria and endoplasmic reticulum (ER) (Fig. 9.1). The mature autophagosome eventually fuses with the lysosome, forming an autolysosome, in which its contents are degraded by lysosomal enzymes. Autophagic cell death can be accompanied by membrane blebbing and partial chromatin condensation, yet DNA fragmentation and caspase activation do not have an active part in the process.

One salient question that has emerged from the recent studies on autophagic cell death is whether autophagy suppresses tumorigenesis, as does its better known counterpart, apoptosis. Although it seems obvious that any block in any cell death pathway would promote cancer growth, for autophagy the question is not so simple. Unlike apoptosis, autophagy has homeostatic functions as a catabolic process by which cellular components are recycled. During times of cell stress, such as starvation, autophagic degradation of cellular organelles and proteins provides the cell with essential nutrients and biochemical building blocks that are not available through external supply or de novo biosynthesis. Autophagy can also be used to remove damaged organelles, such as depolarized mitochondria, which, rather than killing the cell, prevents further damage and release of proapoptotic factors, thereby blocking cellular demise. In these scenarios, autophagy serves a prosurvival role. There are in fact, many examples in which inhibition of autophagy enhances cell death (see below for details). However, other scenarios clearly indicate that beyond some unknown threshold, too much self-eating and destruction of cellular contents can be lethal and contribute to cell death. Furthermore, several death-inducing stimuli

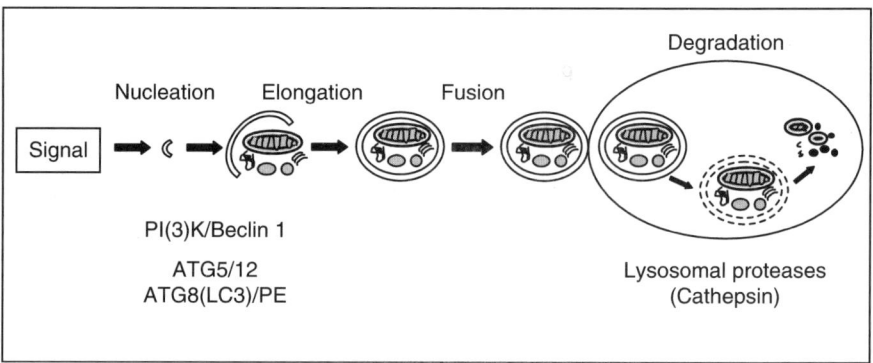

Fig. 9.1 Stages of autophagosome formation and the protein complexes that regulate them

have been shown to induce characteristics of autophagy, in addition to, or instead of, apoptosis. There is still some debate in the literature whether these signs of autophagy are causative to cell death, or merely accompany it, and may actually reflect a futile attempt at rescuing the cell. The recent identification of several mammalian autophagic genes has enabled researchers to elegantly block the autophagic pathway through genetic knockout and RNA interference (RNAi)-based knockdown experiments. Results of these studies has indicated that in certain circumstances (i.e., depending on the type of stimulus and the genetic makeup of the cell), autophagy, does in fact, contribute to the death of the cell. This may be one reason why several genes which regulate and/or execute autophagy have been implicated as tumor suppressors, in much the same way that apoptotic genes have been so characterized. These include Beclin 1 and DAP-kinase. In addition, several prominent oncogenes and tumor suppressor genes more commonly known to play a role in apoptotic signaling, such as p53, PI(3)K, PTEN, Bcl-2, and p19ARF, have now been shown to regulate autophagy.

This chapter will briefly present a summary of what is known about the molecular machinery that mediates and regulates mammalian autophagy, with particular emphasis on the components that have been linked to tumorigenesis. It will describe research indicating the contribution of autophagy to both cell survival and cell death pathways. Furthermore, it will also explore the possibility of harnessing autophagy as a means of destroying tumor cells.

2   The Molecular Basis of Autophagy

2.1   The Basic Machinery of Autophagosome Formation

Much of the known molecular mechanisms that control and/or execute autophagy were originally deciphered in yeast, although, recently, many of the relevant mammalian orthologues were identified (Tsukada and Ohsumi, 1993; Thumm et al., 1994 Harding et al., 1995). The yeast genes, now referred to by common consensus as the *ATG* genes (autophagy-related genes), encode 27 proteins that are necessary for the various stages of autophagic vesicle formation, fusion to the lysosome, and degradation of autophagosome contents (Fig. 9.1) (Klionsky et al., 2003). Prominent among these are several proteins that form a complex with the class III phosphatidylinositol 3-kinase (PI(3)K) Vps34, to produce phosphatidylinositol 3-phosphate (PI3-P), a lipid signaling molecule that is critical in the early stages of autophagosome nucleation (Petiot et al., 2000). Vps34 forms a complex with, and is regulated by, ATG6 (Beclin 1 in mammalians) and myristylated serine kinase Vps15/p150 (Stack et al., 1995). A fourth component of the yeast complex, ATG14, directs the complex to organizing centers of prevacuolar structures known as pre-autophagosomal structures (PAS) (Kim et al., 2002). The mammalian equivalent of ATG14 has yet to be discovered, and the PAS has not been observed in mammalian cells. Yet, in these cells, the PAS may be mimicked by sites on the trans-Golgi network and the ER to which Beclin 1 localizes, which serve as foci of PI3-P formation (Liang et al., 1998; Kihara et al., 2001; Pattingre et al., 2005). PI3-P is necessary for the nucleation of nascent membranes that will form the autophagosome. The exact mechanism is not yet known, but it presumably involves the docking to these nucleation sites of autophagy-specific proteins containing the domains FYVE (conserved in Fab1, YOTB, Vac1, and EEA1) or PX (Phox homology), which have a high affinity for PI3-P (Gillooly et al., 2001; Wishart et al., 2001). These proteins are predicted to control membrane formation and elongation.

The next stages of autophagic vesicle membrane recruitment and elongation utilize two ubiquitin-like pathways (Fig. 9.2). ATG12, a ubiquitin-like protein, is covalently conjugated to ATG5 in a constitutive manner via the sequential E1-ligase and E2-like activities of ATG7 and ATG10 (Mizushima et al., 1998; Tanida et al., 1999; Shintani et al., 1999). The ATG12/ATG5 dimer binds ATG16, which, through its ability to homo-oligomerize, leads to the formation of larger complexes of 800 kDa in mammals (Mizushima et al., 1999, 2003). This complex associates with the outer membrane of the elongating vesicle until completion of the autophagosome (Mizushima et al., 2001, 2003). The ATG12/ATG5 conjugation system is necessary for the second ubiquitin-like pathway (Mizushima et al., 2001; Suzuki et al., 2001). In this pathway, ATG7 and a second E2-like protein, ATG3, mediate the conjugation of ATG8 (or its mammalian counterpart, microtubule-associated protein 1 light chain 3, or LC3), not to a ubiquitin-like molecule, but rather to the lipid phosphatidylethanolamine (PE) (Ichimura et al., 2000). This is a critical step in the recruitment of lipid molecules for the expansion of the autophagic vesicle. The conjugation occurs via an amide bond formed between the amino

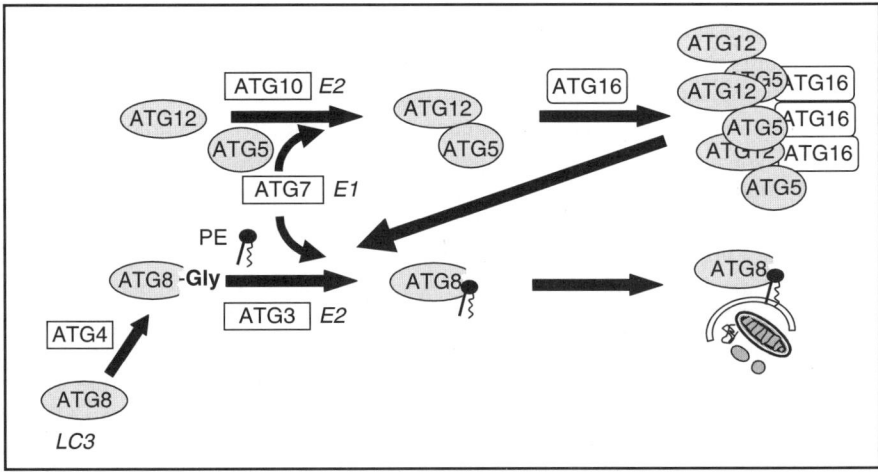

Fig. 9.2 Two ubiquitin-like conjugating systems mediate vesicle nucleation

group of the lipid molecule and the carboxyl-terminal glycine residue of ATG8, which is exposed following the cleavage of ATG8's C-terminus by the ATG4 cysteine protease (Ichimura et al., 2000; Kirisako et al., 2000). Although the proteolysis takes place immediately following translation of the protein, its lipidation occurs only upon stimulation of autophagy, and converts it from a soluble, cytosolic protein of 18 kDa (LC3-I) to a vesicle-associated form that migrates more rapidly on sodium dodecyl sulfate polyacrylamide gel electrophoresis (SDS-PAGE) (LC3-II). These properties of LC-3 have been used extensively as a marker for autophagy in mammalian cells (Kirisako et al., 1999; Kamada et al., 2000).

Once formed, the mature autophagosome fuses with the lysosome, generating the autolysosome. An intact microtubular network is required for this fusion step, at least in some cell types (Webb et al., 2004; Kochl et al., 2006), while actin microfilaments are involved in earlier stages of autophagosome formation (Aplin et al., 1992). Within the autolysosomal compartment, the engulfed organelles and cytosolic proteins are degraded by resident lysosomal enzymes, such as cathepsins.

2.2   Regulators of Autophagy Signaling

Many complex signaling pathways regulate the activity of the ATG proteins and autophagosome formation (Fig. 9.3). One prime regulator is target of rapamycin (TOR) kinase, a sensor of growth factor, nutrient, and energy availability, which converts these signals to cell growth and proliferative responses. TOR, whose activity is associated with inhibition of autophagy, is active in nutrient-rich conditions and upon growth factor stimulation. The mammalian TOR (mTOR) is regulated by several survival signals emanating from signaling molecules such as Akt/PKB (Inoki et al., 2002; Hahn-Windgassen et al., 2005), ERK (Ma et al., 2005), RSK1 (Roux et al., 2004), and the small GTP-binding protein, Rheb (Inoki et al., 2003; Fingar and Blenis, 2004). In yeast, TOR phosphorylates ATG13, which reduces its affinity to the ATG1 kinase (Kamada et al., 2000; Scott et al., 2000). ATG1 kinase activity is necessary for autophagy induction, and this activity requires tight association with ATG13. When TOR activity is blocked, such as during nutrient deprivation or upon treatment with rapamycin, ATG13 is rapidly dephosphorylated and binds to and activates ATG1 (Kamada et al., 2000; Abeliovich et al., 2003). The precise function of ATG1 and the mammalian counterpart of this pathway are not yet known.

mTOR is an important determinant of cell survival vs growth. Two critical mTOR substrates are 4E-BP1 and p70S6K. Phosphorylation of 4E-BP1 enhances cap-dependent translation. p70S6K in turn phosphorylates proteins that are involved in transcription, protein synthesis, and RNA splicing (Fingar and Blenis, 2004; Wang et al., 2001). For example, p70S6K phosphorylates eEF-2 kinase, inhibiting its activity (Wang et al., 2001). The active form of eEF-2 kinase, through phosphorylation of the translation factor eEF-2, blocks the elongation phase of translation. Thus, activation of mTOR leads ultimately to derepression of eEF-2

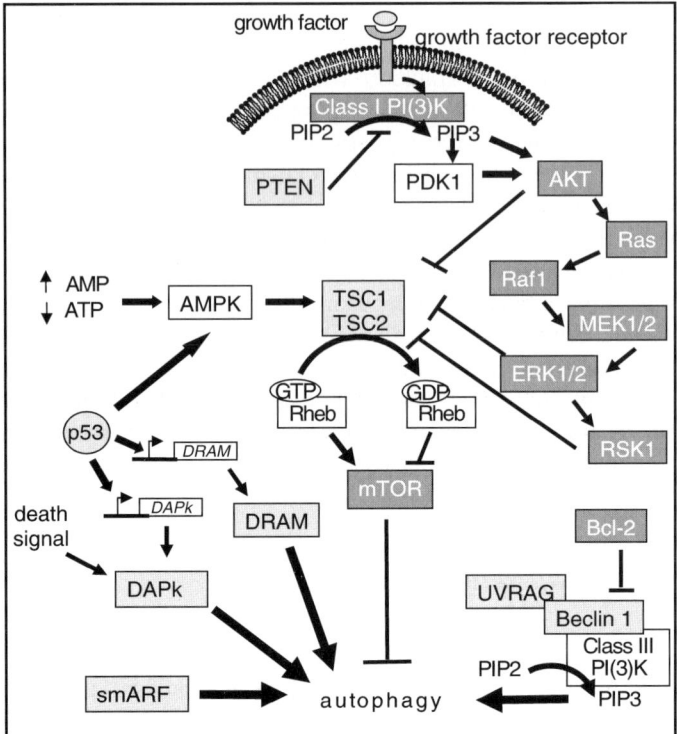

Fig. 9.3 Signaling pathways that regulate mammalian autophagy. Lightly shaded boxes represent tumor suppressor genes and darkly shaded genes are known oncogenes

and the promotion of translation. Under these circumstances, autophagy is repressed. In scenarios when cap-dependent protein synthesis is turned off, such as during amino acid starvation, ER-stress, and viral infection, certain proteins involved in the transcriptional regulation of autophagy-related genes are nevertheless upregulated (Natarajan et al., 2001), and autophagy results.

2.3 Autophagy Regulators Linked to Cancer

The most powerful genetic evidence linking autophagy to cancer development and progression is the emerging notion that genes which positively control autophagy display tumor suppressive functions when assessed in human tumors or in cancer model systems. Examples of such genes include Beclin 1, DAPk, p19ARF (via a novel isoform, SmARF [For short, mitochondrial ARF]), p53, PTEN, and TSC1/2. Conversely, several oncogenes, such as AKT, ERK1/2, and class I PI(3)K, have been shown to antagonize autophagy. This section will discuss these genes and their contributions to the promotion and suppression of autophagy, respectively.

2.3.1   Beclin 1

The cloning of Beclin 1, the human orthologue of yeast ATG6, provided the first link between an autophagic gene and tumor suppression. Beclin 1 was originally identified in a screen for Bcl-2-interacting proteins (Liang et al., 1998). The *Beclin 1* gene showed a high incidence of haploinsufficeincy in numerous breast cancer cell lines, and was downregulated in more than 50% of breast tumors analyzed in one initial study (Liang et al., 1999). In fact, the Beclin 1 monoallelic deletion on chromosome 17q21 is common not only to breast cancer, but also to ovarian and prostate cancers (Aita et al., 1999). Experimental deletion of Beclin 1 in mice confirmed the findings in human tumors; heterozygous knockout mice showed an increase in the preponderance of spontaneous lung cancer, lymphoma, and hepatocellular carcinoma (Qu et al., 2003; Yue et al., 2003). These tumor-suppressive activities of Beclin 1 were attributed to its role as an inducer of autophagy. Expression of Beclin-1 in MCF-7 breast carcinoma cells induced autophagy and blocked tumor formation in nude mice (Liang et al., 1999).

These studies provide strong evidence that autophagy, like apoptosis, is tumor suppressive, and that its downregulation provides tumor cells with a distinct advantage that promotes tumor growth. Furthermore, they may provide an additional functional explanation to Bcl-2's oncogenic properties. Bcl-2 antagonizes Beclin 1's autophagic activity, by binding Beclin 1 and blocking its association with Vps34, thus inhibiting PI(3)K activity (Pattingre et al., 2005). In this manner, overexpression of Bcl-2, a common occurrence in cancer, may lead to tumor growth as a result of its antiapoptotic properties as well as its ability to inhibit autophagy (Pattingre et al., 2005; Cardenas-Aguayo Mdel et al., 2003; Saeki et al., 2000).

A second Beclin 1-binding protein is UVRAG (for UV irradiation resistance-associated gene), which has the opposite effect on the Beclin 1/PI(3)K complex from Bcl-2 (Liang et al., 2006). UVRAG was isolated as part of a multiprotein complex containing Bcl-2, Beclin 1, and PI(3)K. Expression of UVRAG in a colon cancer cell line in which the endogenous gene is downregulated due to a heterozygous mutation in the *UVRAG* gene led to increases in both basal and starvation-induced autophagy. This was dependent on the presence of Beclin 1 and on the ability of UVRAG to interact with Beclin 1. Conversely, UVRAG was necessary for Beclin 1 and starvation-induced autophagy in MCF-7 cells. UVRAG binding to Beclin 1 directly enhanced both the Beclin 1/PI(3)K complex and PI(3)K activity. As a consequence of the increased autophagy, UVRAG expression suppressed proliferation, anchorage-independent growth, and tumor formation in vivo, but paradoxically, did not enhance cell death or otherwise reduce cell number. Thus, UVRAG-mediated autophagy is tumor suppressive, but not simply as a consequence of increased cell death. UVRAG appears to be a bona fide tumor suppressor; it maps to chromosome 11q13, a locus frequently associated with breast and colon cancer, and is found to be monoallelically deleted in multiple human cancer cell lines and tumors.

2.3.2 DAPk and Family Members

Death-associated protein kinase (DAP-kinase, DAPk) was isolated as a gene whose function was necessary for interferon-γ (IFN-γ)-induced death in HeLa cervical cancer cells (Deiss et al., 1995). It was later recognized that IFN-γ induced a caspase-independent death in these cells that bore evidence of autophagy (Inbal et al., 2002). In fact, overexpression of DAPk, as well as its closely related family members, DRP-1 (DAPK2) and ZIPk (DAPk3), can induce autophagosome formation and cell death in numerous cell lines (Inbal et al., 2002; Shani et al., 2004). DRP-1, too, was shown to be necessary for autophagic cell death in MCF-7 cells starved of amino acids or treated with tamoxifen (Inbal et al., 2002). The interest in these kinases further increased once it became apparent that they can be linked to both apoptotic and autophagic cell deaths, suggesting that they may function as molecular switches or integrators of both pathways. Hence, it became important to define the exact cellular settings and the underlying molecular mechanisms which dictate the choice between apoptosis and autophagy when triggered by these kinases.

DAPk is a Ca^{2+}/calmodulin-dependent Ser/Thr protein kinase and the founding member of a family of death-associated kinases, all of which share significant homology within the common kinase domain (for review, see Bialik and Kimchi, 2006). The family members' extra-catalytic domains differ, and reflect their divergent regulation and cellular localizations. While DAPk, and sometimes ZIPk, localize to the actin cytoskeleton, DRP-1 is a soluble protein, which has been found inside autophagosomes upon overexpression (Cohen et al., 1997; Bialik et al., 2004; Inbal et al., 2002; Page et al., 1999; Vetterkind et al., 2005; Komatsu and Ikebe, 2004). Their kinase activity is necessary for both apoptotic and autophagic cell deaths. A wide range of death stimuli have been reported to activate these kinases, and furthermore, require their activities for completion of the death process (Bialik and Kimchi, 2006). The kinases are regulated mostly by posttranslational modifications, including phosphorylation, and DAPk has also been shown to be regulated at the transcriptional level by p53 and TGFβ (Martoriati et al., 2005; and see Bialik and Kimchi, 2006).

Several substrates and downstream pathways have been identified over the years that may explain some of the death-inducing capabilities of these kinases. For example, one substrate common to DAPk, DRP-1, and ZIPk is the regulatory light chain of myosin II, phosphorylation of which mediates membrane blebbing in both apoptotic and autophagic cells (Bialik et al., 2004; Kuo et al., 2003; Vetterkind et al., 2005; Komatsu and Ikebe, 2004; Murata-Hori et al., 2001; Inbal et al., 2002). Another DAPk substrate is syntaxin-1A, a component of the SNARE complex, which mediates docking and fusion of synaptic vesicles with the plasma membrane (Tian et al., 2003). This, combined with the fact that RNAi-based knockdown of both DAPk and DRP-1 blocked clathrin-mediated endocytosis (Pelkmans et al., 2005), indicates a potential role in membrane fusion events that may be related to their ability to induce autophagy.

DAPk has been shown to be a tumor suppressor, whose activities have been directly linked to its ability to promote cell death. It functions at several stages of tumor development. It can block initial cellular transformation by growth-promoting oncogenes, by activating a p53/p19^{ARF}-dependent apoptotic checkpoint (Raveh et al., 2001). Specifically, expression of DAPk in primary mouse embryonic fibroblasts suppressed the oncogenic properties of E2F-1 and c-Myc by inducing caspase-dependent apoptosis, provided that functional p53 and p19^{ARF} were present in these cells. Moreover, DAPk expression led to increases in p53 and p53 transcriptional activity. The apoptotic response to oncogenes was attenuated in DAPk knockout cells, as was the induction of p53 and p19^{ARF} (Raveh et al., 2001). In addition to this role in an early apoptotic checkpoint, DAPk has also been shown to block tumor metastasis. Highly metastatic lung carcinoma cells lacked DAPk expression. Reintroduction of DAPk to these cells at physiological levels resulted in a reduced metastatic activity in mouse models of metastasis compared to the parental clones (Inbal et al., 1997). This was attributed to the ability of DAPk to sensitize the tumor cells to various death stimuli. Significantly, loss of DAPk expression due mainly to promoter methylation, but also to loss of heterozygosity, has been documented in a wide range of tumors, including B- and T-cell malignancies, breast cancer, lung carcinoma, head and neck cancer, gastric cancer, cervical, and prostate cancer (see Bialik and Kimchi (2004) for review, and also supplementary Table 9.1 in Bialik and Kimchi (2006) for details). In fact, DAPk promoter methylation has been used as a diagnostic tool for cancer detection in tumor and blood samples. Furthermore, DAPk loss of expression has been associated in some cases with disease progression and severity, metastatic rates and disease recurrence (Bialik and Kimchi, 2004).

As mentioned earlier, DAPk can modulate both apoptotic and autophagic cell deaths. It is not known the degree by which each of these mechanisms contributes to its tumor suppressive capabilities. This may very well be dependent on cell type and the individual signaling environment present.

2.3.3 The ARF Tumor Suppressor and smARF

The INK4a/ARF locus is commonly deleted in many cancers (Lowe and Sherr, 2003). It encodes two tumor suppressors: the p16^{INK4a} inhibitor of the retinoblastoma gene (*Rb*) and the ARF protein (p14^{ARF} in human and p19^{ARF} in mouse), translated from an alternative leading frame. ARF's tumor suppressive capabilities stem in most part, from its ability to activate p53 by negatively antagonizing its inhibitor, Mdm2. Yet, p53- and Mdm2-independent functions have also been ascribed to ARF. These include inhibition of rRNA processing in the nucleolus by binding of ARF to nucleoplasmin/B23 (Bertwistle et al., 2004; Sugimoto et al., 2003). Recently, it has been reported that the p53-independent effects of ARF on cell death may be attributed to a short, mitochondrial ARF isoform (known as smARF) that is produced from internal translation of both the human and the mouse mRNAs (Reef et al., 2006). This novel isoform lacks the N-terminal domains that mediate nuclear localization and Mdm2 binding. smARF localizes to the mitochondria in a compartment which is resistant to proteinase K and induces mitochondrial depolarization, without causing cytochrome C release or caspase activation. At the cellular level, smARF expression led to pronounced autophagy and caspase-independent cell death. smARF-induced cell death was partially attenuated by knockdown of Beclin 1 or ATG5, implying that, in this case, autophagy was causative to cell death. smARF is an unstable, short-lived protein that was upregulated by oncogene expression. This suggests that the autophagic function of ARF, mediated by its short form, may serve to counteract hyperproliferative signals generated by oncogenes. The dual nature of ARF, to induce apoptosis via the long p19ARF nucleolar isoform, or autophagy via mitochondrial smARF, may enable a choice of death pathways whose execution will depend on the particular genetic environment. In cases when p53-mediated apoptosis is blocked, as occurs in many tumor cells, smARF may provide a convenient back up plan that ensures cell death and maintains ARF's tumor suppressive function.

2.3.4 Tumor Suppressors and Oncogenes that Regulate the mTOR Pathway

Many of the signaling molecules that regulate the mTOR pathway are known oncogenes or tumor suppressors (see Fig. 9.3). mTOR activity is controlled by Rheb, a small GTPase of the Ras superfamily that activates mTOR in its GTP-bound form (see Sarbassov dos et al., 2005 for review). Rheb's GTPase activity is enhanced by a GTP-activating protein complex comprised of TSC1 and TSC2. Thus, TSC1/TSC2 negatively regulate mTOR by converting active Rheb-GTP to inactive Rheb-GDP. Significantly, TSC1 and TSC2 are tumor suppressors, mutations in which lead to tuberous sclerosis syndrome, a disease manifested by the occurrence of benign tumors in multiple organs, especially in the brain, leading to severe neuropathologies (reviewed in Kwiatkowski and Manning, 2005).

The TSC1/TSC2 complex is regulated by phosphorylation of TSC2 by either AKT, ERK1/2, or RSK, in response to growth factors, or by the AMP kinase (AMPK), which senses energy and nutrient deprivation and is activated by high AMP/ATP ratios. AKT/ERK/RSK-mediated phosphorylation serves to inhibit TSC's activity, and thus activates the mTOR pathway, while phosphorylation by AMPK has the reverse effects, leading to inactivation of mTOR. AKT, ERK1/2 and RSK, and class I PI(3)K, which is an upstream activator of AKT, are all oncogenic and are associated with proliferative growth (reviewed in Samuels and Ericson, 2006). Interestingly, RSK, in addition to its role in activating mTOR through phosphorylation of TSC2, was also recently shown to phosphorylate DAPk on a site known to antagonize its functions (Anjum et al., 2005). This may suggest a second mechanism by which RSK can block autophagy. Conversely, ERK1/2 was shown to activate DAPk through phosphorylation, and DAPk, in turn, suppressed ERK nuclear functions through its sequestration to the cytoplasm (Chen et al., 2005a, b). Whether these inhibitory and activating phosphorylations affect DAPk's autophagic properties has not yet been assessed. Thus, how these events are integrated with ERK and RSK's established roles in suppression of autophagy, through modulation of the mTOR pathway, remains to be seen.

Upstream of the class I PI(3)K lies in the dual protein- and phosphoinositide-phosphatase PTEN, which antagonizes the PI(3)K/AKT pathway by dephosphorylating the second messenger PIP3. PTEN is also a known tumor suppressor, located on chromosome 10q23. It is subject to deletion and/or mutation in numerous cancers (reviewed in Kim and Mak, 2006). Thus, many of the signaling molecules that negatively regulate the mTOR pathway are known tumor suppressors, while activators of the pathway have been described as oncogenes. While these factors have multiple targets and affect many cellular signaling and survival pathways, their modulation of autophagy through regulation of mTOR may contribute to their tumor suppressive and oncogenic tendencies.

2.3.5   p53

The p53 tumor suppressor is mutated in 50% of all human cancers. Its antitumor properties stem from its function as the pivotal controller of cell cycle checkpoints, inducing, as appropriate, cell cycle arrest, cellular senescence, or apoptosis (also see Chapter 10). Now autophagy can be added to its numerous tumor-suppressive functions, as two recent reports have linked p53 to signaling pathways that mediate autophagy.

p53 can modulate autophagy through regulation of the mTOR pathway. Activation of p53 by DNA damage resulting from etoposide treatment, or p53 over-expression, led to inhibition of mTOR and reduced phosphorylation of its downstream substrates (Feng et al., 2005). This was accompanied by induction of autophagy. p53's effects were mediated by TSC1 and TSC2; deletion of TSC1 and TSC2 blocked p53's inactivation of mTOR, as did chemical inhibition of AMPK. AKT, on the other hand, seemed not to be affected by p53. Inactivation of mTOR by p53 may be an important component of its growth-suppressive functions. Upon sensing genotoxic or cytotoxic stresses, such as oncogene activation, DNA damage, or hypoxia, p53's first line of defense is to induce cell cycle arrest to enable repair, or at the very least, to prevent passage of the damage to daughter cells. Through inactivation of mTOR and subsequent suppression of protein synthesis, p53 achieves a halt not only in cell cycle progression, but also in cell growth. At the same time, autophagy is induced. This may provide nutrients and energy to the cell in its time of stress, or, in more extreme circumstances, may join p53's apoptotic responses in eliminating the damaged cell once and for all.

p53 has also been recently linked to autophagy through the upregulation of a novel transcriptional target, DRAM (for damage-regulated autophagy modulator) (Crighton et al., 2006). DRAM has a p53-response element in its promoter, and is induced by p53 expression or DNA-damaging agents. Knockdown experiments indicated that DRAM is necessary for p53-induced death and autophagy. DRAM may be a specific stress-related regulator of autophagy, and not part of the general autophagic machinery. This is based on the observation that while knockdown of ATG5 inhibits clonogenicity even in the absence of any outside signal, knockdown of DRAM enhances clonogenic growth of cells treated with DNA-damaging agents, but has no effect on basal growth of untreated cells. Interestingly, expression of

DRAM alone is not sufficient to induce death, but does lead to enhanced autophagy. DRAM localizes to the lysosomal membrane, yet its exact functional activity is not yet known. Significantly, DRAM has characteristics of a tumor suppressor. It was found to be downregulated at the mRNA level in nearly 50% of primary squamous cell carcinomas, but not in breast tumors. CpG island methylation-mediated suppression of gene expression accounted for 28% of the cases. Suggestively, *DRAM* tended to be lost more frequently from tumors with intact *p53* compared with mutant *p53*, indicating that the two proteins operate in overlapping pathways to suppress tumorigenesis.

These studies suggest several mechanisms by which p53 can induce autophagy: transactivation of DRAM and inhibition of the mTOR pathway through AMPK and TSC1/2. However, these results are recent and rather preliminary, and require further investigation in order to fully understand their contributions to the regulation of autophagy and to p53's tumor suppressive activity. Considering the multifaceted nature of p53, its numerous transcriptional targets, and transcriptional-independent functions, future research is likely to reveal further mechanisms by which p53 modulates the autophagic pathway.

3   The Cell Death vs Cell Survival Paradox: How Does Autophagy Contribute to Malignancy?

It is clear from the abundance of tumor suppressors and oncogenes that serve some role in the autophagic process that autophagy has a strong link to the development of cancer. Yet, the exact nature of this link has been subject to debate. Autophagy for the most part has been shown to suppress tumor growth and cellular transformation, and several mechanisms have been proposed to explain this phenomenon. Yet, there have also been studies showing that autophagy positively contributes to tumorigenesis. This controversy stems from the bifunctionality of autophagy, which can display either a cytoprotective or cytotoxic role, depending on the nature of the stress conditions and the genetic milieu of the cells which are exposed to these stimuli.

Autophagy, as a fundamental process that controls protein and organelle recycling, has been shown to play an essential cellular survival role. This role is particularly apparent in the phenotype of ATG knockout mice. While Beclin 1 heterozygotes developed spontaneous cancer, homozygous deletion of the gene was embryonic lethal (Yue et al., 2003). ATG7 deficiency in the central nervous system resulted in death within 28 weeks of birth. Severe behavioral defects and neuropathies were observed and the accumulation of inclusion bodies containing uncleared ubiquitinated proteins led to the death of cerebral and cerebellar neurons (Komatsu et al., 2006). Similar accumulation of abnormal ubiquitinated proteins and inclusions were observed in mice deficient for ATG5 in neurons, resulting in neurodegeneration and loss of motor function (Hara et al., 2006). Thus, in these scenarios, autophagy as a means for normal clearance of cytosolic proteins is essential for cell survival.

Early studies on autophagy were performed in yeast, where it is activated by limiting nutrient conditions to provide the cell with energy and amino acids through self-catabolism, thereby ensuring cell survival (Huang and Klionsky, 2002). Although unicellular organisms like yeast may more frequently face a changing extracellular environment that requires autophagic adaptation, multicellular organisms at times are also subjected to stresses that induce autophagy. An elegant example of physiologically relevant starvation-induced autophagy was demonstrated in neonatal mice (Kuma et al., 2004). Immediately after birth, neonates undergo an adaptation phase to the loss of placental blood supply until they learn to suckle and receive nutrients through mother's milk. During this initial period of starvation, autophagy is observed to increase in neonatal tissue, peaking at 3–12 h post-birth, and returning to normal low levels within 1–2 days. This autophagic phase is essential for providing nutrients, and inhibition of autophagy at this stage is lethal. ATG5 knockout mice, though born healthy, die within 12 h of birth unless force-fed, due to the lack of the alternate nutrient source (Kuma et al., 2004). Autophagy is also required for the starvation response in adult mice, as demonstrated by the observance of cell swelling, accumulation of abnormal membrane structures, and damaged mitochondria in the liver of starved ATG7 conditional knockout mice (Komatsu et al., 2005).

Other studies focused on starvation-induced autophagy in cell culture models. For example, removal of IL-3 from cultures of Bax/Bak double knockout bone marrow cells resulted in starvation as a result of an impairment in nutrient uptake (Lum et al., 2005). In the absence of an intact apoptotic pathway, cells remained viable for as long as 24 weeks, but failed to proliferate and showed signs of atrophy and autophagy. The phenotype was reversible; restoration of IL-3 growth factor until 12 weeks after its removal led to the resumption of glycolysis, and eventually, to normal cell growth and proliferation rates. Thus, autophagy in this model was not associated with a point of no return. In fact, inhibition of autophagy through the knockdown of ATG5 or ATG7 accelerated cell death to 2–3 days. Likewise, blockage of autophagy through the use of chemical inhibitors or through RNAi-mediated knockdown of Beclin 1, ATG5, ATG10, or ATG12 enhanced the apoptotic cell death of amino acid- and serum-starved HeLa cells (Boya et al., 2005). In all of these examples, autophagy, rather than killing the cell, supports cell survival through the provision of otherwise lacking nutrients. Similarly, in tumors, cells present in the poorly vascularized tumor core are deprived of oxygen and nutrients, and may utilize autophagy-based recycling to offset starvation. In fact, autophagy was observed in the ischemic, unvascularized central portions of tumors derived from epithelial cells that could not undergo apoptosis due to deletion of Bax and Bak (Degenhardt et al., 2006). In such a scenario, autophagy would contribute to cell survival, and hence, tumorigenesis (Cuervo, 2004).

The very definitive and elegant studies presented earlier were utilized as proof by an adamant school of thought that autophagy is not a death-inducing process. However, equally convincing data has recently emerged that questions this one-sided approach, and it is now clear, that in certain cellular settings, autophagy can, in fact, lead to death. Cell death scenarios have long been observed to be accompanied by signs of autophagy. For example, autophagy was observed in developmental cell death such as that which occurs during insect metamorphosis, limb bud morphogenesis in birds, and palatal closure in mammals (Schweichel and Merker, 1973; Clarke, 1990; Bursch, 2001). As far back as the 1970s, ultrastructural morphologies that were consistent with autophagic cell death were present upon treatment with certain toxins (Schweichel and Merker, 1973). With the rise in popularity of autophagic cell death, many researchers have now specifically examined cell death morphologies for signs of autophagy, and have observed that common death stimuli thought previously to induce apoptosis can also induce autophagy in certain cell types. For example, both apoptosis and autophagy are observed upon lumen formation of acinar MCF-10A cells, and the death ligand TRAIL was shown to regulate the autophagic pathway (Mills et al., 2004). Etoposide, a known inducer of p53-dependent apoptosis, was shown to induce autophagy in a p53-dependent manner in primary mouse embryo fibroblasts (MEFs) (Feng et al., 2005). Ionizing radiation, inhibition of platelet-derived growth factor (PDGF) signaling, and arsenic trioxide induced autophagy, but not apoptosis, in malignant glioma cells (Kanzawa et al., 2003; Takeuchi et al., 2005a, b; Ito et al., 2005). Autophagy accompanied apoptosis in response to treatment of bovine mammary gland epithelial cells with TGFβ, as a model for mammary gland involution (Gajewska et al., 2005). It was also observed in MCF-7 cells treated with novel analogs of paclitaxel (Gorka et al., 2005) and in gastric cancer and glioma cell lines exposed to oncogenic Ras (Chi et al., 1999).

These basically correlative studies were further supported by results from directed intervention experiments which definitively established a causal relationship between autophagy and subsequent cellular demise in certain scenarios. This latter approach included the use of chemical inhibitors of autophagy, such as 3-methyl-adenine (3-MA) and wortmannin, which block PI(3)K activity, and more elegantly, specific genetic inhibition of autophagy through the use of RNAi-mediated knock-down of autophagy regulators. For example, anti-estrogen treatment of MCF-7 cells led to death which was accompanied by autophagic vesicle accumulation and was blocked by 3-MA (Bursch et al., 1996). Autophagy was evident upon nerve growth factor deprivation of primary sympathetic neurons, and again, death was blocked by 3-MA (Xue et al., 1999). Treatment of Bax/Bak double knockout fibroblasts with etoposide, thapsigargin, or staurosporine killed the cells despite the absence of apoptosis, and autophagy was evident. Death was attenuated upon addition of 3-MA, ATG5 RNAi, and in Beclin 1 -/- cells, indicating that autophagy contributed to death when apoptosis was blocked (Shimizu et al., 2004). Autophagy was the cause of cell death upon amino acid starvation of PC12 cells, since 3-MA, but not caspase inhibitors, blocked cell death. smARF-induced cell death was like-wise shown to be a result of autophagy, as RNAi to Beclin 1 and ATG5 blocked death (Reef et al., 2006). TNFα induced autophagy in Ewing sarcoma cells in the absence of NF-κB signaling, which activates the mTOR pathway. In this system, autophagy enhanced apoptotic cell death, since apoptotic morphologies were inhib-ited when autophagy was blocked through knockdown of Beclin 1 and ATG7 (Djavaheri-Mergny et al., 2006). A similar phenomenon was observed upon treatment of T-lymphoblast cell lines with TNFα (Jia et al., 1997). Macrophage cell death, triggered by lipopolysaccharides (LPS) in the presence of zVAD, showed evidence of autophagy and was blocked by chemical autophagy inhibitors and by RNAi to Beclin 1 (Xu et al., 2006). Likewise, autophagy was activated upon inhibition of caspases in L929 mouse fibroblast cells and U937 monocytes (Yu et al., 2004). Cell death, caused by the accumulation of ROS and the selective autophagy-mediated degradation of catalase, was attenuated by knockdown of LC3 and ATG7 (Yu et al., 2006). Furthermore, the involvement of several known death genes in the induction of autophagy, such as DAPk and DRP1 (Inbal et al., 2002), and BNIP3, a BH3-only member of the Bcl-2 family (Vande Velde et al., 2000), is further proof that autophagy promotes death.

In addition to inducing cell death and the removal of unwanted cells, autophagy may suppress tumor growth by other means. Even in circumstances when autophagy does not lead to cell death, it is counterproductive to cell growth. Cancer cells often show lower rates of autophagy and long-lived protein turnover compared to non-transformed cells (e.g., Gunn et al., 1977; Gronostajski and Pardee, 1984; Knecht et al., 1984; Kisen et al., 1993; Toth et al., 2002). Highly proliferative cells require a general increase in protein synthesis to keep up with the high demand for cell mass that must be divided among the ever-increasing number of daughter cells. Blocking the degradation of long-lived proteins through downregulation of autophagy helps favor the balance towards increased cell mass, thus providing a selective advantage to highly proliferative cells (Ng and Huang, 2005). Basal autophagy also serves to eliminate damaged organelles such as depolarized mitochondria that are a source of genotoxic free radicals. In the absence of this important scavenger mechanism, DNA mutations that may lead to cellular transformation can accumulate more readily. Thus, in this manner too, the lack of autophagy promotes tumorigenesis.

4   Autophagy as a Target for Therapeutic Intervention

In light of the data presented earlier, autophagy presents itself as a target for therapeutic intervention in the treatment of malignancies. In fact, several studies have demonstrated that use of chemotherapeutic drugs induces autophagic cell death, in addition to apoptosis, and even in cells that are resistant to apoptosis. For example, the vitamin D analog EB1089 used in cancer treatment induces autophagic cell death in MCF-7 cells, which leads to caspase-independent nuclear apoptosis (Hoyer-Hansen et al., 2005). The proteasome inhibitor MG132 kills PC3 prostate cancer cells by means of a caspase-dependent apoptotic pathway, and at the same time, leads to the upregulation of several autophagic genes. In fact, the cell death response is attenuated by addition of 3-MA. Autophagy was likewise responsible for the toxicity observed in non-small-cell lung cancer cells upon treatment with the rare earth element, neodymium oxide (Chen et al., 2005a, b). An innovative therapy involving a conditionally replicating adenovirus that targets telomerase-positive cancer cells was shown to kill malignant glioma, cervical cancer, and prostate cancer cells by means of autophagy, most likely through downregulation of mTOR signaling (Ito et al., 2006). The adenovirus treatment also slowed the growth of subcutaneous gliomas in nude mice and prolonged survival of the mice. Both a rapid apoptotic response and a slower autophagic one were observed upon treatment of MCF-7 cells with camptothecin (Lamparska-Przybysz et al., 2005). When apoptotic death was blocked, however, through disruption of Bax and Bid function, autophagy increased. Thus, here, autophagy serves as a backup to the disabled apoptotic pathway. In a similar vein, migrating glioblastoma multiform cells that were refractory to apoptosis-inducing drugs were induced to die via autophagy. The AKT/mTOR signaling pathway was constitutively active in these cells, providing a significant survival advantage. However, inhibition of this pathway with drugs such as temozolomide stimulated autophagy and cell death (Lefranc and Kiss, 2006). Temozolomide was also shown to induce autophagy, but not apoptosis, in malignant glioma cell lines, leading to cytotoxicity (Kanzawa et al., 2004). Histone deacetylase inhibitors, such as butyrate and suberoylanilide hydroxamic acid, can also trigger autophagic cell death in cells that have lost the ability to undergo apoptosis (Shao et al., 2004).

As stated earlier, however, autophagy can also act as a prosurvival pathway in certain cell environments and can thwart the induction of apoptosis. This may be especially true in hypoxic regions of tumors, where autophagy may serve as the only means to provide nutrients and energy to the starved tumor cell. Furthermore, one study of carcinogen-induced pancreatic cancer showed that rates of autophagy, although lower in advanced adenocarcinoma, actually were increased in early-stage premalignant nodules and adenomas (Toth et al., 2002). In these cases, drug treatments that affect autophagy may have opposing effects than those described earlier. For example, sulforaphane induced pronounced autophagy in prostate cancer cell lines, inhibition of which led to the rapid induction of apoptosis (Herman-Antosiewicz et al., 2006). Likewise, inhibition of autophagy enhanced the apoptotic response and, specifically, cytochrome C release, of the colon cancer cell line HT-29 to sulindac sulfide (Bauvy et al., 2001). Crotoxin, a neurotoxin derived from the venom of a South American rattlesnake, induced both apoptosis and autophagy in chronic myeloid leukemia cell lines (Yan et al., 2006). While caspase inhibition blocked cell death, inhibition of autophagy enhanced it. The authors of these papers concluded that the autophagic response of these cells to the chemotherapeutic agent served to suppress the apoptotic response. Similarly, ionizing radiation induced autophagy in breast, prostate, and colon cancer cells, and in malignant glioma cells, yet its inhibition sensitized cells to radiotherapy (Paglin et al., 2001; Ito et al., 2005). These results are consistent with the hypothesis that autophagy actually blocks the damaging affects of radiation by eliminating damaged organelles before they induce apoptosis. Altogether, it appears that effective treatment of malignancies in these specific cases would entail the combinatorial use of chemotherapeutic drugs that induce apoptosis and inhibit autophagy.

Despite the logic behind this approach, one should be wary of drawing conclusions based on responses of cancer cells in culture. An important recent paper addressed this limitation by assessing the effects of manipulations of the cell death programs, not just on tumor cells in culture, but also in the intact tumor in vivo (Degenhardt et al., 2006). In fact, the authors come to the opposite conclusion of those studies performed exclusively in vitro. Metabolic stress induced by ischemia was utilized to kill epithelial cells. Through genetic manipulation, immortalized baby mouse kidney epithelial cells (iBMK) that were defective in apoptosis, autophagy, or both, were generated. Cells exposed to metabolic stress died primarily by a rapid apoptotic cell death (within 24–72 h). When apoptosis was blocked (i.e., by Bax/Bak deficiency or Bcl-2/X$_L$ expression), autophagy was apparent, but enabled cell survival rather than death. Autophagic cells survived for long periods of time in culture, with normal proliferation rates initially, but eventually stopped dividing and moving, and exhibited signs of cell condensation. These defects, although resembling a death phenotype, were reversible upon restoration of oxygen and nutrients. When autophagy, too, was deficient (i.e., AKT expression or reduction in Beclin 1), cells exhibited a slow, inefficient death defined as necrosis. Significantly, the most aggressive, fastest-growing solid tumors developed from those cells that were deficient in both apoptosis and autophagy. These tumors contained large necrotic areas in ischemic regions, with macrophage infiltration and induction of an innate immune response. Necrotic tumors are known to be particularly aggressive, possibly due to proliferative signals generated by infiltrating immune cells, which encourage cell growth and angiogenesis in regions of the tumor that border the necrotic area.

Based on this study, it seems that effective treatment of solid tumors should involve strategies that encourage both apoptosis and autophagy. The latter could include inhibitors of the mTOR/AKT signaling pathways such as rapamycin and PI(3)K inhibitors (Takeuchi et al., 2005a, b). The benefit of triggering autophagy is somewhat of a paradox, as here, in contrast to the examples cited in the beginning of this section, autophagy enhanced cell survival. Yet, the survival of nonproliferative, dormant cells is preferable to the induction of necrosis, which has severe repercussions on overall tumor growth. Furthermore, treatments that activate autophagy may overcome the safe threshold under which autophagy promotes survival, and actually drive the cells towards autophagic cell death, providing an additional advantage to this strategy.

5   Conclusions and Future Perspectives

The dual nature of autophagy, and the dichotomy created by its contradictory effects on tumorigenesis, translates itself into a debate on whether one should be inhibiting or activating autophagy to treat cancer. More studies on the effects of manipulations of the autophagic program, in conjunction with the apoptotic pathway, on tumor growth in vivo are necessary. The role that autophagy plays, as either a cell survival mechanism or a cell death inducer, may be cancer-type specific, i.e., influenced by the genetic makeup of the corresponding tumor cells and the nature of the external stresses to which they are exposed. As a consequence, strategies for treatment will require a tumor-by-tumor genetic and environmental analysis. In order to accomplish such an analysis, it is necessary to acquire a complete understanding of autophagy, its molecular regulation, and its cellular effects. In the past several years, the field of mammalian autophagy has advanced in leaps and bounds; hopefully, the continued advances will translate to concrete clinical benefits in the near future.

References

Abeliovich, H., Zhang, C., Dunn, W. A., Jr., Shokat, K. M., and Klionsky, D. J. (2003). Chemical genetic analysis of Apg1 reveals a non-kinase role in the induction of autophagy. Mol Biol Cell 14, 477–490.

Aita, V. M., Liang, X. H., Murty, V. V., Pincus, D. L., Yu, W., Cayanis, E., Kalachikov, S., Gilliam, T. C., and Levine, B. (1999). Cloning and genomic organization of beclin 1, a candidate tumor suppressor gene on chromosome 17q21. Genomics 59, 59–65.

Anjum, R., Roux, P. P., Ballif, B. A., Gygi, S. P., and Blenis, J. (2005). The tumor suppressor DAP kinase is a target of RSK-mediated survival signaling. Curr Biol 15, 1762–1767.

Aplin, A., Jasionowski, T., Tuttle, D. L., Lenk, S. E., and Dunn, W. A., Jr. (1992). Cytoskeletal elements are required for the formation and maturation of autophagic vacuoles. J Cell Physiol 152, 458–466.

Bauvy, C., Gane, P., Arico, S., Codogno, P., and Ogier-Denis, E. (2001). Autophagy delays sulindac sulfide-induced apoptosis in the human intestinal colon cancer cell line HT-29. Exp Cell Res 2682, 139–149.

Bertwistle, D., Sugimoto, M., and Sherr, C. J. (2004). Physical and functional interactions of the arf tumor suppressor protein with nucleophosmin/b23. Mol Cell Biol 24, 985–996.

Bialik, S. and Kimchi, A. (2004). DAP-kinase as a target for drug design in cancer and diseases associated with accelerated cell death, Semin. Cancer Biol 14, 283–294.

Bialik, S. and Kimchi, A. (2006). The death-associated protein kinases: structure, function, and beyond. Annu Rev Biochem 75, 189–210.

Bialik, S., Bresnick, A. R., and Kimchi, A. (2004). DAP-kinase-mediated morphological changes are localization dependent and involve myosin-II phosphorylation. Cell Death Differ 11, 631–644.

Boya, P., Gonzalez-Polo, R. A., Casares, N., Perfettini, J. L., Dessen, P., Larochette, N., Metivier, D., Meley, D., Souquere, S., Yoshimori, T., Pierron, G., Codogno, P., and Kroemer, G. (2005). Inhibition of macroautophagy triggers apoptosis. Mol Cell Biol 25, 1025–1040.

Bursch, W. (2001). The autophagosomal-lysosomal compartment in programmed cell death. Cell Death Differ 8, 569–581.

Bursch, W., Ellinger, A., Kienzl, H., Torok, L., Pandey, S., Sikorska, M., Walker, R., and Hermann, R. S. (1996). Active cell death induced by the anti-estrogens tamoxifen and ICI 164 384 in human mammary carcinoma cells (MCF-7) in culture: the role of autophagy. Carcinogenesis 17, 1595–1607.

Cardenas-Aguayo Mdel, C., Santa-Olalla, J., Baizabal, J. M., Salgado, L. M., and Covarrubias, L. (2003). Growth factor deprivation induces an alternative non-apoptotic death mechanism that is inhibited by Bcl2 in cells derived from neural precursor cells. J Hematother Stem Cell Res 12, 735–748.

Chen, C. H., Wang, W. J., Kuo, J. C., Tsai, H. C., Lin, J. R., Chang, Z. F., Chen, R. H. (2005a). Bidirectional signals transduced by DAPK-ERK interaction promote the apoptotic effect of DAPK. EMBO J 24, 294–304.

Chen, Y., Yang, L., Feng, C., and Wen, L. P. (2005b). Nano neodymium oxide induces massive vacuolization and autophagic cell death in non-small cell lung cancer NCI-H460 cells. Biochem Biophys Res Commun 337, 52–60.

Chi, S., Kitanaka, C., Noguchi, K., Mochizuki, T., Nagashima, Y., Shirouzu, M., Fujita, H., Yoshida, M., Chen, W., Asai, A., Himeno, M., Yokoyama, S., and Kuchino, Y. (1999). Oncogenic Ras triggers cell suicide through the activation of a caspase-independent cell death program in human cancer cells. Oncogene 18, 2281–2290.

Clarke, P. G. (1990). Developmental cell death: morphological diversity and multiple mechanisms. Anat Embyol 181, 195–213.

Cohen, O., Feinstein, E., and Kimchi, A. (1997). DAP-kinase is a Ca2+/calmodulin-dependent, cytoskeletal-associated protein kinase, with cell death-inducing functions that depend on its catalytic activity. EMBO J 16, 998–1008.

Crighton, D., Wilkinson, S., O'Prey, J., Syed, N., Smith, P., Harrison, P. R., Gasco, M., Garrone, O., Crook T, and Ryan, K. M. (2006). DRAM, a p53-induced modulator of autophagy, is critical for apoptosis. Cell 126, 121–134.

Cuervo, A. M. (2004). Autophagy: in sickness and in health. Trends Cell Biol 14, 70–77.

Degenhardt, K., Mathew, R., Beaudoin, B., Bray, K., Anderson, D., Chen, G., Mukherjee, C., Shi, Y., Gelinas, C., Fan, Y., Nelson, D. A., Jin, S., and White, E. (2006). Autophagy promotes tumor cell survival and restricts necrosis, inflammation, and tumorigenesis. Cancer Cell 10, 51–64.

Deiss, L., Feinstein, E., Berissi, H., Cohen, O., and Kimchi, A. (1995). Identification of a novel serine/threonine kinase and a novel 15-kD protein as potential mediators of the gamma interferon-induced cell death. Genes Dev 9, 15–30.

Djavaheri-Mergny, M., Amelotti, M., Mathieu, J., Besancon, F., Bauvy, C., Souquere, S., Pierron, G., and Codogno, P. (2006). NF-kappa B activation represses TNF alpha-induced autophagy. J Biol Chem 281, 34870–34879.

Feng, Z., Zhang, H., Levine, A. J., and Jin, S. (2005). The coordinate regulation of the p53 and mTOR pathways in cells. Proc Natl Acad Sci USA 102, 8204–8209.

Fingar, D. C. and Blenis, J. (2004). Target of rapamycin (TOR): an integrator of nutrient and growth factor signals and coordinator of cell growth and cell cycle progression. Oncogene 23, 3151–3171.

Gajewska, M., Gajkowska, B., and Motyl, T. (2005). Apoptosis and autophagy induced by TGF-B1 in bovine mammary epithelial BME-UV1 cells. J Physiol Pharmacol 6S3, 143–157.

Gillooly, D. J., Simonsen, A., and Stenmark, H. (2001). Cellular functions of phosphatidylinositol 3-phosphate and FYVE domain proteins. Biochem J 355, 249–258.

Gorka, M., Daniewski, W. M., Gajkowska, B., Lusakowska, E., Godlewski, M. M., and Motyl, T. (2005). Autophagy is the dominant type of programmed cell death in breast cancer MCF-7 cells exposed to AGS 115 and EFDAC, new sesquiterpene analogs of paclitaxel. Anticancer Drugs 16, 777–788.

Gozuacik, D. and Kimchi, A. (2004). Autophagy as a cell death and tumor suppressor mechanism. Oncogene 2, 2891–2906.

Gronostajski, R. M. and Pardee, A. B. (1984). Protein degradation in 3T3 cells and tumorigenic transformed 3T3 cells. J Cell Physiol 119, 127–132.

Gunn, J. M., Clark, M. G., Knowles, S. E., Hopgood, M. F., and Ballard, F. J. (1977). Reduced rates of proteolysis in transformed cells. Nature 266, 58–60.

Hahn-Windgassen, A., Nogueira, V., Chen, C. C., Skeen, J. E., Sonenberg, N., and Hay, N. (2005). Akt activates the mammalian target of rapamycin by regulating cellular ATP level and AMPK activity. J Biol Chem 280, 32081–32089.

Hara, T., Nakamura, K., Matsui, M., Yamamoto, A., Nakahara, Y., Suzuki-Migishima, R., Yokoyama, M., Mishima, K., Saito, I., Okano, H., and Mizushima, N. (2006). Suppression of basal autophagy in neural cells causes neurodegenerative disease in mice. Nature 441, 885–889.

Harding, T. M., Morano, K. A., Scott, S. V., and Klionsky, D. J. (1995). Isolation and characterization of yeast mutants in the cytoplasm to vacuole protein targeting pathway. J Cell Biol 131, 591–602.

Herman-Antosiewicz, A., Johnson, D. E., and Singh, S. V. (2006). Sulforaphane causes autophagy to inhibit release of cytochrome C and apoptosis in human prostate cancer cells. Cancer Res 66, 5828–5835.

Hoyer-Hansen, M., Bastholm, L., Mathiasen, I. S., Elling, F., and Jaattela, M. (2005). Vitamin D analog EB1089 triggers dramatic lysosomal changes and Beclin 1-mediated autophagic cell death. Cell Death Differ 12, 1297–1309.

Huang, W. P. and Klionsky, D. J. (2002). Autophagy in yeast: a review of the molecular machinery. Cell Struct Funct 27, 409–420.

Ichimura, Y., Kirisako, T., Takao, T., Satomi, Y., Shimonishi, Y., Ishihara, N., Mizushima, N., Tanida, I., Kominami, E., Ohsumi, M., Noda, T., and Ohsumi, Y. (2000). A ubiquitin-like system mediates protein lipidation. Nature 408, 488–492.

Inbal, B., Cohen, O., Polak-Charcon, S., Kopolovic, J., Vadai, E., Eisenbach, L., and Kimchi, A. (1997). DAP kinase links the control of apoptosis to metastasis. Nature 390, 180–184.

Inbal, B., Bialik, S., Sabanay, I., Shani, G., and Kimchi, A. (2002). DAP kinase and DRP-1 mediate membrane blebbing and the formation of autophagic vesicles during programmed cell death. J Cell Biol 157, 455–468.

Inoki, K., Li, Y., Zhu, T., Wu, J., and Guan, K. L. (2002). TSC2 is phosphorylated and inhibited by Akt and suppresses mTOR signaling. Nat Cell Biol 4, 648–657.

Inoki, K., Zhu, T., and Guan, K. L. (2003). TSC2 mediates cellular energy response to control cell growth and survival. Cell 115, 577–590.

Ito, H., Daido, S., Kanzawa, T., Kondo, S., and Kondo, Y. (2005). Radiation-induced autophagy is associated with LC3 and its inhibition sensitizes malignant glioma cells. Int J Oncol 26, 1401–1410.

Ito, H., Aoki, H., Kuhnel, F., Kondo, Y., Kubicka, S., Wirth, T., Iwado, E., Iwamaru, A., Fujiwara, K., Hess, K. R., Lang, F. F., Sawaya, R., and Kondo, S. (2006). Autophagic cell death of malignant glioma cells induced by a conditionally replicating adenovirus. J Natl Cancer Inst 98, 625–636.

Jia, L., Dourmashkin, R. R., Allen, P. D., Gray, A. B., Newland, A. C., and Kelsey, S. M. (1997). Inhibition of autophagy abrogates tumour necrosis factor alpha induced apoptosis in human T-lymphoblastic leukaemic cells. Br J Haematol 98, 673–685.

Kamada, Y., Funakoshi, T., Shintani, T., Nagano, K., Ohsumi, M., and Ohsumi, Y. (2000). Tor-mediated induction of autophagy via an Apg1 protein kinase complex. J Cell Biol 150, 1507–1513.

Kanzawa, T., Kondo, Y., Ito, H., Kondo, S., and Germano, I. (2003). Induction of autophagic cell death in malignant glioma cells by arsenic trioxide. Cancer Res 63, 2103–2108.

Kanzawa, T., Zhang, L., Xiao, L., Germano, I. M., Kondo, Y., and Kondo, S. (2004). Arsenic trioxide induces autophagic cell death in malignant glioma cells by upregulation of mitochondrial cell death protein BNIP3. Cell Death Differ 11, 448–457.

Kihara, A., Kabeya, Y., Ohsumi, Y., and Yoshimori, T. (2001). Beclin-phosphatidylinositol 3-kinase complex functions at the trans-Golgi network. EMBO Rep 2, 330–335.

Kim, R. H. and Mak, T. W. (2006). Tumours and tremors: how PTEN regulation underlies both. Br J Cancer 94, 620–624.

Kim, J., Huang, W. P., Stromhaug, P. E., and Klionsky, D. J. (2002). Convergence of multiple autophagy and cytoplasm to vacuole targeting components to a perivacuolar membrane compartment prior to de novo vesicle formation. J Biol Chem 277, 763–773.

Kirisako, T., Baba, M., Ishihara, N., Miyazawa, K., Ohsumi, M., Yoshimori, T., Noda, T., and Ohsumi, Y. (1999). Formation process of autophagosome is traced with Apg8/Aut7p in yeast. J Cell Biol 147, 435–446.

Kirisako, T., Ichimura, Y., Okada, H., Kabeya, Y., Mizushima, N., Yoshimori, T., Ohsumi, M., Takao, T., Noda, T., and Ohsumi, Y. (2000). The reversible modification regulates the membrane-binding state of Apg8/Aut7 essential for autophagy and the cytoplasm to vacuole targeting pathway. J Cell Biol 151, 263–276.

Kisen, G. O., Tessitore, L., Costelli, P., Gordon, P. B., Schwarze, P. E., Baccino, F. M., and Seglen, P. O. (1993). Reduced autophagic activity in primary rat hepatocellular carcinoma and ascites hepatoma cells. Carcinogenesis 14, 2501–2505.

Klionsky, D. J., Cregg, J. M., Dunn, W. A., Jr., Emr, S. D., Sakai, Y., Sandoval, I. V., Sibirny, A., Subramani, S., Thumm, M., Veenhuis, M., and Ohsumi, Y. (2003). A unified nomenclature for yeast autophagy-related genes. Dev Cell 5, 539–545.

Knecht, E., Hernandez-Yago, J., and Grisolia, S. (1984). Regulation of lysosomal autophagy in transformed and non-transformed mouse fibroblasts under several growth conditions. Exp Cell Res 154, 224–232.

Kochl, R., Hu, X. W., Chan, E. Y., and Tooze, S. A. (2006). Microtubules facilitate autophagosome formation and fusion of autophagosomes with endosomes. Traffic 7, 129–145.

Komatsu, S. and Ikebe, M. (2004). ZIP kinase is responsible for the phosphorylation of myosin II and necessary for cell motility in mammalian fibroblasts. J Cell Biol 165, 243–254.

Komatsu, M., Waguri, S., Ueno, T., Iwata, J., Murata, S., Tanida, I., Ezaki, J., Mizushima, N., Ohsumi, Y., Uchiyama, Y., Kominami, E., Tanaka, K., and Chiba, T. (2005). Impairment of starvation-induced and constitutive autophagy in Atg7-deficient mice. J Cell Biol 169, 425–434.

Komatsu, M., Waguri, S., Chiba, T., Murata, S., Iwata, J., Tanida, I., Ueno, T., Koike, M., Uchiyama, Y., Kominami, E., and Tanaka, K. (2006). Loss of autophagy in the central nervous system causes neurodegeneration in mice. Nature 441, 880–884.

Kuma, A., Hatano, M., Matsui, M., Yamamoto, A., Nakaya, H., Yoshimori, T., Ohsumi, Y., Tokuhisa, T., and Mizushima, N. (2004). The role of autophagy during the early neonatal starvation period. Nature 432, 1032–1036.

Kuo, J. C., Lin, J. R., Staddon, J. M., Hosoya, H., and Chen, R. H. (2003). Uncoordinated regulation of stress fibers and focal adhesions by DAP kinase. J Cell Sci 116, 4777–4790.

Kwiatkowski, D. J. and Manning, B. D. (2005). Tuberous sclerosis: a GAP at the crossroads of multiple signaling pathways. Hum Mol Genet 14, R251–R258.

Lamparska-Przybysz, M., Gajkowska, B., and Motyl, T. (2005). Cathepsins and BID are involved in the molecular switch between apoptosis and autophagy in breast cancer MCF-7 cells exposed to camptothecin. J Physiol Pharmacol 56 (Suppl 3), 159–179.

Lefranc, F. and Kiss, R. (2006). Autophagy, the Trojan horse to combat glioblastomas, Neurosurg. Focus 20, E7.

Liang, X. H., Kleeman, L. K., Jiang, H. H., Gordon, G., Goldman, J. E., Berry, G., Herman, B., and Levine, B. (1998). Protection against fatal Sindbis virus encephalitis by beclin, a novel Bcl-2-interacting protein. J Virol 72, 8586–8596.

Liang, X. H., Jackson, S., Seaman, M., Brown, K., Kempkes, B., Hibshoosh, H., and Levine, B. (1999). Induction of autophagy and inhibition of tumorigenesis by beclin 1. Nature 402, 672–676.

Liang, C., Feng, P., Ku, B., Dotan, I., Canaani, D., Oh, B. H., and Jung, J. U. (2006). Autophagic and tumour suppressor activity of a novel Beclin1-binding protein UVRAG. Nature Cell Biol 8, 688–699.

Lowe, S. W. and Sherr, C. J. (2003). Tumor suppression by Ink4a-Arf: progress and puzzles. Curr Opin Genet Dev 13, 77–83.

Lum, J. J., Bauer, D. E., Kong, M., Harris, M. H., Li, C., Lindsten, T., and Thompson, C. B. (2005). Growth factor regulation of autophagy and cell survival in the absence of apoptosis. Cell 120, 237–248.

Ma, L., Chen, Z., Erdjument-Bromage, H., Tempst, P., and Pandolfi, P. P. (2005). Phosphorylation and functional inactivation of TSC2 by Erk implications for tuberous sclerosis and cancer pathogenesis. Cell 121, 179–193.

Martoriati, A., Doumont, G., Alcalay, M., Bellefroid, E., Pelicci, P. G., and Marine, J. C. (2005). Dapk1, encoding an activator of a p19ARF-p53-mediated apoptotic checkpoint, is a transcription target of p53. Oncogene 24, 1461–1466.

Mills, K. R., Reginato, M., Debnath, J., Queenan, B., and Brugge, J. S. (2004). Tumor necrosis factor-related apoptosis-inducing ligand (TRAIL) is required for induction of autophagy during lumen formation in vitro. Proc Natl Acad Sci USA 101, 3438–3443.

Mizushima, N., Noda, T., Yoshimori, T., Tanaka, Y., Ishii, T., George, M. D., Klionsky, D. J., Ohsumi, M., and Ohsumi, Y. (1998). A protein conjugation system essential for autophagy. Nature 395, 395–398.

Mizushima, N., Noda, T., and Ohsumi, Y. (1999). Apg16p is required for the function of the Apg12p-Apg5p conjugate in the yeast autophagy pathway. EMBO J 18, 3888–3896.

Mizushima, N., Yamamoto, A., Hatano, M., Kobayashi, Y., Kabeya, Y., Suzuki, K., Tokuhisa, T., Ohsumi, Y., and Yoshimori, T. (2001). Dissection of autophagosome formation using Apg5-deficient mouse embryonic stem cells. J Cell Biol 152, 657–668.

Mizushima, N., Kuma, A., Kobayashi, Y., Yamamoto, A., Matsubae, M., Takao, T., Natsume, T., Ohsumi, Y., and Yoshimori, T. (2003). Mouse Apg16L, a novel WD-repeat protein, targets to the autophagic isolation membrane with the Apg12-Apg5 conjugate. J Cell Sci 116, 1679–1688.

Murata-Hori, M., et al. (2001). HeLa ZIP kinase induces diphosphorylation of myosin II regulatory light chain and reorganization of actin filaments in nonmuscle cells. Oncogene 20, 8175–8183.

Natarajan, K., Meyer, M. R., Jackson, B. M., Slade, D., Roberts, C., Hinnebusch, A. G., and Marton, M. J. (2001). Transcriptional profiling shows that Gcn4p is a master regulator of gene expression during amino acid starvation in yeast. Mol Cell Biol 21, 4347–4368.

Ng, G. and Huang, J. (2005). The significance of autophagy in cancer. Mol Carcinogenesis 43, 183–187.

Page, G., Kogel, D., Rangnekar, V., and Scheidtmann, K. H. (1999). Interaction partners of Dlk/ZIP kinase: co-expression of Dlk/ZIP kinase and Par-4 results in cytoplasmic retention and apoptosis. Oncogene 18, 7265–7273.

Paglin, S., Hollister, T., Delohery, T., Hackett, N., McMahill, M., Sphicas, E., Domingo, D., and Yahalom, J. (2001). A novel response of cancer cells to radiation involves autophagy and formation of acidic vesicles. Cancer Res. 61, 439–444.

Pattingre, S., Tassa, A., Qu, X., Garuti, R., Liang, X. H., Mizushima, N., Packer, M., Schneider, M. D., and Levine, B. (2005). Bcl-2 antiapoptotic proteins inhibit beclin 1-dependent autophagy. Cell 122, 927–939.

Pelkmans, L, Pelkmans, L., Fava, E., Grabner, H., Hannus, M., Habermann, B., Krausz, E., and Zerial, M. (2005). Genome-wide analysis of human kinases in clathrin- and caveolae/raft-mediated endocytosis. Nature 436, 78–86.

Petiot, A., Ogier-Denis, E., Blommaart, E. F., Meijer, A. J., and Codogno, P. (2000). Distinct classes of phosphatidylinositol 3'-kinases are involved in signaling pathways that control macroautophagy in HT-29 cells. J Biol Chem 275, 992–998.

Qu, X., Yu, J., Bhagat, G., Furuya, N., Hibshoosh, H., Troxel, A., Rosen, J., Eskelinen, E. L., Mizushima, N., Ohsumi, Y., Cattoretti, G., and Levine, B. (2003). Promotion of tumorigenesis by heterozygous disruption of the beclin 1 autophagy gene. J Clin Invest 112, 1809–1820.

Raveh, T., Droguett, G., Horwitz, M. S., DePinho, R. A., and Kimchi, A. (2001). DAP kinase activates a p19ARF/p53-mediated apoptotic checkpoint to suppress oncogenic transformation. Nat Cell Biol 3, 1–7.

Reef, S., Zalckvar, E., Shifman, O., Bialik, S., Sabanay, H., Oren, M., and Kimchi, A. (2006). A short mitochondrial form of p19ARF induces autophagy and caspase-independent cell death. Mol Cell 22, 463–475.

Roux, P. P., Ballif, B. A., Anjum, R., Gygi, S. P., and Blenis, J. (2004). Tumor-promoting phorbol esters and activated Ras inactivate the tuberous sclerosis tumor suppressor complex via p90 ribosomal S6 kinase. Proc Natl Acad Sci USA 101, 13489–13494.

Saeki, K., You, A., Okuma, E., Yazaki, Y., Susin, S. A., Kroemer, G., Takaku, F. (2000). Bcl-2 down-regulation causes autophagy in a caspase-independent manner in human leukemic HL60 cells. Cell Death Differ 7, 1263–1269.

Samuels, Y. and Ericson, K. (2006). Oncogenic PI3K and its role in cancer. Curr Opin Oncol 18, 77–82.

Sarbassov dos, D., Ali, S. M., and Sabatini, D. M. (2005). Growing roles for the mTOR pathway. Curr Opin Cell Biol 17, 596–603.

Schweichel, J. U. and Merker, H. J. (1973). The morphology of various types of cell death in prenatal tissues. Teratology 7, 253–266.

Scott, S. V., Nice, D. C., III, Nau, J. J., Weisman, L. S., Kamada, Y., Keizer-Gunnink, I., Funakoshi, T., Veenhuis, M., Ohsumi, Y., and Klionsky, D. J. (2000). Apg13p and Vac8p are part of a complex of phosphoproteins that are required for cytoplasm to vacuole targeting. J Biol Chem 275, 25840–25849.

Shani, G., Marash, L., Gozuacik, D., Bialik, S., Teitelbaum, L., Shohat, G., and Kimchi, A. (2004). Death-associated protein kinase phosphorylates ZIP kinase, forming a unique kinase hierarchy to activate its cell death functions. Mol Cell Biol 24, 8611–8626.

Shao, Y., Gao, Z., Marks, P. A., and Jiang, X. (2004). Apoptotic and autophagic cell death induced by histone deacetylase inhibitors. Proc Natl Acad Sci USA 101, 18030–18035.

Shimizu, S., Kanaseki, T., Mizushima, N., Mizuta, T., Arakawa-Kobayashi, S., Thompson, C. B., and Tsujimoto, Y. (2004). Role of Bcl-2 family proteins in a non-apoptotic programmed cell death dependent on autophagy genes. Nature Cell Biol 6, 1221–1228.

Shintani, T., Mizushima, N., Ogawa, Y., Matsuura, A., Noda, T., and Ohsumi, Y. (1999). Apg10p, a novel protein-conjugating enzyme essential for autophagy in yeast. EMBO J 18, 5234–5241.

Stack, J. H., DeWald, D. B., Takegawa, K., and Emr, S. D. (1995). Vesicle-mediated protein transport: regulatory interactions between the Vps15 protein kinase and the Vps34 PtdIns 3-kinase essential for protein sorting to the vacuole in yeast. J Cell Biol 129, 321–334.

Sugimoto, M., Kuo, M. L., Roussel, M. F., and Sherr, C. J. (2003). Nucleolar Arf tumor suppressor inhibits ribosomal RNA processing. Mol Cell 11, 415–424.

Suzuki, K., Kirisako, T., Kamada, Y., Mizushima, N., Noda, T., and Ohsumi, Y. (2001). The pre-autophagosomal structure organized by concerted functions of APG genes is essential for autophagosome formation. EMBO J 20, 5971–5981.

Takeuchi, H., Kondo, Y., Fujiwara, K., Kanzawa, T., Aoki, H., Mills, G. B., and Kondo, S. (2005a). Synergistic augmentation of rapamycin-induced autophagy in malignant glioma cells by phosphatidylinositol 3-kinase/protein kinase B inhibitors. Cancer Res 65, 3336–3346.

Takeuchi, H., Kanzawa, T., Kondo, Y., and Kondo, S. (2005b). Inhibition of platelet-derived growth factor signalling induces autophagy in malignant glioma cells. Br J Cancer 90, 1069–1075.

Tanida, I., Mizushima, N., Kiyooka, M., Ohsumi, M., Ueno, T., Ohsumi, Y., and Kominami, E. (1999). Apg7p/Cvt2p: A novel protein-activating enzyme essential for autophagy. Mol Biol Cell 10, 1367–1379.

Thumm, M., Egner, R., Koch, B., Schlumpberger, M., Straub, M., Veenhuis, M., and Wolf, D. H. (1994). Isolation of autophagocytosis mutants of Saccharomyces cerevisiae. FEBS Lett 349, 275–280.

Tian, J. H., Das, S., and Sheng, Z. H. (2003). Ca2+-dependent phosphorylation of syntaxin-1A by the death-associated protein (DAP) kinase regulates its interaction with Munc18. J Biol Chem 278, 26265–26274.

Toth, S., Nagy, K., Palfia, Z., and Rez, G. (2002). Cellular autophagic capacity changes during azaserine-induced tumour progression in the rat pancreas. Up-regulation in all premalignant stages and down-regulation with loss of cycloheximide sensitivity of segregation along with malignant transformation. Cell Tissue Res 309, 409–416.

Tsukada, M. and Ohsumi, Y. (1993). Isolation and characterization of autophagy-defective mutants of Saccharomyces cerevisiae. FEBS Lett 333, 169–174.

Vande Velde, C., Cizeau, J., Dubik, D., Alimonti, J., Brown, T., Israels, S., Hakem, R., and Greenberg, A. H. (2000). BNIP3 and genetic control of necrosis-like cell death through the mitochondrial permeability transition pore. Mol Cell Biol 20, 5454–5468.

Vetterkind, S., Illenberger, S., Kubicek, J., Boosen, M., Appel, S., Naim, H. Y., Scheidtmann, K. H., and Preuss, U. (2005). Binding of Par-4 to the actin cytoskeleton is essential for Par-4/Dlk-mediated apoptosis. Exp Cell Res 305, 392–408.

Wang, X., Li, W., Williams, M., Terada, N., Alessi, D. R., and Proud, C. G. (2001). Regulation of elongation factor 2 kinase by p90(RSK1) and p70 S6 kinase. EMBO J 20, 4370–4379.

Webb, J. L., Ravikumar, B., and Rubinsztein, D. C. (2004). Microtubule disruption inhibits autophagosome-lysosome fusion: implications for studying the roles of aggresomes in polyglutamine diseases. Int J Biochem Cell Biol 36, 2541–2550.

Wishart, M. J., Taylor, G. S., and Dixon, J. E. (2001). Phoxy lipids: revealing PX domains as phosphoinositide binding modules. Cell 105, 817–820.

Xu, Y., Kim, S. O., Li, Y., and Han, J. (2006). Autophagy contributes to caspase-independent macrophage cell death. J Biol Chem 281, 19179–19187.

Xue, L., Fletcher, G. C., and Tolkovsky, A. M. (1999). Autophagy is activated by apoptotic signalling in sympathetic neurons: an alternative mechanism of death execution. Mol Cell Neurosci 114, 180–198.

Yan, C. H., Liang, Z. Q., Gu, Z. L., Yang, Y. P., Reid, P., and Qin, Z. H. (2006). Contributions of autophagic and apoptotic mechanisms to CrTX-induced death of K562 cells. Toxicon 47, 521–530.

Yu, L., Alva, A., Su, H., Dutt, P., Freundt, E., Welsh, S., Baehrecke, E. H., and Lenardo, M. J. (2004). Regulation of an ATG7-beclin 1 program of autophagic cell death by caspase-8. Science 304, 1500–1502.

Yu, L., Wan, F., Dutta, S., Welsh, S., Liu, Z., Freundt, E., Baehrecke, E. H., and Lenardo, M. (2006). Autophagic programmed cell death by selective catalase degradation. Proc Natl Acad Sci USA 103, 4952–4957.

Yue, Z., Jin, S., Yang, C., Levine, A. J., and Heintz, N. (2003). Beclin 1, an autophagy gene essential for early embryonic development, is a haploinsufficient tumor suppressor. Proc Natl Acad Sci USA 100, 15077–15082.

Chapter 10
Regulation of Programmed Cell Death by the P53 Pathway

Kageaki Kuribayashi and Wafik S. El-Deiry*

Abstract The p53 pathway is targeted for inactivation in most human cancers either directly or indirectly, highlighting its critical function as a tumor suppressor gene. p53 is normally activated by cellular stress and mediates a growth-suppressive response that involves cell cycle arrest and apoptosis. In the case of cell cycle arrest, p21 appears sufficient to block cell cycle progression out of G1 until repair has occurred or the cellular stress has been resolved. The p53-dependent apoptotic response is more complex and involves transcriptional activation of multiple proapoptotic target genes, tissue, and signal specificity, as well as additional events that are less well understood. In this chapter, we summarize the apoptosis pathway regulated by p53 and include some open questions in this field.

Keywords p53, apoptosis, transcription, TRAIL receptors, p53-dependent cell death.

1 Introduction

The p53 pathway is inactivated in most human tumors. It is inactivated directly as a result of mutations, with substitution mutations being common, indirectly by binding to viral or cellular proteins, or as a consequence of alterations in proteins regulating its functions (Vogelstein et al., 2000). p53 function is usually switched off, although when the cells get exposed to stress such as DNA damage induced by ionizing radiation or ultraviolet rays, activation of oncogenic signaling, hypoxia, or nucleotide depletion, p53 is accumulated in the nucleus in a tetrameric form (Bode and Dong, 2004). Upon activation, p53 mediates a growth-suppressive effect on cells by blocking the cell cycle or it can lead the cells to undergo programmed cell

Kageaki Kuribayashi and Wafik S. El-Deiry
Laboratory of Molecular Oncology and Cell Cycle Regulation, Departments of Medicine (Hematology/Oncology), Genetics, and Pharmacology, The Institute for Translational Medicine and Therapeutics and the Abramson Comprehensive Cancer Center, University of Pennsylvania School of Medicine, Philadelphia, PA

* To whom correspondence should be addressed: e-mail: wafik@mail.med.upenn.edu

R. Khosravi-Far and E. White (eds.), *Programmed Cell Death in Cancer Progression and Therapy*.
© Springer 2008

death primarily by binding to particular DNA sequences and activating transcription of specific genes (El-Deiry, 2003).

Programmed cell death, frequently referred to as apoptosis, is induced by either intracellular or extracellular stimuli. In addition to the toxic stresses mentioned earlier, serum deprivation, ligand–receptor interactions between FAS ligand (FasL)–FAS/APO1, tumor necrosis factor (TNF)–TNF receptors, and TRAIL–TRAIL receptors will also induce apoptosis (Ozoren and El-Deiry, 2003). In CD95-mediated apoptosis, there are two cell-type-specific signaling pathways, so-called type I and type II pathways (Scaffidi et al., 1998). In the type I (extrinsic) pathway, caspase-8 activation is sufficient to kill cells as a direct consequence of death receptor ligation with subsequent activation of effector caspase-3, caspase-6, and caspase-7. This death is independent of the mitochondria and is not blocked by overexpression of Bcl-2 or treatment of cells by a caspase-9 inhibitor. On the other hand, the type II (intrinsic) pathway amplifies a cell membrane-initiated death signal via the mitochondria and this form of death can be blocked by Bcl-2 or treatment of cells by a caspase-9 inhibitor.

p53 regulates these classical cell death pathways (Fig. 10.1) by either upregulating proapoptotic genes or by associating with proapoptotic genes in a transcription-independent manner. Understanding of apoptosis is very important as its dysregulation leads to variety of human diseases including cancer, autoimmune diseases, and neurodegenerative disorders. Greater insight into the pathways of apoptosis and their deregulation in disease in fundamental to understanding pathophysiology and to developing novel therapeutic agents.

2   Stabilization and Activation of P53

p53 is normally maintained at low levels in unstressed mammalian cells. The amount of p53 is determined by the rate of its degradation rather than its transcription, as blocking of its interaction with its main negative regulator Mdm2 (also known as HDM2) is sufficient to induce accumulation of the protein in cells (Michael and Oren, 2003; Vassilev et al., 2004). The primary structure of the p53 cDNA can be subdivided into three functional domains. The N-terminal region consists of a transactivation and Src homology 3-like domain, as well as a proline-rich domain. The central core consists primarily of the DNA-binding domain, where contains hot spots for various missense mutations found in human tumors. Several of the hot spots represent contact points between p53 protein and its DNA-response element. The C-terminal domain contains a nuclear localization signal, a nuclear export signal, and a tetramerization domain. The C-terminus provides a regulatory domain whose conformation and acetylation state may impact on p53 DNA binding and transactivation activity.

Mdm2 inactivates p53 by binding to its N-terminal transactivation domain to inhibit its transcriptional activity and by ligating ubiquitin at its C-terminal lysines thereby ultimately targeting p53 for proteasome-mediated degradation

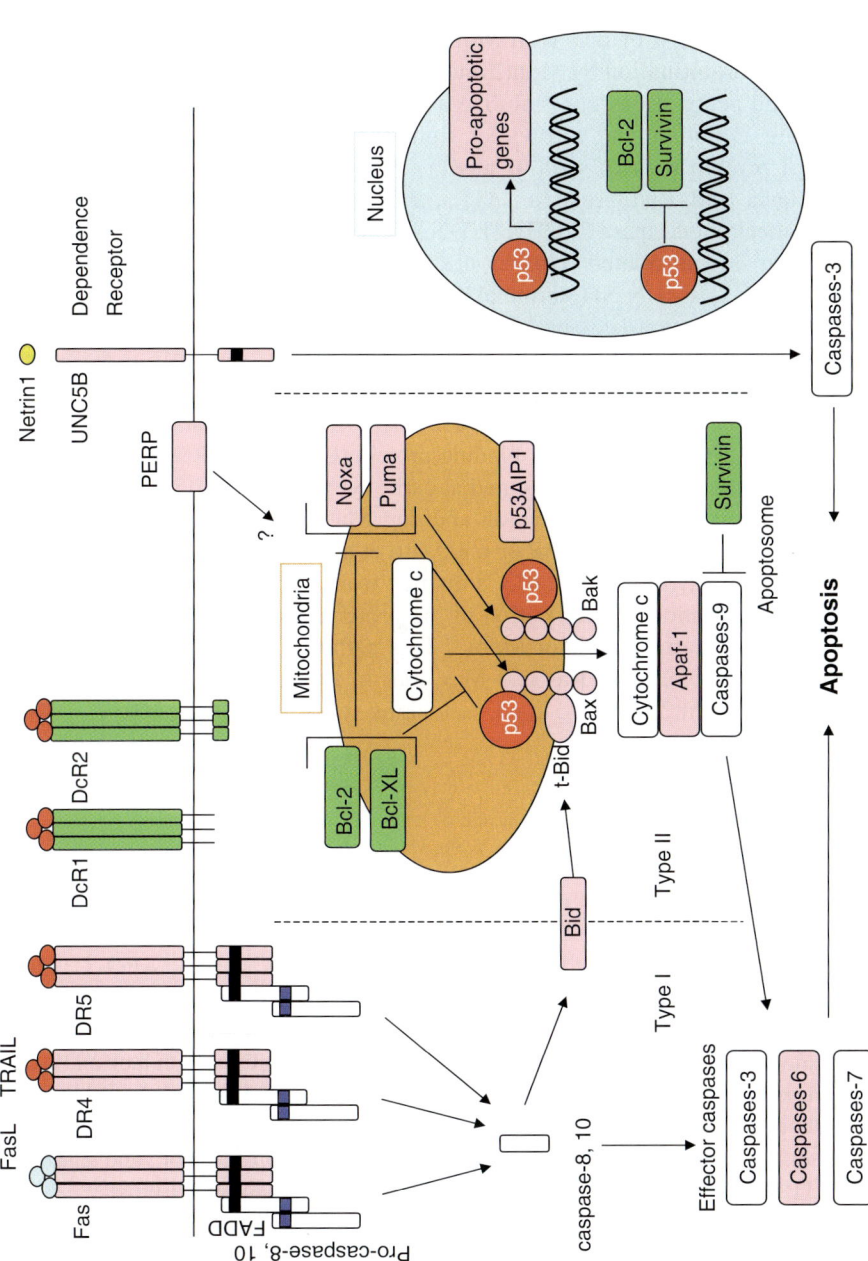

Fig. 10.1 Apoptotic pathways regulated by p53. Please see text for details of the genes, pathways, and mechanisms involved in cell death regulation and the complex signaling networks governed by p53 activity

(Rodriguez et al., 2000). Recent work by the laboratory of Wei Gu has documented monoubiquitination of p53 by Mdm2 leading to nuclear export, polyubiquitination, and degradation. As phosphorylation of N-terminal serines (particularly serine 20) blocks the interaction of p53 with Mdm2 and acetylation of C-terminal lysines prevent p53 ubiquitination by Mdm2 and subsequent degradation, these phosphorylations and acetylations can stabilize and activate p53. DNA damage induced by ionizing radiation and ultraviolet light induce p53 phosphorylation by a number of protein kinases such as ataxia telangiectasia mutated (ATM), ataxia telangiectasia and Rad3 related (ATR), casein kinases, checkpoint kinase 1 (CHK1), checkpoint kinase 2 (CHK2), DNA-dependent protein kinase (DNA-PK), extracellular signal-related kinase (ERK), homeodomain-interacting protein kinase 2 (HIPK2), c-JUN NH_2-terminal kinase (JNK), and p38 kinase in a stimulant/kinase/phosphorylation-site-specific manner (Bode and Dong, 2004). It has also been reported and is well known that phosphorylation of serine 46 is associated with apoptosis induced by p53AIP1 (Oda et al., 2000a) and that exogenous expression of p53 mutant that has defect in serine 46 shows resistance to apoptosis (Ichwan et al., 2006). p300/CREB-binding protein (CBP) and p300/CBP-associated factor (PCAF) acetylate lysines located at carboxyl terminus of p53. Further more, Mdm2 mediates PCAF ubiquitination and degradation (Jin et al., 2004), and can inhibit acetylation of p53 by CBP or p300 (Ito et al., 2001).

Another pathway to stabilize p53 is distinct from the first two mechanisms that involve posttranslational modification of the protein, and rather acts to inhibit the activity of the negative regulator Mdm2. This pathway is activated by oncogenic signals, for example, from Ras and Myc that in turn activate p14[ARF] leading to inactivation of Mdm2 resulting in p53 activation (Lowe and Sherr, 2003).

3   Type I Pathway

Type I pathway is initiated by ligand binding to its cognate death receptors. Overall eight receptors possessing death domains (DD) have been identified and all belong to the TNF family of receptors, including TNF-R1, Fas (CD95, APO-1), DR3, TRAIL-R1 (DR4), TRAIL-R2 (KILLER/DR5), DR6, p75[NTR], and EDAR (Ozoren and El-Deiry, 2003). Fas has a secretory decoy receptor (DcR3) that lacks a transmembrane domain. There are two decoy receptors for TRAIL, which lack a DD and these are known as TRAIL-R3 (DcR1, TRID) and TRAIL-R4 (DcR2, TRUNDD). These decoy receptors act as negative regulators of the death pathway. Till date, Fas (Muller et al., 1998), DR4 (Liu et al., 2004), KILLER/DR5 (Takimoto and El-Deiry, 2000), TRID (Ruiz de Almodovar et al., 2004), and TRUNDD (Liu et al., 2005) are reported to contain p53-specific binding sequences in intron 1 and are transcriptionally regulated by p53.

p53 target proteins Fas, DR4, and KILLER/DR5 contain cysteine-rich extracellular domains that bind their cognate ligands and intracellular portions consisting of approximately 80 amino acid DDs that transduce apoptosis-inducing signals.

As death ligands such as FasL or TRAIL exist in a homotrimeric form, binding to their respective receptors leads to receptor trimerization. The ligand/receptor interaction triggers formation of the death-inducing signaling complex (DISC) which contains Fas-activating DD (FADD) and initiator caspases, pro-caspase-8 or pro-caspase-10. FADD is the adaptor protein which links receptors and pro-caspases through its two distinct domains, a DD that binds with the DD of the receptors and a death effector domain (DED) which binds with the DED of pro-caspase-8. Association of pro-caspase-8 with the DISC generates a p20 fragment from the caspase by cleavage and further processing leads to a p10 fragment for its full activation (Medema et al., 1997). This mature caspase can cleave downstream effector caspase-3, caspase-6, and caspase-7 (Riedl and Shi, 2004). The extrinsic pathway can cross talk with the intrinsic pathway via BID. When BID is cleaved by caspase-8, truncated BID is myristoylated, translocates to mitochondria, releases proapoptotic proteins, and further activates death signaling and execution events (Zha et al., 2000).

3.1 Fas (APO-1, CD95)

The Fas receptor is a type I membrane protein expressed abundantly in various tissues. The *Fas* gene is located on human chromosome 10q24.1 and on chromosome 11 in mice (Nagata, 1999). The human and mouse *Fas* genes contain a p53 DNA-binding site in intron 1 and through this site the expression of the *Fas* gene can be transcriptionally upregulated by p53 (Muller et al., 1998; Munsch et al., 2000). Fas is also transcriptionally regulated by Sp1 and NF-κB (Chan et al., 1999; Xiao et al., 2001). The FasL–Fas interaction plays an important role in immune homeostasis, especially maintaining immune privilege in the eye and testis. As tumor cells can evade the host immune surveillance system by overexpressing FasL and can induce apoptosis in the T-cells responsible for the immune response, a phenomenon known as Fas counterattack, an understanding of the Fas-mediated apoptotic pathway is important for understanding tumor biology, It has been reported that CMT93 colon carcinoma cell downregulation of FasL has no effect on cell growth in vitro, but results in reduced tumorigenicity in vivo, possibly by the mechanism of loss of the Fas counterattack (Ryan et al., 2005).

Fas is transcriptionally upregulated by 5-FU and mediates apoptosis in a p53-dependent manner in MCF7 and HCT116 cells (Longley et al., 2004). It has also been reported that p53 relocalizes Fas to the cell surface (Bennett et al., 1998), providing a role of p53 in the Fas apoptotic pathway independent of its transcriptional activity. Wild-type p53 transduction in p53-mutant non-small-cell lung carcinoma cells induces Fas expression and the cells become susceptible to cytotoxic T lymphocyte-mediated killing (Thiery et al., 2005). There are certain p53 mutants, which can induce cell cycle arrest, but not apoptosis, so-called discriminatory mutants. Munsch et al. reported that discriminatory mutants Pro-175 and Ala-143 have activity to induce Fas transcription, but not apoptosis, suggesting upregulation of Fas is not enough to induce apoptosis in some circumstances (Munsch et al.,

2000). Furthermore, Fas does not appear to be required for p53-dependent apoptosis in response to DNA damage by irradiation (Fuchs et al., 1997).

3.2 Trail Receptors

In humans, there are four homologous TRAIL receptors including DR4, KILLER/DR5, TRID, and TRUNDD, as well as a fifth soluble receptor osteoprotegerin (Wang and El-Deiry, 2003a). The extracellular cysteine-rich domains of DR4, KILLER/DR5, TRID, and TRUNDD are 52–69% identical to each other and the DD of DR4 and KILLER/DR5 are 64% identical to each other (Ozoren and El-Deiry, 2003). As these genes are clustered on human chromosome 8p21–22, they might have arisen from a common ancestral gene (Degli-Esposti et al., 1997a). TRAIL seems to be promising for cancer therapeutics as many cancer cells are sensitive to TRAIL while normal cells are not (Wang and El-Deiry, 2003a).

3.2.1 DR4

DR4 protein is a 445 amino acid-containing type I transmembrane receptor, which is generated through the cleavage of a signal sequence of 23 amino acids from a primary protein (Pan et al., 1997a). The protein has three cysteine-rich repeats in the extracellular domain. The DD of DR4 is 30 and 19% identical with that of TNF-R1 and Fas, respectively. It has been reported that nucleotide substitutions in the extracellular domain of DR4 were correlated with increased risk of lung, and head and neck cancers (Fisher et al., 2001). Somatic mutations of DR4 have been found in non-Hodgkin's lymphoma (Lee et al., 2001), breast cancers (Shin et al., 2001), and osteosarcoma (Dechant et al., 2004). The DR4 expression level is known to be the one of the determinants to TRAIL sensitivity in many cancer cell lines (Kim et al., 2000), and homozygous deletion of the *DR4* gene has been reported in the FaDu nasopharyngeal cancer cell line and this is associated with TRAIL resistance (Ozoren et al., 2000). Approximately 20% of the normal population carries the polymorphic DR4 variant that contains adenine to guanine alteration in the DD (K441R). When a DR4 K441R expressing plasmid was transfected into human cells, it acted as a dominant negative TRAIL receptor resulting in decreased sensitivity to TRAIL (Kim et al., 2000). From these observations, DR4 seems to be a major factor determining TRAIL sensitivity. p53 overexpression by adenovirus-p53 induces upregulation of DR4 and DR5 resulting in increased apoptosis by TRAIL treatment in myeloma cells (Liu et al., 2001). As, wild-type p53 is not required for TRAIL sensitivity in many cancer cell lines (Kim et al., 2000), it is the open question that to what extent p53 is involved in DR4-mediated apoptosis.

3.2.2   KILLER/DR5

KILLER/DR5 is a 411 amino acid containing protein that includes a 51 amino acid signal peptide sequence. Other than p53, a recent study demonstrated that NF-κB can also upregulate KILLER/DR5 transcription in the presence of p53 (Shetty et al., 2005). Germline or somatic mutations of the *KILLER/DR5* gene have been reported in head and neck cancer (Pai et al., 1998), non-Hodgkin's lymphoma (Lee et al., 2001), and breast cancer (Shin et al., 2001).

Compared to DR4, many studies were conducted to elucidate the role of KILLER/DR5 in the p53 pathway, as it was first found as a DNA damage-inducible p53-regulated gene in doxorubicin-treated cell lines (Wu et al., 1997). Comparison of the apoptotic response of p53+/+ and p53−/− mice after ionizing radiation is a good in vivo model to study DNA damage-induced p53-dependent apoptosis in the context of studying p53 target gene tissue specificity. While thymus, spleen, and small intestine underwent p53-dependent apoptosis in the mouse model, among p21, E124/PIG8, Bax, Fas, and KILLER/DR5, KILLER/DR5 was the only upregulated gene after γ-irradiation in a p53-dependent manner to induce apoptosis in the spleen and small intestine, implicating a critical role of KILLER/DR5 in the radiation response (Burns et al., 2001). Recent results using DR5 knockout mice further support the importance of DR5 in the p53 pathway (Finnberg et al., 2005). DR5-null mice showed a slightly larger thymus than wild-type mice. As DR5 is the only known TRAIL receptor in mice, the results suggest that negative selection in thymocytes might be in part controlled through the DR5 receptor. In these mice, there was no evidence of spontaneous autoimmune disease as reported in TRAIL knockout mice (Lamhamedi-Cherradi et al., 2003). E1A stabilizes p53 and transactivates its target genes. The result that DR5-null mouse embryo fibroblasts (MEFs) expressing E1A did not undergo apoptosis after TRAIL treatment suggests that there are no other TRAIL receptors in mice besides DR5, which can be transactivated by p53 in mice. DR5-null tissues showed reduced amounts of apoptosis compared to wild-type thymus, spleen, Peyer's patches, and the white matter of the brain. However, because gene targeting of DR5 failed to nullify all death in these organs, it is likely that DR5 is only one of the several p53 target genes that are important in this response. In the colon, DR5 wild type and null mice showed approximately the same amount of radiation-induced cell death. However, in the human colon cancer cell line HCT116 silencing of DR5 induces accelerated growth of tumor xenografts (Wang and El-Deiry, 2004a, b) and also DR5 is required for p53-dependent TRAIL sensitivity in mismatch repair deficient Bax−/− HCT116 cells (Wang and El-Deiry, 2003b). From these observations, even in the colon, DR5 is important in apoptosis within the p53 pathway that suppresses tumor formation or progression. Taken together, DR5 seems to play a critical role in DNA damage-induced programmed cell death in the p53 pathway.

3.2.3 TRID/DCR1 and TRUNDD/DCSR2

TRID and TRUNDD contain extracellular cysteine-rich domains. TRID consists of an extracellular TRAIL-binding domain linked to the membrane through a glycosylphosphatidylinositol (GPI) anchor and completely lacks an intracellular domain, whereas TRUNDD contains an intracellular domain that has a truncated DD which can transduce NF-κB signal (Degli-Esposti et al., 1997b). Over the extracellular domain, TRID is 69 and 52% identical with DR4 and DR5, and TRUNDD is 70%, 57%, and 58% identical with TRID, DR5, and DR4, respectively. TRID mRNA is expressed in normal tissues but not in many tumor cells, giving a rationale for TRAIL on its tumor-specific apoptosis-inducing activity (Sheridan et al., 1997; Pan et al., 1997b).

Although these decoy receptors are regulated by p53, little is known about their role in the p53-regulated apoptotic pathway. TRID is overexpressed by genotoxic stress in p53 intact cells and is overexpressed in gastrointestinal tumors (Sheikh et al., 1999). TRUNDD is induced by adenovirus-p53 overexpression and TRUNDD can delay TRAIL-, p53-, and KILLER/DR5-dependent apoptosis in colon cancer cells (Meng et al., 2000). It has also been reported that silencing of TRUNDD enhances doxorubicin-induced apoptosis in HCT116 cells (Liu et al., 2005). It therefore seems that these decoy receptors are forming a negative feedback loop to dampen p53-mediated apoptotic signaling.

4 Type II Pathway

The type II death pathway is evoked through intrinsic stimuli such as DNA damage, cytotoxic drugs, hypoxia, oncogenic signaling, or even extrinsic death receptor signals in type II cells. Mitochondrial factors are crucial in the efficiency of cell death mediated by this pathway. Cytochrome c released from mitochondria assembles a cytosolic caspase-activating complex called apoptosome which consists of Apaf-1, caspase-9, and cytochrome c, while release of Smac/DIABLO and Htra2/Omi inactivate inhibitor of apoptosis proteins (IAPs), the inhibitors of caspases, enhance apoptosis (Danial and Korsmeyer, 2004). Bcl-2 family members are the key components in this process. They are categorized into three groups according to their function and numbers of Bcl-2 homology (BH) domains. The first group includes antiapoptotic members such as Bcl-2, Bcl-X_L, and MCL-1, which contains four BH domains. Their BH1–3 domains are in close spatial proximity and create a hydrophobic pocket, which can mask a BH3 domain of proapoptotic members, blocking their proapoptotic functions (Muchmore et al., 1996; Sattler et al., 1997). Multidomain proapoptotic members of the family are Bax and Bak, which are thought to form a pore in the mitochondrial membrane and release cytochrome c into the cytosol. These two molecules are thought to be required in the type II pathway as cells lacking both Bax and Bak, but not cells lacking one of them are completely resistant to tBid-induced cytochrome c release and apoptosis (Wei et al.,

2001). The last members of the family are BH3-only proapoptotic proteins, which are Bid, Noxa, Puma, Bad, Bik, and Bim. Bid provides the only known connection between the extrinsic and intrinsic pathways, while the others are thought to act upstream of Bax and Bak. Cartron et al. (2004) showed that Bid and Puma specifically bind to the first α-helix of Bax leading to its activation. Bad and Bik cannot directly activate Bax but promote apoptosis by binding to Bcl-2 to inhibit their antiapoptotic functions (Letai et al., 2002).

4.1 Bid

Bid gene is transcriptionally regulated by p53 and contains a functional p53 DNA-binding site in the first large intron (Sax et al., 2002). Bid−/− MEFs are resistant to adriamycin and 5-FU as compared to Bid+/+ MEFs, showing its role as a chemosensitivity determinant (Sax et al., 2002). Recently, Bid was shown to be a sentinel for DNA damage, and it was reported to be phosphorylated by ATM (Zinkel et al., 2005; Kamer et al., 2005). It was also shown that other than the proapoptotic function of the protein, when it is phosphorylated following exposure to low dose of ionizing radiation or the DNA-damaging agent etoposide, Bid can block the cell cycle in the G2 phase. Even though it does not induce cell cycle arrest in the G1/S phase as is brought about by p21, Bid might be one of the regulators that determines cell fate after DNA damage, i.e., whether the cells should live or die.

4.2 Puma and Noxa

Puma (bbc3) and Noxa are p53 target genes belonging to the BH-3 only proteins, contain p53 DNA-binding sequences in the first intron and can induce apoptosis by p53 overexpression or exposure to DNA-damaging stimuli (Oda et al., 2000a; Yu et al., 2001; Nakano and Vousden, 2001; Han et al., 2001). Serum starvation and glucocorticoid treatment also induce Puma and virus infection and interferon induce Noxa expression independent of p53 activation, respectively (Han et al., 2001; Sun and Leaman, 2005). In hematopoietic progenitor cells, Slug represses p53-mediated transcription of Puma in turn protecting the cells from γ-radiation-induced cell death, and it was also found that Slug itself was upregulated by p53 (Wu et al., 2005). Furthermore, it was recently demonstrated that p53 family member p73 transactivates Puma and Noxa expression independent of p53, and the other member delta p63 acts as their repressor inhibiting head and neck tumor cells from apoptosis (Rocco et al., 2006). Noxa and Puma are tightly regulated genes with redundancy in their stimulation and regulation by transcription factors, suggesting their important role in apoptosis.

These p53 targets seem to have tissue specificity. By ionizing radiation, Noxa was expressed in the red pulp, where as Puma was induced in the white pulp of the spleen in a p53-dependent fashion (Fei et al., 2005). Puma−/− MEFs and Noxa−/− MEFs showed increased resistance in apoptosis induced by etoposide treatment or γ-irradiation. Although single gene, *Puma* or *Noxa*, knockdown could not attain the resistant level to that of p53−/− MEFs, it was suggested that Puma and Noxa have redundancy in inducing apoptosis in these cells (Villunger et al., 2003). On the other hand, Puma knockout nullified nearly all of the cell death attributed to p53 in primary hematopoietic cells and the developing central nervous system in response to γ-radiation or oncogenic signals from c-Myc, i.e., it has indispensable role in apoptosis in these tissues (Jeffers et al., 2003). Yu et al. (2003) reported that targeted deletion of *Puma* gene in HCT116 cells completely blocked apoptosis induced by p53 overexpression, adriamycin exposure, or a hypoxic environment (Yu et al., 2003). Another important notion of the study is that in the presence of p21, cellular stresses lead to cell cycle arrest, whereas deprivation of p21 by gene targeting results in enhanced apoptosis induced by the same stimuli. From such observations, it has been proposed that cell fate between cell cycle arrest and apoptosis is determined by the balance between p21 and Puma. Another study showed Noxa and Bax doubly knocked out MEFs were more resistant to apoptosis induced by adriamycin or oncogenic signals as compared to single knockouts of these genes (Shibue et al., 2003). It was suggested from the result that Noxa and Bax carry out different functions in the apoptosis pathway.

4.3 Bax and Bak

The gene-encoding BAX is a transcriptional target of p53 (Miyashita and Reed, 1995). BAK has also been reported to be upregulated by p53 (Pearson et al., 2000; Pohl et al., 1999). Bax and Bak appear to have some overlapping roles in apoptosis, as either thymocytes from Bak−/− or Bax−/− null mice do not show radiation-induced apoptosis, although thymocytes from Bak and Bax double-knockout mice show resistance to γ-radiation or etoposide treatment (Lindsten et al., 2000). Furthermore, Bak and Bax doubly deficient MEFs show resistance to multiple intrinsic death-inducing stimuli such as staurosporine, ultraviolet radiation, and growth factor deprivation (Wei et al., 2001). Bax−/− HCT116 cells are resistant to TRAIL treatment, but etoposide and camptothecin treatment of the cells restores their sensitivity to TRAIL by upregulating Bak and DR5 (LeBlanc et al., 2002). However, recent studies with DR5 or Bak knockdown suggests that this restored sensitivity relies more on DR5 upregulation and a conversion of cells from type II to type I signaling, in the case of TRAIL and chemotherapy treatment (Wang et al., 2003b).

Recent studies revealed that p53 may have a transcription-independent activity in the mitochondrial death pathway involving Bak and Bax. After DNA damage induced by irradiation or chemotherapeutic agents, p53 has been reported to translocate to the mitochondria and to activate either Bax- (Chipuk et al., 2005) or Bak- (Leu et al., 2004) dependent mitochondrial outer membrane permeabilization (MOMP) and release of cytochrome c into the cytosol. However, translocation of p53 to the mitochondria is not sufficient to induce cell death (Essmann et al., 2005), as p53 is sequestered by Bcl-X$_L$ at mitochondria and its activity to induce MOMP is blocked (Mihara et al., 2003; Chipuk et al., 2005). Puma has been reported to act on the complex of p53–Bcl-X$_L$ thereby releasing p53 from Bcl-X$_L$ to allow for the MOMP-inducing activity (Chipuk et al., 2005). However, whether the p53–Bcl-X$_L$ or p53–Bcl2 complexes act as positive or negative regulators of cytochrome c release is still under study (Mihara et al., 2003; Chipuk et al., 2005; Tomita et al., 2006).

4.4   P53AIP1

The *p53AIP1* gene is induced following severe DNA damage associated with p53 ser-46 phosphorylation and localization of p53AIP1 at mitochondria (Oda et al., 2000b). Phosphorylation of p53 and subsequent p53AIP1 induction is also regulated by the p53-inducible protein p53DINP1 (Okamura et al., 2001). p53AIP1 has been reported to have potential to release cytochrome c from mitochondria into the cytosol and to induce apoptosis, although its precise mechanism and relation with other apoptotic factors is not clarified yet.

4.5   Apaf-1

The *Apaf-1* gene is a transcriptional target of p53 and it is also transcriptionally induced by E2F (Moroni et al., 2001). The study comparing p53–/– and wild-type mice showed that Apaf-1 expression was p53-dependent in the spleen and heart (Ho et al., 2003). Apaf-1–/– MEFs were resistant to p53-dependent cell death-induced by oncogenic Myc and Ras signaling (Soengas et al., 1999). Apaf-1 was found to be silenced in metastatic melanomas by hypermethylation and restoration of Apaf-1 expression led to efficient caspase-9 activation and adriamycin-induced cell death (Soengas et al., 2001), supporting its role as a chemosensitivity determinant.

5   Dependency Receptor Pathway

There is unique apoptotic pathway called dependency receptor pathway. In the absence of ligand, expression-dependent receptors induce apoptosis, whereas binding of cognate ligands to their receptors blocks apoptosis and this apoptotic pathway seems to be independent of mitochondria (Arakawa, 2004; Bredesen et al., 2004). Examples for the receptor/ligand are p75[NTR]/neurotrophin, UNC5B (p53RDL1)/Netrin-1, and deleted in colorectal cancer (DCC)/Netrin-1. These receptors are involved in axon guidance during neuronal development and among these receptors

UNC5B was shown to be a p53 transcriptional target, which is implicated in p53-dependent apoptosis (Tanikawa et al., 2003). Loss of DCC has been reported in many cancers, and binding of Netrin-1 to UNC5B has been reported to repress the p53 target genes *Bax* and *p21*. This newly found pathway might be also involved in p53-related tumorigenesis.

6   PERP

The *PERP* gene is transcriptionally upregulated by p53 (Attardi et al., 2000) as well as p63 (Ihrie et al., 2005). It is a membrane protein involved in apoptosis induced by p53 overexpression and Bcl-2 reduces the cell death, suggesting that the mitochondria are involved in its signaling (Attardi et al., 2000). PERP localizes specifically to desmosomes, adhesion junctions important for tissue integrity. Numerous structural defects in desmosomes are observed in skin of PERP–/– mice (Ihrie et al., 2005). It was recently reported that PERP-null mice are not tumor-prone as compared to wild-type mice (Ihrie et al., 2006). As p53-null mice are tumor prone, whereas single knockout of the other p53 targets such as Puma, Bak, or Bax do not produce tumor-prone mice, the observation does not imply that PERP is not important in the p53 apoptotic pathway.

7   PIGs

The PIGs are "p53-induced genes," identified by transducing p53 into the human colorectal cancer line DLD-1 that undergoes apoptosis in response to p53 expression (Polyak et al., 1997). As many of these genes were capable of producing or responding to reactive oxygen species, the importance of reactive oxygen species in the p53 pathway was suggested. One of the PIGs, EI24/PIG8, has also been identified as the gene upregulated by etoposide treatment in murine NIH3T3 cells (Lehar et al., 1996). It was recently shown that EI24/PIG8 colocalizes at the endoplasmic reticulum with Bcl-2 and loss of EI24/PIG8 is positively related with invasiveness of breast cancers (Zhao et al., 2005).

8   Caspase-6

Caspases, the cysteine proteases that cleave after an aspartate residue in their substrate, are the central components of the apoptotic pathway. They are usually divided into two classes, the initiator caspase-2, caspase-8, caspase-9, and caspase-10, and the effectors, caspase-3, caspase-6, and caspase-7 (Riedl and Shi, 2004). Caspase-6 is a transcriptional target of p53 in the apoptotic response (MacLachlan and El-Deiry, 2002). Caspase-1 is also a p53 transcriptional target, although it is involved in inflammatory response rather than apoptotic pathway (Gupta et al., 2001). p53 seems to have potential to activate caspase-6 and sensitize cells to chemotherapeutic drugs leading them to apoptosis by the mechanism other than its transcriptional upregulation (MacLachlan and El-Deiry, 2002). We have also previously reported that caspase-10 is directly induced by p53.

9   P53-Dependent Apoptosis Under Hypoxic Conditions

Solid tumors acquire regions of hypoxia as a result of insufficient blood supply. Cells containing wild-type but not mutant p53 undergo apoptosis in hypoxic regions (Graeber et al., 1996), leading to a powerful selection pressure to promote tumor progression and therapeutic resistance (Harris, 2002). p53 shows an altered behavior under hypoxia. Under hypoxia, p53 does accumulate in cells, although it does not upregulate most of the known p53 target genes such as *p21, Bax, GADD45, DR5,* or *Puma* (Koumenis et al., 2001; Fei et al., 2004). We have recently identified that Bnip3L is playing a role in apoptosis during hypoxia in some human tumor cell lines (Fei et al., 2004). Bnip3L was found to be a direct transcriptional target of p53 as well as hypoxia-inducible factor 1 (HIF1). p53-dependent apoptosis during hypoxia was reduced after knocking down Bnip3L. Furthermore, nontumorigenic U2OS cells were converted into a tumorigenic state in mouse xenograft experiments following stable Bnip3L knockdown.

10   Transcriptional Repression of Antiapoptotic Genes

IAPs and Bcl2 block apoptosis by inhibiting caspase activation and MOMP. In addition to transcriptional activation activity, p53 exerts its apoptosis-promoting effects by repressing antiapoptotic gene transactivation (Murphy et al., 1999; Wu et al., 2001; Hoffman et al., 2002). Its mechanism appears to involve association of p53 with histone deacetylases (HDACs) and its interaction is mediated by corepressor mSin3a. DNA damage induces the p53–mSin3a interaction and targets HDACs to the promoters of the p53-repressed genes, where HDACs deacetylate histones and create a chromatin environment that is unfavorable for transcription.

11   P53 as a Therapeutic Target

A number of strategies have been developed to target p53 in cancer therapy. For about half of human cancers, which possess wild-type p53, the Mdm2–p53 interaction could be a major target to prevent p53 from degradation. Nutlin-3 is an example of a small molecule that specifically disrupts the p53–Mdm2 interaction. It was recently demonstrated that administration of Nutlin-3 suppressed xenograft growth in a dose-dependent manner (Tovar et al., 2006). As Mdm2 downregulation and subsequent p53 upregulation is reported to bring lymphocytopenia as a side effect in hypomorphic Mdm2 mice, further study may help to compare its benefit to disadvantage or advantages over standard chemotherapy.

Histone deacetylase inhibitors (HDACIs) have been shown to exert various antitumor effects and they are presently in clinical trials (see Chapter 13). p53 is one of the targets of HDACIs, as HDACIs inhibit deacetylation of the C-terminal lysines and induce apoptosis in gastric cancer (Terui et al., 2003) and prostate cancer cells (Roy et al., 2005). It has also been demonstrated that HDACIs enhance the tumoricidal effects of p53 adenovirally transferred gene therapy (Takimoto et al., 2005).

The status of an intact p53 pathway positively correlates with the response to the majority of chemotherapeutic drugs, most, although not all, of them being DNA-damaging agents (Weinstein et al., 1997; O'Connor et al., 1997). However, there are some clinically useful agents such as the antimitotic agent taxol, which was found to be more effective in tumor cells with mutant p53 (Weinstein et al., 1997). In this context, we have identified the Polo-like kinase family member serum-inducible kinase (Snk/Plk2) as a p53 target and its silencing by siRNA leads to mitotic catastrophe after taxol treatment, suggesting p53-dependent activation of Snk/Plk prevents mitotic catastrophe following spindle damage (Burns et al., 2003).

Much effort has been devoted to overcome mutant p53 by small molecules that can restore the wild-type functions to mutant p53. CP-31398, a strylquinazoline, was identified from a screen of the library containing more than 10,000 synthetic compounds (Foster et al., 1999). The molecule not only promotes the stability of wild-type p53, but also allows mutant p53 to maintain an active conformation, enabling transcription and subsequent tumor growth suppression. CP-31398 can cause either cell cycle arrest or cell death in tumor cell lines carrying mutant p53, and combination of CP-3198 with chemotherapy or TRAIL exhibit synergistic effects enhancing cell killing (Takimoto et al., 2002). It has been shown that stabilization of p53 by CP-31398 involves a mechanism targeting blockade of ubiquitination of p53 and its further degradation (Wang et al., 2003). Neither phosphorylation of p53 at serine 15, 20, or interaction between Mdm2 was inhibited by CP-31398, highlighting a unique mechanism by which it can activate p53. PRIMA-1 also induces apoptosis in tumor cells (Bykov et al., 2002) and it has synergistic effects with chemotherapeutic drugs (Bykov et al., 2005).

A peptide derived from the C-terminus of p53 is known to activate its specific binding to DNA including several p53 DNA contact mutants (Hupp et al., 1995). Several cationic peptides such as TAT and polyArg can penetrate into the cells through a mechanism called macropinocytosis (Wang and El-Deiry, 2004b). Utilizing this technology, Snyder et al. (2004) showed that the C-terminal peptide of p53 fused with TAT induced cell cycle arrest and apoptosis in a peritoneal carcinomatosis model and prolonged survival of the mice.

12   Future Directions

Already a quarter century has passed since the discovery of p53 and we have learned much about its important role as a tumor suppressor gene as well as its complicated network governing programmed cell death. However, there are still important problems left to be solved. There are numerous genes known to be involved in the p53 pathway, but are they all equally important? Which genes are involved in which tissues? No single gene so far can account for p53-mediated apoptosis alone, and it might be possible that there is no such gene. The principle question is that we still do not know how p53 determines cell fate. Progress towards this understanding as well as efforts to develop therapies targeting this p53 pathway and its family members represent important future directions.

References

Arakawa, H. (2004). Netrin-1 and its receptors in tumorigenesis. Nat Rev Cancer 4, 978–987.

Attardi LD, Reczek EE, Cosmas C, Demicco EG, McCurrach ME, Lowe SW, Jacks T. PERP, an apoptosis-associated target of p53, is a novel member of the PMP-22/gas3 family. Genes Dev. 2000 14:704–18.

Bennett, M., Macdonald, K., Chan, S. W., Luzio, J. P., Simari, R., and Weissberg, P. (1998). Cell surface trafficking of Fas: a rapid mechanism of p53-mediated apoptosis. Science 282, 290–293.

Bode, A. M. and Dong, Z. (2004). Post-translational modification of p53 in tumorigenesis. Nat Rev Cancer 4, 793–805.

Bredesen, D. E., Mehlen, P., and Rabizadeh, S. (2004). Apoptosis and dependence receptors: a molecular basis for cellular addiction. Physiol Rev 84, 411–430.

Burns, T. F., Bernhard, E. J., and El-Deiry, W. S. (2001). specific expression of p53 target genes suggests a key role for KILLER/DR5 in p53-dependent apoptosis in vivo. Oncogene 20, 4601–4612.

Burns, T. F., Fei, P., Scata, K. A., Dicker, D. T., and El-Deiry, W. S. (2003). Silencing of the novel p53 target gene Snk/Plk2 leads to mitotic catastrophe in paclitaxel (taxol)-exposed cells. Mol Cell Biol 23, 5556–5571.

Bykov, V. J., Issaeva, N., Shilov, A., Hultcrantz, M., Pugacheva, E., Chumakov, P., Bergman, J., Wiman, K. G., and Selivanova, G. (2002). Restoration of the tumor suppressor function to mutant p53 by a low-molecular-weight compound. Nat Med 8, 282–288.

Bykov, V. J., Zache, N., Stridh, H., Westman, J., Bergman, J., Selivanova, G., and Wiman, K. G. (2005). PRIMA-1(MET) synergizes with cisplatin to induce tumor cell apoptosis. Oncogene 24, 3484–3491.

Cartron, P. F., Gallenne, T., Bougras, G., Gautier, F., Manero, F., Vusio, P., Meflah, K., Vallette, F. M., and Juin, P. (2004). The first alpha helix of Bax plays a necessary role in its ligand-induced activation by the BH3-only proteins Bid and PUMA. Mol Cell 16, 807–818.

Chan, H., Bartos, D. P., and Owen-Schaub, L. B. (1999). Activation-dependent transcriptional regulation of the human Fas promoter requires NF-kappaB p50-p65 recruitment. Mol Cell Biol 19, 2098–2108.

Chipuk, J. E., Bouchier-Hayes, L., Kuwana, T., Newmeyer, D. D., and Green, D. R. (2005). PUMA couples the nuclear and cytoplasmic proapoptotic function of p53. Science 309, 1732–1735.

Danial, N. N. and Korsmeyer, S. J. (2004). Cell death: critical control points. Cell 116, 205–219.

Dechant, M. J., Fellenberg, J., Scheuerpflug, C. G., Ewerbeck, V., and Debatin, K. M. (2004). Mutation analysis of the apoptotic "death-receptors" and the adaptors TRADD and FADD/MORT-1 in osteosarcoma tumor samples and osteosarcoma cell lines. Int J Cancer 109, 661 667.

Degli-Esposti, M. A., Smolak, P. J., Walczak, H., Waugh, J., Huang, C. P., DuBose, R. F., Goodwin, R. G., and Smith, C. A. (1997a). Cloning and characterization of TRAIL-R3, a novel member of the emerging TRAIL receptor family. J Exp Med 186, 1165–1170.

Degli-Esposti, M. A., Dougall, W. C., Smolak, P. J., Waugh, J. Y., Smith, C. A., and Goodwin, R. G. (1997b). The novel receptor TRAIL-R4 induces NF-kappaB and protects against TRAIL-mediated apoptosis, yet retains an incomplete death domain. Immunity 7, 813–820.

El-Deiry, W. S. (2003). The role of p53 in chemosensitivity and radiosensitivity. Oncogene 22, 7486–7495.

Essmann, F., Pohlmann, S., Gillissen, B., Daniel, P, T., Schulze,-Osthoff. K., and Janicke, R. U. (2005). Irradiation-induced translocation of p53 to mitochondria in the absence of apoptosis. J Biol Chem 280, 37169–37177.

Fei, P., Wang, W., Kim, S. H., Wang, S., Burns, T. F., Sax, J. K., Buzzai, M., Dicker, D. T., McKenna, W. G., Bernhard, E. J., and El-Deiry, W. S. (2004). Bnip3L is induced by p53 under hypoxia, and its knockdown promotes tumor growth. Cancer Cell 6, 597–609.

Fei, P., Bernhard, E. J., and El-Deiry, W. S. (2005). Tissue-specific induction of p53 targets in vivo. Cancer Res 62, 7316–7327.

Finnberg, N., Gruber, J. J., Fei, P., Rudolph, D., Bric, A., Kim, S. H., Burns, T. F., Ajuha, H., Page, R., Wu, G. S., Chen, Y., McKenna, W. G., Bernhard, E., Lowe, S., Mak, T., and El-Deiry, W. S. (2005). DR5 knockout mice are compromised in radiation-induced apoptosis. Mol Cell Biol 25, 2000–2013.

Fisher, M. J., Virmani, A. K., Wu, L., Aplenc, R., Harper, J. C., Powell, S. M., Rebbeck, T. R., Sidransky, D., Gazdar, A. F., and El-Deiry, W. S. (2001). Nucleotide substitution in the ecto-domain of trail receptor DR4 is associated with lung cancer and head and neck cancer. Clin Cancer Res 7, 1688–1697.

Foster, B. A., Coffey, H. A., Morin, M. J., and Rastinejad, F. (1999). Pharmacological rescue of mutant p53 conformation and function. Science 286, 2507–2510.

Fuchs, E. J., McKenna, K. A., Bedi, A. (1997). p53-dependent DNA damage-induced apoptosis requires Fas/APO-1-independent activation of CPP32beta. Cancer Res 57, 2550–2554.

Graeber, T. G., Osmanian, C., Jacks, T., Housman, D. E., Koch, C. J., Lowe, S. W., and Giaccia, A. J. (1996). Hypoxia-mediated selection of cells with diminished apoptotic potential in solid tumours. Nature 379, 88–91.

Gupta, S., Radha, V., Furukawa, Y., and Swarup, G. (2001). Direct transcriptional activation of human caspase-1 by tumor suppressor p53. J Biol Chem 276, 10585–10588.

Han, J., Flemington, C., Houghton, A. B., Gu, Z., Zambetti, G. P, Lutz, R. J., Zhu, L., and Chittenden, T. (2001). Expression of bbc3, a pro-apoptotic BH3-only gene, is regulated by diverse cell death and survival signals. Proc Natl Acad Sci USA 98, 11318–11323.

Harris, A. L. (2002). Hypoxia – a key regulatory factor in tumour growth. Nat Rev Cancer 2, 38–47.

Ho, C. K., Bush, J. A., and Li, G. (2003). Tissue-specific regulation of Apaf-1 expression by p53. Oncol Rep 10, 1139–1143.

Hoffman, W. H., Biade, S., Zilfou, J. T., Chen, J., and Murphy, M. (2002). Transcriptional repression of the anti-apoptotic survivin gene by wild type p53. J Biol Chem 277, 3247–3257.

Ichwan, S. J., Yamada, S., Sumrejkanchanakij, P., Ibrahim-Auerkari, E., Eto, K., and Ikeda, M. A. (2006). Defect in serine 46 phosphorylation of p53 contributes to acquisition of p53 resistance in oral squamous cell carcinoma cells. Oncogene 25, 1216–1224.

Hupp, T. R, Sparks, A., and Lane, D. P. (1995). Small peptides activate the latent sequence-specific DNA binding function of p53. Cell 83, 237–245.

Ihrie, R. A., Marques, M. R., Nguyen, B. T., Horner, J. S., Papazoglu, C., Bronson, R. T., Mills, A. A., and Attardi, L. D. (2005). Perp is a p63-regulated gene essential for epithelial integrity. Cell 120, 843–856.

Ihrie, R. A., Bronson, R. T., and Attardi, L. D. (2006). Adult mice lacking the p53/p63 target gene Perp are not predisposed to spontaneous tumorigenesis but display features of ectodermal dysplasia syndromes. Cell Death Differ 13(9), 1614–1618.

Ito, A., Lai, C. H., Zhao, X., Saito, S., Hamilton, M. H., Appella, E., and Yao, T. P. (2001). p300/CBP-mediated p53 acetylation is commonly induced by p53-activating agents and inhibited by MDM2. EMBO J 20, 1331–1340.

Jeffers, J. R., Parganas, E., Lee, Y., Yang, C., Wang, J., Brennan, J., MacLean, K. H., Han, J., Chittenden, T., Ihle, J. N., McKinnon, P. J., Cleveland, J. L., and Zambetti, G. P. (2003). Puma is an essential mediator of p53-dependent and -independent apoptotic pathways. Cancer Cell 4, 321–328.

Jin, Y., Zeng, S. X., Lee, H., and Lu, H. (2004). MDM2 mediates p300/CREB-binding protein-associated factor ubiquitination and degradation. J Biol Chem 279, 20035–20043.

Kamer, I., Sarig, R., Zaltsman, Y., Niv, H., Oberkovitz, G., Regev, L., Haimovich, G., Lerenthal, Y., Marcellus, R. C., and Gross, A. (2005). Proapoptotic BID is an ATM effector in the DNA-damage response. Cell 122, 593–603.

Kim, K., Fisher, M. J, Xu, S. Q., and El-Deiry, W. S. (2000). Molecular determinants of response to TRAIL in killing of normal and cancer cells. Clin Cancer Res 6, 335–346.

Lamhamedi-Cherradi, S. E., Zheng, S. J., Maguschak, K. A., Peschon, J., and Chen, Y. H. (2003). Defective thymocyte apoptosis and accelerated autoimmune diseases in TRAIL-/- mice. Nat Immunol 4, 255–260.

Koumenis, C., Alarcon, R., Hammond, E., Sutphin, P., Hoffman, W., Murphy, M., Derr, J., Taya, Y., Lowe, S. W., Kastan, M., and Giaccia, A. (2001). Regulation of p53 by hypoxia: dissociation of transcriptional repression and apoptosis from p53-dependent transactivation. Mol Cell Biol 21, 1297–1310.

LeBlanc, H., Lawrence, D., Varfolomeev, E., Totpal, K., Morlan, J, Schow, P., Fong, S., Schwall, R., Sinicropi, D., and Ashkenazi, A. (2002). Tumor-cell resistance to death receptor–induced apoptosis through mutational inactivation of the proapoptotic Bcl-2 homolog Bax. Nat Med 8, 274–281.

Lee, S. H., Shin, M. S., Kim, H. S., Lee, H. K., Park, W. S., Kim, S. Y., Lee, J. H., Han, S. Y., Park, J. Y., Oh, R. R., Kang, C. S., Kim, K. M., Jang, J. J., Nam, S. W., Lee, J. Y., and Yoo, N. J. (2001). Somatic mutations of TRAIL-receptor 1 and TRAIL-receptor 2 genes in non-Hodgkin's lymphoma. Oncogene 20, 399–403.

Lehar, S. M., Nacht, M., Jacks, T., Vater, C. A., Chittenden, T., and Guild, B. C. (1996). Identification and cloning of EI24, a gene induced by p53 in etoposide-treated cells. Oncogene 12, 1181–1187.

Letai, A., Bassik, M. C., Walensky, L. D., Sorcinelli, M. D., Weiler, S., and Korsmeyer, S. J. (2002). Distinct BH3 domains either sensitize or activate mitochondrial apoptosis, serving as prototype cancer therapeutics. Cancer Cell 2, 183–192.

Leu, J. I., Dumont, P., Hafey, M., Murphy, M. E., and George. D. L. (2004). Mitochondrial p53 activates Bak and causes disruption of a Bak-Mcl1 complex. Nat Cell Biol 6, 443–450.

Lindsten, T., Ross, A. J., King, A., Zong, W. X., Rathmell, J. C., Shiels, H. A., Ulrich, E., Waymire, K. G., Mahar, P., Frauwirth, K., Chen, Y., Wei, M., Eng, V. M., Adelman, D. M., Simon, M. C., Ma, A., Golden, J. A., Evan, G., Korsmeyer, S. J., MacGregor, G. R., and Thompson, C. B. (2000). The combined functions of proapoptotic Bcl-2 family members bak and bax are essential for normal development of multiple tissues. Mol Cell 6, 1389–1399.

Liu, Q., El-Deiry, W. S., and Gazitt, Y. (2005). Additive effect of Apo2L/TRAIL and Adeno-p53 in the induction of apoptosis in myeloma cell lines. Exp Hematol 29, 962–970.

Liu, X., Yue, P., Khuri, F. R., and Sun, S. Y. (2004). p53 upregulates death receptor 4 expression through an intronic p53 binding site. Cancer Res 64, 5078–5083.

Liu, X., Yue, P., Khuri, F. R., and Sun, S. Y. (2005). Decoy receptor 2 (DcR2) is a p53 target gene and regulates chemosensitivity. Cancer Res 65, 9169–9175.

Longley, D. B., Allen, W. L., McDermott, U., Wilson, T. R., Latif, T., Boyer, J., Lynch, M., and Johnston, P. G. (2004). The roles of thymidylate synthase and p53 in regulating Fas-mediated apoptosis in response to antimetabolites. Clin Cancer Res 10, 3562–3571.

Lowe, S. W. and Sherr, C. J. (2003). Tumor suppression by Ink4a-Arf: progress and puzzles. Curr Opin Genet Dev 13, 77–83.

MacLachlan, T. K. and El-Deiry, W. S. (2002). Apoptotic threshold is lowered by p53 transactivation of caspase-6. Proc Natl Acad Sci USA 99, 9492–9497.

Medema, J. P, Scaffidi, C., Kischkel, F. C., Shevchenko, A., Mann M,, Krammer, P. H, and Peter, M. E. (1997). FLICE is activated by association with the CD95 death-inducing signaling complex (DISC). EMBO J 16, 2794–2804.

Meng, R. D., McDonald, E. R. III, Sheikh, M. S., Fornace, A. J., Jr., and El-Deiry, W. S. (2000). The TRAIL decoy receptor TRUNDD (DcR2, TRAIL-R4) is induced by adenovirus-p53 over-expression and can delay TRAIL-, p53-, and KILLER/DR5-dependent colon cancer apoptosis. Mol Ther 1, 130–144.

Michael, D. and Oren, M. (2003). The p53-Mdm2 module and the ubiquitin system. Semin Cancer Biol 13, 49–58.

Mihara, M., Erster, S., Zaika, A., Petrenko, O., Chittenden, T., Pancoska, P., and Moll, U. M. (2003). p53 has a direct apoptogenic role at the mitochondria. Mol Cell 11, 577–590.

Muller, M., Wilder, S., Bannasch, D., Israeli, D., Lehlbach, K., Li-Weber, M., Friedman, S. L., Galle, P. R., Stremmel, W., Oren, M., and Krammer, P. H. (1998). p53 activates the CD95 (APO-1/Fas) gene in response to DNA damage by anticancer drugs. J Exp Med 188, 2033–2045.

Miyashita, T. and Reed, J. C. (1995). Tumor suppressor p53 is a direct transcriptional activator of the human bax gene. Cell 80, 293–299.

Moroni, M. C., Hickman, E. S., Lazzerini Denchi, E., Caprara, G., Colli, E., Cecconi, F., Muller, H., and Helin, K. (2001). Apaf-1 is a transcriptional target for E2F and p53. Nat Cell Biol 3, 552–558.

Muchmore, S. W., Sattler, M., Liang, H., Meadows, R. P., Harlan, J. E., Yoon, H. S., Nettesheim, D., Chang, B. S., Thompson, C. B., Wong, S. L., Ng, SL., and Fesik, S. W. (1996). X-ray and NMR structure of human Bcl-xL, an inhibitor of programmed cell death. Nature 381, 335–341.

Munsch, D., Watanabe-Fukunaga, R., Bourdon, J. C., Nagata, S., May, E., Yonish-Rouach, E., and Reisdorf, P. (2000). Human and mouse Fas (APO-1/CD95) death receptor genes each contain a p53-responsive element that is activated by p53 mutants unable to induce apoptosis. J Biol Chem 275, 3867–3872.

Murphy, M., Ahn, J., Walker, K. K., Hoffman, W. H., Evans, R. M., Levine, A. J., and George, D. L. (1999). Transcriptional repression by wild-type p53 utilizes histone deacetylases, mediated by interaction with mSin3a. Genes Dev 13, 2490–2501.

Nagata, S. (1999). Fas ligand-induced apoptosis. Annu Rev Genet. 33:29–55.

Nakano, K. and Vousden, K. (2001). H. PUMA, a novel proapoptotic gene, is induced by p53. Mol Cell 7, 683–694.

O'Connor, P. M., Jackman, J., Bae, I., Myers, T. G., Fan, S., Mutoh, M., Scudiero, D. A., Monks, A., Sausville, E. A., Weinstein, J. N., Friend, S., Fornace, A. J., Jr., and Kohn, K. W. (1997). Characterization of the p53 tumor suppressor pathway in cell lines of the National Cancer Institute anticancer drug screen and correlations with the growth-inhibitory potency of 123 anticancer agents. Cancer Res 57, 4285–4300.

Oda, E., Ohki, R., Murasawa, H., Nemoto, J., Shibue, T., Yamashita, T., Tokino, T., Taniguchi, T., and Tanaka, N. (2000a). Noxa, a BH3-only member of the Bcl-2 family and candidate media-tor of p53-induced apoptosis. Science 288, 1053–1058.

Oda, K., Arakawa, H., Tanaka, T., Matsuda, K., Tanikawa, C., Mori, T., Nishimori, H., Tamai, K., Tokino, T., Nakamura, Y., and Taya, Y. (2000b). p53AIP1, a potential mediator of p53-dependent apoptosis, and its regulation by Ser-46-phosphorylated p53. Cell 102, 849–862.

Okamura, S., Arakawa, H., Tanaka, T., Nakanishi, H., Ng, C. C., Taya, Y., Monden, M., Nakamura, Y. (2001). p53DINP1, a p53-inducible gene, regulates p53-dependent apoptosis. Mol. Cell. 8, 85–94.

Owen-Schaub, L. B., Zhang, W., Cusack, J. C., Angelo, L. S., Santee, S. M., Fujiwara, T., Roth, J. A., Deisseroth, A. B., Zhang, W. W., Kruzel, E., et al. (1995). Wild-type human p53 and a temperature-sensitive mutant induce Fas/APO-1 expression. Mol. Cell Biol. 15, 3032–3040.

Ozoren, N. and El-Deiry, W. S. (2003). Cell surface Death Receptor signaling in normal and cancer cells. Semin Cancer Biol 13, 135–1347.

Ozoren, N., Fisher, M. J., Kim, K., Liu, C. X., Genin, A., Shifman, Y., Dicker, D. T., Spinner, N. B., Lisitsyn, N. A., and El-Deiry, W. S. (2000). Homozygous deletion of the death receptor DR4 gene in a nasopharyngeal cancer cell line is associated with TRAIL resistance. Int J Oncol 16, 917–925.

Pai, S. I., Wu, G. S., Ozoren, N., Wu, L., Jen, J., Sidransky, D., and El-Deiry, W. S. (1998). Rare loss-of-function mutation of a death receptor gene in head and neck cancer. Cancer Res 58, 3513–3518.

Pan, G., O'Rourke, K., Chinnaiyan, A. M., Gentz, R., Ebner, R., Ni, J., and Dixit, V. M. (1997a). The receptor for the cytotoxic ligand TRAIL. Science 276, 111–113.

Pan, G., Ni, J., Wei, Y. F., Yu, G., Gentz, R., and Dixit, V. M. (1997b). An antagonist decoy receptor and a death domain-containing receptor for TRAIL. Science 277, 815–818.

Riedl, S. J. and Shi, Y. (2004). Molecular mechanisms of caspase regulation during apoptosis. Nat Rev Mol Cell Biol 5, 897–907.

Pearson, A. S., Spitz, F. R., Swisher, S. G., Kataoka, M., Sarkiss, M. G., Meyn, R. E., McDonnell, T. J., Cristiano, R. J., and Roth, J. A. (2000). Up-regulation of the proapoptotic mediators Bax and Bak after adenovirus-mediated p53 gene transfer in lung cancer cells. Clin Cancer Res 6, 887–890.

Pohl, U., Wagenknecht, B., Naumann, U., and Weller, M. (1999). p53 enhances BAK and CD95 expression in human malignant glioma cells but does not enhance CD95L-induced apoptosis. Cell Physiol Biochem 9, 29–37.

Polyak, K., Xia, Y., Zweier, J.L., Kinzler, K.W., Vogelstein, B. (1997). A model for p53-induced poptosis. Nature. 389:300–5.

Rocco, J. W., Leong, C. O., Kuperwasser, N., DeYoung, M. P., and Ellisen, L. W. (2006). p63 mediates survival in squamous cell carcinoma by suppression of p73-dependent apoptosis. Cancer Cell 9, 45–56.

Rodriguez, M. S., Desterro, J. M., Lain, S., Lane, D. P., and Hay, R. T. (2000). Multiple C-terminal lysine residues target p53 for ubiquitin-proteasome-mediated degradation. Mol Cell Biol 20, 8458–8467.

Roy, S., Packman, K., Jeffrey, R., and Tenniswood, M. (2005). Histone deacetylase inhibitors differentially stabilize acetylated p53 and induce cell cycle arrest or apoptosis in prostate cancer cells. Cell Death Differ 12, 482–491.

Ruiz de Almodovar, C., Ruiz-Ruiz, C., Rodriguez, A., Ortiz-Ferron, G., Redondo, J. M., and Lopez-Rivas, A. (2004). Tumor necrosis factor-related apoptosis-inducing ligand (TRAIL) decoy receptor TRAIL-R3 is up-regulated by p53 in breast tumor cells through a mechanism involving an intronic p53-binding site. J Biol Chem 279, 4093–4101.

Ryan, A. E., Shanahan, F., O'Connell, J., and Houston, A. M. (2005). Addressing the "Fas counterattack" controversy: blocking fas ligand expression suppresses tumor immune evasion of colon cancer in vivo. Cancer Res 65, 9817–9823.

Sattler, M., Liang, H., Nettesheim, D., Meadows, R. P., Harlan, J. E., Eberstadt, M., Yoon, H. S., Shuker, S. B., Chang, B. S., Minn, A. J., Thompson, C. B., and Fesik, S. W. (1997). Structure of Bcl-xL-Bak peptide complex: recognition between regulators of apoptosis. Science 275, 983–986.

Sax, J. K., Fei, P., Murphy, M. E., Bernhard, E., Korsmeyer, S. J., and El-Deiry, W. S. (2002). BID regulation by p53 contributes to chemosensitivity. Nat Cell Biol 4, 842–849.

Scaffidi, C., Fulda, S., Srinivasan, A., Friesen, C., Li, F., Tomaselli, K. J., Debatin, K. M., Krammer, P. H., and Peter, M. E. (1998). Two CD95 (APO-1/Fas) signaling pathways. EMBO J 17, 1675–1687.

Sheikh, M. S., Huang, Y., Fernandez-Salas, E. A., El-Deiry, W. S., Friess, H., Amundson, S., Yin, J., Meltzer, S. J., Holbrook, N. J., and Fornace, A. J., Jr. (1999). The antiapoptotic decoy receptor TRID/TRAIL-R3 is a p53-regulated DNA damage-inducible gene that is overexpressed in primary tumors of the gastrointestinal tract. Oncogene 18, 4153–4159.

Sheridan, J. P., Marsters, S. A., Pitti, R. M., Gurney, A., Skubatch, M., Baldwin, D., Ramakrishnan, L., Gray, C. L., Baker, K., Wood, W. I., Goddard, A. D., Godowski, P., and Ashkenazi, A. (1997).

Control of TRAIL-induced apoptosis by a family of signaling and decoy receptors. Science 277, 818–821.

Shetty, S., Graham, B. A., Brown, J. G., Hu, X., Vegh-Yarema, N., Harding, G., Paul, J. T., and Gibson, S. B. (2005). Transcription factor NF-kappaB differentially regulates death receptor 5 expression involving histone deacetylase 1. Mol Cell Biol 25, 5404–5416.

Shibue, T., Takeda, K., Oda, E., Tanaka, H., Murasawa, H., Takaoka, A., Morishita, Y., Akira, S., Taniguchi, T., and Tanaka, N. (2003). Integral role of Noxa in p53-mediated apoptotic response. Genes Dev 17, 2233–2238.

Shin, M. S., Kim, H. S., Lee, S. H., Park, W. S., Kim, S. Y., Park, J. Y., Lee, J. H., Lee, S. K., Lee, S. N., Jung. S. S., Han, J. Y., Kim, H., Lee, J. Y., and Yoo, N. J. (2001). Mutations of tumor necrosis factor-related apoptosis-inducing ligand receptor 1 (TRAIL-R1) and receptor 2 (TRAIL-R2) genes in metastatic breast cancers. Cancer Res 61, 4942–4946.

Snyder, E. L., Meade, B. R., Saenz, C. C., and Dowdy, S. F. (2004). Treatment of terminal peritoneal carcinomatosis by a transducible p53-activating peptide. PLoS Biol E36.

Soengas, M. S., Alarcon, R. M., Yoshida, H., Giaccia, A. J., Hakem, R., Mak, T. W., and Lowe, S. W. (1999). Apaf-1 and caspase-9 in p53-dependent apoptosis and tumor inhibition. Science 284, 156–159.

Soengas, M. S., Capodieci, P., Polsky, D., Mora, J., Esteller, M., Opitz-Araya, X., McCombie, R., Herman, J. G., Gerald, W. L., Lazebnik, Y. A., Cordon-Cardo, C., and Lowe, S. W. (2001). Inactivation of the apoptosis effector Apaf-1 in malignant melanoma. Nature 409, 207–211.

Sun, Y. and Leaman, D. W. (2005). Involvement of Noxa in cellular apoptotic responses to interferon, double-stranded RNA, and virus infection. J Biol Chem 280(16), 15561–15568.

Takimoto, R. and El-Deiry, W. S. (2000). Wild-type p53 transactivates the KILLER/DR5 gene through an intronic sequence-specific DNA-binding site. Oncogene 19, 1735–1743.

Takimoto, R., Wang, W., Dicker, D. T., Rastinejad, F., Lyssikatos, J., and El-Deiry, W. S. (2002). The mutant p53-conformation modifying drug, CP-31398, can induce apoptosis of human cancer cells and can stabilize wild-type p53 protein. Cancer Biol Ther 1, 47–55.

Takimoto, R., Kato, J., Terui, T., Takada, K., Kuroiwa, G., Wu, J., Ohnuma, H., Takahari, D., Kobune, M., Sato, Y., Takayama, T., Matsunaga, T., and Niitsu, Y. (2005). Augmentation of Antitumor Effects of p53 Gene Therapy by Combination with HDAC Inhibitor. Cancer Biol. Ther. 4, 421–428.

Tanikawa, C., Matsuda, K., Fukuda, S., Nakamura, Y., and Arakawa, H. (2003). p53RDL1 regulates p53-dependent apoptosis. Nat Cell Biol 5, 216–222.

Thiery, J., Abouzahr, S., Dorothee, G., Jalil, A., Richon, C., Vergnon, I., Mami-Chouaib, F., and Chouaib, S. (2005). p53 potentiation of tumor cell susceptibility to CTL involves Fas and mitochondrial pathways. J Immunol 174, 871–878.

Terui, T., Murakami, K., Takimoto, R., Takahashi, M., Takada, K., Murakami, T., Minami, S., Matsunaga, T., Takayama, T., Kato, J., and Niitsu, Y. (2003). Induction of PIG3 and NOXA through acetylation of p53 at 320 and 373 lysine residues as a mechanism for apoptotic cell death by histone deacetylase inhibitors. Cancer Res 63, 8948–8954.

Tomita, Y., Marchenko, N., Erster, S., Nemajerova, A., Dehner, A, Klein, C., Pan, H., Kessler, H., Pancoska, P., and Moll, U. M. (2006). WTp53 but not tumor-derived mutants bind to BCL2 via the DNA binding domain and induce mitochondrial permeabilization. J Biol Chem 281, 8600.

Tovar, C., Rosinski, J., Filipovic, Z., Higgins, B., Kolinsky, K., Hilton, H., Zhao, X., Vu, B. T., Qing, W., Packman, K., Myklebost, O., Heimbrook, D. C., and Vassilev, L. T. (2006). Small-molecule MDM2 antagonists reveal aberrant p53 signaling in cancer: Implications for therapy. Proc Natl Acad Sci USA 103, 1888–1893.

Vassilev, L. T., Vu, B. T., Graves, B., Carvajal, D., Podlaski, F., Filipovic, Z., Kong, N., Kammlott, U., Lukacs, C., Klein, C., Fotouhi, N., and Liu, E. A. (2004). In vivo activation of the p53 pathway by small-molecule antagonists of MDM2. Science 303, 844–848.

Villunger, A., Michalak, E. M., Coultas, L., Mullauer, F., Bock, G., Ausserlechner, M. J., Adams, J. M., and Strasser, A. (2003). p53- and drug-induced apoptotic responses mediated by BH3-only proteins puma and noxa. Science 302, 1036–1038.

Vogelstein, B., Lane, D., and Levine, A. J. (2000). Surfing the p53 network. Nature 408, 307–310.

Wang, S. and El-Deiry, W. S. (2003a). TRAIL and apoptosis induction by TNF-family death receptors. Oncogene 22, 8628–8633.

Wang, S. and El-Deiry, W. S. (2003b). Requirement of p53 targets in chemosensitization of colonic carcinoma to death ligand therapy. Proc Natl Acad Sci USA 100, 15095–15100.

Wang, S. and El-Deiry, W. S. (2004a). Inducible silencing of KILLER/DR5 in vivo promotes bio-luminescent colon tumor xenograft growth and confers resistance to chemotherapeutic agent 5-fluorouracil. Cancer Res 64, 6666–7662.

Wang, W. and El-Deiry, W. S. (2004b). Targeting p53 by PTD-mediated transduction. Trends Biotechnol 22, 431–434.

Wang, W., Takimoto, R., Rastinejad, F., and El-Deiry, W. S. (2003). Stabilization of p53 by CP-31398 inhibits ubiquitination without altering phosphorylation at serine 15 or 20 or MDM2 binding. Mol Cell Biol 23, 2171–2181.

Wei, M. C., Zong, W. X., Cheng, E. H., Lindsten, T., Panoutsakopoulou, V., Ross, A. J., Roth, K. A., MacGregor, G. R., Thompson, C. B., and Korsmeyer, S. J. (2001). Proapoptotic BAX and BAK: a requisite gateway to mitochondrial dysfunction and death. Science 292, 727–730.

Weinstein, J. N., Myers, T. G., O'Connor, P. M., Friend, S. H., Fornace, A. J. Jr., Kohn, K. W., Fojo, T., Bates, S. E., Rubinstein, L. V., Anderson, N. L., Buolamwini, J. K., van Osdol, W. W., Monks, A. P., Scudiero, D. A., Sausville, E. A., Zaharevitz, D. W., Bunow, B., Viswanadhan, V. N., Johnson, G. S., Wittes, R. E., and Paull, K. D. (1997). An information-intensive approach to the molecular pharmacology of cancer. Science 275, 343–349.

Wu, G. S., Burns, T. F., McDonald, E. R., III, Jiang, W., Meng, R., Krantz, I. D., Kao, G., Gan, D. D., Zhou, J. Y., Muschel, R., Hamilton, S. R., Spinner, N. B., Markowitz, S., Wu, G., and El-Deiry, W. S. (1997). KILLER/DR5 is a DNA damage-inducible p53-regulated death receptor gene. Nat Genet 17, 141–143.

Wu, W. S., Heinrichs, S., Xu, D., Garrison, S. P., Zambetti, G. P., Adams, J. M., and Look, A. T. (2005). Slug antagonizes p53-mediated apoptosis of hematopoietic progenitors by repressing puma. Cell 123, 641–653.

Wu, Y., Mehew, J. W., Heckman, C. A., Arcinas, M., and Boxer, L. M. (2001). Negative regulation of bcl-2 expression by p53 in hematopoietic cells. Oncogene 20, 240–251.

Xiao, S., Marshak-Rothstein, A., and Ju, S. T. (2001). Sp1 is the major fasl gene activator in abnormal CD4(−)CD8(−)B220(+) T cells of lpr and gld mice. Eur J Immunol 31, 3339–3348.

Yu, J., Zhang, L., Hwang, P. M., Kinzler, K. W., and Vogelstein, B. (2001). PUMA induces the rapid apoptosis of colorectal cancer cells. Mol Cell 7: 673–682.

Yu, J., Wang, Z., Kinzler, K. W., Vogelstein, B., and Zhang, L. (2003). PUMA mediates the apoptotic response to p53 in colorectal cancer cells. Proc Natl Acad Sci USA 100, 1931–1936.

Zha, J., Weiler, S., Oh, K. J., Wei, M. C., and Korsmeyer, S. J. (2000). Posttranslational N-myristoylation of BID as a molecular switch for targeting mitochondria and apoptosis. Science 290, 1761–1765.

Zinkel, S. S., Hurov, K. E., Ong, C., Abtahi, F. M., Gross, A., and Korsmeyer, S. J. (2005). A role for proapoptotic BID in the DNA-damage response. Cell 122, 579–591.

Zhao, X., Ayer, R. E., Davis, S. L., Ames, S. J., Florence, B., Torchinsky, C., Liou, J. S., Shen, L., and Spanjaard, R. A. (2005). Apoptosis factor EI24/PIG8 is a novel endoplasmic reticulum-localized Bcl-2-binding protein which is associated with suppression of breast cancer invasiveness. Cancer Res 65, 2125–2129.

Chapter 11
Regulation of Programmed Cell Death by NF-κB and its Role in Tumorigenesis and Therapy

Yongjun Fan, Jui Dutta, Nupur Gupta, Gaofeng Fan, and Céline Gélinas*

Abstract The Rel/NF-κB transcription factors are key regulators of programmed cell death (PCD). Their activity has significant physiological relevance for normal development and homeostasis in various tissues and important pathological consequences are associated with aberrant NF-κB activity, including hepatocyte apoptosis, neurodegeneration, and cancer. While NF-κB is best characterized for its protective activity in response to proapoptotic stimuli, its role in suppressing programmed necrosis has come to light more recently. NF-κB most commonly antagonizes PCD by activating the expression of antiapoptotic proteins and antioxidant molecules, but it can also promote PCD under certain conditions and in certain cell types. It is therefore important to understand the pathways that control NF-κB activation in different settings and the mechanisms that regulate its anti- vs pro-death activities. Here, we review the role of NF-κB in apoptotic and necrotic PCD, the mechanisms involved, and how its activity in the cell death response impacts cancer development, progression, and therapy. Given the role that NF-κB plays both in tumor cells and in the tumor microenvironment, recent findings underscore the NF-κB signaling pathway as a promising target for cancer prevention and treatment.

Keywords Rel/NF-κB, apoptosis, necrosis, transcription factor, cancer, therapy

Yongjun Fan
Center for Advanced Biotechnology and Medicine

Jui Dutta, Nupur Gupta, and Gaofeng Fan
Center for Advanced Biotechnology and Medicine
Graduate Program in Biochemistry and Molecular Biology

Céline Gélinas
Center for Advanced Biotechnology and Medicine
Department of Biochemistry and Cancer Institute of New Jersey, Robert Wood Johnson Medical School, University of Medicine and Dentistry of New Jersey, Piscataway, NJ 08854-5638, USA

*To whom correspondence should be addressed: CABM, 679 Hoes Lane, Piscataway, NJ 08854-5638, USA. Tel.: (732) 235–5035; fax: (732) 235–4466; e-mail: gelinas@cabm.rutgers.edu

R. Khosravi-Far and E. White (eds.), *Programmed Cell Death in Cancer Progression and Therapy*.
© Springer 2008

223

1 Introduction

The Rel/NF-κB family of proteins is comprised of homologous transcription factors that mediate the cellular response to various exogenous or endogenous stimuli including infection, inflammation, stress, or injury (reviewed in Bonizzi and Karin, 2004; Hayden and Ghosh, 2004). This multimember family consists of the vertebrate c-Rel, RelA, RelB, p105/p50 NF-κB1, and p100/p52 NF-κB2 subunits, the viral oncoprotein v-Rel, Xenopus X-Rell, and the Dorsal, Dif, and Relish factors from *Drosophila*. Rel/NF-κB proteins share a highly conserved Rel homology domain (RHD) at their N-terminus that allows them to engage in homodimer or heterodimer formation, enter the nucleus, and bind to consensus GGGRNNYYCC NF-κB DNA sites. It also enables them to associate with inhibitory IκB molecules that act in an autoregulatory feedback fashion to terminate the activation process. The C-terminal domains of NF-κB factors are more divergent across the family and impart transcriptional activation properties to c-Rel, RelA, RelB and v-Rel proteins, or inhibitory properties to p105/NF-κB1 and p100/NF-κB2 that contain ankyrin-repeats akin to those found in IκB proteins.

In resting cells, cytosolic NF-κB dimers are inactive and typically bound to IκB proteins that prevent their nuclear translocation and binding to consensus NF-κB DNA-binding sites. Two distinct NF-κB activation cascades that respond to different stimuli have been documented (Fig. 11.1). The canonical (or classical) NF-κB pathway is activated by proinflammatory and mitogenic stimuli such as cytokines, bacterial lipopolysaccharides (LPS), interleukin-1 (IL-1), and antigens. This pathway commonly converges upon activation of the IκB kinase complex (IKK complex), a large multisubunit entity comprised of the catalytic IKKα and IKKβ subunits and the regulatory subunit IKKγ/NEMO. Phosphorylation of IκBα on serines 32 and 36 targets it for ubiquitination at lysines 21 and 22 by the E3 ligase SCF-βTrCP. Degradation of polyubiquitinated IκBα by the 26S proteasome frees NF-κB dimers, like the classical p50/p65 complex, enabling their entry into the nucleus where they bind to NF-κB DNA sites. This commonly results in the transcriptional activation of genes important for immune and inflammatory responses, cell proliferation, and/or suppression of apoptosis. Among the many genes that NF-κB regulates, transcriptional activation of its inhibitor IκBα generates an autoregulatory feedback loop that terminates the activation process. Consequently, activation of the NF-κB pathway is normally a regulated and transient process that is important for normal innate and adaptive immunity, inflammatory and acute phase responses and for embryonic development, organogenesis, and homeostasis. In contrast, sustained activation of the NF-κB pathway is implicated in a wide variety of pathological conditions including immune system disorders, chronic inflammation, and cancer.

The noncanonical (or alternative) NF-κB signaling cascade is characterized by the tightly regulated processing of the p100/NF-κB2 precursor protein into a mature p52 subunit and is commonly involved in the preferential activation of RelB/p52

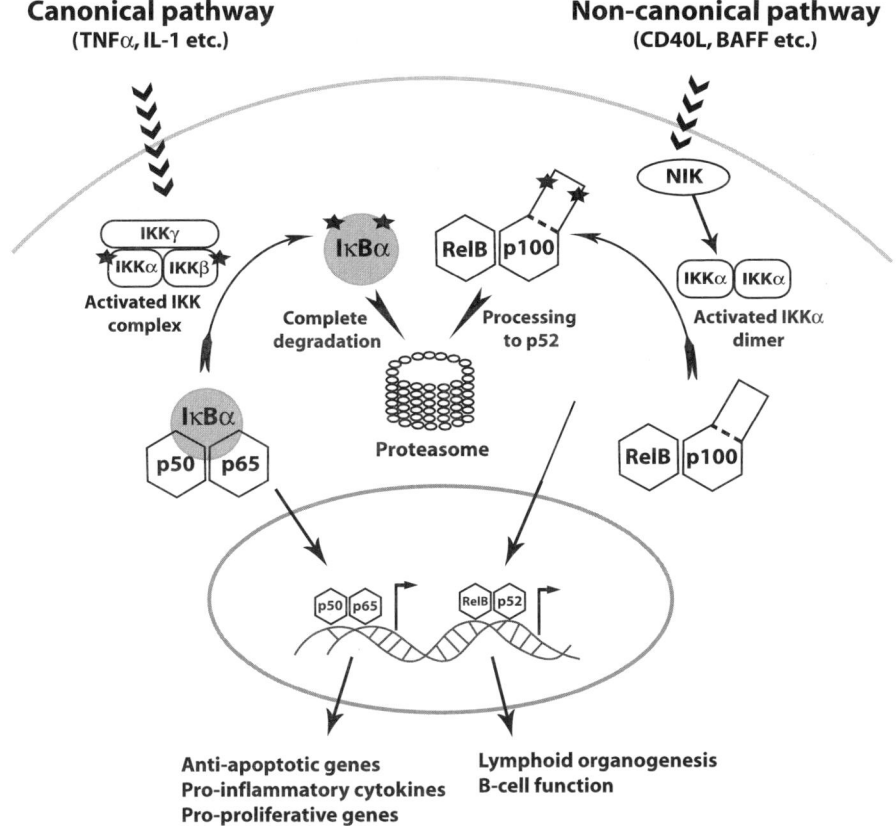

Fig. 11.1 The canonical (classical) and noncanonical (alternative) NF-κB signaling pathways. In the canonical NF-κB pathway, binding of cytokines, LPS, IL-1, or antigen T-cell surface receptors leads to activation of the IKK kinase complex that induces phosphorylation of IκBα and promotes its ubiquitin-dependent degradation via the proteasome. Cytosolic NF-κB dimers (e.g., p50/p65 complexes) are then free to translocate to the nucleus where they bind to consensus NF-κB DNA sites and activate gene expression. This pathway is commonly involved in the activation of antiapoptotic genes, inflammatory cytokines and genes that promote cell proliferation, angiogenesis, and metastasis. The noncanonical NF-κB cascade is activated in response to cell stimulation with BAFF, LTβ, or CD40L and leads to activation of the kinase NIK. NIK phosphorylates IKKα to induce phosphorylation of the C-terminus of p100/NF-κB2. This targets p100 for ubiquitination and partial proteasome-mediated degradation to generate a mature p52/NF-κB2 form. This commonly results in nuclear translocation of RelB/p52 complexes, their binding to NF-κB DNA sites and the activation of gene expression. This pathway is important for lymphoid organogenesis and B-cell function

dimers (Qing and Xiao, 2005; Senftleben et al., 2001; Xiao et al., 2001, 2004); (Fig. 11.1). Activation of this pathway occurs predominantly in B cells stimulated with BAFF, lymphotoxinβ (LTβ), or CD40L and is important for B-cell function and lymphoid organogenesis. In this cascade receptor stimulation leads to activation of the NF-κB-inducing kinase NIK that activates IKKα complexes, independently of

IKKβ and IKKγ/NEMO, to phosphorylate serines 866 and 870 in the C-terminus of p100/NF-κB2 (Liang et al., 2006; Qing et al., 2005). Consequent ubiquitination of p100 by SCF-βTrCP results in the cleavage and partial degradation of p100 to the mature p52 form via the proteasome (Rape and Jentsch, 2004). Some reported that p100/NF-κB2 could also undergo cotranslational processing by the proteasome (Heusch et al., 1999). The functional consequences of NF-κB activation via the canonical or noncanonical pathways are many, but for the purpose of this review, we will focus on those associated with its role in programmed cell death (PCD) via apoptosis or necrosis, and on the mechanisms by which it operates in these different contexts.

2   Role of NF-κB in Apoptosis and Necrosis

NF-κB can protect cells from apoptosis induced by many different death-inducing stimuli including antigen receptor cross-linking in B cells, chemotherapeutic agents, radiation, and the proinflammatory cytokine TNFα, although in some instances it can behave in a proapoptotic manner (Grumont et al., 1998; Owyang et al., 2001; Van Antwerp et al., 1998; Wang et al., 1998; Wu et al., 1996; reviewed in Kucharczak et al., 2003). Studies focusing on the activity of NF-κB in cells treated with TNFα, UV radiation, or chemotherapeutic agents have provided important insights into the mechanisms that underlie its antiapoptotic vs pro-death effects and on those that govern this decision, as reviewed below.

2.1   Choosing between Life and Death Downstream of Activated TNFR1

Detailed analysis of the signaling cascade initiated by TNFα revealed important clues regarding the role that NF-κB plays in the apoptotic response, and recently in necrosis (see Section 2.2). These studies also illustrated that NF-κB plays a crucial role in tipping the balance in favor of survival following Tumor necrosis factor receptor (TNFR) activation.

Although activation of TNF receptor 1 (TNFR1) by TNFα can initiate PCD, TNFα is usually not cytotoxic, as concomitant activation of the NF-κB pathway confers efficient protection. In fact, cell killing by TNFα is seen only under conditions where NF-κB activity is suppressed, or if RNA or protein synthesis is inhibited. Binding of TNFα to TNFR1 triggers trimerization of the receptor and initiates three different cascades that can differentially affect the fate of the cells (Fig. 11.2; Micheau and Tschopp, 2003; reviewed in Jaattela and Tschopp, 2003). The first involves the cooperative recruitment of the adaptor molecule TNFR-associated death domain (TRADD) and the receptor-interacting protein kinase 1 (RIP1), along with TNFR-associated factor 2 (TRAF2). This promotes IKK-dependent activation

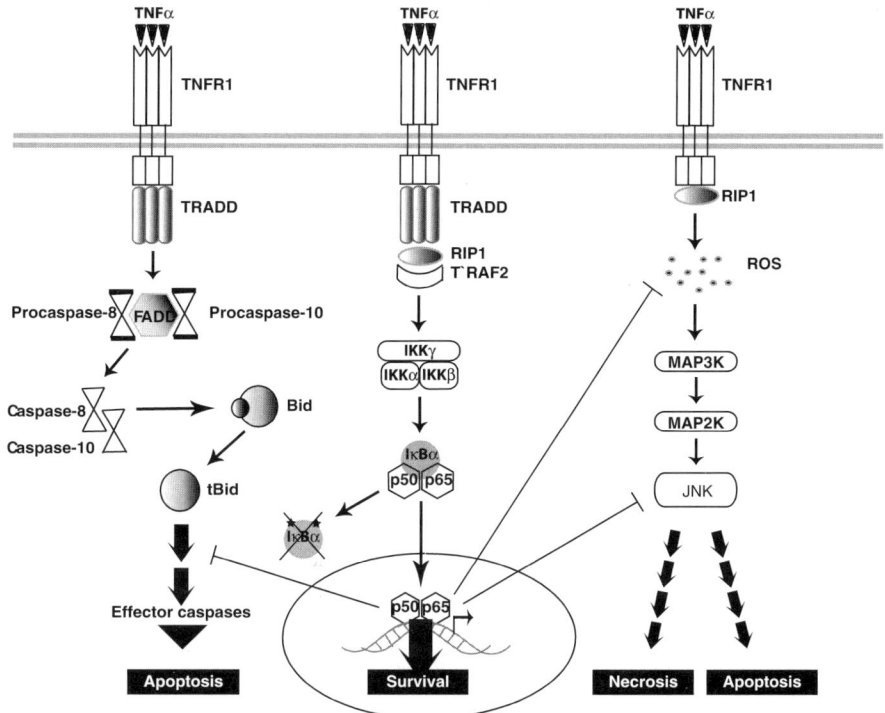

Fig. 11.2 Activation of TNFR1 initiates three different signaling cascades that differentially affect the fate of the cells. Cooperative recruitment of TRADD, RIP1, and TRAF2 to TNFR1 promotes IKK-dependent activation of NF-κB and the activation of antiapoptotic genes, leading to cell survival (*center*). Recruitment of TRADD to TNFR1 in absence of RIP1 engages FADD, caspase-8, and caspase-10 leading to cleavage-mediated activation of Bid into tBid. Translocation of tBid to mitochondria provokes the release of cytochrome C and Smac/Diablo, and activation of effector caspases resulting in apoptosis (*left*). Ligand binding to TNFR1 can also trigger recruitment of RIP1 to the TNFR1 complex in absence of TRADD. This leads to production of reactive oxygen species (ROS), activation of the JNK signaling cascade, and results in PCD via necrosis or apoptosis (*right*). Efficient NF-κB-dependent synthesis of antiapoptotic proteins and antioxidant molecules is thus necessary to block apoptosis and necrosis triggered by TNFα

of NF-κB and that of antiapoptotic proteins like the cellular FLICE/caspase-8 inhibitor protein (c-FLIP). Ubiquitination of TNFR1 and TRADD promotes initiation of a second cascade that engages FLICE-associated death domain (FADD) and caspase-8/FLICE along with caspase-10, that leads to the cleavage-mediated activation of the BH3-only protein Bid into tBid. tBid translocates to mitochondria and associates with the proapoptotic BH1–3 factors Bax and Bak. This provokes the mitochondrial release of cytochrome C and Smac/Diablo, activation of caspase-9, and downstream effector caspases resulting in apoptosis. Thus, efficient NF-κB-dependent synthesis of antiapoptotic proteins like c-FLIP by the first cascade is necessary to block apoptosis induced by the second cascade. Consequently, cells deficient for NF-κB readily undergo apoptosis in response to TNFα, as do cells in which inhibition of RNA or protein synthesis precludes activation of prosurvival NF-κB target genes (Yeh et al., 2000).

Ligand binding to TNFR1 can also trigger a third cascade that leads to necrotic cell death, a mode of PCD that is morphologically distinct from apoptosis and is independent of caspases. This cascade depends on the recruitment of RIP1 to the TNFR1 complex, in absence of TRADD, on RIP1 kinase activity and its ability to induce production of reactive oxygen species (ROS) and activate the JNK signaling cascade (see Section 2.2; Zheng et al., 2006; reviewed in Jaattela and Tschopp, 2003; Leist and Jaattela, 2001; Papa et al., 2006). While it remains to be determined how cells decide to die by apoptosis vs necrosis, their metabolic state appears to be an important factor in this decision (Edinger and Thompson, 2004). Furthermore, recent evidence that recruitment of RIP1 to TNFR1 precludes engagement of TRADD in this cascade suggests that the joint vs exclusive engagement of these molecules by TNFR1 may also help to determine whether the cell will take the NF-κB survival path, the apoptotic path or will undergo death via necrosis (Zheng et al., 2006). An important distinction between cells dying by apoptosis or by necrosis is that contrary to apoptosis, cells dying by necrosis trigger a strong inflammatory response due to the release of potent proinflammatory factors such as the chromatin-associated HMGB1 protein. HMGB1 binds to the receptor for advanced glycation end products (RAGE), Toll-like receptors TLR2 or TLR4 on macrophages to signal production of proinflammatory cytokines (reviewed in Lotze and Tracey, 2005; Zeh and Lotze, 2005), although some have recently argued that HMGB1 binds preferentially TLR2 and TLR4, but not RAGE as determined by fluorescence resonance energy transfer (FRET) analysis (Park et al., 2006).

2.2   Interplay Between NF-κB and JNK: Additional Insights into NF-κB's Protective Activity Toward Apoptosis and Necrosis

The cross talk between the NF-κB- and JNK-signaling pathways and its impact on the outcome of the cells has been the subject of several excellent reviews (Luo et al., 2005a; Nakano et al., 2006; Papa et al., 2006). Here, we briefly outline how their interplay can result in cell survival, apoptosis, or necrosis.

In addition to promoting activation of NF-κB, binding of TNFα to TNFR1 triggers activation of the MAPK-related stress-activated Jun kinase (JNK), as illustrated in the third cascade (Fig. 11.2). Detailed analyses with cells defective for either JNK or NF-κB outlined a key role for JNK in TNF-induced cell death and showed that the ability of NF-κB to antagonize JNK signaling is an important component of NF-κB's arsenal against the cytotoxic effects of TNFα (reviewed in Luo et al., 2005a; Papa et al., 2004b, 2006; see below). Studies showed that NF-κB is responsible for the transient activation of JNK in response to TNFα, and that suppression of NF-κB activity results in sustained JNK activation, aberrant ROS accumulation, and cell death (De Smaele et al., 2001; Javelaud and Besancon, 2001; reviewed in Papa et al., 2004,

2006; Tang et al., 2001). In turn, ROS can induce sustained JNK activity by inactivating MAPK phosphatases (MKPs), thus allowing TNFα to kill cells in which NF-κB is active (Kamata and Hirata, 1999; Kamata et al., 2005; Sakon et al., 2003). However, this may not be the only mechanism that activates JNK following ROS accumulation, as others showed that ROS activate ASK1/MEKK5 that leads to prolonged JNK activation downstream of TNFR1 (Davis, 2000; Matsuzawa and Ichijo, 2005).

There is still debate in the field regarding the extent to which NF-κB-mediated suppression of JNK signaling blunts apoptosis vs necrosis (Ventura et al., 2004; reviewed in Papa et al., 2006). Differences in the metabolic state of the cells are likely to sway which form of cell death will prevail (Ventura et al., 2004; reviewed in Papa et al., 2006). It was suggested that actively dividing cells that depend on glycolysis are more likely to die by necrosis, whereas quiescent cells that undergo oxidative phosphorylation predominantly die by apoptosis (reviewed in Edinger and Thompson, 2004). Clearly, the interplay between the JNK and NF-κB signaling cascades is an important factor in dictating the fate of the cells be it survival, apoptosis or necrosis.

3   A Role for NF-κB in Autophagy

Autophagy is a form of PCD distinct from apoptosis and necrosis that has come under increasing scrutiny lately, particularly as it relates to cancer (reviewed in Edinger and Thompson, 2004; Hait et al., 2006; Levine and Yuan, 2005) (also see Chapter 9). Cells undergo autophagy in response to nutrient and growth factor deficiency as a temporary means of survival. They do so by undergoing self-digestion under conditions where adequate nutrient supplies are limited, as would be the case for cancer cells lacking an adequate blood supply. However, a prolonged state of autophagy ultimately results in metabolic cell death. The process itself involves assembly of an autophagosome in which a cell's organelles and cytoplasm are swallowed. Its contents are then degraded by lysosomes, which allow salvation of amino acids and fatty acids for energy generation. Although studies are only beginning to explore a possible role of NF-κB in autophagy, its protective activity in ventricular myocytes was recently shown to involve transcriptional repression of the hypoxia-inducible BH3-only protein BNIP3 that was demonstrated to induce autophagy (Baetz et al., 2005; Daido et al., 2004; Kanzawa et al., 2005). Future studies will surely shed more light on this subject and on whether it is implicated in NF-κB-associated cancers.

4   Mechanisms that NF-κB Employs to Suppress PCD

NF-κB utilizes several different means to suppress PCD. NF-κB most commonly suppresses apoptosis by activating the transcription of antiapoptotic genes (reviewed in Kucharczak et al., 2003; Luo et al., 2005b; Papa et al., 2006). Among them are antiapoptotic Bcl-2 family members Bcl-2, Bcl-x$_L$, Bfl-1/A1, and NR13 that antagonize the activity of proapoptotic Bcl-2 family proteins and thus blunt the release of proapoptotic cytochrome c and Smac/Diablo from mitochondria. The cellular inhibitor of apoptosis molecules XIAP, c-IAP1, and c-IAP2 also contribute to its protective activity (reviewed in Wright and Duckett, 2005). While some IAPs like XIAP directly block cleavage-mediated activation of pro-caspase-9 and the activity of caspases 3 and 7, others are less potent in this regard (Deveraux et al., 1997; Liston et al., 2003). Recent studies showing that the baculoviral IAP protein (OpIAP) promotes ubiquitination of the IAP antagonist Smac/DIABLO uncovered a novel mechanism whereby cytoprotective IAPs can block apoptosis in a caspase-independent manner (Duckett, 2005; Wilkinson et al., 2004). Of late, the zinc finger protein A20 was shown to suppress cell death by promoting degradation of the TNFR1 complex component RIP1, via its deubiquitinating (DUB) and E3 ligase activities (Wertz et al., 2004). Other NF-κB-regulated candidates include decoy TRAIL receptor 1 (DcR1) (Bernard et al., 2001a) and c-FLIP that interferes with activation of pro-caspases 8 and 10 (Kreuz et al., 2004; Micheau et al., 2001). cFLIP can also work with caspase-8 to enhance NF-κB activation via the B-cell lymphoma 10 (BCL-10) and mucosa-associated-lymphoid-tissue lymphoma-translocation gene 1 (MALT1) that act as E3 ligases for IKKγ/NEMO, along with RIP1 (Zhou et al., 2004; reviewed in Budd et al., 2006). NF-κB-mediated induction of the serine protease inhibitor 2A (Spi2A), that inhibits cathepsin B, was shown to suppress cell killing by TNFα by blocking lysosome-mediated PCD (Liu et al., 2003).

In the antagonistic relationship between the NF-κB and JNK signaling cascades, GADD45β/Myd118 and XIAP were among the first NF-κB targets proposed to block sustained JNK activation (De Smaele et al., 2001; Tang et al., 2001). GADD45β associates with and blocks the catalytic activity of the JNK-activating kinase MKK7/JNKK2 (Kaur et al., 2005; Papa et al., 2004a, b). How XIAP blocks prolonged JNK activation is still not clear, but a recent study suggests that it can inhibit TGF-β1-induced JNK activation and apoptosis by ubiquitinating the kinase TAK1, leading to its degradation (Kaur et al., 2005). It should be noted, however, that homozygous deletion of *gadd45β* or *xiap* had no significant effect on JNK activation in vivo (Amanullah et al., 2003; Sanna et al., 2002), suggesting that compensatory mechanisms may exist or that another NF-κB-dependent inhibitor(s) of proapoptotic JNK signaling remain to be identified. Relevant candidates in this regard are the antioxidant molecules manganese superoxide dismutase (MnSOD), and ferritin heavy chain (FHC) that inhibits JNK by suppressing ROS accumulation through iron sequestration (Bernard et al., 2001b, 2002; Delhalle et al., 2002; Pham et al., 2004; Tanaka et al., 2002; reviewed in Papa et al., 2004b).

Other means have been described to explain the antiapoptotic effects of NF-κB in certain contexts. One of them involves NF-κB-induced destabilization of tumor suppressor p53 as a result of increased expression of Mdm2 (Egan et al., 2004; Tergaonkar et al., 2002). RelA-dependent suppression of caspase-8 and TRAIL receptors DR4 and DR5 was shown to confer survival to TRAIL along with induction of c-IAP1 and c-IAP2 (Chen et al., 2003). The peptidyl prolyl-isomerase Pin1 was reported to enhance nuclear accumulation of RelA/p65 by blocking its association with IκBα and to also lead to p65 stabilization by interfering with its interaction with the ubiquitin ligase SOCS-1 (Ryo et al., 2003). Although direct evidence is still lacking that Pin1 enhances the protective activity of RelA, Pin1 is frequently upregulated in breast cancer compared to normal mammary glands (Currier et al., 2005; Ryo et al., 2003).

5   Mechanisms that Underlie the Pro-Death Activity of NF-κB

Although NF-κB is best known for its ability to antagonize PCD, it should be noted that it can be proapoptotic in certain cells and in response to certain stimuli (reviewed in Kucharczak et al., 2003; and see below). Some NF-κB transcriptional targets that were implicated in this effect include factors that modulate the mitochondrial and death receptor apoptotic pathways including the p53 tumor suppressor, death receptor Fas and its ligand FasL, TNFα, TRAIL receptors DR4, DR5, DR6, TRAIL itself, and the proapoptotic Bcl-2 family members Bcl-xS and Bax.

Work in recent years uncovered an interesting new way whereby the typically antiapoptotic NF-κB subunit RelA can behave as a pro-death factor in response to certain stimuli (reviewed in Perkins and Gilmore, 2006). Cell treatment with atypical activators of NF-κB such as UV-C radiation or the chemotherapeutic drugs daunorubicin and doxorubicin switches RelA from a transcriptional activator into a gene-specific transcriptional repressor of antiapoptotic genes (like Bcl-x_L, $XIAP$, and $A20$), but not IκBα by promoting association of RelA with histone deacetylase HDAC1, resulting in cell death (Campbell et al., 2004). This occurs in a RelA phosphorylation-independent manner. Tumor suppressor alternative reading frame (ARF) can also suppress the protective activity of RelA by using a slightly different mechanism, i.e., by directing ATR- and Chk1-dependent phosphorylation of the RelA transactivation domain (Thr 505). This creates a potential docking site for HDAC1 to suppress expression of antiapoptotic genes and sensitize cells to TNF-induced killing (Rocha et al., 2003, 2005). Lately the DNA cross-linking chemotherapeutic drug cisplatin was identified to imitate ARF's activity, by promoting Chk1-dependent phosphorylation of RelA to repress expression of Bcl-x_L (Campbell et al., 2006). It should be noted, however, that not all genotoxic drugs convert RelA into a transcriptional repressor, as etoposide promotes RelA-dependent activation of the antiapoptotic genes Bcl-x_L and $XIAP$ (Campbell et al., 2006).

Targeting of RelA to the nucleolus was recently suggested as a novel means to antagonize its transcriptional and antiapoptotic activities in colorectal cancer cells treated with aspirin, serum deprivation, or UV-C radiation (Stark and Dunlop, 2005), although others previously reported that aspirin suppresses NF-κB activation by interfering with the activity of the IKK complex (Kopp and Ghosh, 1994; Yamamoto et al., 1999; Yin et al., 1998).

6   NF-κB's Role in PCD has Important Developmental and Physiological Consequences

The phenotypes of mice deficient for individual Rel/NF-κB subunits highlighted the crucial contribution of NF-κB in the control of apoptosis during development and/or homeostasis in the hepatic, epidermal, immune, and nervous systems (reviewed in Kucharczak et al., 2003; Li and Verma, 2002). For example, homozygous inactivation of RelA or of its upstream activating kinase IKKβ, alone or together with IKKα, is embryonic lethal due to massive liver apoptosis (Beg and Baltimore, 1996; Li et al., 1999a, b, 2000; Rudolph et al., 2000; Tanaka et al., 1999). That this phenotype is rescued by concerted deletion of RelA with TNFα, or of IKKβ with TNFR1, indicates that developing hepatocytes undergo apoptosis induced by circulating TNFα (Alcamo et al., 2001; Doi et al., 1999). Although RelA is believed to protect developing hepatocytes from TNF-induced killing by upregulating the expression of antiapoptotic genes, expression of Bcl-2 recently failed to rescue fatal liver apoptosis in RelA-deficient mice (Gugasyan et al., 2006). It therefore appears that the protective activity of RelA against physiological levels of TNF requires activation of other NF-κB targets in developing hepatocytes. The protective role of NF-κB in hepatocytes is also evident in cells treated with transforming growth factor β (TGF-β), which induces cell death by promoting synthesis and stabilization of the NF-κB inhibitor IκBα (Arsura et al., 2003; Cavin et al., 2003). Induction of apoptosis in this context coincides with suppression of the prosurvival NF-κB targets Bcl-x_L and XIAP, as well as alpha-fetoprotein that suppresses TNF-induced cell death by inhibiting TNFR1 signaling (Cavin et al., 2004).

The protective role of NF-κB in the immune system is also well documented. NF-κB orchestrates survival and differentiation during early lymphopoiesis, where RelA suppresses apoptosis of precursor cells in presence of high levels of TNF (Prendes et al., 2003); reviewed in Claudio et al., 2006; Gerondakis and Strasser, 2003; Siebenlist et al., 2005). Later in development, NF-κB activation via the pre-B-cell receptor (pre-BCR) is key for suppressing apoptosis and promoting proliferation and developmental progression. Combined inactivation of c-Rel and RelA impairs maturation to the IgM(lo)IgD(hi) stage and causes premature cell death (Feng et al., 2004; Grossmann et al., 2000; reviewed in Gilmore et al., 2004). NF-κB activation downstream of the BCR is also crucial for survival and proliferation of mature peripheral B cells. Homozygous deletion of c-Rel renders primary B cells exquisitely susceptible to apoptosis following stimulation with mitogens, as does B-lineage-specific inactivation of IKKβ or IKKγ/NEMO (Grumont et al., 1998, 1999; Kontgen et al., 1995; Leitges et al., 2001; Li et al., 2003; Martin et al., 2002; Owyang et al., 2001; Pasparakis et al., 2002b; Petro and Khan, 2001; Petro et al., 2000; Tan et al., 2001; Tumang et al., 1998). The recent analysis of mice deficient for the B-cell adaptor for phosphoinositide 3-kinase (BCAP), that signals through c-Rel, is consistent with this (Yamazaki and Kurosaki, 2003). NF-κB also reduces apoptosis induced by the cytokine BlyS/BAFF that is involved in peripheral B-cell development. Its protective activity in this context coincides with induction of Bcl-2, Bcl-x$_L$, and Bfl-1/A1 (Do et al., 2000; Hsu et al., 2002; Schiemann et al., 2001). Combined, these results highlight a crucial role for NF-κB in B-cell survival, maturation, and function.

NF-κB is also prominent in determining the fate of T cells, in which it serves either in an antiapoptotic or a proapoptotic fashion. NF-κB activation following T-cell receptor (TCR) engagement together with CD28 costimulation fosters survival and proliferation of naïve T cells (Khoshnan et al., 2000). Both p50/NF-κB1 and c-Rel were implicated in inducing expression of cell death inhibitors Bcl-x$_L$, Bcl-2, and Bfl-1/A1 (Verschelde et al., 2003; Zheng et al., 2003). Despite its protective role, it appears that only an appropriate dose of NF-κB activity is tolerated as survival of B and T lymphocytes is compromised in mice deficient for both inhibitory subunits IκBα and IκBε in which NF-κB is highly activated, akin to the phenotype of mice lacking NF-κB activity (Goudeau et al., 2003). There is also evidence suggesting that NF-κB can be proapoptotic in double-positive thymocytes (Hettmann et al., 1999). NF-κB-dependent induction of Fas ligand (FasL) in mature T cells undergoing activation-induced cell death (AICD) has also been reported (Kasibhatla et al., 1999; Lin et al., 1999; Zheng et al., 2001).

Inhibition of apoptosis by NF-κB is also important for the development of most ectodermal appendages, as tissue-specific suppression of NF-κB activity leads to impaired development of hair follicles and exocrine glands due to increased apoptosis (Headon et al., 2001; Pasparakis et al., 2002a; Schmidt-Supprian et al., 2000; Schmidt-Ullrich et al., 2001; Yan et al., 2002). In this regard tumor suppressor cylindromatosis (CYLD), whose loss predisposes patients to tumors of hair follicles, sweat, and scent glands acts as a deubiquitinating enzyme for IKKγ/NEMO and TRAF2, and suppresses NF-κB activation of the TNFR family members CD40, XEDAR, and EDAR (Brummelkamp et al., 2003; Kovalenko et al., 2003; Trompouki et al., 2003). It thus seems that the antiapoptotic activity of NF-κB may contribute to cancer development in these tissues.

In the nervous system too, NF-κB can either block or induce apoptosis depending on the cell context and the stimulus. It is neuroprotective in response to injury as illustrated in experimental models of stroke or seizure, where it induces expression of the prosurvival genes *IAP*, *Bcl-2*, *Bcl-x$_L$*, and *MnSOD* (reviewed in Kucharczak et al., 2003; Mattson and Camandola, 2001). While increased NF-κB activity is observed in neurodegenerative disorders like Alzheimer's and Parkinson's diseases, amyotrophic lateral sclerosis, epilepsy, and stroke, it was postulated that it helps to protect against oxidative stress and mitochondrial dysfunction (reviewed in Mattson and Camandola, 2001). This is supported by the increased susceptibility of p50/NF-κB1-deficient mice to neuronal damage following treatment with a mitochondrial toxin in an experimental model of Huntington's disease (Yu et al., 2000). However, others reported that NF-κB promotes cell death in models of neuronal injury following ischemia/reperfusion and excitotoxic insult in which tumor suppressor p53 was implicated as a harmful downstream effector (Crumrine et al., 1994; Morrison et al., 1996; Xiang et al., 1996). Interestingly, recent work indicates that K$^+$ loss in cortical neurons subjected to serum withdrawal leads to increased levels of NF-κB and that apoptosis is associated with upregulation of the pro-death factor Bcl-xS (Tao et al., 2006).

The cell type in which NF-κB is activated appears to significantly influence whether NF-κB is neuroprotective or neurodegenerative. While its activation in neurons is often cytoprotective, NF-κB activation in microglia promotes neuronal cell death (Mattson and Camandola, 2001). In this regard, it was suggested that NF-κB activation in glial cells might induce neuronal apoptosis by promoting production of proinflammatory cytokines, ROS, and excitotoxins (John et al., 2003; Mattson and Meffert, 2006). Consistent with this idea, inactivation of astroglial NF-κB by transgenic expression of a superrepressor IκBα was recently shown to reduce production of proinflammatory cytokines and to dramatically improve recovery after spinal cord injury (Brambilla et al., 2005).

7   NF-κB's Role in PCD Fosters Cancer Development and Progression

Constitutive activation of NF-κB contributes to the pathogenesis of a large number of human cancers (reviewed in Rayet and Gelinas, 1999; Kim et al., 2006). Many tumor cells, including those derived from activated B cell-like diffuse large B-cell lymphoma (ABC-DLBCL), primary mediastinal B-cell lymphomas (PMBL), classical Hodgkin's lymphoma (cHL), acute lymphoblastic leukemia (ALL), chronic myelogenous leukemia (CML), adult T-cell leukemia (ATL), breast, lung, or head and neck cancer show constitutively high levels of nuclear Rel/NF-κB factors and depend upon them for survival (Alizadeh et al., 2000; Bargou et al., 1997; Davis et al., 2001; Hinz et al., 2001; Kordes et al., 2000; Shipp et al., 2002). Suppression of NF-κB activity using a degradation-resistant form of IκB blocks tumor cell proliferation and sensitizes them to apoptosis (reviewed in Baldwin, 2001; Barkett and Gilmore, 1999; Kucharczak et al., 2003; Sonenshein, 1997). This agrees with the acute oncogenicity of the viral NF-κB oncoprotein v-Rel of reticuloendotheliosis virus strain T that causes fatal leukemia/lymphoma in animal models (reviewed in Fan et al., 2006; Gilmore, 1999). NF-κB activation is also implicated in malignant cell transformation by many viruses, as reviewed previously (Fan et al., 2006; Hiscott et al., 2001; Kucharczak et al., 2003; Santoro et al., 2003). These include Epstein-Barr virus (EBV) implicated in Burkitt's lymphoma, human Herpesvirus 8/Kaposi's sarcoma-associated Herpes virus (HHV8/KHSV) associated with Kaposi's sarcoma, and primary effusion lymphoma (PEL) and human T-cell leukemia virus type-1 (HTLV-1) associated with ATL.

In a majority of human cancers, persistent NF-κB activity results from constitutive activation of the IKK complex, although the mechanisms responsible for IKK activation in these tumors have remained elusive. Using an RNA interference screen Staudt's group recently uncovered that CARD11, that signals though MALT1 and BCL10, is a key factor responsible for the constitutive activation of IKK in ABC-DLBCL (Ngo et al., 2006). In other instances, constitutively high levels of nuclear Rel/NF-κB proteins are due to chromosomal rearrangement, amplification and/or overexpression of the *rel/nf-κb* genes, or in some cases due to mutations in IκB (reviewed in Fan et al., 2006; Gilmore et al., 2002; Karin et al., 2002; Rayet and Gelinas, 1999). For example, *c-rel* is overexpressed in PMBL, certain follicular large cell lymphoma, and in cHL (Barth et al., 1998, 2003; Feuerhake et al., 2005; Houldsworth et al., 1996; Joos et al., 1996, 2002; Lu et al., 1991; Rao et al., 1998; Savage et al., 2003; Wessendorf et al., 2003). In several of these cases, this was correlated with accumulation of nuclear c-Rel protein (Barth et al., 2003; Savage et al., 2003). There is emerging evidence that NF-κB may also contribute to brain cancer, as constitutive NF-κB activity coincides with expression of a novel TrkA splice variant (trkAIII) in neuroblastoma cell lines (Tacconelli et al., 2004). In addition, reduced expression of the candidate tumor suppressor ING4 is correlated with increased expression of NF-κB target genes that foster survival, growth, and angiogenesis of brain tumors, and ING4 was proposed to regulate NF-κB activity by directly interacting with RelA/p65 (Garkavtsev et al., 2004).

A large number of studies have delineated a cell autonomous role for NF-κB in tumor cell survival, but recent publications provided compelling evidence that activation of NF-κB in the tumor microenvironment plays a vital role in promoting tumor cell growth (Fig. 11.3; reviewed in de Visser and Coussens, 2005). In a

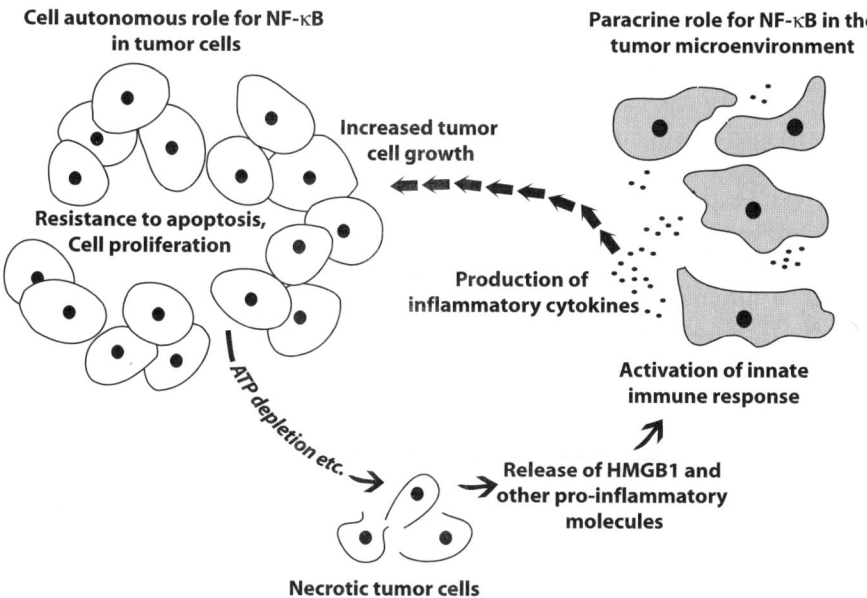

Fig. 11.3 NF-κB plays an essential role in tumor cells and in the tumor microenvironment. Activation of NF-κB in tumor cells acts in a cell autonomous fashion to increased cell resistance to apoptosis, cell proliferation, and metastatic capacity. Rapidly dividing tumor cells that depend on glycolysis can undergo PCD via necrosis under conditions where ATP is depleted. Necrotic cells release potent proinflammatory factors like HMGB1, which trigger activation of the innate immune response in the tumor microenvironment, resulting in NF-κB-dependent production of proinflammatory cytokines that promote tumor cell growth in a paracrine fashion

mouse model of colitis-associated cancer, abrogation of NF-κB activity either in intestinal epithelial cells or in myeloid cells significantly reduced tumor incidence following administration of a carcinogen that promotes colonic tumor formation together with dextran sulfate sodium (DSS) salt to induce inflammation and accelerate tumor growth (Greten et al., 2004). Inactivation of IKKβ in intestinal epithelial cells decreased tumor incidence due to increased apoptosis, coincident with decreased expression of antiapoptotic proteins like Bcl-x$_L$, but it had no effect on tumor cell proliferation. In contrast, ablation of IKKβ in myeloid cells decreased tumor incidence by inhibiting epithelial cell proliferation due to reduced expression of proinflammatory genes, but had no effect on tumor cell survival. Consequently, it seems that NF-κB activation promotes tumor cell survival, whereas its activation in myeloid cells promotes production of cytokines that accelerate tumor cell growth in a paracrine fashion (Greten et al., 2004). A similar correlation has emerged between NF-κB activation and inflammation-associated tumor growth in a mouse model of chronic hepatitis that evolves into hepatocellular carcinoma (HCC) (Pikarsky et al., 2004). In this system, chronic liver inflammation triggers production of TNFα by endothelial and inflammatory cells that leads to chronic activation of NF-κB in hepatocytes. While inactivation of NF-κB in hepatocytes had no effect on the onset of early neoplastic events, its inactivation at later stages increased hepatocyte apoptosis and blunted progression to carcinoma (Pikarsky et al., 2004).

Further evidence that NF-κB plays a prominent role in inflammation-associated tumor growth and metastasis came to light in studies in which administration of bacterial LPS induced systemic inflammation and production of TNFα by cells in the tumor microenvironment that accelerated the growth and metastasis of colon and breast cancer cell lines (Luo et al., 2004) Inhibition of NF-κB in the tumor cells themselves prompted tumor regression in response to LPS-induced inflammation, where reduced tumor cell proliferation and increased apoptosis resulted from induction of TRAIL receptor DR5 on NF-κB-deficient tumor cells and of its ligand TRAIL on surrounding immune cells. Together these studies highlight a critical role for NF-κB in inflammation-associated tumor promotion, progression, and metastasis.

Exposure to carcinogens is an important contributing factor to the onset of sporadic human cancer. NF-κB plays an important role in this scenario as well, as evidenced in various mouse models of chemically induced cancer. An interesting link between inflammation and chemical carcinogenesis was unveiled in a mouse model of diethylnitrosamine (DEN)-induced HCC, in which IKK-mediated NF-κB activation plays a critical role both in hepatocytes and in hematopoietic-derived Kupffer cells (Maeda et al., 2005). A surprising finding was that hepatocyte-specific deletion of IKKβ noticeably enhanced tumor development, as hepatocyte apoptosis was offset by proliferation of surviving hepatocytes, coincident with increased ROS production, and JNK activation. Administration of antioxidant or compound inactivation of IKKβ in both hepatocytes and hematopoietic-derived Kupffer cells reduced the incidence of HCC in this model. Although the mechanism whereby DEN triggers this inflammatory response is unclear, it was proposed that hepatocytes undergoing necrosis release factors like HMGB1 that can trigger a strong inflammatory response in the microenvironment, that in turn promotes the growth and tumorigenesis of surviving hepatocytes (Scaffidi et al., 2002). This model is supported by: (1) the observation that supernatant from necrotic hepatocytes can activate NF-κB in primary macrophages (Maeda et al., 2005); (2) an increasing number of studies showing that inflammation and necrosis support tumor growth (Vakkila and Lotze, 2004; reviewed in Lotze and Tracey, 2005; Zeh and Lotze, 2005); and (3) work indicating that HMGB1 released from necrotic cells is an important mediator of inflammation (reviewed in Lotze and Tracey, 2005; Zeh and Lotze, 2005; Fig. 11.3).

In contrast to its well-documented growth-promoting effects in most cell types, NF-κB inhibits cell growth in the epidermis and loss of NF-κB activity promotes epidermal cell proliferation and hyperplasia (Seitz et al., 1998, 2000; van Hogerlinden et al., 1999). Furthermore, suppression of NF-κB activity in epidermal keratinocytes in conjunction with expression of oncogenic Ras promotes invasive neoplasia reminiscent of squamous cell carcinoma (SCC; Dajee et al., 2003). The growth suppressive effects of NF-κB in epidermal homeostasis were recently shown to result from suppression of the G1 cell cycle kinase CDK4 (Zhang et al., 2005). Suppression of NF-κB in the epidermis was accompanied by upregulation of CDK4 in a TNFR1- and JNK-dependent manner and CDK4 was necessary for epidermal cell hyperplasia under conditions in which NF-κB activity was inhibited (Zhang et al., 2005). This highlights an important tumor suppressor function for NF-κB in certain cells.

Interestingly, NF-κB was found to be preferentially activated in epithelial cells of ER-negative breast tumors and particularly in ER-negative and ErbB2-positive tumors (86%; Biswas et al., 2004). Interestingly, in ER-negative and ErbB2-negative breast cancer samples, nuclear NF-κB was predominantly found in the stroma (Biswas et al., 2004).

8 Approaches for Prevention and Therapy

Many dietary and natural agents that show chemopreventive activity can block NF-κB activity (reviewed in Yamamoto and Gaynor, 2001). These include the green tea polyphenol epigallocatechin-3 gallate, resveratrol, and curcumin that can block tumor initiation and progression by suppressing tumor cell proliferation and by inducing apoptosis (Hofmann and Sonenshein, 2003; Bharti et al., 2003; reviewed in Signorelli and Ghidoni, 2005).

Not only is NF-κB important for the inherent resistance of tumor cells to PCD and for promoting tumor cell growth, but it is also a central figure in the resistance of many tumors to anticancer treatment. Compounds that interrupt NF-κB signaling counteract the growth and survival of many tumor cells in which NF-κB is implicated and can potentiate the efficacy of anticancer drugs (Wang et al., 1996; reviewed in Baldwin, 2001; Karin et al., 2002; Yamamoto and Gaynor, 2001). Since several signaling molecules and posttranslational modifications are necessary to mediate NF-κB activation, different steps in the pathway can be used as potential therapeutic targets. One of them involves suppression of the proteasome-dependent degradation of IκBα. The proteasome inhibitor Velcade/bortezomib is currently used for the treatment of advanced multiple myeloma (also see Chapter 12) and also shows promise in preclinical models of breast, colon, lung, prostate, and pancreatic cancer (reviewed in Kim et al., 2006; Richardson et al., 2004). Since the proteasome is involved in the turnover of many cellular factors, Velcade's effectiveness does not solely derive from inhibition of the NF-κB pathway, as illustrated by recent evidence that it can also affect mitochondrial function and blunt activation of JNK (Landowski et al., 2005; Small et al., 2004). Moreover, its proapoptotic effects for melanoma cells do not seem to coincide with widespread inhibition of NF-κB, suggesting the need to identify more specific inhibitors of NF-κB (Fernandez et al., 2005).

Another valuable approach is to target phosphorylation events critical for NF-κB activation. In this regard, there is a growing inventory of compounds that can suppress IKK activity and promote apoptosis in tumor-derived cells (reviewed in Kim et al., 2006). A few examples include nonsteroidal anti-inflammatory drugs (NSAIDs) like celecoxib, sulfasalazine or aspirin (e.g. Ashikawa et al., 2004; Robe et al., 2004; Subhashini et al., 2005; Takada et al., 2004). Incidentally, prolonged use of NSAIDs has been linked with a decreased incidence of colon cancer (reviewed in Li et al., 2005). Thalidomide and arsenic, respectively show efficacy in combined therapy for relapsed or refractory multiple myeloma, and in the treatment of acute promyelocytic leukemia (reviewed in Kim et al., 2006; Mathas et al., 2003). However, since the activity of these agents is not selective for NF-κB, there is a great deal of interest in identifying small molecule inhibitors specific for IKK subunits. Among them, the β-carboline derivative PS-1145 was shown to specifically kill ABC- DLBCL and PMBL-derived tumor cell lines that rely on NF-κB for growth and survival (Lam et al., 2005). Other specific and potent IKK inhibitors are undergoing preclinical testing. These include the quinazoline analogue SPC839 and the imidazoquinoxaline derivative BMS-345541 (reviewed in Karin et al., 2004). Their safety and effectiveness in combination therapy is currently under investigation (reviewed in Nakanishi and Toi, 2005).

Lately, there has been a significant new development in the quest to identify new molecular targets critical for the pathogenesis of NF-κB-associated cancers. Using an inducible RNA interference library to identify genes important for tumor cell proliferation and apoptosis resistance, Staudt's group uncovered that CARD11/CARMA1 is responsible for the constitutive activation of IKK in ABC-DLBCL-derived tumor cells (Ngo et al., 2006). CARD11/CARMA1 lies downstream of the BCR and TCR and engages MALT1 and BCL10 to promote ubiquitination of IKKγ/NEMO. This new finding opens the possibility to develop strategies to inhibit signaling by CARD11, an approach that may have limited side effects since CARD11 expression is restricted to lymphoid cells. Moreover, the work provides compelling evidence that RNA interference screens might be particularly useful to uncover new therapeutic targets that are crucial for tumor cell survival and proliferation (Ngo et al., 2006).

Although compounds that suppress NF-κB activity offer promising avenues to antagonize tumor development or progression and to enhance the efficacy of existing therapeutic agents, it is important to remember that NF-κB can be proapoptotic in certain cell types and in response to certain stimuli. This suggests that the particular cell context may be important for the therapeutic outcome. For example, the chemotherapeutic drug doxorubicin triggers apoptosis in colon cancer cells by activating NF-κB (Ashikawa et al., 2004), but others found that NF-κB protects HeLa cells from apoptosis induced by this agent (Baldwin, 2001; Nakanishi and Toi, 2005). The tumor suppressor activity of NF-κB in the epidermis is another consideration, as a possible adverse effect of long-term inhibition of NF-κB might be an increased susceptibility to develop certain tumors associated with suppression of NF-κB, such as SCC. Lastly, recent work indicating that certain chemotherapeutic drugs can convert RelA into a transcriptional repressor of antiapoptotic genes suggests that the response of tumor cells to particular chemotherapeutic regimens may differ significantly depending on the tumor type, the status of endogenous tumor suppressors, and the stage of tumor development (Perkins, 2004; Perkins and Gilmore, 2006). Ongoing efforts to clarify the mechanisms that govern the anti- vs pro-death effects of NF-κB in different cell contexts will certainly be very informative to help predict the impact of NF-κB inhibition in different tumor cell contexts and the outcome of therapy.

Acknowledgments   We apologize to the many investigators whose work could not be cited due to space limitations. Research performed in this laboratory on the roles of Rel/NF-κB in apoptosis and oncogenesis and on its antiapoptotic target Bfl-1/A1 was supported by grants from the National Institutes of Health CA54999 and CA83937 to CG.

References

Alcamo, E., Mizgerd, J. P., Horwitz, B. H., Bronson, R., Beg, A. A., Scott, M., Doerschuk, C. M., Hynes, R. O., and Baltimore, D. (2001). Targeted mutation of TNF receptor I rescues the RelA-deficient mouse and reveals a critical role for NF-kappa B in leukocyte recruitment. J Immunol 167, 1592–1600.

Alizadeh, A., Eisen, M., Davis, R., Ma, C., Lossos, I., Rosenwald, A., Boldrick, J., Sabet, H., Tran, T., Yu, X., Powell, J., Yang, L., Marti, G., Moore, T., Hudson, J. J., Lu, L., Lewis, D., Tibshirani, R., Sherlock, G., Chan, W., Greiner, T., Weisenburger, D. D., Armitage, J. O., Warnke, R., Levy, R., Wilson, W. H., Greyer, M. R., Byrd, J., Botstein, D., Brown, P. O., and Staudt, L. M. (2000). Distinct types of diffuse large B-cell lymphoma identified by gene expression profiling. Nature 403, 503–511.

Amanullah, A., Azam, N., Balliet, A., Hollander, C., Hoffman, B., Fornace, A., and Liebermann, D. (2003). Cell signalling: cell survival and a Gadd45-factor deficiency. Nature 424, 741; discussion 742.

Arsura, M., Panta, G. R., Bilyeu, J. D., Cavin, L. G., Sovak, M. A., Oliver, A. A., Factor, V., Heuchel, R., Mercurio, F., Thorgeirsson, S. S., and Sonenshein, G. E. (2003). Transient activation of NF-kappaB through a TAK1/IKK kinase pathway by TGF-beta1 inhibits AP-1/SMAD signaling and apoptosis: implications in liver tumor formation. Oncogene 22, 412–425.

Ashikawa, K., Shishodia, S., Fokt, I., Priebe, W., and Aggarwal, B. B. (2004). Evidence that activation of nuclear factor-kappaB is essential for the cytotoxic effects of doxorubicin and its analogues. Biochem Pharmacol 67, 353–364.

Baetz, D., Regula, K. M., Ens, K., Shaw, J., Kothari, S., Yurkova, N., and Kirshenbaum, L. A. (2005). Nuclear factor-kappaB-mediated cell survival involves transcriptional silencing of the mitochondrial death gene BNIP3 in ventricular myocytes. Circulation 112, 3777–3785.

Baldwin, A. S. (2001). Control of oncogenesis and cancer therapy resistance by the transcription factor NF-kappaB. J Clin Invest 107, 241–246.

Bargou, R., Emmerich, F., Krappmann, D., Bommert, K., Mapara, M., Arnold, W., Royer, H., Grinstein, E., Greiner, A., Scheidereit, C., and Dorken, B. (1997). Constitutive nuclear factor-kappaB-RelA activation is required for proliferation and survival of Hodgkin's disease tumor cells. J Clin Invest 100, 2961–2969.

Barkett, M. and Gilmore, T. D. (1999). Control of apoptosis by Rel/NF-kappaB transcription factors. Oncogene 18, 6910–6924.

Barth, T. F., Dohner, H., Werner, C. A., Stilgenbauer, S., Schlotter, M., Pawlita, M., Lichter, P., Moller, P., and Bentz, M. (1998). Characteristic pattern of chromosomal gains and losses in primary large B-cell lymphomas of the gastrointestinal tract. Blood 91, 4321–4330.

Barth, T. F., Martin-Subero, J. I., Joos, S., Menz, C. K., Hasel, C., Mechtersheimer, G., Parwaresch, R. M., Lichter, P., Siebert, R., and Moeller, P. (2003). Gains of 2p involving the *REL* locus correlate with nuclear c-Rel protein accumulation in neoplastic cells of classical Hodgkin lymphoma. Blood 101, 3681–3686.

Beg, A. A. and Baltimore, D. (1996). An essential role for NF-κB in preventing TNF-α-induced cell death. Science 274, 782–784.

Bernard, D., Quatannens, B., Vandenbunder, B., and Abbadie, C. (2001a). Rel/NF-kappaB transcription factors protect against tumor necrosis factor (TNF)-related apoptosis-inducing ligand (TRAIL)-induced apoptosis by up-regulating the TRAIL decoy receptor DcR1. J Biol Chem 276, 27322–27328.

Bernard, D., Slomianny, C., Vandenbunder, B., and Abbadie, C. (2001b). cRel induces mitochondrial alterations in correlation with proliferation arrest. Free Radic Biol Med 31, 943–953.

Bernard, D., Monte, D., Vandenbunder, B., and Abbadie, C. (2002). The c-Rel transcription factor can both induce and inhibit apoptosis in the same cells via the upregulation of MnSOD. Oncogene 21, 4392–4402.

Bharti, A. C., Donato, N., Singh, S., and Aggarwal, B. B. (2003). Curcumin (diferuloylmethane) down-regulates the constitutive activation of nuclear factor-kappa B and IkappaBalpha kinase in human multiple myeloma cells, leading to suppression of proliferation and induction of apoptosis. Blood 101, 1053–1062.

Biswas, D. K., Shi, Q., Baily, S., Strickland, I., Ghosh, S., Pardee, A. B., and Iglehart, J. D. (2004). NF-kappa B activation in human breast cancer specimens and its role in cell proliferation and apoptosis. Proc Natl Acad Sci USA 101, 10137–10142.

Bonizzi, G. and Karin, M. (2004). The two NF-kappaB activation pathways and their role in innate and adaptive immunity. Trends Immunol 25, 280–288.

Brambilla, R., Bracchi-Ricard, V., Hu, W. H., Frydel, B., Bramwell, A., Karmally, S., Green, E. J., and Bethea, J. R. (2005). Inhibition of astroglial nuclear factor kappaB reduces inflammation and improves functional recovery after spinal cord injury. J Exp Med 202, 145–156.

Brummelkamp, T. R., Nijman, S. M., Dirac, A. M., and Bernards, R. (2003). Loss of the cylindromatosis tumour suppressor inhibits apoptosis by activating NF-kappaB. Nature 424, 797–801.

Budd, R. C., Yeh, W. C., and Tschopp, J. (2006). cFLIP regulation of lymphocyte activation and development. Nat Rev Immunol 6, 196–204.

Campbell, K. J., Rocha, S., and Perkins, N. D. (2004). Active repression of antiapoptotic gene expression by RelA(p65) NF-kappa B. Mol Cell 13, 853–865.

Campbell, K. J., Witty, J. M., Rocha, S., and Perkins, N. D. (2006). Cisplatin mimics ARF tumor suppressor regulation of RelA (p65) nuclear factor-kappaB transactivation. Cancer Res 66, 929–935.

Cavin, L. G., Romieu-Mourez, R., Panta, G. R., Sun, J., Factor, V. M., Thorgeirsson, S. S., Sonenshein, G. E., and Arsura, M. (2003). Inhibition of CK2 activity by TGF-beta1 promotes IkappaB-alpha protein stabilization and apoptosis of immortalized hepatocytes. Hepatology 38, 1540–1551.

Cavin, L. G., Venkatraman, M., Factor, V. M., Kaur, S., Schroeder, I., Mercurio, F., Beg, A. A., Thorgeirsson, S. S., and Arsura, M. (2004). Regulation of alpha-fetoprotein by nuclear factor-kappaB protects hepatocytes from tumor necrosis factor-alpha cytotoxicity during fetal liver development and hepatic oncogenesis. Cancer Res 64, 7030–7038.

Chen, X., Kandasamy, K., and Srivastava, R. K. (2003). Differential roles of RelA (p65) and c-Rel subunits of nuclear factor kappa B in tumor necrosis factor-related apoptosis-inducing ligand signaling. Cancer Res 63, 1059–1066.

Claudio, E., Brown, K., and Siebenlist, U. (2006). NF-kappaB guides the survival and differentiation of developing lymphocytes. Cell Death Differ 13, 697–701.

Crumrine, R. C., Thomas, A. L., and Morgan, P. F. (1994). Attenuation of p53 expression protects against focal ischemic damage in transgenic mice. J Cereb Blood Flow Metab 14, 887–891.

Currier, N., Solomon, S. E., Demicco, E. G., Chang, D. L., Farago, M., Ying, H., Dominguez, I., Sonenshein, G. E., Cardiff, R. D., Xiao, Z. X., Sherr, D. H., and Seldin, D. C. (2005). Oncogenic signaling pathways activated in DMBA-induced mouse mammary tumors. Toxicol Pathol 33, 726–737.

Daido, S., Kanzawa, T., Yamamoto, A., Takeuchi, H., Kondo, Y., and Kondo, S. (2004). Pivotal role of the cell death factor BNIP3 in ceramide-induced autophagic cell death in malignant glioma cells. Cancer Res 64, 4286–4293.

Dajee, M., Lazarov, M., Zhang, J. Y., Cai, T., Green, C. L., Russell, A. J., Marinkovich, M. P., Tao, S., Lin, Q., Kubo, Y., and Khavari, P. A. (2003). NF-kB blockade and oncogenic Ras trigger invasive human epidermal neoplasia. Nature 421, 639–643.

Davis, R. E., Brown, K. D., Siebenlist, U., and Staudt, L. M. (2001). Constitutive nuclear factor kappaB activity is required for survival of activated B cell-like diffuse large B cell lymphoma cells. J Exp Med 194, 1861–1874.

Davis, R. J. (2000). Signal transduction by the JNK group of MAP kinases. Cell 103, 239–252.

De Smaele, E., Zazzeroni, F., Papa, S., Nguyen, D. U., Jin, R., Jones, J., Cong, R., and Franzoso, G. (2001). Induction of gadd45beta by NF-kappaB downregulates pro-apoptotic JNK signalling. Nature 414, 308–313.

de Visser, K. E. and Coussens, L. M. (2005). The interplay between innate and adaptive immunity regulates cancer development. Cancer Immunol Immunother 54, 1143–1152.

Delhalle, S., Deregowski, V., Benoit, V., Merville, M. P., and Bours, V. (2002). NF-kappaB-dependent MnSOD expression protects adenocarcinoma cells from TNF-alpha-induced apoptosis. Oncogene 21, 3917–3924.

Deveraux, Q. L., Takahashi, R., Salvesen, G. S. and Reed, J. C. (1997). X-linked IAP is a direct inhibitor of cell-death proteases. Nature 388, 300–304.

Do, R. K., Hatada, E., Lee, H., Tourigny, M. R., Hilbert, D., and Chen-Kiang, S. (2000). Attenuation of apoptosis underlies B lymphocyte stimulator enhancement of humoral immune response. J Exp Med 192, 953–964.

Doi, T. S., Marino, M. W., Takahashi, T., Yoshida, T., Sakakura, T., Old, L. J., and Obata, Y. (1999). Absence of tumor necrosis factor rescues RelA-deficient mice from embryonic lethality. Proc Natl Acad Sci USA 96, 2994–2999.

Duckett, C. S. (2005). IAP proteins: sticking it to Smac. Biochem J 385, e1– e2.

Edinger, A. L. and Thompson, C. B. (2004). Death by design: apoptosis, necrosis and autophagy. Curr Opin Cell Biol 16, 663–669.

Egan, L. J., Eckmann, L., Greten, F. R., Chae, S., Li, Z. W., Myhre, G. M., Robine, S., Karin, M., and Kagnoff, M. F. 2004. IkappaB-kinasebeta-dependent NF-kappaB activation provides radioprotection to the intestinal epithelium. Proc Natl Acad Sci USA 101, 2452–2457.

Fan, Y., Dutta, J., Gupta, N., and Gélinas, C. (2006). Molecular basis of oncogenesis by NF-kB: from a bird's eye view to a RELevant role in cancer. In: NF-kB/Rel Transcription Factor Family, ed. Liou, H. C. Landes Bioscience Publishers, Georgetown, TX, pp. 112–130.

Feng, B., Cheng, S., Hsia, C. Y., King, L. B., Monroe, J. G., and Liou, H. C. (2004). NF-kappaB inducible genes BCL-X and cyclin E promote immature B-cell proliferation and survival. Cell Immunol 232, 9–20.

Fernandez, Y., Verhaegen, M., Miller, T. P., Rush, J. L., Steiner, P., Opipari, A. W., Jr., Lowe, S. W., and Soengas, M. S. (2005). Differential regulation of noxa in normal melanocytes and melanoma cells by proteasome inhibition: therapeutic implications. Cancer Res 65, 6294–6304.

Feuerhake, F., Kutok, J. L., Monti, S., Chen, W., LaCasce, A. S., Cattoretti, G., Kurtin, P., Pinkus, G. S., de Leval, L., Harris, N. L., Savage, K. J., Neuberg, D., Habermann, T. M., Dalla-Favera, R., Golub, T. R., Aster, J. C., and Shipp, M. A. (2005). NFkappaB activity, function, and target-gene signatures in primary mediastinal large B-cell lymphoma and diffuse large B-cell lymphoma subtypes. Blood 106, 1392–1399.

Garkavtsev, I., Kozin, S. V., Chernova, O., Xu, L., Winkler, F., Brown, E., Barnett, G. H., and Jain, R. K. (2004). The candidate tumour suppressor protein ING4 regulates brain tumour growth and angiogenesis. Nature 428, 328–332.

Gerondakis, S. and Strasser, A. (2003). The role of Rel/NF-kappaB transcription factors in B lymphocyte survival. Semin Immunol 15, 159–166.

Gilmore, T., Gapuzan, M. E., Kalaitzidis, D., and Starczynowski, D. (2002). Rel/NF-kB/IkB signal transduction in the generation and treatment of human cancer. Cancer Lett 181, 1–9.

Gilmore, T. D. (1999). Multiple mutations contribute to the oncogenicity of the retroviral oncoprotein v-Rel. Oncogene 18, 6925–6937.

Gilmore, T. D., Kalaitzidis, D., Liang, M. C., and Starczynowski, D. T. (2004). The c-Rel transcription factor and B-cell proliferation: a deal with the devil. Oncogene 23, 2275–2286.

Goudeau, B., Huetz, F., Samson, S., Di Santo, J. P., Cumano, A., Beg, A., Israel, A., and Memet, S. (2003). IkappaBalpha/IkappaBepsilon deficiency reveals that a critical NF-kappaB dosage is required for lymphocyte survival. Proc Natl Acad Sci USA 100, 15800–15805.

Greten, F. R., Eckmann, L., Greten, T. F., Park, J. M., Li, Z. W., Egan, L. J., Kagnoff, M. F., and Karin, M. (2004). IKKbeta links inflammation and tumorigenesis in a mouse model of colitis-associated cancer. Cell 118, 285–296.

Grossmann, M., O'Reilly, L. A., Gugasyan, R., Strasser, A., Adams, J. M., and Gerondakis, S. (2000). The anti-apoptotic activities of Rel and RelA required during B-cell maturation involve the regulation of Bcl-2 expression. EMBO J 19, 6351–6360.

Grumont, R. J., Rourke, I. J., O'Reilly, L. A., Strasser, A., Miyake, K., Sha, W., and Gerondakis, S. (1998). B lymphocytes differentially use the Rel and nuclear factor kB1 (NF-kB1) transcription factors to regulate cell cycle progression and apoptosis in quiescent and mitogen-activated cells. J Exp Med 187, 663–674.

Grumont, R. J., Rourke, I. J., and Gerondakis, S. (1999). Rel-dependent induction of A1 transcription is required to protect B cells from antigen receptor ligation-induced apoptosis. Genes Dev 13, 400–411.

Gugasyan, R., Christou, A., O'Reilly L, A., Strasser, A., and Gerondakis, S. (2006). Bcl-2 transgene expression fails to prevent fatal hepatocyte apoptosis induced by endogenous TNFalpha in mice lacking RelA. Cell Death Differ 13(7), 1235–1237.

Hait, W. N., Jin, S., and Yang, J. M. (2006). A matter of life or death (or both): understanding autophagy in cancer. Clin Cancer Res 12, 1961–1965.

Hayden, M. S. and Ghosh, S. (2004). Signaling to NF-kappaB. Genes Dev 18, 2195–2224.

Headon, D. J., Emmal, S. A., Ferguson, B. M., Tucker, A. S., Justice, M. J., Sharpe, P. T., Zonana, J., and Overbeek, P. A. (2001). Gene defect in ectodermal dysplasia implicates a death domain adapter in development. Nature 414, 913–916.

Hettmann, T., DiDonato, J., Karin, M., and Leiden, J. M. (1999). An essential role for nuclear factor kappaB in promoting double positive thymocyte apoptosis. J Exp Med 189, 145–158.

Heusch, M., Lin, L., Geleziunas, R., and Greene, W. C. (1999). The generation of nfkb2 p52: mechanism and efficiency. Oncogene 18, 6201–6208.

Hinz, M., Loser, P., Mathas, S., Krappmann, D., Dorken, B., and Scheidereit, C. (2001). Constitutive NF-kappaB maintains high expression of a characteristic gene network, including CD40, CD86, and a set of antiapoptotic genes in Hodgkin/Reed-Sternberg cells. Blood 97, 2798–2807.

Hiscott, J., Kwon, H., and Genin, P. (2001). Hostile takeovers: viral appropriation of the NF-kappaB pathway. J Clin Invest 107, 143–151.

Hofmann, C. S. and Sonenshein, G. E. (2003). Green tea polyphenol epigallocatechin-3 gallate induces apoptosis of proliferating vascular smooth muscle cells via activation of p53. FASEB J 17, 702–704.

Houldsworth, J., Mathew, S., Rao, P. H., Dyomina, K., Louie, D. C., Parsa, N., Offit, K., and Chaganti, R. S. K. (1996). REL proto-oncogene is frequently amplified in extranodal diffuse large cell lymphoma. Blood 87, 25–29.

Hsu, B. L., Harless, S. M., Lindsley, R. C., Hilbert, D. M., and Cancro, M. P. (2002). Cutting edge: BLyS enables survival of transitional and mature B cells through distinct mediators. J Immunol 168, 5993–5996.

Jaattela, M. and Tschopp, J. (2003). Caspase-independent cell death in T lymphocytes. Nat Immunol 4, 416–423.

Javelaud, D. and Besancon, F. (2001). NF-kappa B activation results in rapid inactivation of JNK in TNF alpha-treated Ewing sarcoma cells: a mechanism for the anti-apoptotic effect of NF-kappa B. Oncogene 20, 4365–4372.

John, G. R., Lee, S. C., and Brosnan, C. F. (2003). Cytokines: powerful regulators of glial cell activation. Neuroscientist 9, 10–22.

Joos, S., Otano-Joos, M. I., Ziegler, S., Brüderlein, S., du Manoir, S., Bentz, M., Möller, P., and Lichter, P. (1996). Primary mediastinal (thymic) B-cell lymphoma is characterized by gains of chromosomal material including 9p and amplification of the REL gene. Blood 87, 1571–1578.

Joos, S., Menz, C. K., Wrobel, G., Siebert, R., Gesk, S., Ohl, S., Mechtersheimer, G., Trumper, L., Moller, P., Lichter, P., and Barth, T. F. (2002). Classical Hodgkin lymphoma is characterized by recurrent copy number gains of the short arm of chromosome 2. Blood 99, 1381–1387.

Kamata, H. and Hirata, H. (1999). Redox regulation of cellular signalling. Cell Signal 11, 1–14.

Kamata, H., Honda, S., Maeda, S., Chang, L., Hirata, H., and Karin, M. (2005). Reactive oxygen species promote TNFalpha-induced death and sustained JNK activation by inhibiting MAP kinase phosphatases. Cell 120, 649–661.

Kanzawa, T., Zhang, L., Xiao, L., Germano, I. M., Kondo, Y., and Kondo, S. (2005). Arsenic trioxide induces autophagic cell death in malignant glioma cells by upregulation of mitochondrial cell death protein BNIP3. Oncogene 24, 980–991.

Karin, M., Cao, Y., Greten, F. R., and Li, Z. W. (2002). NF-kappaB in cancer: from innocent bystander to major culprit. Nat Rev Cancer 2, 301–310.

Karin, M., Yamamoto, Y., and Wang, Q. M. (2004). The IKK NF-kappa B system: a treasure trove for drug development. Nat Rev Drug Discov 3, 17–26.

Kasibhatla, S., Genestier, L., and Green, D. R. (1999). Regulation of fas-ligand expression during activation-induced cell death in T lymphocytes via nuclear factor kB. J Biol Chem 274, 987–992.

Kaur, S., Wang, F., Venkatraman, M., and Arsura, M. (2005). X-linked inhibitor of apoptosis (XIAP) inhibits c-Jun N-terminal kinase 1 (JNK1) activation by transforming growth factor beta1 (TGF-beta1) through ubiquitin-mediated proteosomal degradation of the TGF-beta1-activated kinase 1 (TAK1). J Biol Chem 280, 38599–38608.

Khoshnan, A., Bae, D., Tindell, C. A., and Nel, A. E. (2000). The physical association of protein kinase C theta with a lipid raft-associated inhibitor of kappa B factor kinase (IKK) complex plays a role in the activation of the NF-kappa B cascade by TCR and CD28. J Immunol 165, 6933–6940.

Kim, H. J., Hawke, N., and Baldwin, A. S. (2006). NF-kappaB and IKK as therapeutic targets in cancer. Cell Death Differ 13, 738–747.

Kontgen, F., Grumont, R. J., Strasser, A., Metcalf, D., Li, R., Tarlinton, D., and Gerondakis, S. (1995). Mice lacking the c-rel proto-oncogene exhibit defects in lymphocyte proliferation, humoral immunity, and interleukin-2 expression. Genes Dev 9, 1965–1977.

Kopp, E. and Ghosh, S. (1994). Inhibition of NF-κB by sodium salycilate and aspirin. Science 265, 956.

Kordes, U., Krappmann, D., Heissmeyer, V., Ludwig, W. D., and Scheidereit, C. (2000). Transcription factor NF-kB is constitutively activated in acute lymphoblastic leukemia cells. Leukemia 14, 399–402.

Kovalenko, A., Chable-Bessia, C., Cantarella, G., Israel, A., Wallach, D., and Courtois, G. (2003). The tumour suppressor CYLD negatively regulates NF-kappaB signalling by deubiquitination. Nature 424, 801–805.

Kreuz, S., Siegmund, D., Rumpf, J. J., Samel, D., Leverkus, M., Janssen, O., Hacker, G., Dittrich-Breiholz, O., Kracht, M., Scheurich, P., and Wajant, H. (2004). NFkappaB activation by Fas is mediated through FADD, caspase-8, and RIP and is inhibited by FLIP. J Cell Biol 166, 369–380.

Kucharczak, J. F., Simmons, M. J., Fan, Y., and Gelinas, C. (2003). To be, or not to be: NF-kappaB is the answer – role of Rel/NF-kappaB in the regulation of apoptosis. Oncogene 22, 8961–8982.

Lam, L. T., Davis, R. E., Pierce, J., Hepperle, M., Xu, Y., Hottelet, M., Nong, Y., Wen, D., Adams, J., Dang, L., and Staudt, L. M. (2005). Small molecule inhibitors of IkappaB kinase are selectively toxic for subgroups of diffuse large B-cell lymphoma defined by gene expression profiling. Clin Cancer Res 11, 28–40.

Landowski, T. H., Megli, C. J., Nullmeyer, K. D., Lynch, R. M., and Dorr, R. T. (2005). Mitochondrial-mediated disregulation of Ca2+ is a critical determinant of Velcade (PS-341/bortezomib) cytotoxicity in myeloma cell lines. Cancer Res 65, 3828–3836.

Leist, M. and Jaattela, M. (2001). Four deaths and a funeral: from caspases to alternative mechanisms. Nat Rev Mol Cell Biol 2, 589–598.

Leitges, M., Sanz, L., Martin, P., Duran, A., Braun, U., Garcia, J. F., Camacho, F., Diaz-Meco, M. T., Rennert, P. D., and Moscat, J. (2001). Targeted disruption of the zetaPKC gene results in the impairment of the NF-kappaB pathway. Mol Cell 8, 771–780.

Levine, B. and Yuan, J. (2005). Autophagy in cell death: an innocent convict? J Clin Invest 115, 2679–2688.

Li, Q. and Verma, I. M. (2002). NF-kappaB regulation in the immune system. Nat Rev Immunol 2, 725–734.

Li, Q., Van Antwerp, D., Mercurio, F., Lee, K. F., and Verma, I. M. (1999a). Severe liver degeneration in mice lacking the IkappaB kinase 2 gene. Science 284, 321–325.

Li, Q., Estepa, G., Memet, S., Israel, A., and Verma, I. M. (2000). Complete lack of NF-kappaB activity in IKK1 and IKK2 double-deficient mice: additional defect in neurulation. Genes Dev 14, 1729–1733.

Li, Q., Withoff, S., and Verma, I. M. (2005). Inflammation-associated cancer: NF-kappaB is the lynchpin. Trends Immunol 26, 318–325.

Li, Z. W., Chu, W., Hu, Y., Delhase, M., Deerinck, T., Ellisman, M., Johnson, R., and Karin, M. (1999b). The IKKbeta subunit of IkappaB kinase (IKK) is essential for nuclear factor kappaB activation and prevention of apoptosis. J Exp Med 189, 1839–1845.

Li, Z. W., Omori, S. A., Labuda, T., Karin, M., and Rickert, R. C. (2003). IKK beta is required for peripheral B cell survival and proliferation. J Immunol 170, 4630–4637.

Liang, C., Zhang, M., and Sun, S. C. (2006). beta-TrCP binding and processing of NF-kappaB2/p100 involve its phosphorylation at serines 866 and 870. Cell Signal 18(8), 1309–1317.

Lin, B., Williams-Skipp, C., Tao, Y., Schleicher, M. S., Cano, L. L., Duke, R. C., and Scheinman, R. I. (1999). NF-kappaB functions as both a proapoptotic and antiapoptotic regulatory factor within a single cell type. Cell Death Differ 6, 570–582.

Liston, P., Fong, W. G., and Korneluk, R. G. (2003). The inhibitors of apoptosis: there is more to life than Bcl2. Oncogene 22, 8568–8580.

Liu, N., Raja, S. M., Zazzeroni, F., Metkar, S. S., Shah, R., Zhang, M., Wang, Y., Bromme, D., Russin, W. A., Lee, J. C., Peter, M. E., Froelich, C. J., Franzoso, G., and Ashton-Rickardt, P. G. (2003). NFkappaB protects from the lysosomal pathway of cell death. EMBO J 22, 5313–5322.

Lotze, M. T. and Tracey, K. J. (2005). High-mobility group box 1 protein (HMGB1): nuclear weapon in the immune arsenal. Nat Rev Immunol 5, 331–342.

Lu, D., Thompson, J. D., Gorski, G. K., Rice, N. R., Mayer, M. G., and Yunis, J. J. (1991). Alterations at the *rel* locus in human lymphoma. Oncogene 6, 1235–1241.

Luo, J. L., Maeda, S., Hsu, L. C., Yagita, H., and Karin, M. (2004). Inhibition of NF-kappaB in cancer cells converts inflammation- induced tumor growth mediated by TNFalpha to TRAIL-mediated tumor regression. Cancer Cell 6, 297–305.

Luo, J. L., Kamata, H., and Karin, M. (2005a). The anti-death machinery in IKK/NF-kappaB signaling. J Clin Immunol 25, 541–550.

Luo, J. L., Kamata, H., and Karin, M. (2005b). IKK/NF-kappaB signaling: balancing life and death – a new approach to cancer therapy. J Clin Invest 115, 2625–2632.

Maeda, S., Kamata, H., Luo, J. L., Leffert, H., and Karin, M. (2005). IKKbeta couples hepatocyte death to cytokine-driven compensatory proliferation that promotes chemical hepatocarcinogenesis. Cell 121, 977–990.

Martin, P., Duran, A., Minguet, S., Gaspar, M. L., Diaz-Meco, M. T., Rennert, P., Leitges, M., and Moscat, J. (2002). Role of zeta PKC in B-cell signaling and function. EMBO J 21, 4049–4057.

Mathas, S., Lietz, A., Janz, M., Hinz, M., Jundt, F., Scheidereit, C., Bommert, K., and Dorken, B. (2003). Inhibition of NF-kappaB essentially contributes to arsenic-induced apoptosis. Blood 102, 1028–1034.

Matsuzawa, A. and Ichijo, H. (2005). Stress-responsive protein kinases in redox-regulated apoptosis signaling. Antioxid Redox Signal 7, 472–481.

Mattson, M. P. and Camandola, S. (2001). NF-kappaB in neuronal plasticity and neurodegenerative disorders. J Clin Invest 107, 247–254.

Mattson, M. P. and Meffert, M. K. (2006). Roles for NF-kappaB in nerve cell survival, plasticity, and disease. Cell Death Differ 13, 852–860.

Micheau, O. and Tschopp, J. (2003). Induction of TNF receptor I-mediated apoptosis via two sequential signaling complexes. Cell 114, 181–190.

Micheau, O., Lens, S., Gaide, O., Alevizopoulos, K., and Tschopp, J. (2001). NF-kappaB signals induce the expression of c-FLIP. Mol Cell Biol 21, 5299–5305.

Morrison, R. S., Wenzel, H. J., Kinoshita, Y., Robbins, C. A., Donehower, L. A., and Schwartzkroin, P. A. (1996). Loss of the p53 tumor suppressor gene protects neurons from kainate-induced cell death. J Neurosci 16, 1337–1345.

Nakanishi, C. and Toi, M. (2005). Nuclear factor-kappaB inhibitors as sensitizers to anticancer drugs. Nat Rev Cancer 5, 297–309.

Nakano, H., Nakajima, A., Sakon-Komazawa, S., Piao, J. H., Xue, X., and Okumura, K. (2006). Reactive oxygen species mediate crosstalk between NF-kappaB and JNK. Cell Death Differ 13, 730–737.

Ngo, V. N., Davis, R. E., Lamy, L., Yu, X., Zhao, H., Lenz, G., Lam, L. T., Dave, S., Yang, L., Powell, J., and Staudt, L. M. (2006). A loss-of-function RNA interference screen for molecular targets in cancer. Nature 441, 106–110.

Owyang, A. M., Tumang, J. R., Schram, B. R., Hsia, C. Y., Behrens, T. W., Rothstein, T. L., and Liou, H. C. (2001). c-Rel is required for the protection of B cells from antigen receptor-mediated, but not Fas-mediated, apoptosis. J Immunol 167, 4948–4956.

Papa, S., Zazzeroni, F., Bubici, C., Jayawardena, S., Alvarez, K., Matsuda, S., Nguyen, D. U., Pham, C. G., Nelsbach, A. H., Melis, T., De Smaele, E., Tang, W. J., D'Adamio, L., and Franzoso, G. (2004a). Gadd45 beta mediates the NF-kappa B suppression of JNK signalling by targeting MKK7/JNKK2. Nat Cell Biol 6, 146–153.

Papa, S., Zazzeroni, F., Pham, C. G., Bubici, C., and Franzoso, G. (2004b). Linking JNK signaling to NF-kappaB: a key to survival. J Cell Sci 117, 5197–5208.

Papa, S., Bubici, C., Zazzeroni, F., Pham, C. G., Kuntzen, C., Knabb, J. R., Dean, K., and Franzoso, G. (2006). The NF-kappaB-mediated control of the JNK cascade in the antagonism of programmed cell death in health and disease. Cell Death Differ 13, 712–729.

Park, J. S., Gamboni-Robertson, F., He, Q., Svetkauskaite, D., Kim, J. Y., Strassheim, D., Sohn, J. W., Yamada, S., Maruyama, I., Banerjee, A., Ishizaka, A., and Abraham, E. (2006). High mobility group box 1 protein interacts with multiple Toll-like receptors. Am J Physiol Cell Physiol 290, C917–924.

Pasparakis, M., Courtois, G., Hafner, M., Schmidt-Supprian, M., Nenci, A., Toksoy, A., Krampert, M., Goebeler, M., Gillitzer, R., Israel, A., Krieg, T., Rajewsky, K., and Haase, I. (2002a). TNF-mediated inflammatory skin disease in mice with epidermis-specific deletion of IKK2. Nature 417, 861–866.

Pasparakis, M., Schmidt-Supprian, M., and Rajewsky, K. (2002b). IkappaB kinase signaling is essential for maintenance of mature B cells. J Exp Med 196, 743–752.

Perkins, N. D. (2004). NF-kappaB: tumor promoter or suppressor? Trends Cell Biol 14, 64–69.

Perkins, N. D. and Gilmore, T. D. (2006). Good cop, bad cop: the different faces of NF-kappaB. Cell Death Differ 13, 759–772.

Petro, J. B. and Khan, W. N. (2001). Phospholipase C-gamma 2 couples Bruton's tyrosine kinase to the NF-kappaB signaling pathway in B lymphocytes. J Biol Chem 276, 1715–1719.

Petro, J. B., Rahman, S. M., Ballard, D. W., and Khan, W. N. (2000). Bruton's tyrosine kinase is required for activation of IkappaB kinase and nuclear factor kappaB in response to B cell receptor engagement. J Exp Med 191, 1745–1754.

Pham, C. G., Bubici, C., Zazzeroni, F., Papa, S., Jones, J., Alvarez, K., Jayawardena, S., De Smaele, E., Cong, R., Beaumont, C., Torti, F. M., Torti, S. V., and Franzoso, G. (2004). Ferritin heavy chain upregulation by NF-kappaB inhibits TNFalpha-induced apoptosis by suppressing reactive oxygen species. Cell 119, 529–542.

Pikarsky, E., Porat, R. M., Stein, I., Abramovitch, R., Amit, S., Kasem, S., Gutkovich-Pyest, E., Urieli-Shoval, S., Galun, E., and Ben-Neriah, Y. (2004). NF-kappaB functions as a tumour promoter in inflammation-associated cancer. Nature 431, 461–466.

Prendes, M., Zheng, Y., and Beg, A. A. (2003). Regulation of developing B cell survival by RelA-containing NF-kappa B complexes. J Immunol 171, 3963–3969.

Qing, G. and Xiao, G. (2005). Essential role of IkappaB kinase alpha in the constitutive processing of NF-kappaB2 p100. J Biol Chem 280, 9765–9768.

Qing, G., Qu, Z., and Xiao, G. (2005). Stabilization of basally translated NF-kappaB-inducing kinase (NIK) protein functions as a molecular switch of processing of NF-kappaB2 p100. J Biol Chem 280, 40578–40582.

Rao, P. H., Houldsworth, J., Dyomina, K., Parsa, N. Z., Cigudosa, J. C., Louie, D. C., Popplewell, L., Offit, K., Jhanwar, S. C., and Chaganti, R. S. (1998). Chromosomal and gene amplification in diffuse large B-cell lymphoma. Blood 92, 234–240.

Rape, M. and Jentsch, S. (2004). Productive RUPture: activation of transcription factors by proteasomal processing. Biochim Biophys Acta 1695, 209–213.

Rayet, B. and Gelinas, C. (1999). Aberrant rel/nfkb genes and activity in human cancer. Oncogene 18, 6938–6947.

Richardson, P. G., Hideshima, T., Mitsiades, C., and Anderson, K. (2004). Proteasome inhibition in hematologic malignancies. Ann Med 36, 304–314.

Robe, P. A., Bentires-Alj, M., Bonif, M., Rogister, B., Deprez, M., Haddada, H., Khac, M. T., Jolois, O., Erkmen, K., Merville, M. P., Black, P. M., and Bours, V. (2004). In vitro and in vivo activity of the nuclear factor-kappaB inhibitor sulfasalazine in human glioblastomas. Clin Cancer Res 10, 5595–5603.

Rocha, S., Campbell, K. J., and Perkins, N. D. (2003). p53- and Mdm2-independent repression of NF-kappa B transactivation by the ARF tumor suppressor. Mol Cell 12, 15–25.

Rocha, S., Garrett, M. D., Campbell, K. J., Schumm, K., and Perkins, N. D. (2005). Regulation of NF-kappaB and p53 through activation of ATR and Chk1 by the ARF tumour suppressor. EMBO J 24, 1157–1169.

Rudolph, D., Yeh, W. C., Wakeham, A., Rudolph, B., Nallainathan, D., Potter, J., Elia, A. J., and Mak, T. W. (2000). Severe liver degeneration and lack of NF-kappaB activation in NEMO/IKKgamma-deficient mice. Genes Dev 14, 854–862.

Ryo, A., Suizu, F., Yoshida, Y., Perrem, K., Liou, Y. C., Wulf, G., Rottapel, R., Yamaoka, S., and Lu, K. P. (2003). Regulation of NF-kappaB signaling by Pin1-dependent prolyl isomerization and ubiquitin-mediated proteolysis of p65/RelA. Mol Cell 12, 1413–1426.

Sakon, S., Xue, X., Takekawa, M., Sasazuki, T., Okazaki, T., Kojima, Y., Piao, J. H., Yagita, H., Okumura, K., Doi, T., and Nakano, H. (2003). NF-kappaB inhibits TNF-induced accumulation of ROS that mediate prolonged MAPK activation and necrotic cell death. EMBO J 22, 3898–3909.

Sanna, M. G., da Silva Correia, J., Ducrey, O., Lee, J., Nomoto, K., Schrantz, N., Deveraux, Q. L., and Ulevitch, R. J. (2002). IAP suppression of apoptosis involves distinct mechanisms: the TAK1/JNK1 signaling cascade and caspase inhibition. Mol Cell Biol 22, 1754–1766.

Santoro, M. G., Rossi, A., and Amici, C. (2003). NF-kappaB and virus infection: who controls whom. EMBO J 22, 2552–2560.

Savage, K. J., Monti, S., Kutok, J. L., Cattoretti, G., Neuberg, D., De Leval, L., Kurtin, P., Dal Cin, P., Ladd, C., Feuerhake, F., Aguiar, R. C., Li, S., Salles, G., Berger, F., Jing, W., Pinkus, G. S., Habermann, T., Dalla-Favera, R., Harris, N. L., Aster, J. C., Golub, T. R., and Shipp, M. A. (2003). The molecular signature of mediastinal large B-cell lymphoma differs from that of other diffuse large B-cell lymphomas and shares features with classical Hodgkin lymphoma. Blood 102, 3871–3879.

Scaffidi, P., Misteli, T., and Bianchi, M. E. (2002). Release of chromatin protein HMGB1 by necrotic cells triggers inflammation. Nature 418, 191–195.

Schiemann, B., Gommerman, J. L., Vora, K., Cachero, T. G., Shulga-Morskaya, S., Dobles, M., Frew, E., and Scott, M. L. (2001). An essential role for BAFF in the normal development of B cells through a BCMA-independent pathway. Science 293, 2111–2114.

Schmidt-Supprian, M., Bloch, W., Courtois, G., Addicks, K., Israel, A., Rajewsky, K., and Pasparakis, M. (2000). NEMO/IKK gamma-deficient mice model incontinentia pigmenti. Mol Cell 5, 981–992.

Schmidt-Ullrich, R., Aebischer, T., Hulsken, J., Birchmeier, W., Klemm, U., and Scheidereit, C. (2001). Requirement of NF-kappaB/Rel for the development of hair follicles and other epidermal appendices. Development 128, 3843–3853.

Seitz, C. S., Deng, H., Hinata, K., Lin, Q., and Khavari, P. A. (2000). Nuclear factor kappaB subunits induce epithelial cell growth arrest. Cancer Res 60, 4085–4092.

Seitz, C. S., Lin, Q., Deng, H. and Khavari, P. A. (1998). Alterations in NF-kappaB function in transgenic epithelial tissue demonstrate a growth inhibitory role for NF-kappaB. Proc Natl Acad Sci USA 95, 2307–2312.

Senftleben, U., Cao, Y., Xiao, G., Greten, F. R., Krahn, G., Bonizzi, G., Chen, Y., Hu, Y., Fong, A., Sun, S. C., and Karin, M. (2001). Activation by IKKalpha of a second, evolutionary conserved, NF-kappa B signaling pathway. Science 293, 1495–1499.

Shipp, M., Ross, K., Tamayo, P., Weng, A. P., Kutok, J. L., Aguiar, R. C. T., Gaasenbeek, M., Angelo, M., Reich, M., Pinkus, G. S., Ray, T. S., Koval, M. A., Last, K. W., Norton, A., Lister, A., Mesirov, J., Neuberg, D. S., Lander, E. S., Aster, J. C., and Golub, T. R. (2002). Diffuse large B-cell lymphoma outcome prediction by gene expression profiling and supervised machine learning. Nat Med 8, 68–74.

Siebenlist, U., Brown, K., and Claudio, E. (2005). Control of lymphocyte development by nuclear factor-kappaB. Nat Rev Immunol 5, 435–445.

Signorelli, P. and Ghidoni, R. (2005). Resveratrol as an anticancer nutrient: molecular basis, open questions and promises. J Nutr Biochem 16, 449–466.

Small, G. W., Shi, Y. Y., Edmund, N. A., Somasundaram, S., Moore, D. T., and Orlowski, R. Z. (2004). Evidence that mitogen-activated protein kinase phosphatase-1 induction by proteasome inhibitors plays an antiapoptotic role. Mol Pharmacol 66, 1478–1490.

Sonenshein, G. E. (1997). Rel/NF-kappa B transcription factors and the control of apoptosis. Semin Cancer Biol 8, 113–119.

Stark, L. A. and Dunlop, M. G. (2005). Nucleolar sequestration of RelA (p65) regulates NF-kappaB-driven transcription and apoptosis. Mol Cell Biol 25, 5985–6004.

Subhashini, J., Mahipal, S. V., and Reddanna, P. (2005). Anti-proliferative and apoptotic effects of celecoxib on human chronic myeloid leukemia in vitro. Cancer Lett 224, 31–43.

Tacconelli, A., Farina, A.R., Cappabianca, L., Desantis, G., Tessitore, A., Vetuschi, A., Sferra, R., Rucci, N., Argenti, B., Screpanti, I., Gulino, A., Mackay, A.R. (2004). TrkA alternative splicing: a regulated tumor-promoting switch in human neuroblastoma. Cancer Cell 6, 347–360.

Takada, Y., Bhardwaj, A., Potdar, P., and Aggarwal, B. B. (2004). Nonsteroidal anti-inflammatory agents differ in their ability to suppress NF-kappaB activation, inhibition of expression of cyclooxygenase-2 and cyclin D1, and abrogation of tumor cell proliferation. Oncogene 23, 9247–9258.

Tan, J. E., Wong, S. C., Gan, S. K., Xu, S., and Lam, K. P. (2001). The adaptor protein BLNK is required for b cell antigen receptor-induced activation of nuclear factor-kappa B and cell cycle entry and survival of B lymphocytes. J Biol Chem 276, 20055–20063.

Tanaka, H., Matsumura, I., Ezoe, S., Satoh, Y., Sakamaki, T., Albanese, C., Machii, T., Pestell, R. G. and Kanakura, Y. (2002). E2F1 and c-Myc potentiate apoptosis through inhibition of NF-kappaB activity that facilitates MnSOD-mediated ROS elimination. Mol Cell 9, 1017–1029.

Tanaka, M., Fuentes, M. E., Yamaguchi, K., Durnin, M. H., Dalrymple, S. A., Hardy, K. L., and Goeddel, D. V. (1999). Embryonic lethality, liver degeneration, and impaired NF-kappa B activation in IKK-beta-deficient mice. Immunity 10, 421–429.

Tang, G., Minemoto, Y., Dibling, B., Purcell, N. H., Li, Z., Karin, M., and Lin, A. (2001). Inhibition of JNK activation through NF-kappaB target genes. Nature 414, 313–317.

Tao, Y., Yan, D., Yang, Q., Zeng, R., and Wang, Y. (2006). Low K+ promotes NF-kappaB/DNA binding in neuronal apoptosis induced by K+ loss. Mol Cell Biol 26, 1038–1050.

Tergaonkar, V., Pando, M., Vafa, O., Wahl, G., and Verma, I. (2002). p53 stabilization is decreased upon NFkappaB activation: a role for NFkappaB in acquisition of resistance to chemotherapy. Cancer Cell 1, 493–503.

Trompouki, E., Hatzivassiliou, E., Tsichritzis, T., Farmer, H., Ashworth, A., and Mosialos, G. (2003). CYLD is a deubiquitinating enzyme that negatively regulates NF-kappaB activation by TNFR family members. Nature 424, 793–796.

Tumang, J. R., Owyang, A., Andjelic, S., Jin, Z., Hardy, R. R., Liou, M. L., and Liou, H. C. (1998). c-Rel is essential for B lymphocyte survival and cell cycle progression. Eur J Immunol 28, 4299–4312.

Vakkila, J. and Lotze, M. T. (2004). Inflammation and necrosis promote tumour growth. Nat Rev Immunol 4, 641–648.

Van Antwerp, D. J., Martin, S. J., Verma, I. M., and Green, D. R. (1998). Inhibition of TNF-induced apoptosis by NF-kB. Cell Biol 8, 107–111.

van Hogerlinden, M., Rozell, B. L., Ahrlund-Richter, L., and Toftgard, R. (1999). Squamous cell carcinomas and increased apoptosis in skin with inhibited Rel/nuclear factor-kappaB signaling. Cancer Res 59, 3299–3303.

Ventura, J. J., Cogswell, P., Flavell, R. A., Baldwin, A. S., Jr., and Davis, R. J. (2004). JNK potentiates TNF-stimulated necrosis by increasing the production of cytotoxic reactive oxygen species. Genes Dev 18, 2905–2915.

Verschelde, C., Walzer, T., Galia, P., Biemont, M. C., Quemeneur, L., Revillard, J. P., Marvel, J., and Bonnefoy-Berard, N. (2003). A1/Bfl-1 expression is restricted to TCR engagement in T lymphocytes. Cell Death Differ 10, 1059–1067.

Wang, C. Y., Mayo, M. W., and Baldwin, A. S., Jr. (1996). TNF- and cancer therapy-induced apoptosis: potentiation by inhibition of NF-kappaB. Science 274, 784–787.

Wang, C. Y., Mayo, M. W., Korneluk, R. G., Goeddel, D. V., and Baldwin, A. S., Jr. (1998). NF-kappaB antiapoptosis: induction of TRAF1 and TRAF2 and c-IAP1 and c- IAP2 to suppress caspase-8 activation. Science 281, 1680–1683.

Wertz, I. E., O'Rourke, K. M., Zhou, H., Eby, M., Aravind, L., Seshagiri, S., Wu, P., Wiesmann, C., Baker, R., Boone, D. L., Ma, A., Koonin, E. V., and Dixit, V. M. (2004). De-ubiquitination and ubiquitin ligase domains of A20 downregulate NF-kappaB signalling. Nature 430, 694–699.

Wessendorf, S., Schwaenen, C., Kohlhammer, H., Kienle, D., Wrobel, G., Barth, T. F., Nessling, M., Möller, P., Döhner, H., Lichter, P., and Bentz, M. (2003). Hidden gene amplifications in aggressive B-cell non-Hodgkin lymphomas detected by microarray-based comparative genomic hybridization. Oncogene 22, 1425–1429.

Wilkinson, J. C., Wilkinson, A. S., Scott, F. L., Csomos, R. A., Salvesen, G. S., and Duckett, C. S. (2004). Neutralization of Smac/Diablo by inhibitors of apoptosis (IAPs). A caspase-independent mechanism for apoptotic inhibition. J Biol Chem 279, 51082–51090.

Wright, C. W. and Duckett, C. S. (2005). Reawakening the cellular death program in neoplasia through the therapeutic blockade of IAP function. J Clin Invest 115, 2673–2678.

Wu, M., Arsura, M., Bellas, R. E., Fitzgerald, M. J., Lee, H., Schauer, S. L., Sherr, D. H., and Sonenshein, G. E. (1996). Inhibition of c-myc expression induces apoptosis of WEHI 231 murine B cells. Mol Cell Biol 16, 5015–5025.

Xiang, H., Hochman, D. W., Saya, H., Fujiwara, T., Schwartzkroin, P. A., and Morrison, R. S. (1996). Evidence for p53-mediated modulation of neuronal viability. J Neurosci 16, 6753–6765.

Xiao, G., Harhaj, E., and Sun, S. C. (2001). NF-kB-inducing kinase regulates the processing of NF-kB2 p100. Mol Cell 7, 401–409.

Xiao, G., Fong, A., and Sun, S. C. (2004). Induction of p100 processing by NF-kappaB-inducing kinase involves docking IkappaB kinase alpha (IKKalpha) to p100 and IKKalpha-mediated phosphorylation. J Biol Chem 279, 30099–30105.

Yamamoto, K. and Gaynor, R. B. (2001). Therapeutic potential of inhibition of the NF-kappaB pathway in the treatment of inflammation and cancer. J Clin Invest 107, 135–142.

Yamamoto, Y., Yin, M. J., Lin, K. M., and Gaynor, R. B. (1999). Sulindac inhibits activation of the NF-kappaB pathway. J Biol Chem 274, 27307–27314.

Yamazaki, T. and Kurosaki, T. (2003). Contribution of BCAP to maintenance of mature B cells through c-Rel. Nat Immunol 4, 780–786.

Yan, M., Zhang, Z., Brady, J. R., Schilbach, S., Fairbrother, W. J., and Dixit, V. M. (2002). Identification of a novel death domain-containing adaptor molecule for ectodysplasin-A receptor that is mutated in crinkled mice. Curr Biol 12, 409–413.

Yeh, W. C., Itie, A., Elia, A. J., Ng, M., Shu, H. B., Wakeham, A., Mirtsos, C., Suzuki, N., Bonnard, M., Goeddel, D. V., and Mak, T. W. (2000). Requirement for Casper (c-FLIP) in regulation of death receptor-induced apoptosis and embryonic development. Immunity 12, 633–642.

Yin, M.-J., Yamamoto, Y., and Gaynor, R. B. (1998). The anti-inflammatory agents aspirin and salicylate inhibit the activity of IkB kinase-b. Nature 396, 77–80.

Yu, Z., Zhou, D., Cheng, G., and Mattson, M. P. (2000). Neuroprotective role for the p50 subunit of NF-kappaB in an experimental model of Huntington's disease. J Mol Neurosci 15, 31–44.

Zeh, H. J., III and Lotze, M. T. (2005). Addicted to death: invasive cancer and the immune response to unscheduled cell death. J Immunother 28, 1–9.

Zhang, J. Y., Tao, S., Kimmel, R., and Khavari, P. A. (2005). CDK4 regulation by TNFR1 and JNK is required for NF-kappaB-mediated epidermal growth control. J Cell Biol 168, 561–566.

Zheng, L., Bidere, N., Staudt, D., Cubre, A., Orenstein, J., Chan, F. K., and Lenardo, M. (2006). Competitive control of independent programs of tumor necrosis factor receptor-induced cell death by TRADD and RIP1. Mol Cell Biol 26, 3505–3513.

Zheng, Y., Ouaaz, F., Bruzzo, P., Singh, V., Gerondakis, S., and Beg, A. A. (2001). NF-kappa B RelA (p65) is essential for TNF-alpha-induced fas expression but dispensable for both TCR-induced expression and activation-induced cell death. J Immunol 166, 4949–4957.

Zheng, Y., Vig, M., Lyons, J., Van Parijs, L., and Beg, A. A. (2003). Combined deficiency of p50 and cRel in CD4+ T cells reveals an essential requirement for nuclear factor kappaB in regulating mature T cell survival and in vivo function. J Exp Med 197, 861–874.

Zhou, H., Wertz, I., O'Rourke, K., Ultsch, M., Seshagiri, S., Eby, M., Xiao, W., and Dixit, V. M. (2004). Bcl10 activates the NF-kappaB pathway through ubiquitination of NEMO. Nature 427, 167–171.

Chapter 12
Targeting Proteasomes as Therapy in Multiple Myeloma

Dharminder Chauhan, Teru Hideshima, and Kenneth C. Anderson*

Abstract The Ubiquitin-proteasome pathway (UPP) regulates normal intracellular protein degradation processes essential for cell cycle progression, inflammation, transcription, DNA replication, and apoptosis. Blockade of UPP using proteasome inhibitor Bortezomib (Velcade) is an effective therapy for relapsed/refractory multiple myeloma (MM). Both oligonucleotide microarrays and proteomic studies are delineating the molecular mechanisms mediating Bortezomib-induced cytotoxicity, defining targets of sensitivity vs resistance, allowing for the development of next generation therapies, and providing the rationale for combination therapies.

Keywords proteasomes, apoptosis, drug resistance, myeloma

1 Introduction

The proteasome is a multisubunit complex with catalytic activities mediating proteolysis of ubiquitinated intracellular proteins (Adams, 2004; Goldberg and Rock, 2002) The 26S proteasome complex consist of 19S units flanking a barrel-shaped 20S proteasome core; the 19S units regulate entry only of those proteins marked for degradation into the 20S core chamber (Adams, 2004; Goldberg and Rock, 2002). Proteasomal protein degradation is a multistep process: protein is first earmarked with a chain of ubiquitin molecules; E1 ubiquitin enzyme then activates ubiquitin and links it to the ubiquitin-conjugating enzyme E2 in an ATP-dependent manner; E3 ubiquitin ligase then links the ubiquitin molecule to the protein; a long polypeptide chain of ubiquitin moieties is formed; and finally, proteasomes degrade the protein into small fragments and free ubiquitin for recycling (Goldberg and

Dharminder Chauhan, Teru Hideshima, and Kenneth C. Anderson
The Jerome Lipper Multiple Myeloma Center, Department of Medical Oncology, Dana Farber Cancer Institute, Harvard Medical School, Boston, MA 02115, USA

*To whom correspondence should be addressed: Fax: 617 632-2140;
e-mail: kenneth_anderson@dfci.harvard.edu

R. Khosravi-Far and E. White (eds.), *Programmed Cell Death in Cancer Progression and Therapy*.
© Springer 2008

Rock, 2002; Pickart, 2004). Protein degradation is predominantly regulated by caspase-like (CT-L) (beta-5), trypsin-like (beta-1), and caspase-like (beta-2) proteolytic activities residing within the 20S proteasome core. Importantly, the substrates of proteasomes include many cellular proteins that maintain normal cell cycle progression, growth, and survival. Most of the damaged or misfolded, short or long-lived, proteins in the cell are eliminated by UPP; conversely, blockade of protein degradation by proteasome inhibitors (PIs) causes intracellular accumulation of redundant proteins, resulting in induction of heat-shock response and apoptosis (Adams, 2004; Goldberg and Rock, 2002).

Since the proteasome regulates normal cellular functions, its value as a possible therapeutic target was viewed with skepticism due to cytotoxicity to normal cells. However, various studies suggest that PIs are more cytotoxic to proliferating malignant cells than quiescent normal cells, thereby providing an acceptable therapeutic index (Adams, 2004). The mechanism whereby cancer cells are more susceptible to PIs than normal cellular counterparts is unclear. One possibility is that cancer cells have altered cell cycle machinery, leading to an increase in their proliferation rate. These cells therefore accumulate damaged proteins at a much higher rate than do normal cells, which in turn increases dependency on proteasomal degradation. In contrast, quiescent cancer cells may be more susceptible to proteasome inhibition than normal cells. PIs also inhibit prosurvival signaling pathways. For example, nuclear factor-kappa B (NF-κB) is linked to proliferation and drug resistance in cancer cells (Haefner, 2002); conversely, PIs downregulate NF-κB activation, thereby enhancing the cytotoxic effects of chemotherapy. Together, these findings support the notion of targeting proteasomes in novel therapeutics.

Naturally occurring and synthetic inhibitors of the ubiquitin-proteasome pathway (UPP) include peptide aldehydes, peptide boronates, nonpeptide inhibitors, peptide vinyl sulfones, and peptide epoxyketones (Adams, 2004). All of these PIs differentially affect proteasome activities and also show activity against other proteases. For example, peptide aldehydes (MG-132, MG-115, ALLN, or PSI) potently, but reversibly, block the chymotrypsin-like (CT-L) (beta-5) activity of the proteasome; they also inhibit lysosomal cysteine and serine proteases, as well as calpains, thereby limiting their clinical utility. Lactacystin is a natural, irreversible, nonpeptide inhibitor; the clasto-lactacystin beta-lactone, an analog of its active metabolite, is currently in phase I clinical trails. Importantly, studies by Adams et al. led to the development of peptide boronic acid PIs (Adams, 2004). The dipeptidyl boronic acid Bortezomib/PS-341 is a potent and reversible inhibitor of CT-L (beta-5) activity. Moreover, our recent study using radiolabeled active site-directed probe specific for proteasome catalytic subunits showed that Bortezomib targets beta-5 (CT-L activity) and beta-1 (C-L activity), as well as beta-5i and beta-1i catalytic subunits of the immunoproteasome (Berkers et al., 2005). Initial NCI screening showed remarkable antitumor activity of Bortezomib in a panel of 60 tumor cell lines. We have shown that Bortezomib induces MM cell apoptosis, downregulates adhesion molecules, inhibits constitutive and MM cell adhesion-induced cytokine secretion, and blocks angiogenesis in the BM milieu (Chauhan et al., 2005b). It also inhibits human MM cell growth and prolongs host survival in a severe combined immunodeficient (SCID) mouse model of human MM (Chauhan et al., 2005b). Phase I trials showed safety and acceptable toxicity, as well as early signs of anti-MM activity (Richardson, 2004; Voorhees and Orlowski, 2006) Phase II clinical trails demonstrated durable responses (including complete responses) with associated clinical benefits, providing the basis for the Food and Drug Administration (FDA) approval to treat relapsed refractory MM (Richardson, 2004). A randomized phase III trial showed higher responses as well as prolonged time to progression and survival in patients treated with Bortezomib vs Dexamethasone (Richardson et al., 2005), providing the basis for FDA approval extended to include relapsed MM. Although Bortezomib is a major advance, treatment is associated with toxicity and the development of drug-resistance in most patients. Recent studies have therefore delineated the mechanisms mediating Bortezomib-induced cytotoxicity and drug resistance, in order to design novel therapeutic strategies.

2 Bortezomib-Triggered Signaling Pathways

The proteasome is the primary target of PIs; however, PIs also affect growth/survival and apoptotic molecules. A major mechanism whereby PIs inhibit growth and survival of cancer cells is by blocking prosurvival NF-κB signaling (Adams, 2004). Constitutive activation of NF-κB, associated with growth/proliferation and drug resistance, occurs via these sequential events: IκB-a kinase (IKK) activation; IκB phosphorylation; ubiquitination and degradation of IκB; and nuclear translocation of p50/65 NF-κB. Nuclear localization of NF-κB induces transcription of genes-encoding cytokines (IL-6, TNF-α), survival factors (inhibitors of apoptosis proteins [IAPs], Bcl-x$_L$), and cell adhesion molecules (intracellular adhesion molecule [ICAM], vascular cell adhesion molecule [VCAM], and E-selectin). NF-κB activation is also associated with growth and survival of MM cells; specifically, adhesion of MM cells to bone marrow stromal cells (BMSCs) triggers NF-κB-mediated transcription and secretion of IL-6 and insulin-like growth factor-I (IGF-I) (Chauhan et al., 2005b), both of which promote survival and conventional drug resistance in MM cells in the BM milieu. Moreover, patient MM cells and BMSCs have upregulated NF-κB activity relative to normal cells; within the tumor cell population, drug-sensitive MM cells have lower NF-κB activity than drug-resistant MM cells. Importantly, treatment of MM cells with Bortezomib inhibits NF-κB activation and related cytokine production, thereby overcoming the survival advantage for MM cells conferred by BMSCs. Our work also shows that NF-κB inhibition alone is unlikely to account for the total anti-MM activity of Bortezomib. Both PS-1145, a specific inhibitor of IκB, and Bortezomib block TNF-α-induced NF-κB activation by inhibiting phosphorylation and degradation of IκB-α; in contrast to Bortezomib, however, PS-1145 only partially inhibits MM cell growth (Hideshima et al., 2002).

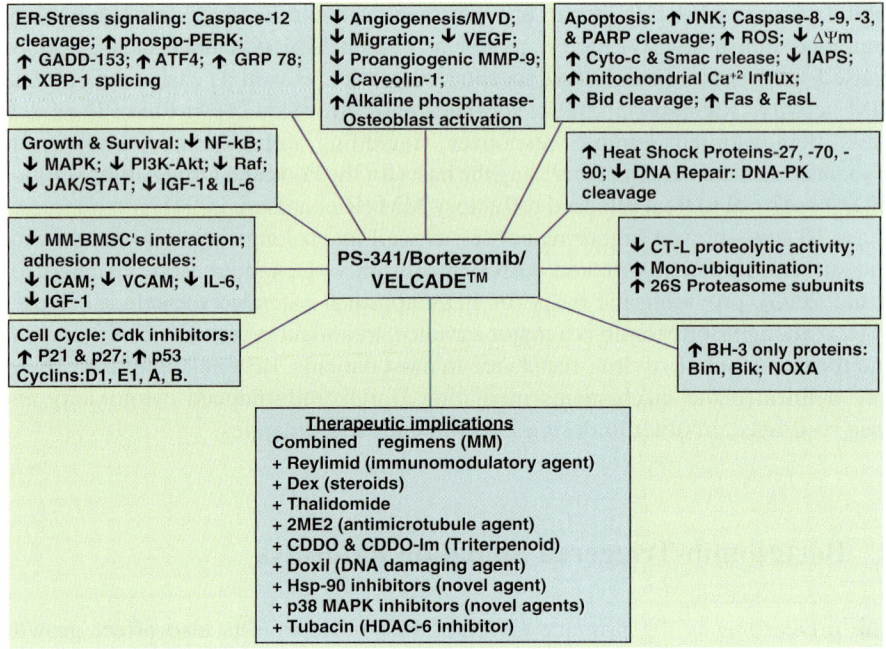

Fig. 12.1 Identification of Bortezomib/Velcade/PS-341-triggered molecular mechanisms mediating growth/survival, apoptosis, and drug resistance in tumor cells, including host-BM microenvironment and angiogenesis ("↑" arrow: induction/upregulation; "↓" arrow: reduction /inhibition/downregulation). Delineation of the Bortezomib signaling profile allow us to combine it with agents that utilize either similar or additional apoptotic pathways to enhance its tumor cytotoxicity, reduce toxicity to normal cells, prevent development of drug-resistance, and improve patient outcome. Shown are the ongoing therapeutic strategies in MM using the combination of Bortezomib with various conventional and novel agents

Recent oligonucleotide microarray and proteomic studies show that Bortezomib affects various signaling pathways (Figure 12.1). For example, Bortezomib-induced apoptosis is associated with: (1) activation of stress response proteins such as heat shock proteins, Hsp-27, Hsp-70, and Hsp-90 (Chauhan et al., 2003; Mitsiades et al., 2002); (2) upregulation of proapoptotic c-Jun-NH2-terminal kinase (JNK) (Chauhan and Anderson, 2003); (3) alteration of mitochondrial membrane potential (MMP) and generation of reactive oxygen species (ROS); (4) induction of the intrinsic cell death pathway via the release of mitochondrial proteins cytochrome-c/Smac into cytosol, resulting in activation of caspase-9 > caspase-3 cascade; (5) activation of extrinsic apoptotic signaling through Bid and caspase-8 cleavage (Mitsiades et al., 2002); (6) upregulation of ubiquitin/proteasome pathway members (Mitsiades et al., 2002); (7) inactivation of DNA-dependent protein kinase (DNA-PK) (Mitsiades et al., 2003), which is essential for the repair of DNA double-strand breaks; and (8) inhibition of MM cell growth factor-triggered MAPK and PI3-kinase/Akt signaling (Hideshima et al., 2003). Our studies using dominant negative strategies and knockout cell line models have established a direct role for JNK and Bax/Bak during Bortezomib-induced apoptosis (Chauhan et al., 2005b). Stress stimuli that induce mitochondrial outer membrane permeabilization (MOMP) use BH3-only proteins to facilitate Bax/Bak translocation to mitochondria; treatment of various cell types with Bortezomib induces stabilization of the BH3-only proteins Bim and Bik, whereas Bik or Bim and Bik-deficient MEFs are less susceptible to Bortezomib-induced killing. Moreover, Ca^{2+} influx into mitochondria triggers cyto-c and caspase-9-mediated apoptosis; conversely, treatment of MM cells with mitochondrial Ca^{2+} uptake inhibitor abrogates Bortezomib-triggered apoptosis (Landowski et al., 2005). These findings suggest that Bortezomib also affects signaling events upstream of mitochondria.

Recent studies link endoplasmic reticulum (ER)-related stress signaling to Bortezomib-induced death in MM cells (Landowski et al., 2005; Mitsiades et al., 2002). Oligonucleotide microarrays show a predominant induction of gene products associated with endoplasmic reticulum secretory pathways in MM cell lines following short-term exposure to high-dose Bortezomib (Landowski et al., 2005). Bortezomib triggers: expression of proteins associated with ER secretory pathways; activation of ER-resident caspase-12; and dysregulation of Ca^{2+} homeostasis, thereby resulting in cell death. Specifically, Bortezomib activates ER membrane-resident stress kinase PERK, accompanied by steady-state levels of ER protein-folding chaperone GRP-78; as well as proapoptotic ATF-4 and CHOP/GADD153, coupled with a simultaneous decrease in general protein synthesis. Another study in head and neck squamous cell carcinoma cells suggests that Bortezomib enhances efficacy of chemotherapeutic drugs via activation of the proapoptotic ER stress-ROS pathway (Fribley et al., 2004). Both caspase-12 and caspase-4 have been implicated in ER stress-induced apoptosis; however, neither caspase-12 nor caspase-4 are required for ER stress-induced apoptosis (Obeng and Boise, 2005), and it remains unclear how ER stress causes caspase activation. Nonetheless, Bortezomib-induced apoptosis involves activation of ER-related stress pathways, including activation of caspase-12. Together, these findings suggest that inhibition of growth/survival signaling cascades and concurrent activation of apoptotic signaling pathways mediate overall Bortezomib-induced cytotoxicity in MM cells.

Although Bortezomib triggers remarkable antitumor activity in MM cells, intrinsic or acquired drug resistance occurs in most cases. Recent studies have therefore focused on defining mechanisms mediating Bortezomib resistance. Our study shows that treatment with Bortezomib induces apoptosis in SUDHL6 (DHL6), but not SUDHL4 (DHL4), lymphoma cells (Chauhan et al., 2003). Microarray analysis demonstrates high RNA levels for heat shock protein-27 (Hsp27) in DHL4 vs DHL6 cells, correlating with increased Hsp27 protein expression. Blockade of Hsp27 in DHL-4 cells using antisense (AS) strategy restores the apoptotic response to Bortezomib; conversely, overexpression of Hsp27 renders Bortezomib-sensitive DHL6 cells resistant to Bortezomib. These data suggest that Hsp27, at least in part, accounts for Bortezomib resistance. MM cells obtained from patient's refractory to Bortezomib treatment show elevated Hsp-27 levels, further supporting this view. Of note, Hsp-27 negatively regulates the mitochondrial release of cytoc-c and Smac, thereby preventing activation of intrinsic cell death-signaling cascade. Upregulated expression of IAPs, such as XIAP, may also contribute to Bortezomib resistance (Mitsiades et al., 2002); conversely, inhibition of these prosurvival molecules may sensitize tumor cells to Bortezomib and even overcome Bortezomib resistance.

3   Therapeutic Implications

It is unlikely that one specific mechanism accounts for Bortezomib-induced cytotoxicity or the development of resistance, suggesting that combinations of Bortezomib with other conventional and/or novel agents may enhance its cytotoxicity and overcome drug resistance. For example, combined Bortezomib and irinotecan treatment triggers apoptosis in pancreatic tumor xenografts and enhances chemosensitivity in colorectal cancer xenograft models (Cusack et al., 2001). Preclinical studies in MM demonstrate that combining Bortezomib with conventional agents such as Dex, Doxorubicin, Melphalan, or Mitoxantrone induces additive or synergistic antitumor activity (Mitsiades et al., 2003). Treatment of MM cells with Bortezomib and novel agents Relvimid or triterpenoids CDDO-Imidazolide also induces synergistic anti-MM activity and overcomes Bortezomib resistance by targeting both intrinsic and extrinsic apoptotic signaling, thereby providing the basis for clinical protocols using combination regimens. Importantly, gene profiling studies show that Bortezomib induces Hsp-90 in MM cells; conversely, blockade of Hsp-90 with 17-AAG enhances sensitivity and even overcomes Bortezomib resistance (Mitsiades et al., 2002). Clinical trials already show promise of combined therapy in Bortezomib refractory MM. Our laboratory has recently demonstrated the significance of the alternative aggresome cascade for protein catabolism in MM cells; identified histone deacetylase-6 (HDAC-6) to be essential in the chaperoning of ubiquitinated proteins for aggresomal degradation; and validated the preclinical anti-MM activity of HDAC-6 inhibitor Tubacin (Hideshima et al., 2005). Importantly, dual inhibition of proteasomes and aggresomes with Bortezomib and tubacin, respectively, triggers synergistic cytotoxicity, setting the stage again for clinical translation of this new class of cancer therapeutics. Finally, correlative science studies of samples from patients on Bortezomib treatment protocols show that resistance is associated with upregulation of Hsp-27; already preclinical and clinical studies of p38 MAPK inhibitors to downregulate Hsp-27 and thereby overcome Bortezomib resistance have been completed.

4   Novel Proteasome Inhibitor Npi-0052 and its Clinical Utility Vis-À-Vis Bortezomib

Besides the combination therapeutic strategies, our recent study also shows that a novel proteasome inhibitor NPI-0052 can overcome Bortezomib resistance in MM cells. NPI-0052 is a small molecule derived from fermentation of Salinospora, a new marine gram-positive actinomycete (Chauhan et al., 2005a; Feling et al., 2003;

Macherla et al., 2005). NPI-0052 is a nonpeptide PI with structural similarity to Omuralide (Feling et al., 2003; Groll et al., 2006; Macherla et al., 2005), a beta-lactone derived from naturally occurring lactacystin. NPI-0052, in contrast to Omuralide, possess a uniquely methylated C3 ring juncture, chlorinated alkyl group at C2, and cyclohexene ring at C5 (Macherla et al., 2005), which accounts for its higher antitumor activity than omuralide. Initial screening of NPI-0052 against the NCI panel of 60 tumor cell lines showed GI_{50} of <110 nM in all cases. Our data showed that (1) NPI-0052 triggers apoptosis in MM cells sensitive and resistant not only to conventional, but also to Bortezomib therapies; and (2) The IC_{50} of NPI-0052 for MM cells is within the low nanomolar concentration (Chauhan et al., 2005a). Importantly, NPI-0052 similarly triggered apoptosis in purified tumor cells from several MM patients relapsing after various prior therapies including Bortezomib and thalidomide.

The mechanism whereby NPI-0052 overcomes Bortezomib resistance in MM cells is unclear; however, this may be due to its differential mode of action than Bortezomib. For example, NPI-0052 and Bortezomib differentially affect 20S proteasomal activities: (1) NPI-0052 inhibits CT-L and T-L activities at much lower concentrations than Bortezomib, and (2) higher concentrations of NPI-0052 than Bortezomib are required to inhibit C-L activity (Chauhan et al., 2005a). Animal studies using whole blood lysates showed that NPI-0052 blocked CT-L activity, which was recoverable to near basal levels by day 7; whereas inhibition of CT-L activity is significantly restored at 24 h after Bortezomib. NPI-0052 inhibits 50% of T-L activity, which is restored by day 7; whereas Bortezomib enhances T-L activity, which remains upregulated even at day 7. Interestingly, both NPI-0052 and Bortezomib inhibited C-L activity, which recovered only at day 7. The comparative kinetics of proteosomal activities suggest that NPI-0052, in contrast to Bortezomib, triggers a sustained inhibition of CT-L, T-L, and C-L (up to 7 days), which may therefore allow for a less-frequent administration schedule in patients. In this context, previous studies showed that CT-L activity is inhibited in peripheral blood cells of patients within 1 h of Bortezomib administration, and recoverable before the next dose (Adams, 2002; Hamilton et al., 2005).

A recent study showed that simultaneous inhibition of multiple proteasome activities is a prerequisite for significant (i.e., >50%) proteolysis (Kisselev et al., 2006). Another study showed that 50% inhibition of cystic fibrosis transmembrane conductance regulator degradation in reticulocytes extracts required concurrent blockade of CT-L and C-L proteasome activities (Oberdorf et al., 2001). Importantly, our study showed that MM cells exhibit higher constitutive levels of T-L proteasome activity than either CT-L or C-L activities (Crawford et al., 2006). These data, together with the results that NPI-0052, but not Bortezomib, efficiently inhibits CT-L + T-L activities (Chauhan et al., 2005a), suggest that NPI-0052 may block more protein breakdown than Bortezomib in MM cells. Moreover, mechanisms conferring Bortezomib resistance may not be effective against NPI-0052. Importantly, our study suggests that NPI-0052 is a potent inducer of MM cells apoptosis in tumor cells obtained from Bortezomib-refractory MM patients.

Another distinction between NPI-0052 and Bortezomib is their toxicity profile against normal cells. NPI-0052 does not significantly decrease normal lymphocyte viability at the IC_{50} doses for MM cells, with only modest effects at higher concentrations. By contrast, Bortezomib decreased the survival of lymphocytes at concentrations close to the IC_{50} doses for MM cells. NPI-0052 inhibits CT-L activity at doses which does not trigger apoptosis in MM cells. Previous observations that Bortezomib inhibits 20S proteasome activity in murine WBCs at 1 h postinjection, and that a similar degree of proteasome inhibition was noted in blood from responders vs nonresponders to Bortezomib therapy (Adams, 2002; Richardson, 2004), suggest that inhibition of proteasome activity in blood may not correlate to tumor cell cytotoxicity. Nonetheless, the above data suggest that (1) NPI-0052, in contrast to Bortezomib, is likely to have less toxic effects than Bortezomib on normal cells; and (2) NPI-0052 has a larger therapeutic index, which may allow for dose escalation therapy.

In vivo efficacy of NPI-0052 was shown using a human plasmacytoma xenograft mouse. Model (LeBlanc et al., 2002). Specifically, NPI-0052 inhibited MM tumor growth and prolongs survival of these mice at concentrations which were well tolerated and without significant weight loss or any neurological behavioral changes. Analysis at day 300 showed no recurrence of tumor in 57% of NPI-0052-treated mice.

Examination of signal transduction pathways showed that (1) NPI-0052 is a more potent inhibitor of NF-κB and related cytokine transcription and secretion than Bortezomib; (2) NPI-0052-induced MM cell death is predominantly mediated by caspase-8; and (3) Bortezomib-induced apoptosis requires both caspase-8 and caspase-9 activation. These findings further confirm differential actions of NPI-0052 vs Bortezomib in MM cells. The mechanistic differences between NPI-0052 and Bortezomib, i.e., their effect on proteasome activities and their dependence on specific apoptotic signal transduction pathway, provide a rationale for combination regimens for the treatment of MM. Indeed, the combination of NPI-0052 with Bortezomib induced synergistic anti-MM activity, without significantly affecting the viability of normal lymphocytes. The mechanisms mediating enhanced cytotoxicity of the combination regimen may simply reflect higher levels of proteasome inhibition with the two-drug regimens and/or activation of differential apoptotic signaling pathways. These data provide the framework for clinical trials of combined PIs to improve patient outcome in MM.

Acknowledgments This research study was supported by NIH grants CA 50947, CA 78373, CA100707-01, a Doris Duke Distinguished Clinical Research Scientist Award (KCA), The Cure for Myeloma Fund, and Multiple Myeloma Research Foundation Senior Research Award (DC).

References

Adams, J. (2002). Preclinical and clinical evaluation of proteasome inhibitor PS-341 for the treatment of cancer. Curr Opin Chem Biol 6, 493–500.

Adams, J. (2004). The proteasome: a suitable antineoplastic target. Nat Rev Cancer 4, 349–360.

Berkers, C. R., Verdoes, M., Lichtman, E., Fiebiger, E., Kessler, B. M., Anderson, K. C., Ploegh, H. L., Ovaa, H., and Galardy, P. J. (2005). Activity probe for in vivo profiling of the specificity of proteasome inhibitor bortezomib. Nat Methods 2, 357–362.

Chauhan, D. and Anderson, K. C. (2003). Mechanisms of cell death and survival in multiple myeloma (MM): Therapeutic implications. Apoptosis 8, 337–343.

Chauhan, D., Li, G., Shringarpure, R., Podar, K., Ohtake, Y., Hideshima, T. and Anderson, K. C. (2003). Blockade of Hsp27 overcomes Bortezomib/proteasome inhibitor PS-341 resistance in lymphoma cells. Cancer Res 63, 6174–6177.

Chauhan, D., Catley, L., Li, G., Podar, K., Hideshima, T., Velankar, M., Mitsiades, C., Mitsiades, N., Yasui, H., Letai, A., et al. (2005a). A novel orally active proteasome inhibitor induces apoptosis in multiple myeloma cells with mechanisms distinct from Bortezomib. Cancer Cell 8, 407–419.

Chauhan, D., Hideshima, T., and Anderson, K. C. (2005b). Proteasome inhibition in multiple myeloma: therapeutic implication. Annu Rev Pharmacol Toxicol 45, 465–476.

Crawford, L., Walker, B., Ovaa, H., Chauhan, D., Anderson, K., Morris, T., and Irvine, T. (2006). Comparative selectivity and specificity of the proteasome inhibitor BzLLLCOCHO, PS-341, and MG-132. Cancer Res 66, 6379–6386.

Cusack, J. C., Jr., Liu, R., Houston, M., Abendroth, K., Elliott, P. J., Adams, J., and Baldwin, A. S., Jr. (2001). Enhanced chemosensitivity to CPT-11 with proteasome inhibitor PS-341: implications for systemic nuclear factor-kappaB inhibition. Cancer Res 61, 3535–3540.

Feling, R. H., Buchanan, G. O., Mincer, T. J., Kauffman, C. A., Jensen, P. R., and Fenical, W. (2003). Salinosporamide A: a highly cytotoxic proteasome inhibitor from a novel microbial source, a marine bacterium of the new genus salinospora. Angew Chem Int Ed Engl 42, 355–357.

Fribley, A., Zeng, Q., and Wang, C. Y. (2004). Proteasome inhibitor PS-341 induces apoptosis through induction of endoplasmic reticulum stress-reactive oxygen species in head and neck squamous cell carcinoma cells. Mol Cell Biol 24, 9695–9704.

Goldberg, A. L. and Rock, K. (2002). Not just research tools – proteasome inhibitors offer therapeutic promise. Nat Med 8, 338–340.

Groll, M., Huber, R., and Potts, B. C. (2006). Crystal structures of salinosporamide A (NPI-0052) and B (NPI-0047) in complex with the 20S proteasome reveal important consequences of beta-lactone ring opening and a mechanism for irreversible binding. J Am Chem Soc 128, 5136–5141.

Haefner, B. (2002). NF-kappa B: arresting a major culprit in cancer. Drug Discov Today 7, 653–663.

Hamilton, A. L., Eder, J. P., Pavlick, A. C., Clark, J. W., Liebes, L., Garcia-Carbonero, R., Chachoua, A., Ryan, D. P., Soma, V., Farrell, K., et al. (2005). Proteasome inhibition with bortezomib (PS-341): a phase I study with pharmacodynamic end points using a day 1 and day 4 schedule in a 14-day cycle. J Clin Oncol 23, 6107–6116.

Hideshima, T., Chauhan, D., Richardson, P., Mitsiades, C., Mitsiades, N., Hayashi, T., Munshi, N., Dong, L., Castro, A., Palombella, V., et al. (2002). NF-kappa B as a therapeutic target in multiple myeloma. J Biol Chem 28, 28.

Hideshima, T., Mitsiades, C., Akiyama, M., Hayashi, T., Chauhan, D., Richardson, P., Schlossman, R., Podar, K., Munshi, N. C., Mitsiades, N., and Anderson, K. C. (2003). Molecular mechanisms mediating antimyeloma activity of proteasome inhibitor PS-341. Blood 101, 1530–1534.

Hideshima, T., Bradner, J. E., Wong, J., Chauhan, D., Richardson, P., Schreiber, S. L. and Anderson, K. C. (2005). Small-molecule inhibition of proteasome and aggresome function induces synergistic antitumor activity in multiple myeloma. Proc Natl Acad Sci USA 102, 8567–8572.

Kisselev, A. F., Callard, A., and Goldberg, A. L. (2006). Importance of the different proteolytic sites of the proteasome and the efficacy of inhibitors varies with the protein substrate. J Biol Chem 281, 8582–8590.

Landowski, T. H., Megli, C. J., Nullmeyer, K. D., Lynch, R. M., and Dorr, R. T. (2005). Mitochondrial-mediated disregulation of Ca2+ is a critical determinant of Velcade (PS-341/bortezomib) cytotoxicity in myeloma cell lines. Cancer Res 65, 3828–3836.

LeBlanc, R., Catley, L. P., Hideshima, T., Lentzsch, S., Mitsiades, C. S., Mitsiades, N., Neuberg, D., Goloubeva, O., Pien, C. S., Adams, J., et al. (2002). Proteasome inhibitor PS-341 inhibits human myeloma cell growth in vivo and prolongs survival in a murine model. Cancer Res 62, 4996–5000.

Macherla, V. R., Mitchell, S. S., Manam, R. R., Reed, K. A., Chao, T. H., Nicholson, B., Deyanat-Yazdi, G., Mai, B., Jensen, P. R., Fenical, W. F., et al. (2005). Structure-activity relationship studies of salinosporamide A (NPI-0052), a novel marine derived proteasome inhibitor. J Med Chem 48, 3684–3687.

Mitsiades, N., Mitsiades, C. S., Poulaki, V., Chauhan, D., Fanourakis, G., Gu, X., Bailey, C., Joseph, M., Libermann, T. A., Treon, S. P., et al. (2002). Molecular sequelae of proteasome inhibition in human multiple myeloma cells. Proc Natl Acad Sci USA 99, 14374–14379.

Mitsiades, N., Mitsiades, C. S., Richardson, P. G., Poulaki, V., Tai, Y. T., Chauhan, D., Fanourakis, G., Gu, X., Bailey, C., Joseph, M., et al. (2003). The proteasome inhibitor PS-341 potentiates sensitivity of multiple myeloma cells to conventional chemotherapeutic agents: therapeutic applications. Blood 101, 2377–2380.

Obeng, E. A. and Boise, L. H. (2005). Caspase-12 and caspase-4 are not required for caspase-dependent endoplasmic reticulum stress-induced apoptosis. J Biol Chem 280, 29578–29587.

Oberdorf, J., Carlson, E. J., and Skach, W. R. (2001). Redundancy of mammalian proteasome beta subunit function during endoplasmic reticulum associated degradation. Biochemistry 40, 13397–13405.

Pickart, C. M. (2004). Back to the future with ubiquitin. Cell 116, 181–190.

Richardson, P. G. (2004). A review of the proteasome inhibitor bortezomib in multiple myeloma. Expert Opin Pharmacother 5, 1321–1331.

Richardson, P. G., Sonneveld, P., Schuster, M. W., Irwin, D., Stadtmauer, E. A., Facon, T., Harousseau, J. L., Ben-Yehuda, D., Lonial, S., Goldschmidt, H., et al. (2005). Bortezomib or high-dose dexamethasone for relapsed multiple myeloma. N Engl J Med 352, 2487–2498.

Voorhees, P. M. and Orlowski, R. Z. (2006). The proteasome and proteasome inhibitors in cancer therapy. Annu Rev Pharmacol Toxicol 46, 189–213.

Chapter 13
Histone Deacetylase Inhibitors: Mechanisms and Clinical Significance in Cancer

HDAC Inhibitor-Induced Apoptosis

Sharmila Shankar and Rakesh K. Srivastava*

Abstract Epigenic modifications, mainly DNA methylation and acetylation, are recognized as the main mechanisms contributing to the malignant phenotype. Acetylation and deacetylation are catalyzed by specific enzymes, histone acetyltransferases (HATs) and histone deacetylases (HDACs), respectively. While histones represent a primary target for the physiological function of HDACs, the antitumor effect of HDAC inhibitors might also be attributed to transcription-independent mechanisms by modulating the acetylation status of a series of non-histone proteins. HDAC inhibitors may act through the transcriptional reactivation of dormant tumor suppressor genes. They also modulate expression of several other genes related to cell cycle, apoptosis, and angiogenesis. Several HDAC inhibitors are currently in clinical trials both for solid and hematologic malignancies. Thus, HDAC inhibitors, in combination with DNA-demethylating agents, chemopreventive, or classical chemotherapeutic drugs, could be promising candidates for cancer therapy. Here, we review the molecular mechanisms and therapeutic potential of HDAC inhibitors for the treatment of cancer.

Keywords HDAC inhibitors, HAT, SAHA, MS-275, TSA, TRAIL, apoptosis, caspase

1 Introduction

Recent years have seen major advances in elucidating the complexity of chromatin and its role as an epigenetic regulator of gene expression in eukaryotes. Epigenic modifications, mainly DNA methylation and acetylation, are recognized as additional

Sharmila Shankar and Rakesh K. Srivastava
Department of Biochemistry, The University of Texas Health Center at Tyler, 11937 US Highway 271, Tyler, TX 75708-3154, USA.

* To whom correspondence should be addressed: e-mail: rakesh.srivastava@uthct.edu
Tel.: 903-877-7559; fax: 903-877-5320

R. Khosravi-Far and E. White (eds.), *Programmed Cell Death in Cancer Progression and Therapy*.
© Springer 2008

mechanisms contributing to the malignant phenotype (Jones, 2002; Plass, 2002). Acetylation and deacetylation of histones play an important role in the regulation of gene expression (Grunstein, 1997). Histone acetylation is a reversible process whereby histone acetyltransferase (HAT) transfers the acetyl moiety from acetyl coenzyme A to the lysine; histone deacetylase (HDAC) removes the acetyl groups, reestablishing the positive charge in the histones. HATs and HDACs have recently been shown to regulate cell proliferation, differentiation, and apoptosis in various hematological and solid malignancies (Kouzarides, 1999). Altered HAT or HDAC activity is associated with cancer by changing the expression pattern of selected genes (Grignani et al., 1998; Lin et al., 1998). Hyperacetylation of histones correlates with gene activation, whereas deacetylation mediates eukaryotic chromatin condensation and gene expression silencing (Johnstone and Licht, 2003; Strahl and Allis, 2000). Recently, new roles of histone acetylation have been uncovered, not only in transcription, but also in DNA replication, repair, and heterochromatin formation (Kurdistani and Grunstein, 2003).

2   Histone Deacetylases

HDACs catalyze the removal of an acetyl group from the ε-amino group of lysine side chains of the core nucleosomal histones (H2A, H2B, H3, and H4), thereby reconstituting the positive charge on the lysine. Recent studies have revealed 12 human HDAC enzymes, HDAC1-11 (Emiliani et al., 1998; Gao et al., 2002; Grozinger et al., 1999; Taunton et al., 1996; Yang et al., 1996) and HDAC-A (Fischle et al., 1999). Based on the structural properties, HDACs can be divided into three classes (Gray and Ekstrom, 2001). Class I members (HDAC 1, 2, 3, 8, and 11) are transcriptional corepressors homologous to yeast RPD3 and have a single deacetylase domain at the N-termini and diversified C-terminal regions (de Ruijter et al., 2003). Class II members (HDAC 4, 5, 6, 7, 9, and 10) have domains similar to yeast HDA1 with a deacetylase domain at a C-terminal position (Verdin et al., 2003). In addition, HDAC 6 contains a second N-terminal deacetylase domain, which can function independently of its C-terminal counterpart. Class III HDACs are distinct from class I and II and are homologous of the yeast silent information regulator 2 (Sir2). All of these HDACs apparently exist in the cell as subunits of multiprotein complexes. Class II HDACs translocate from the cytoplasm to the nucleus in response to external stimuli, whereas class I HDACs are constitutively nuclear and play important roles in dynamic gene regulation (McKinsey and Olson, 2005).

Sir2 enzymes (or sirtuins) are NAD(+)-dependent deacetylases that modulate gene silencing, aging, and energy metabolism. Previous work has implicated several transcription factors as Sir2 targets. Sir2 silences transcription at silent mating loci, telomerese, and ribosomal DNA (rDNA), and this also suppresses recombination in rDNA. Earlier experiments have shown that the overexpression of Sir2 in yeast induced the global deacetylation of histones, indicating that Sir2 was an

HDAC (Braunstein et al., 1993). Later, it was shown that *cobB*, a bacterial homologue of Sir2, had ribosyltransferase activity, leading to experiments showing that Sir2 was also able to transfer adenosine diphosphate-ribose (ADP-ribose) from nicotinamide adeninedinucleotide (NAD) (Frye, 1999). Subsequently, it was confirmed that Sir2 was an NAD-dependent HDAC (Imai et al., 2000). The ADP-ribosylation of an acetylated lysine residue is an intermediate state of the enzymatic reaction catalyzed by Sir2. Only class III enzymes use NAD as a cofactor. Therefore, they are known as NAD-dependent HDACs.

Recently, Sir2 has attracted much attention, because it is related to longevity (Bordone and Guarente, 2005). The overexpression of Sir2 extends the life span of budding yeast, while its knockout shortens the life span by about 50% (Kaeberlein et al., 1999). Sir2 is conserved from bacteria to humans. In the nematodes, the gene most homologous to yeast *Sir2* gene is Sir-2.1. A duplication containing the *Sir-2.1* gene confers a life span that is extended by up to 50% (Tissenbaum and Guarente, 2001). The mammalian homologues consist of seven members, Sirt1–Sirt7. In mammalian cells, Sirt1 downregulates stress-induced p53 and FOXO pathways for apoptosis, thus favoring survival under stress. In the absence of applied stress, Sirt1 silencing induces growth arrest and/or apoptosis in human epithelial cancer cells (Ford et al., 2005). In contrast, normal human epithelial cells and normal human diploid fibroblasts seem to be refractory to Sirt1 silencing. Further studies have revealed that the Sirt1-regulated pathway is independent of p53, Bax, and caspase-2. Alternatively, Sirt1 may suppress apoptosis downstream from these apoptotic factors. FOXO4 (but not FOXO3) is required as proapoptotic mediator. Caspase-3 and caspase-7 act as downstream executioners of Sirt1/FOXO4-regulated apoptosis. These data suggest that Sirt1 as a novel target for selective killing of cancer vs noncancer epithelial cells. Upregulation of Sirt1 may be a double-edged sword that both promotes survival of aging cells and increases cancer risk in mammals.

Histones are part of the core proteins of nucleosomes. The recruitment of HATs and HDACs plays an important role in proliferation, differentiation and apoptosis (Glass and Rosenfeld, 2000; Kouzarides, 1999). Altered HAT or HDAC activity is associated with the development of cancer by changing the expression of several genes (Grignani et al., 1998; Lin et al., 1998). Treatment of malignant cells with HDAC inhibitors regulates only a small number (1–2%) of genes, as examined by DNA microarray studies (Van Lint et al., 1996). HDAC1 interacts directly with other transcription repressors, including all three of the pocket proteins, Rb, p107 and p130, and YY1. HDAC1 causes transcription repression by locally deacetylating histones, leading to a compact nucleosomal structure that prevents transcription factors from accessing DNA to promote transcription. Furthermore, HDAC1 knockout mice were embryonic lethal, possibly due to a proliferative defect upon unrestricted expressions of the cell cycle inhibitors p21[WAF1/CIP1] and p27[KIP1] (Lagger et al., 2002). Overexpression of HDAC I confers resistance to sodium butyrate-mediated apoptosis in melanoma cells through a p53-mediated pathway (Bandyopadhyay et al., 2004). We and others have shown that inhibition of HDAC activity induces apoptosis in various types of cancer (Fandy et al., 2005; Fang, 2005; Marks et al., 2003; Rosato et al., 2001; Singh et al., 2005).

Stability of HDACs is an important factor in determining the biological activity. HDAC4 is unusually unstable, with a half-life of less than 8 h (Liu et al., 2004). Consistent with the instability of HDAC4 protein, its mRNA was also highly unstable (with a half-life of less than 4 h). The exposure of cells to ultraviolet (UV) irradiation resulted in the degradation of HDAC4. This degradation was not dependent on proteasome or CRM1-mediated export activity but instead was caspase-dependent and was detectable in diverse human cancer lines. Of two potential caspase consensus motifs in HDAC4, both lying within a region containing proline, glutamic acid-, serine-, and threonine-rich (PEST) sequences, Asp-289 as the prime cleavage site was identified by site-directed mutagenesis (Liu et al., 2004). Notably, this residue is not conserved among other class IIa members, HDAC5, HDAC7, and HDAC9. Finally, the induced expression of caspase-cleavable HDAC4 led to markedly increased apoptosis. These results therefore link the regulation of HDAC4 protein stability to caspases, enzymes that are important for controlling cell death and differentiation.

3   Histone Deacetylase Inhibitors

It is well established that hyperacetylation of the N-terminal tails of histones H3 and H4 correlates with gene activation, whereas deacetylation mediates transcriptional repression (Strahl and Allis, 2000). Revived interest in these enzymatic pathways and how they modulate eukaryotic transcription has led to the identification of multiple cofactors whose complex interplay with HDAC affects gene expression. Concurrent with these discoveries, screening of natural product libraries yielded new small molecules that were subsequently identified as potent inhibitors of HDAC. While predominantly identified by using antiproliferative assays, the biological activity of these new HDAC inhibitors also encompasses significant antiprotozoal, antifungal, phytotoxic, and antiviral applications. During the past decade, a number of HDAC inhibitors have been shown to induce growth arrest, differentiation, and/or apoptosis in cancer cells (Boyle et al., 2005; Fandy et al., 2005; Kwon et al., 2002b; Marks et al., 2004; Singh et al., 2005), and inhibit tumor growth in various xenograft models (Bordin et al., 2004; Butler et al., 2000; Park et al., 2004; Sakajiri et al., 2005; Shao et al., 2004; Takimoto et al., 2005; Tang et al., 2004; Zhang et al., 2004c). HDAC inhibitors induce expression of cell cycle regulatory (e.g., $p21^{/WAF1/CIP1}$) and apoptotic proteins (e.g., Bax, PUMA, and Noxa), downregulate survival signaling pathways (e.g., Raf/MAPkinase/ERK), and disrupt cellular redox state (e.g., reactive oxygen species, ROS). Therefore, HDAC inhibitors are considered candidate drugs in cancer therapy (Johnstone, 2002; Marks et al., 2001b; McLaughlin and La Thangue, 2004).

Seven classes of HDAC inhibitors have been characterized and include short-chain fatty acids (e.g., sodium butyrate and phenylbutyrate); hydroxamic acids (e.g., suberoylanilide hydroxamic acid [SAHA], LAQ824, and trichostatin A [TSA]); benzamides (e.g., MS-275, CI994); cyclic tetrapeptide containing a 2-amino-8-oxo-9,

10-epoxy-decanoyl (AOE) moiety (e.g., trapoxin A); cyclic peptides without the AOE moiety (e.g., FK228/depsipeptide, apicidin); and epoxides (e.g., depudecin). These inhibitors induce a dose-dependent inhibition of either class I or class II HDACs, or both. Newly characterized HDAC inhibitors are now available that preferentially inhibit specific HDAC classes, including SK7041 (inhibits class I HDACs) and splitomicin (inhibits class III HDACs). A wide variety of HDAC inhibitors of both natural and synthetic origin has been reported. Except for dep-sipeptide (FK228), natural HDACs (TSA, depudecin, trapoxins, and apicidins), as well as sodium butyrate, phenylbutyrate, and SAHA, while effective in vivo, are marked by instability and low retention. Subsequently, synthetic analogs isolated from screening libraries (oxamflatin, scriptaid) were discovered as having a com-mon structure with TSA and SAHA: a hydroxamic acid zinc-binding group linked via a spacer (5 or 6 CH2) to a hydrophobic group. Second-generation HDAC inhibitors such as LAQ824 and PDX101 are currently under clinical trials. Synthetic benzamide-containing HDAC inhibitors (e.g., MS-275 and CI-994) are also being evaluated in the clinics.

3.1 Short-Chain Fatty Acid

Butyrate inhibits HDAC activity at micromolar concentrations. It is generated by the fermentation of dietary fibers in the lumen of the large intestine. The aromatic fatty acids phenylbutyrate and phenylacetate, which has been used to treat patients with disorders of urea metabolism, also inhibits HDAC activity and possess anti-cancer activity (Appelskog et al., 2004; Boivin et al., 2002; Pili et al., 2001; Sowa and Sakai, 2000; Warrell et al., 1998; Zhang et al., 2004a). Valproic acid (VPA), an anticonvulsant, has been shown to have HDAC inhibitory activity at relatively high concentrations (Catalano et al., 2005; De Felice et al., 2005; Facchetti et al., 2004; Sakajiri et al., 2005; Shen et al., 2005; Takai et al., 2004a). VPA also inhibits angiogenesis, but displays no toxicity in endothelial cells (Michaelis et al., 2005). VPA increases extracellular signal-regulated kinase 1/2 (ERK 1/2) phosphorylation in human umbilical vein endothelial cells. Moreover, the combination of VPA with PD98059, a pharmacological inhibitor of the mitogen-activated protein kinase kinase 1/2, synergistically inhibited angiogenesis in vitro and in vivo.

3.2 Hydroxamic Acids

Essential characteristics of hydroxamic acid-based inhibitors are the polar hydroximic group – a six-carbon hydrophobic methylene spacer, a second polar site, and a ter-minal hydrophobic group. TSA from *Streptomyces hygroscopicus* was initially identified as an antifungal agent (Tsuji et al., 1976). TSA and SAHA act as noncom-petitive inhibitor of HDAC by mimicking the lysine substrate as well as chelating a zinc atom crucial for enzymatic activity (Yoshida et al., 1990b). TSA and SAHA inhibit both class I and II HDACs. Simple analogs of cyclic tetrapeptides that contain suberic acid linkers and hydroxamate, instead of epoxyketone or ketone functional group, inhibit HDAC activity (Hoffmann et al., 2000). The structurally related hybrid polar compounds (HPCs) were shown to induce differentiation in a wide variety of transformed cells (Marks et al., 1996). The first representative was hexamethylene bisacetamide (HMBA) which induced differentiation of transformed cells in millimolar range (Marks and Rifkind, 1988). HMBA regulates genes that control G1-to-S phase transition, leading to G1 arrest and inhibition of DNA synthesis. Among the inducer-mediated changes, suppression of cyclin-dependent kinase cdk4, which may be required for phosphorylation of the retinoblastoma protein pRB and perhaps p107, is critical in the pathway of terminal differentiation. HMBA induces an increase in the level of p21$^{WAF1/CIP1}$ which inhibits cyclin-dependent kinase activity and, in turn, may cause cells to arrest in G1. p107 complexes with transcription factor E2F, which may alter E2F-dependent gene transcription. HMBA has also been shown to induce differentiation of neoplastic cells in patients. Furthermore, a second generation of HPCs have been synthesized which are up to 1,000-fold more potent than HMBA. Second-generation HPCs such as oxamflatin, SAHA, suberic bishydroxamic acid (SBHA), and m-carboxycinnamic acid bishydroxamide (CBHA) inhibited HDAC activity and induced cancer cell differentiation and apoptosis (Richon et al., 1998; Shankar et al., 2005b). Polyaminohydroxamic acids (PAHAs) represent an important new chemical class of HDAC inhibitors and appear to be more specific than SAHA, TSA, and MS-275, because they are selectively directed to chromatin and associated histones by the positively charged polyamine side chain. Several other analogs of hydroxamic acids are being developed (Hoffmann et al., 2000; Qiu et al., 2000).

These HDAC inhibitors inhibits proliferation, causes cell cycle arrest, and induces differentiation and/or apoptosis in numerous models of lymphoma, leukemia, multiple myeloma, and solid tumors (Fandy et al., 2005; Fronsdal and Saatcioglu, 2005; Inoue et al., 2002; Monneret, 2005; Shankar et al., 2005b; Taghiyev et al., 2005; Toth et al., 2004; Tsatsoulis, 2002; Vanhaecke et al., 2004a; Vanhaecke et al., 2004b; Wang et al., 2002; Yamashita et al., 2003). TSA is also effective in xenograft models (Canes et al., 2005; Touma et al., 2005). TSA attenuates the development of allergic airway inflammation by decreasing expression of the Th2 cytokines, IL-4 and IL-5, and IgE, which results from reduced T-cell infiltration, suggesting that HDAC inhibition may attenuate the development of asthma by a T-cell suppressive effect (Choi et al., 2005). Other analogs of TSA such as oxamflatin, scriptaid, and amide derivatives have been reported to have anticancer activity (Jung et al., 1999; Kim et al., 1999c; Monneret, 2005; Su et al., 2000). Scriptaid induces reticulocytosis and human gamma-globin synthesis (Johnson et al., 2005), suggesting its potential as a treatment option for sickle cell disease. The suppressed RARβ expression in head and neck carcinoma (HNSCC) can be reactivated by TSA (Wang et al., 2005). Additionally, TSA alone or in combination with 5-aza-2′-deoxycytidine (5-AzaC) increases lysine-9 (Lys-9) acetylation and Lys-4 methylation of the first exon at the *RAR*β gene, while decreasing the methylation of Lys-9. Similarly, treatment of gastric carcinoma with 5-aza-C, and/or TSA resulted in reexpressed caspase-1 mRNA (Jee et al., 2005). DNA methylation-mediated repression of eNOS promoter activity was partially reversed by TSA treatment, and combined treatment of TSA and 5-AzaC synergistically induced eNOS expression in nonendothelial cells (Gan et al., 2005). Furthermore, TSA downregulates DNMT3B mRNA and protein expression in human endometrial cancer cells (Xiong et al., 2005). This decrease in DNMT3B mRNA results in a significant reduction in de novo methylation activities, suggesting that TSA may not only modify histone acetylation, but also potentially alter DNA methylation. The above findings suggest that epigenetic events such as DNA methylation and histone deacetylation play important roles in the regulation of cancer-related genes.

3.3 Benzamides

Several benzamides have been found to inhibit HDAC activity in the low micromolar range. A 2′-hydroxy or amino function seems to be essential for the optimum activity (Suzuki et al., 1999). A newly synthesized benzamide derivative with HDAC inhibitory activity, MS-275 is believed to enter the catalytic site and bind the active zinc, inhibits HDAC at micromolar concentrations. MS-275 is the first HDAC inhibitor discovered with oral anticancer activity in several animal models. Pretreatment of human leukemic cells with MS-275 significantly enhances the abrogative capacity of an established nucleoside analogue, fludarabine (Maggio et al., 2004). The study indicates that apart from promoting acetylation of histones and regulation of genes involved in differentiation and apoptosis, MS-275 also induces multiple perturbations in signal transduction, survival and cell cycle regulatory pathways that increase the fludarabine-mediated cell death. CI-994 (N-acetyl dinaline), originally synthesized as an anticonvulsant, does not seem to directly inhibit HDAC, but causes accumulation of acetylated histones by an unknown mechanisms. MS-275, acetyldinaline, and CI-994 are in clinical trials for the treatment of several cancers (Monneret, 2005; Ryan et al., 2005).

3.4 Cyclic Tetrapeptides Containing AOE Moiety

Hydrophobic cyclotetrapeptides contain common amino acid (S)-2-amino-9,10-epoxy-8-xodecanoic acid (L-Aoe) and have been reported to inhibit HDACs (Brosch et al., 1995; Kijima et al., 1993). The epoxyketone was first thought to be essential for activity, as reduction or nucleophoilic attack resulted in inactivation of compounds (Brosch et al., 1995; Kijima et al., 1993). Trapoxin A, a microbially derived cyclotetrapeptide, is an irreversible inhibitor in the low nanomolar range (Kijima et al., 1993). Trapoxin A irreversibly inhibits histone deacetylation in vivo and causes mammalian cells to arrest in the cell cycle (Taunton et al., 1996). On the other hand, related HC toxin (host-selective toxin of *Cochliobolus carbonum*) inhibits maize enzyme activity reversibly (Brosch et al., 1995). K-trap (an analogous of trapoxin A) inhibited HDAC1 activity. A number of derivatives, such as 9-acyloxyapicidins and 9-hydroxy, have been prepared and are under investigation. Trapoxin analogs that combine cyclotetrapeptide and hydroxamic acid moieties have been prepared. The inhibitors of quinolone analogs and the hydroxamic acid analogs of apicidin yielded promising results (Meinke et al., 2000; Meinke and Liberator, 2001). Depudecin, a natural epoxide derivative isolated from the fungus *Alternaria brassicicola*, induces hyperacetylation of histones and morphological reversion in v-ras-transformed NIH 3T3 cells (Kwon et al., 1998).

3.5   *Cyclic Peptides that do not Contain an AOE Moiety*

Cyclic peptides such as depsipeptide (FR901228/FK228) isolated from *Chromobacterium violaceum* inhibits HDAC activity at nanomolar concentrations. Depsipeptide induces differentiation, growth arrest and apoptosis, and inhibits metastasis and angiogenesis (Aron et al., 2003; Doi et al., 2004; Khan et al., 2004; Klisovic et al., 2003a, b, 2005; Kwon et al., 2002a; Mie Lee et al., 2003; Sasakawa et al., 2002, 2003; Sato et al., 2004; Sawa et al., 2004; Vanoosten et al., 2005). Depsipeptide is also very promising antitumor agent against osteosarcoma, inducing apoptosis by the activation of the Fas/FasL system (Imai et al., 2003). A novel fungal metabolite, apicidin (cyclo(*N-O*-methyl-L-tryptophanyl-L-isoleucinyl-D-pipecolinyl-L-2-amino-8-oxodecanoyl)), exhibits potent, broad spectrum antiprotozoal activity in vitro against apicomplexan parasites (Darkin-Rattray et al., 1996). Apicidin's antiparasitic activity appears to be due to low nanomolar inhibition of HDAC, which induces hyperacetylation of histones in treated parasites. Since apicidin and apicidin A possess only a ketone functional group and are active in the low nanomolar concentrations, it appears that the presence of the epoxy group is not essential for activity. Apicidin induces differentiation, cell cycle arrest and apoptosis, and inhibits metastasis and angiogenesis in several cancer models (Cheong et al., 2003; Han et al., 2000, 2001; Hong et al., 2003; Kim et al., 2001a, 2004b, c; Kouraklis and Theocharis, 2002; Kwon et al., 2002b). It promotes histone acetylation and gene transcription. Its activity is enhanced by DNA methyltransferase inhibitors in AML1/ETO-positive leukemic cells (Khan et al., 2004). Preclinical studies with depsipeptide in chronic lymphocytic leukemia (CLL) and acute myeloid leukemia (AML) have demonstrated that it effectively induces apoptosis at concentrations at which HDAC inhibition occurs. A dose-dependent increase in H3 and H4 histone acetylation was noted in depsipeptide-treated AML1/ETO-positive Kasumi-1 cells and blasts from a patient with t(8;21) AML (Klisovic et al., 2003b). A phase I and pharmacodynamic study of depsipeptide in CLL and AML have yielded promising results (Byrd et al., 2005).

Opening of the disulfide bridge leads to a thiol that may be able to enter the active site and complex the zinc ion. In this regard, garlic constituents and their metabolites such as diallylsulfide and allylmercaptan inhibited HDAC activity. Diallyl disulfide caused increased acetylation of H3 and H4 histones in DS19 mouse erythroleukemic cells and K562 human leukemic cells (Lea et al., 1999), suggesting that differentiation in erythroleukemic cells by diallyl disulfide and allyl mercaptan may be mediated through induction of histone acetylation. Acetylation was also induced in rat hepatoma and human breast cancer cells by diallyl disulfide or its metabolite, allyl mercaptan. Diallyl disulfide increased histone acetylation and p21[WAF1/CIP1] expression in human colon tumor cell lines (Druesne et al., 2004).

3.6 Epoxides

The naturally occurring epoxide depudecin (a microbial metabolite containing two epoxide groups) irreversibly binds to HDAC and inhibits its activity at micromolar concentration. Depudecin inhibited embryonic angiogenesis, involving the chorio-allantoic membrane of growing chick embryo (Oikawa et al., 1995). It also affected the growth of vascular endothelial cells, a key event in the process of angiogenesis in vivo. Depudecin reverts the rounded phenotype of NIH 3T3 fibroblasts transformed with v-ras and v-src oncogenes to the flattened phenotype of the nontransformed parental cells (Kwon et al., 1998). These data suggest that depudecin could be promising as an antiangiogenic agent and that its antiangiogenic action involves an inhibitory effect on vascular endothelial cell growth.

3.7 Psammaplins

Psammaplins, isolated from a marine sponge *Pseudoceratina purpurea*, inhibited HDAC and DNA methyltransferase activities (Pina et al., 2003). Psammaplin A (PsA) contains an α-oximatoamide functional group, which inhibits the HDAC activity at the catalytic site. The disulfide group is also an essential feature for HDAC inhibition. PsA showed a potent cytotoxicity against several cancer and endothelial cells (Jiang et al., 1995, 2004; Kim et al., 1999a, b; Nicolaou et al., 2001; Park et al., 2003; Pham et al., 2000; Shim et al., 2004). PsA-induced cyto-toxicity may correlate with its inhibition on DNA replication (Jiang et al., 2004). Furthermore, PsA was found to inhibit mammalian aminopeptidase N (APN) that plays a key role in tumor cell invasion and angiogenesis (Shim et al., 2004). Interestingly, the antiproliferative effect of PsA was dependent on the cellular amount of APN expression. PsA suppressed the invasion and tube formation of endothelial cells stimulated by basic fibroblast growth factor. Several synthetic analogs of PsA are currently being developed as antiangiogenic and anticancer agents.

4   Mechanism of Actions of HDAC Inhibitors

HDAC inhibitors regulate several biological events including cell cycle, differentiation, and apoptosis in vitro and in vivo (Donadelli et al., 2003; Fandy et al., 2005; Fang, 2005; Fenic et al., 2004; Fronsdal and Saatcioglu, 2005; Henderson and Brancolini, 2003; Hu and Colburn, 2005; Imai et al., 2003; Mai et al., 2005; Marks et al., 2001a; Marks and Jiang, 2005; Nome et al., 2005; Sasakawa et al., 2003; Strait et al., 2005; Takimoto et al., 2005; Yoshida et al., 1990a, 2003). The mechanisms by which these inhibitors induce cell cycle arrest, differentiation, and apoptosis appear to involve multiple genes. In addition to inducing growth arrest and apoptosis, they also inhibit metastasis and angiogenesis (Deroanne et al., 2002; Kim et al., 2001b, 2004c; Sasakawa et al., 2003; Sawa et al., 2002; Williams, 2001; Zgouras et al., 2004). These biological processes are described in this section.

Inhibition of ErbB signaling pathway has been an attractive target for cancer therapy. Several studies have shown that HDAC inhibitors decreased expression of ErbB1 and ErbB2 in DU145 and ErbB2 in SKBr3 cancer cell lines (Chinnaiyan et al., 2005b). HDAC inhibitors also inhibited caveolin-1 and hypoxia-inducible factor 1α (HIF-α), and upregulated gelsolin, p19 (INK4D) and Nur77 expressions in DU145 cells (Chinnaiyan et al., 2005b). Synergistic effects of HDAC inhibitor and ErbB blockade have been shown on cell proliferative, apoptosis, and signaling pathways in cancer cells. Thus, anti-ErbB agents and HDAC inhibitors may offer a promising strategy of dual-targeted therapy. The beneficial effects of these agents may not derive solely from modulation of ErbB expression, but may result from effects on other oncogenic processes including angiogenesis, invasion, and cell cycle kinetics.

The ability of HDAC inhibitors to deactivate Akt through the reorganization of PP1 complexes not only provides a unique mode of Akt regulation, but also represent first example of modulating specific PP1-protein interactions by small-molecule agents. HDAC inhibitors have been reported to lower the apoptotic threshold of several molecularly targeted agents in cancer therapy. This therapeutic strategy is illustrated by the synergistic combination of HDAC inhibitors with other therapeutic agents, Hsp-90 antagonist 17-AGG (George et al., 2005; Rahmani et al., 2005), including the Bcr-Abl kinase inhibitor imatinib (Kim et al., 2004b; Nimmanapalli et al., 2003), the purine analog flutarabine (Maggio et al., 2004), the HER2 antibody trastuzumab (Fuino et al., 2003), the receptor tyrosine kinase FLT-3 inhibitor PKC412 (Bali et al., 2004), the proteosome inhibitor Bortezomib (Yu et al., 2003), and tumor necrosis factor-related apoptosis-inducing ligand (TRAIL) (Shankar et al., 2005b; Singh et al., 2005). These chemosensitization effects may be mediated through both histone acetylation-dependent and acetylation-independent effects of HDAC inhibitors, of which the underlying mechanism warrants investigation.

MS-275 upregulates TGFβ signaling pathway via transcriptional activation of the TGFβ type II receptors (TβRII) (Lee et al., 2001), as a result of PCF recruitment to the NF-Y complex on the type II receptor promoter and selective hyperacetylation of histones associated with the TβRII promoter (Park et al., 2002). Thus, MS-275induces TβRII promoter activity by the recruitment of the PCAF protein to the NF-Y complex, interacting with the inverted CCAAT box in the TβRII promoter. TβRII is often inactivated by mutation or transcriptionally repressed in many cancers, and is therefore a potential candidate for reactivation by HDAC inhibitor treatment.

HDAC inhibitor may also enhance tumor-cell immunogenicity through transcriptional activation of MHC class I and II genes, costimulatory molecules (CD40, CD80, and CD86), intercellular adhesion molecule ICAM1, and type I and II interferons (Johnstone, 2002). These proteins play important roles in host defense mechanisms and cell signaling.

Nonepigenic mechanisms of HDAC inhibitors have recently been described. A number of tumor-associated proteins that mediate cell cycle, growth and/or apoptosis, including Ku70 (Cohen et al., 2004a, b; Subramanian et al., 2005), FOXO1 (Yang et al., 2005), p300 (Bouras et al., 2005), androgen receptor (Fu et al., 2003; Gaughan et al., 2002, 2005), Smad7 (Simonsson et al., 2005), Stat3 (O'Shea et al., 2005; Yuan et al., 2005), p53 (Juan et al., 2000; Langley et al., 2002; Luo et al., 2001; Vaziri et al., 2001), Hsp90 (Kovacs et al., 2005), NF-κB/RelA (Greene and Chen, 2004; Quivy and Van Lint, 2004; Yeung et al., 2004), and SRY (Thevenet et al., 2004) have been identified as substrates for various HDACs isoforms. Targeting the acetylation status of these signal mediators might underlie the antiproliferative activities of HDAC inhibitors in cancer cells. Furthermore, various HDACs have been shown to form complexes with cellular proteins including 14-3-3 proteins, α-tubulin, ubiquitin, and PP1(Brush et al., 2004; Canettieri et al., 2003; Grozinger and Schreiber, 2000; Hook et al., 2002; Kawaguchi et al., 2003; Yang and Gregoire, 2005). These protein–protein complexes may be responsible for altering the biological functions. HDACs 1 and 6 formed complexes with PP1 (Brush et al., 2004; Canettieri et al., 2003), of which the combined deacetylase/phosphatase activities underlie the ability of HDAC1 to modulate transcriptional activity of the cAMP-responsive element-binding protein (CREB) and that of HDAC6 to regulate microtubule dynamics. These studies provide new insight into the mechanism by which HDAC inhibitors elicited coordinate changes in cellular protein phosphorylation and acetylation and suggested that changes in these protein modifications at multiple subcellular sites may contribute to HDAC inhibitor's effects to suppress cell growth and transformation.

4.1   Cell Cycle Regulation by HDAC Inhibitors

During the cell-division cycle, chromosomal DNA must initially be precisely duplicated and then correctly segregated to daughter cells. Cell cycle control of transcription seems to be a universal feature of proliferating cells, although relatively little is known about its biological significance and conservation between organisms.

Given the key role of cell cycle integrity in tumor suppression and cancer therapy, a lot of attention has focused on the ability of HDAC inhibitors to alter the levels of cell cycle regulatory proteins. HDAC inhibitors induce growth arrest at both the G1 and G2/M phases of cell cycle and induce differentiation and/or apoptosis of various types of tumor cell lines (Acharya and Figg, 2004; Donadelli et al., 2003; Duan et al., 2005; Fandy et al., 2005; Fang, 2005; Lavelle et al., 2001; Marks and Jiang, 2005; Myzak et al., 2004; Nome et al., 2005; Rocchi et al., 2005; Rosato et al., 2003b; Sakajiri et al., 2005; Sato et al., 2004; Shankar et al., 2005b; Strait et al., 2005). HDAC inhibitors induced both $p21^{WAF1/CIP1}$ and $p27^{KIP1}$ at protein levels, and caused hypophosphorylation of Rb (Fandy et al., 2005; Mitsiades et al., 2005; Nome et al., 2005; Shankar et al., 2005b). Other cell cycle inhibitors that participate in the proliferative arrest elicited by HDAC inhibitors are $p15^{INK4b}$, $p18^{INK4c}$, and $p19^{INK4d}$ (Hitomi et al., 2003; Yokota et al., 2004). Moreover, positive regulators of proliferation, such as cyclins D1 and D2, cMyc, or c-Src, are downregulated by HDAC inhibitors (Dehm and Bonham, 2004; Heruth et al., 1993; Lallemand et al., 1996; Souleimani and Asselin, 1993; Takai et al., 2004b). p53 is activated both by inhibitors of HDACs class I/II, as well as by inhibitors of the Sir2 family (Juan et al., 2000; Luo et al., 2000, 2001; Vaziri et al., 2001). Transcription factor Sp1 regulates $p21^{WAF1/CIP1}$ expression in a p53-independent fashion (Han et al., 2001; Sasakawa et al., 2002; Savickiene et al., 2004; Varshochi et al., 2005). Furthermore, $p21^{WAF1/CIP1}$ expression is also transcriptionally regulated by p53 (Parker et al., 1995).

4.2 Apoptotic Induction by HDAC Inhibitors

HDAC inhibitors induce apoptosis in several types of cancers including breast, prostate, lung and thyroid carcinoma, leukemia, and multiple myeloma (Amin et al., 2001; Chen et al., 2005; de Ruijter et al., 2003; Donadelli et al., 2003; Fandy et al., 2005; Fandy and Srivastava, 2006; Kim et al., 2003; Mitsiades et al., 2005; Mori et al., 2004; Papeleu et al., 2005; Rosato et al., 2003a; Sakajiri et al., 2005; Singh et al., 2005; Vigushin and Coombes, 2002; Zhang et al., 2004d). In addition to TRAIL-R1/DR4 and TRAIL-R2/DR5 receptors, the regulation of Bcl-2 family members is also important for inducing sensitivity by HDAC inhibitors. We and others have shown that HDAC inhibitors selectively induce proapoptotic members such as Bax, Bak, Noxa, Bim and Puma and inhibit antiapoptotic Mcl-1, Bcl-X_L and Bcl-2 expression (Fandy et al., 2005; Fandy and Srivastava, 2006; Khan et al., 2004; Mitsiades et al., 2003; Neuzil et al., 2004; Shankar et al., 2005b; Singh et al., 2005; Zhang et al., 2003, 2004b). Bcl-2 family members mainly exert their apoptotic effects by acting at the level of mitochondria and play a crucial role in cancer development (Green and Reed, 1998). HDAC inhibitors cleave poly(ADP-ribose) polymerase (PARP) and caspase-8, caspase-9, caspase-3, caspase-7, and caspase-2. Transfection of Bcl-2 cDNA partially suppressed SAHA-induced cell death. HDAC inhibitors can also induce TRAIL, suggesting the activation of death receptor pathway without the requirement of exogenous TRAIL. Thus, HDAC inhibitors can induce apoptosis by linking both death receptor and mitochondrial pathways of apoptosis.

Dysregulation in apoptosis has been associated with the development of cancer (Johnstone et al., 2002). Recent studies have shown the involvement of mitochondria in many apoptotic signaling pathways (Kandasamy et al., 2003; Wei et al., 2000). Members of the Bcl-2 family of proteins that regulate apoptotic signaling through mitochondria are key regulators of apoptosis in mammalian development, and their deregulation is associated with disease, particularly cancer (Grimm et al., 1996; Gross et al., 1999). There are three classes of Bcl-2 family members: apoptosis promoters (e.g., Bax and Bak); apoptosis inhibitors (e.g., Bcl-2, Bcl-X_L, and adenoviral E1B 19K); and the BH3-only Bcl-2 family members (e.g., Bid, Puma, Noxa, Bad, and Nbk/Bik) (Gross, 2001). BH-3 only proteins may function as death sensors that mediate activation of the mitochondrial apoptosis pathway in response to oncogenic stress signals or DNA damage. Noxa and PUMA are transcriptionally induced by p53 and mediate apoptosis induced by p53. These proapoptotic activities of certain BH3-only proteins essentially depend on the presence of Bax and Bak. Inactivation of both Bax and Bak was required for tumor growth and was selected for in vivo tumorigenesis (Degenhardt et al., 2002a, b). $Bax^{-/-}$ and $Bak^{-/-}$ double knockout mouse embryo fibroblasts (DKO MEFs) were resistant to death signaling pathway, indicating that they are the required downstream components of mitochondrial signaling pathways (Kandasamy et al., 2003). Bim has been implicated in modulating lymphocyte homeostasis in immune cells. $Bim^{-/-}$ mice succumb to autoimmune kidney disease, accumulation of lymphoid and myeloid cells, and perturbed T-cell development (Bouillet et al., 2002; Bouillet and Strasser, 2002). Therefore, the regulation of Bcl-2 family members by HDAC inhibitor may play important roles on apoptosis by inducing a death activity or by antagonizing a survival activity. Furthermore, HDAC inhibitors can disrupt cellular redox state (e.g., ROS), and damage mitochondria in cells undergoing apoptosis.

Direct inhibitor of apoptosis protein (IAP)-binding protein with low pI/second mitochondrial activator of caspases, HtrA2/Omi and GstPT/eRF3 are mammalian proteins that bind via N-terminal IAP-binding motifs (IBMs) to the baculoviral IAP repeat (BIR) domains of IAPs. These interactions can prevent IAPs from inhibiting caspases, or displace active caspases, thereby promoting cell death (Deveraux and Reed, 1999). IAPs (cIAP-1, cIAP-2, NIAP, Livin/ML-IAP, survivin, and XIAP) protect cells against apoptosis by acting as caspase inhibitors (Deveraux and Reed, 1999). IAPs bind to and directly inhibit caspase-3, caspase-7, and caspase-9 (Deveraux and Reed, 1999; Deveraux et al., 1999). IAP proteins are regulated by interactions with the mitochondrial proteins (e.g., Smac/DIABLO), which may be released into the cytosol upon apoptotic stimulation and through IAP sequestration results in elevated caspase activity (Du et al., 2000; Verhagen et al., 2000). Some IAP proteins are also regulated by proteolysis via the ubiquitin-proteasome pathway and caspase-dependent cleavage of XIAP in cells undergoing apoptosis. The inhibition of XIAP, cIAP1, and cIAP2 expressions by HDAC inhibitors may contribute in sensitization of cells to TRAIL. In this context, we have shown that

TRAIL inhibits the expression of IAPs in breast and prostate cancer cells (Shankar et al., 2005b; Singh et al., 2005). The combination of HDAC inhibitors and TRAIL may further inhibit the expression of some of the IAPs and contribute to the synergistic induction of apoptosis by these agents.

HDAC inhibitors activate the p53 molecule through acetylation of 320 and 373 lysine residues, upregulate PIG3 and NOXA, and induce apoptosis in cancer cells expressing wild and pseudo-wild-type *p53* genes (Terui et al., 2003). SAHA induced polyploidy in human colon cancer cell line HCT116 and human breast cancer cell lines, MCF-7, MDA-MB-231, and MBA-MD-468, but not in normal human embryonic fibroblast SW-38 and normal MEFs (Xu et al., 2005a). The polyploid cells lost the capacity for proliferation and committed to senescence. The induction of polyploidy was enhanced in HCT116 p21$^{WAF1-/-}$ or HCT116 p$^{53-/-}$ cells than in wild-type HCT116. The development of senescence of SAHA-induced polyploidy cells was similar in all colon cell lines (Xu et al., 2005b). The present findings indicate that the HDAC inhibitor could exert antitumor effects by inducing polyploidy, and this effect is more marked in transformed cells with nonfunctioning *p21*$^{WAF1/CIP1}$ or *p53* genes.

In chronic myelocytic leukemia (CML) the activity of the Bcr-Abl tyrosine kinase is known to activate a number of molecular mechanisms, which inhibit apoptosis (Nimmanapalli et al., 2003; Xu et al., 2005b). SAHA markedly decreases protein expression levels of Bcr-Abl, c-Myc, and HDAC3 in CML, suggesting that SAHA exerts its biological activity by inhibiting survival pathway (Xu et al., 2005b). Differential expression of HDAC has been reported in various cancers. To explore the mechanisms of disease-specific HDAC activity in AML, the expression of HDAC in primary AML blasts and in four control cell types (namely CD34+ progenitors from umbilical cord, quiescent or cycling (postculture) cells, cycling CD34+ progenitors from GCSF-stimulated adult donors, and peripheral blood mononuclear cells) was characterized. Only Sirt1 was consistently overexpressed in AML samples compared with all controls, while HDAC6 was overexpressed relative to adult, but not neonatal cells (Bradbury et al., 2005). HDAC5 and SIRT4 were consistently underexpressed. HDAC inhibitors (valproate, butyrate, TSA, and SAHA) caused hyperacetylation of histones in AML blasts and cell lines (Bradbury et al., 2005). Such treatment also modulated the pattern of HDAC expression, with strong induction of HDAC11 in all myeloid cells tested, and lesser, more selective, induction of HDAC9 and SIRT4. The distinct pattern of HDAC expression in AML and its response to HDAC inhibitors is of relevance to the development of HDAC inhibitor-based therapeutic strategies and may contribute to observed patterns of clinical response and development of drug resistance.

4.3   Antiangiogenic Properties of HDAC Inhibitors

Tumor growth requires the development of new vessels that sprout from preexisting normal vessels in a process known as "angiogenesis" (Folkman, 2002). These new vessels arise from local capillaries, arteries, and veins in response to the release of soluble growth factors from the tumor mass, enabling these tumors to grow beyond the diffusion-limited size of approximately 2 mm diameter. Tumor growth and metastasis depend upon the development of a neovasculature in and around the tumor (Folkman, 2002, 2003a, b, d; Folkman and Kalluri, 2004; Liotta et al., 1991). Angiogenesis is regulated by the balance between stimulatory (e.g., bFGF, IL-8, MMP-2, MMP-9, TGFβ1, and vascular endothelial growth factor [VEGF]) and inhibitory (e.g., angiostatin, IL-10, and interferon) factors released by the tumor and its environment (Folkman, 2003b, c). For example, overexpression of bFGF (Allen and Maher, 1993; Ravery et al., 1992) and VEGF (Brown et al., 1993a, b; O'Brien et al., 1995) has been found in the tissue, serum, and urine of patients with bladder cancer and has been associated with cancer progression, suggesting a direct involvement of these proteins in angiogenesis.

HDAC inhibitors also modulate angiogenesis in a potentially therapeutic manner. HDAC1 downregulates expression of p53 and the von Hippel–Lindau tumor suppressor gene and stimulates angiogenesis of human endothelial cells. HDAC inhibitors prevent endothelial cell proliferation and angiogenesis by downregulating angiogenesis-related gene expression (Bapna et al., 2004; Caponigro et al., 2005; Chinnaiyan et al., 2005b; Deroanne et al., 2002; He et al., 2005; Kim et al., 2001b, 2004c; Kwon et al., 2002a; Liu et al., 2003; Michaelis et al., 2004, 2005; Mie Lee et al., 2003; Momparler, 2003; Murakami et al., 2004; Nam and Parang, 2003; Pili et al., 2001; Qian et al., 2004; Rossig et al., 2002; Sasakawa et al., 2003; Sawa et al., 2002; Takimoto et al., 2005; Wang et al., 2003; Wiedmann and Caca, 2005; Williams, 2001; Zgouras et al., 2004). Phenyl butyrate, LBH589, LAQ824, and TSA have antiangiogenic activity both in vitro and in vivo (Pili et al., 2001; Qian et al., 2004, 2006; Williams, 2001). Other HDAC inhibitors such as SAHA, FK228, VPA, and apicidin also have antiangiogenic acitivity (Kim et al., 2001b; Kwon et al., 2002a; Michaelis et al., 2004). Angiogenesis inhibition induced by HDAC inhibitors was associated with modulation of angiogenesis-related genes both in cancer cells (e.g., inhibition of HIF-1α and VEGF) and in endothelial cells (inhibition of Tie-2 and survivin), and inhibition of endothelial cell migration and proliferation (Kim et al., 2001b; Kwon et al., 2002a; Michaelis et al., 2004; Williams, 2001). Furthermore, LBH589 inhibited endothelial tube formation and matrigel invasion (Qian et al., 2006). These data suggest that the effects of HDAC inhibitors on angiogenesis can be further enhanced in the presence of TRAIL.

HDAC inhibitors upregulate p53 and von Hippel–Lindau expression (Kim et al., 2001b). The combination of adenoviral vector carrying wild-type *p53* (*Ad-p53*) gene therapy with sodium butyrate resulted in a complete regression of xenografted human gastric tumor (KATO-III) cells in nude mice (Takimoto et al., 2005). Tumors treated with the combination showed higher numbers of TUNEL-positive cells and lower CD34 staining than those treated with a single modality (Takimoto et al., 2005). This was further supported by the finding that the brain-specific angiogenesis inihibitor-1 (BAI-1), an inhibitor of vascularization, was induced by sodium butyrate treatment in cells transfected with Ad-p53 (Takimoto et al., 2005). These data suggest that HDAC inhibitors can be combined with *p53* gene therapy for the treatment of cancer. The HDAC inhibitors have shown the dual function of targeting both tumor cells and proliferating endothelial cells and to inhibit tumor angiogenesis by gene modulation. Rational clinical testing of these agents either alone or in combination with angiogenesis inhibitors is warranted.

5   Combination of HDAC Inhibitors with Trail/Apo-2L

HDAC inhibitor either alone or in combination with TRAIL can be used in cancer therapy. We and others have shown that several HDAC inhibitors can enhance the apoptosis-inducing potential of TRAIL in TRAIL-sensitive cells and sensitize TRAIL-resistant breast, prostate, and lung cancer cells, and malignant mesothelioma, leukemia, and myeloma cells (Facchetti et al., 2004; Fandy et al., 2005; Goldsmith and Hogarty, 2005; Inoue et al., 2004; Nebbioso et al., 2005; Rosato et al., 2003a; Shetty et al., 2005; Singh et al., 2005; Vanoosten et al., 2005). The sensitization of TRAIL-resistant cells appears to be due to downregulation of the antiapoptotic protein Bcl-2, Bcl-X_L, and Mcl-1, and upregulation of proapoptotic genes *Bax*, *Bak*, *TRAIL*, *Fas*, *FasL*, *DR4*, and *DR5*, and activation of caspases. HDAC inhibitors upregulate proapoptotic genes in cancer cells but not in normal cells (Insinga et al., 2005a, b). Sodium butyrate and TSA enhanced TRAIL-mediated apoptosis to a greater extent than depsipeptide, MS-275, and oxamflatin (Vanoosten et al., 2005). Both sodium butyrate and TSA treatment also increased mRNA and surface expression of TRAIL-R2/DR5 that was dependent on the transcription factor Sp1, thus providing a possible mechanism behind the increased sensitivity to TRAIL. These results show that sensitivity to HDAC inhibitors in cancer cells is a property of the fully transformed phenotype and depends on activation of a specific death pathway. Since HDAC inhibitors sensitize TRAIL-resistant cancer cells to undergo apoptosis by TRAIL, they appear to be promising candidates for combination chemotherapy.

Several studies have demonstrated the engagement of mitochondria during activation of death receptor pathway (Debatin and Krammer, 2004; Sartorius et al., 2001; Shankar et al., 2005b; Suliman et al., 2001). Cross talk between the death-receptor (extrinsic) and mitochondrial (intrinsic) pathways requires caspase-8/caspase-10-dependent cleavage of Bid (Fandy et al., 2005; Shankar et al., 2005b; Singh et al., 2005; Suliman et al., 2001). tBid activates Bax and Bak to release cytochrome c and other mitochondrial proteins (Luo et al., 1998; Wei et al., 2000). Since HDAC inhibitors induced cleavage of Bid, the truncated Bid may trigger activation of mitochondria in the absence of ligand TRAIL. We have shown that the pan-caspase inhibitor z-VAD-fmk completely inhibited TRAIL-induced apoptosis in the presence of HDAC inhibitor (Fandy et al., 2005; Shankar et al., 2005b; Singh et al., 2005). The caspase-8 inhibitor z-IETD and DN-FADD completely inhibited the synergistic interaction between HDAC inhibitor and TRAIL. Furthermore, in the presence of HDAC inhibitors, TRAIL induced caspase-3 and caspase-9 activation and caused cleavage of their substrate poly(ADP-ribose) polymerase (PARP). Antiapoptotic proteins Bcl-2 and Bcl-X$_L$ inhibit HDAC inhibitors and/or TRAIL-induced apoptosis by blocking cytochrome c release. The phosphorylation deficient mutant of Bcl-2 and Bcl-X$_L$ also blocked HDAC inhibitors and/or TRAIL-induced apoptosis. In cell-intrinsic pathway of apoptosis, mitochondria amplify the apoptotic signals leading to activation of caspase-9 (Kandasamy et al., 2003). Caspase-9 in turn activates downstream caspases and the cleavage of apoptotic substrates that finally kill cells. The synergistic effects of HDAC inhibitors and TRAIL on apoptosis occur through activation of downstream caspase-3, which can be activated by both extrinsic and intrinsic pathways (Fandy et al., 2005; Shankar et al., 2005b; Singh et al., 2005).

The sensitization of cancer cells to HDAC inhibitors appears to be p53 independent. We have recently shown that chemotherapeutic drugs (Singh et al., 2003) or irradiation (Shankar et al., 2004a, b) can sensitize breast and prostate cancer cells by upregulating death receptors DR4 and/or DR5 in cells harboring wild-type (MCF-7) and mutated (MDA-MB-231 and MDA-MB-468) p53. Recent studies have shown that HDAC inhibitors induce apoptosis in leukemia in a p53-independent manner but not in normal hematopoietic progenitors (Insinga et al., 2005b; Nebbioso et al., 2005). Other transcription factors such as NF-κB and SP1 have been shown to regulate the expression of death receptors (Chen et al., 2003; Keane et al., 1999; Nagane et al., 2000; Ravi et al., 2001).

Treatment of nude mice with HDAC inhibitors resulted in acetylation of histone H3 and H4, and downregulation of hypoxia-inducible factor 1-alpha and VEGF expression in tumor cells. Furthermore, control mice demonstrating increased rate of tumor growth had increased numbers of CD31-positive or von Willebrand Factor (vWF)-positive blood vessels, and increased circulating vascular VEGFR2-positive endothelial cells compared to HDAC inhibitor and/or TRAIL-treated mice. Sequential treatments of athymic nude mice with HDAC inhibitors followed by TRAIL cause a synergistic apoptotic response through activation of caspase-3 and caspase-7, which is accompanied by regression of tumor growth, inhibition of angiogenesis, and enhancement of survival of xenografted nude mice. Together with our previous studies showing that cancer chemotherapeutic drugs and irradiation upregulate DR4 and/or DR5 expression, thereby enhancing TRAIL-induced apoptosis in vivo (Chinnaiyan et al., 2000; Shankar et al., 2004b, 2005a; Singh et al., 2003), these studies demonstrate the antitumor interactions of HDAC inhibitors with the TRAIL death-receptor pathway. Similarly, several recent studies including ours have demonstrated the additive or synergistic effects of HDAC inhibitors and TRAIL on apoptosis in vitro (Facchetti et al., 2004; Fandy et al., 2005; Goldsmith and Hogarty, 2005; Inoue et al., 2004; Nebbioso et al., 2005; Neuzil et al., 2004; Rosato et al., 2003a; Shankar et al., 2005b; Shetty et al., 2005; Singh et al., 2005; Zhang et al., 2003). The ability of HDAC inhibitors to sensitize cancer cells to TRAIL suggests that HDAC inhibitors can reduce the minimal effective dose or side effects of TRAIL. Thus, these data provide the framework for clinical evaluation of HDAC inhibitors and TRAIL for the treatment of human cancer.

6   Combination of HDAC Inhibitors with Irradiation

HDAC inhibitors have been shown to radiosensitize prostate, breast, and glioma cell lines (Camphausen et al., 2004; Kim et al., 2004a; Nome et al., 2005). TSA has been shown to radiosensitize human glioblastoma U373MG and U87MG cell lines in a dose- and time-dependent manner (Kim et al., 2004a). VPA enhanced the radiosensitivity of brain tumor SF539 and U251 cell lines in vitro and U251 xenografts in vivo, which correlated with the induction of histone hyperacetylation (Camphausen et al., 2005). Similarly, MS-275 can enhance radiosensitivity of DU145 prostate carcinoma and U251 glioma cells suggesting that this effect may involve an inhibition of DNA repair (Camphausen et al., 2004). The combination of HDAC inhibitors with irradiation may be useful for the treatment of cancer and merit further investigation. Given the limited efficacy of standard treatments for patients with cancer, these data provide support for clinical trials integrating HDAC inhibitor with radiation therapy.

Caspase-2 and caspase-3 cleave HDAC4 in vitro, and caspase-3 is critical for HDAC4 cleavage in vivo during UV-induced apoptosis (Paroni et al., 2004). After UV irradiation, GFP-HDAC4 translocates into the nucleus coincidentally/ immediately before the retraction response, but clearly before nuclear fragmentation. Together, these data indicate that caspases could specifically modulate gene repression and apoptosis through the proteolytic processing of HDAC4. Among molecular cell cycle-targeted drugs currently in the pipeline for testing in early-phase clinical trials, HDAC inhibitors may have therapeutic potential as radiosensitizers.

7   Combination of HDAC Inhibitors
with Chemotherapeutic Drugs

Chemotherapeutic treatment with combinations of drugs is frontline therapy for many types of cancer. Combining drugs which target different signaling pathways often lessens adverse side effects while increasing the efficacy of treatment and reducing patient morbidity. It has recently been shown that HDAC inhibitors facilitate the cytotoxic effectiveness of the topoisomerase I inhibitor camptothecin in the killing of tumor cells (Bevins and Zimmer, 2005). SAHA has been shown to act as a chemopreventive agent in mammary tumors in the rat (Cohen et al., 1999) and inhibited the growth of established tumors (Butler et al., 2000; Chinnaiyan et al., 2005a; Cohen et al., 1999). SAHA and sodium butyrate interacted synergistically with camptothecin in inducing apoptosis of breast and lung cancer cell lines. Experiments have shown that cells arrested in G2-M by camptothecin were most sensitive to subsequent addition of HDAC inhibitor. In camptothecin-arrested cells, sodium butyrate decreased cyclin B levels, as well as the levels of the antiapoptotic proteins XIAP and survivin. Overall, these findings suggest that reducing the levels of these critical antiapoptotic factors may increase the efficacy of camptothecin in the clinical setting if given in a sequence that does not prevent or inhibit tumor cell progression through the S phase.

MS-275 also synergistically interacted with fludarabine in inducing apoptosis of human lymphoid and myeloid leukemia cells (Maggio et al., 2004). Prior exposure of Jurkat lymphoblastic leukemia cells to MS-275 increased mitochondrial injury, caspase activation, and apoptosis in response to fludarabine, resulting in highly synergistic antileukemic interactions and loss of clonogenic survival. Simultaneous exposure to MS-275 and fludarabine also led to synergistic effects, but these were not as pronounced as observed with sequential treatment. Similar interactions were noted in the case of (a) other human leukemia cell lines (e.g., U937, CCRF-CEM); (b) other HDAC inhibitors (e.g., sodium butyrate); and (c) other nucleoside analogues (e.g., 1-beta-D-arabinofuranosylcytosine, gemcitabine). Potentiation of fludarabine-induced apoptosis by MS-275 was associated with acetylation of histones H3 and H4, downregulation of the antiapoptotic proteins XIAP and Mcl-1, enhanced cytosolic release of proapoptotic mitochondrial proteins (e.g., cytochrome c, Smac/DIABLO, and AIF), and caspase activation. These events were accompanied by the caspase-dependent downregulation of p27^{KIP1}, cyclins A, E, and D1, and cleavage and diminished phosphorylation of retinoblastoma protein. Prior exposure to MS-275 attenuated fludarabine-mediated activation of MEK1/2, extracellular signal-regulated kinase, and Akt, and enhanced c-Jun NH(2)-terminal kinase phosphorylation; furthermore, inducible expression of constitutively active MEK1/2 or Akt significantly diminished MS-275/fludarabine-induced lethality. Combined exposure of cells to MS-275 and fludarabine was associated with a significant increase in generation of ROS; moreover, both the increase in ROS and apoptosis were largely attenuated by coadministration of the free radical scavenger L-N-acetylcysteine. Finally, prior administration of MS-275 markedly potentiated fludarabine-mediated generation of the proapoptotic lipid second messenger ceramide. Taken together, these findings indicate that MS-275 induces multiple perturbations in signal transduction, survival, and cell cycle regulatory pathways that lower the threshold for fludarabine-mediated mitochondrial injury and apoptosis in human leukemia cells.

A synergistic interaction of retinoic acid and CBHA was shown in a mouse model of neuroblastoma. DNA hypomethylating agents have been found to have synergistic effects with HDAC inhibitors. The combination of TSA with azacytidine caused a dramatic potentiation in the activation of silenced genes (Baylin and Bestor, 2002; Baylin et al., 2001; Chen et al., 1997). Depsipeptide and TSA induced apoptosis in human lung cancer cells. HDAC inhibitor-induced apoptosis was greatly enhanced in the presence of the DNA methyltransferase inhibitor, 5-aza-2′-deoxycytidine, suggesting the DNA methylation status plays an important role on the effectiveness of HDAC inhibitors (Zhu et al., 2001). Furthermore, HDAC inhibitors enhanced paclitaxel-induced cell death in ovarian cancer cell lines independent of p53 status (Chobanian et al., 2004). Similarly, commonly used anticancer drugs doxorubicin and decitabine have been reported to have synergistic effects with HDAC inhibitors (Blagosklonny et al., 2000; Gozzini and Santini, 2005).

Thus, the combination of anticancer drugs with other epigenetic therapies provides potentially safer therapeutic options.

8 Chemoprevention by HDAC Inhibitors

In recent years, the use of naturally occurring chemopreventive agents have attracted many investigators because of their nontoxic effects. The preclinical data on selected chemopreventive agents have been very promising. Evidence indicates that a diet high in fresh fruits and vegetables decreases risk of certain cancers because they contain fiber, folate, and vitamins with antioxidant activity (Howe et al., 1992; Janne and Mayer, 2000). Studies have shown that the dietary fiber provides a protective effect against colon cancer (Howe et al., 1992; Trock et al., 1990). It appears that the fermentation of dietary fiber in the lumen of the colon produces the short chain fatty acid n-butyrate, which has anticarcinogenic activity on a variety of cellular functions, including differentiation, motility, invasion, adhesion, proliferation, and apoptosis. There is a positive correlationship between high fecal butyrate levels and decrease tumor incidence and tumor growth (Cassidy et al., 1994; Hylla et al., 1998; McIntyre et al., 1993). Butyrate is a physiological regulator of colonic epithelial cell proliferation, differentiation, and survival; and it induces histone hyperacetylation and inhibits methylation (de Haan et al., 1986; Riggs et al., 1977). Butyrate induces expression of $p^{21/WAF1/CIP1}$ through a process involving histone hyperacetylation and recruitment of Sp3 to the proximal p21 promoter (Sowa et al., 1999), and p21 is required for butyrate-mediated growth arrest in colon carcinoma cells (Archer et al., 1998). Although p21 is a p53 target gene, p21 induction by butyrate and other HDAC inhibitors is p53-independent (Xiao et al., 1997). Thus, HDAC inhibitors can induce p21-associated growth arrest in the absence of wild-type p53 function.

Sulforaphane (SFN), a compound found at high levels in broccoli and broccoli sprouts, is a potent inducer of phase 2 detoxification enzymes and inhibits tumorigenesis in animal models. SFN also has a marked effect on cell cycle checkpoint controls and cell survival and/or apoptosis in various cancer cells. SFN dose-dependently increased the activity of a β-catenin-responsive reporter, without altering β-catenin or HDAC protein levels (Myzak et al., 2004). SFN inhibits HDAC activity in colon and prostate cancer cells (Myzak et al., 2005). The inhibition of HDAC was accompanied by an increase in acetylated histones. SFN caused enhanced interaction of acetylated histone H4 with the promoter region of the $p^{21/}$ $^{WAF1/CIP1}$ gene and the bax gene. SFN induced cell cycle arrest and apoptosis through caspase activation. These findings provide new insight into the mechanisms of SFN action in benign prostate hyperplasia, and they suggest a novel approach to chemoprotection and chemotherapy of prostate cancer through the inhibition of HDAC.

In summary, several reports have described butyrate, diallyl disulfide, and SFN as HDAC inhibitors, and many other dietary agents likely will be discovered to attenuate HDAC activity. Dietary HDAC inhibitors, as weak ligands, regulate the expression of genes involved in cell growth, differentiation, and apoptosis. An important question is the extent to which dietary HDAC inhibitors, and other dietary agents that affect gene expression via chromatin remodeling, modulate the expression of genes so that cells can respond most effectively to external stimuli and toxic insults.

9 Clinical Trials with HDAC Inhibitors

Phase I and II clinical trials indicate that HDAC inhibitors from several different structural classes are very well tolerated and exhibit clinical activity against a variety of human malignancies; however, the molecular basis for their anticancer selectivity remains largely unknown. Furthermore, HDAC inhibitors have also shown preclinical promise when combined with other therapeutic agents, and innovative drug delivery strategies, including liposome encapsulation, may further enhance their clinical development and anticancer potential. An improved understanding of the mechanistic role of specific HDACs in human tumorigenesis, as well as the identification of more specific HDAC inhibitors, will likely accelerate the clinical development and broaden the future scope and utility of HDAC inhibitors for cancer treatment.

Several HDAC inhibitors (SAHA, MS-275, CI-994, and depsipeptide) are currently undergoing clinical trials (Blanchard and Chipoy, 2005; Hess-Stumpp, 2005; Kelly et al., 2005). HDAC inhibitors represent a relatively new group of targeted anticancer compounds, which are showing significant promise as agents with activity against a broad spectrum of neoplasms, at doses that are well tolerated by cancer patients. SAHA is most advanced in development, currently in phase I and II clinical trials for patients with both hematologic and solid tumors (Kelly et al., 2005). Clinical trials on depsipeptide alone have shown low toxicity and evidence of antitumor activity (Sandor et al., 2002). Additionally, the compound has potential for synergism with radiotherapy, chemotherapy, and biologicals. Second-generation HDAC inhibitors, such as LAQ824 and PDX101, are currently under phase I clinical trials. Simultaneously, synthetic benzamide-containing HDAC inhibitors, CI-994 and MS-275, have reached phase I and II clinical trials, respectively.

10 Conclusions

Epigenetic modifications causing gene transcriptional repression have been associated with malignant transformation and are intriguing new targets in the treatment of cancer. In contrast to genetic deletions causing irreversible loss of gene function, epigenetic gene silencing mediated by DNA methylation and histone deacetylation can be reversed via pharmacologic inhibition of DNA methyltransferases and HDACs, respectively. When this occurs, normal patterns of gene expression, cell differentiation, and apoptosis may be restored and disease response obtained. The HDAC has been considered an attractive target molecule for cancer therapy. The inhibition of HDAC activity by a specific inhibitor induces growth arrest, differentiation, and apoptosis of several cancer cells.

Our studies have shown that HDAC inhibitors upregulate proapoptotic members of Bcl-2 family and death receptors (TRAIL-R1/DR4 and TRAIL-R2/DR5), and downregulate antiapoptotic genes of Bcl-2 family; thus it is possible that sensitization of cancer cells to chemotherapy, irradiation, or TRAIL by HDAC inhibitors may occur at various stages of apoptotic pathways. Furthermore, the ability of HDAC inhibitors to inhibit angiogenesis may further affect tumor growth by regulating angiogenesis-related signaling pathways. Preliminary studies in animal models have revealed a relatively high tumor selectivity of HDAC inhibitors, strengthening their promising potential in cancer chemotherapy. Some of these inhibitors are undergoing phase I and phase II clinical trials. Furthermore, the combination of HDAC inhibitors with commonly used anticancer drugs, irradiation, or TRAIL will be useful for cancer therapy. Since the HDAC inhibitors are frequently used in epigenetic studies and are considered to be promising anticancer drugs, these findings will have implications in both laboratory and clinical settings.

Acknowledgment This work was supported by grants from the Susan G. Komen Breast Cancer Foundation, the National Institutes of Health and the Department of Defense.

References

Acharya, M. R. and Figg, W. D. (2004). Histone deacetylase inhibitor enhances the anti-leukemic activity of an established nucleoside analogue. Cancer Biol Ther 3, 719–720.

Allen, L. E. and Maher, P. A. (1993). Expression of basic fibroblast growth factor and its receptor in an invasive bladder carcinoma cell line. J Cell Physiol 155, 368–375.

Amin, H. M., Saeed, S., and Alkan, S. (2001). Histone deacetylase inhibitors induce caspase-dependent apoptosis and downregulation of daxx in acute promyelocytic leukaemia with t(15;17). Br J Haematol 115, 287–297.

Appelskog, I. B., Ammerpohl, O., Svechnikova, I. G., Lui, W. O., Almqvist, P. M., and Ekstrom, T. J. (2004). Histone deacetylase inhibitor 4-phenylbutyrate suppresses GAPDH mRNA expression in glioma cells. Int J Oncol 24, 1419–1425.

Archer, S. Y., Meng, S., Shei, A., and Hodin, R. A. (1998). p21(WAF1) is required for butyrate-mediated growth inhibition of human colon cancer cells. Proc Natl Acad Sci USA 95, 6791–6796.

Aron, J. L., Parthun, M. R., Marcucci, G., Kitada, S., Mone, A. P., Davis, M. E., Shen, T., Murphy, T., Wickham, J., Kanakry, C., et al. (2003). Depsipeptide (FR901228) induces histone acetylation and inhibition of histone deacetylase in chronic lymphocytic leukemia cells concurrent with activation of caspase-8-mediated apoptosis and down-regulation of c-FLIP protein. Blood 102, 652–658.

Bali, P., George, P., Cohen, P., Tao, J., Guo, F., Sigua, C., Vishvanath, A., Scuto, A., Annavarapu, S., Fiskus, W., et al. (2004). Superior activity of the combination of histone deacetylase inhibitor LAQ824 and the FLT-3 kinase inhibitor PKC412 against human acute myelogenous leukemia cells with mutant FLT-3. Clin Cancer Res 10, 4991–4997.

Bandyopadhyay, D., Mishra, A., and Medrano, E. E. (2004). Overexpression of histone deacetylase 1 confers resistance to sodium butyrate-mediated apoptosis in melanoma cells through a p53-mediated pathway. Cancer Res 64, 7706–7710.

Bapna, A., Vickerstaffe, E., Warrington, B. H., Ladlow, M., Fan, T. P., and Ley, S. V. (2004). Polymer-assisted, multi-step solution phase synthesis and biological screening of histone deacetylase inhibitors. Org Biomol Chem 2, 611–620.

Baylin, S. and Bestor, T. H. (2002). Altered methylation patterns in cancer cell genomes: cause or consequence? Cancer Cell 1, 299–305.

Baylin, S. B., Esteller, M., Rountree, M. R., Bachman, K. E., Schuebel, K., and Herman, J. G. (2001). Aberrant patterns of DNA methylation, chromatin formation and gene expression in cancer. Hum Mol Genet 10, 687–692.

Bevins, R. L. and Zimmer, S. G. (2005). It's about time: scheduling alters effect of histone deacetylase inhibitors on camptothecin-treated cells. Cancer Res 65, 6957–6966.

Blagosklonny, M. V., Robey, R., Bates, S., and Fojo, T. (2000). Pretreatment with DNA-damaging agents permits selective killing of checkpoint-deficient cells by microtubule-active drugs. J Clin Invest 105, 533–539.

Blanchard, F. and Chipoy, C. (2005). Histone deacetylase inhibitors: new drugs for the treatment of inflammatory diseases? Drug Discov Today 10, 197–204.

Boivin, A. J., Momparler, L. F., Hurtubise, A., and Momparler, R. L. (2002). Antineoplastic action of 5-aza-2'-deoxycytidine and phenylbutyrate on human lung carcinoma cells. Anticancer Drugs 13, 869–874.

Bordin, M., D'Atri, F., Guillemot, L., and Citi, S. (2004). Histone deacetylase inhibitors up-regulate the expression of tight junction proteins. Mol Cancer Res 2, 692–701.

Bordone, L. and Guarente, L. (2005). Calorie restriction, SIRT1 and metabolism: understanding longevity. Nat Rev Mol Cell Biol 6, 298–305.

Bouillet, P., Purton, J. F., Godfrey, D. I., Zhang, L. C., Coultas, L., Puthalakath, H., Pellegrini, M., Cory, S., Adams, J. M., and Strasser, A. (2002). BH3-only Bcl-2 family member Bim is required for apoptosis of autoreactive thymocytes. Nature 415, 922–926.

Bouillet, P. and Strasser, A. (2002). BH3-only proteins - evolutionarily conserved proapoptotic Bcl-2 family members essential for initiating programmed cell death. J Cell Sci 115, 1567–1574.

Bouras, T., Fu, M., Sauve, A. A., Wang, F., Quong, A. A., Perkins, N. D., Hay, R. T., Gu, W., and Pestell, R. G. (2005). SIRT1 deacetylation and repression of p300 involves lysine residues 1020/1024 within the cell cycle regulatory domain 1. J Biol Chem 280, 10264–10276.

Boyle, G. M., Martyn, A. C., and Parsons, P. G. (2005). Histone deacetylase inhibitors and malignant melanoma. Pigment Cell Res 18, 160–166.

Bradbury, C. A., Khanim, F. L., Hayden, R., Bunce, C. M., White, D. A., Drayson, M. T., Craddock, C., and Turner, B. M. (2005). Histone deacetylases in acute myeloid leukaemia show a distinctive pattern of expression that changes selectively in response to deacetylase inhibitors. Leukemia 19, 1751–1759.

Braunstein, M., Rose, A. B., Holmes, S. G., Allis, C. D., and Broach, J. R. (1993). Transcriptional silencing in yeast is associated with reduced nucleosome acetylation. Genes Dev 7, 592–604.

Brosch, G., Ransom, R., Lechner, T., Walton, J. D., and Loidl, P. (1995). Inhibition of maize histone deacetylases by HC toxin, the host-selective toxin of Cochliobolus carbonum. Plant Cell 7, 1941–1950.

Brown, L. F., Berse, B., Jackman, R. W., Tognazzi, K., Manseau, E. J., Dvorak, H. F., and Senger, D. R. (1993a). Increased expression of vascular permeability factor (vascular endothelial growth factor) and its receptors in kidney and bladder carcinomas. Am J Pathol 143, 1255–1262.

Brown, L. F., Berse, B., Jackman, R. W., Tognazzi, K., Manseau, E. J., Senger, D. R., and Dvorak, H. F. (1993b). Expression of vascular permeability factor (vascular endothelial growth factor) and its receptors in adenocarcinomas of the gastrointestinal tract. Cancer Res 53, 4727–4735.

Brush, M. H., Guardiola, A., Connor, J. H., Yao, T. P., and Shenolikar, S. (2004). Deactylase inhibitors disrupt cellular complexes containing protein phosphatases and deacetylases. J Biol Chem 279, 7685–7691.

Butler, L. M., Agus, D. B., Scher, H. I., Higgins, B., Rose, A., Cordon-Cardo, C., Thaler, H. T., Rifkind, R. A., Marks, P. A., and Richon, V. M. (2000). Suberoylanilide hydroxamic acid, an inhibitor of histone deacetylase, suppresses the growth of prostate cancer cells in vitro and in vivo. Cancer Res 60, 5165–5170.

Byrd, J. C., Marcucci, G., Parthun, M. R., Xiao, J. J., Klisovic, R. B., Moran, M., Lin, T. S., Liu, S., Sklenar, A. R., Davis, M. E., et al. (2005). A phase 1 and pharmacodynamic study of depsipeptide (FK228) in chronic lymphocytic leukemia and acute myeloid leukemia. Blood 105, 959–967.

Camphausen, K., Burgan, W., Cerra, M., Oswald, K. A., Trepel, J. B., Lee, M. J., and Tofilon, P. J. (2004). Enhanced radiation-induced cell killing and prolongation of gammaH2AX foci expression by the histone deacetylase inhibitor MS-275. Cancer Res 64, 316–321.

Camphausen, K., Cerna, D., Scott, T., Sproull, M., Burgan, W. E., Cerra, M. A., Fine, H., and Tofilon, P. J. (2005). Enhancement of in vitro and in vivo tumor cell radiosensitivity by valproic acid. Int J Cancer 114, 380–386.

Canes, D., Chiang, G. J., Billmeyer, B. R., Austin, C. A., Kosakowski, M., Rieger-Christ, K. M., Libertino, J. A., and Summerhayes, I. C. (2005). Histone deacetylase inhibitors upregulate plakoglobin expression in bladder carcinoma cells and display antineoplastic activity in vitro and in vivo. Int J Cancer 113, 841–848.

Canettieri, G., Morantte, I., Guzman, E., Asahara, H., Herzig, S., Anderson, S. D., Yates, J. R., III, and Montminy, M. (2003). Attenuation of a phosphorylation-dependent activator by an HDAC-PP1 complex. Nat Struct Biol 10, 175–181.

Caponigro, F., Basile, M., de Rosa, V., and Normanno, N. (2005). New drugs in cancer therapy, National Tumor Institute, Naples, 17–18 June 2004. Anticancer Drugs 16, 211–221.

Cassidy, A., Bingham, S. A., and Cummings, J. H. (1994). Starch intake and colorectal cancer risk: an international comparison. Br J Cancer 69, 937–942.

Catalano, M. G., Fortunati, N., Pugliese, M., Costantino, L., Poli, R., Bosco, O., and Boccuzzi, G. (2005). Valproic acid induces apoptosis and cell cycle arrest in poorly differentiated thyroid cancer cells. J Clin Endocrinol Metab 90, 1383–1389.

Chen, C. S., Weng, S. C., Tseng, P. H., Lin, H. P., and Chen, C. S. (2005). Histone acetylation-independent effect of histone deacetylase inhibitors on akt through the reshuffling of protein phosphatase 1 complexes. J Biol Chem 280, 38879–38887.

Chen, W. Y., Bailey, E. C., McCune, S. L., Dong, J. Y., and Townes, T. M. (1997). Reactivation of silenced, virally transduced genes by inhibitors of histone deacetylase. Proc Natl Acad Sci USA 94, 5798–5803.

Chen, X., Kandasamy, K., and Srivastava, R. K. (2003). Differential roles of RelA (p65) and c-Rel subunits of nuclear factor kappa B in tumor necrosis factor-related apoptosis-inducing ligand signaling. Cancer Res 63, 1059–1066.

Cheong, J. W., Chong, S. Y., Kim, J. Y., Eom, J. I., Jeung, H. K., Maeng, H. Y., Lee, S. T., and Min, Y. H. (2003). Induction of apoptosis by apicidin, a histone deacetylase inhibitor, via the activation of mitochondria-dependent caspase cascades in human Bcr-Abl-positive leukemia cells. Clin Cancer Res 9, 5018–5027.

Chinnaiyan, A. M., Prasad, U., Shankar, S., Hamstra, D. A., Shanaiah, M., Chenevert, T. L., Ross, B. D., and Rehemtulla, A. (2000). Combined effect of tumor necrosis factor-related apoptosis-inducing ligand and ionizing radiation in breast cancer therapy. Proc Natl Acad Sci USA 97, 1754–1759.

Chinnaiyan, P., Vallabhaneni, G., Armstrong, E., Huang, S. M., and Harari, P. M. (2005a). Modulation of radiation response by histone deacetylase inhibition. Int J Radiat Oncol Biol Phys 62, 223–229.

Chinnaiyan, P., Varambally, S., Tomlins, S. A., Ray, S., Huang, S., Chinnaiyan, A. M., and Harari, P. M. (2005b). Enhancing the antitumor activity of ErbB blockade with histone deacetylase (HDAC) inhibition. Int. J. Cancer 118(4), 1041–1050.

Chobanian, N. H., Greenberg, V. L., Gass, J. M., Desimone, C. P., Van Nagell, J. R., and Zimmer, S. G. (2004). Histone deacetylase inhibitors enhance paclitaxel-induced cell death in ovarian cancer cell lines independent of p53 status. Anticancer Res 24, 539–545.

Choi, J. H., Oh, S. W., Kang, M. S., Kwon, H. J., Oh, G. T., and Kim, D. Y. (2005). Trichostatin A attenuates airway inflammation in mouse asthma model. Clin Exp Allergy 35, 89–96.

Cohen, H. Y., Lavu, S., Bitterman, K. J., Hekking, B., Imahiyerobo, T. A., Miller, C., Frye, R., Ploegh, H., Kessler, B. M., and Sinclair, D. A. (2004a). Acetylation of the C terminus of Ku70 by CBP and PCAF controls Bax-mediated apoptosis. Mol Cell 13, 627–638.

Cohen, H. Y., Miller, C., Bitterman, K. J., Wall, N. R., Hekking, B., Kessler, B., Howitz, K. T., Gorospe, M., de Cabo, R., and Sinclair, D. A. (2004b). Calorie restriction promotes mammalian cell survival by inducing the SIRT1 deacetylase. Science 305, 390–392.

Cohen, L. A., Amin, S., Marks, P. A., Rifkind, R. A., Desai, D., and Richon, V. M. (1999). Chemoprevention of carcinogen-induced mammary tumorigenesis by the hybrid polar cytodifferentiation agent, suberanilohydroxamic acid (SAHA). Anticancer Res 19, 4999–5005.

Darkin-Rattray, S. J., Gurnett, A. M., Myers, R. W., Dulski, P. M., Crumley, T. M., Allocco, J. J., Cannova, C., Meinke, P. T., Colletti, S. L., Bednarek, M. A., et al. (1996). Apicidin: a novel antiprotozoal agent that inhibits parasite histone deacetylase. Proc Natl Acad Sci USA 93, 13143–13147.

De Felice, L., Tatarelli, C., Mascolo, M. G., Gregorj, C., Agostini, F., Fiorini, R., Gelmetti, V., Pascale, S., Padula, F., Petrucci, M. T., et al. (2005). Histone deacetylase inhibitor valproic acid enhances the cytokine-induced expansion of human hematopoietic stem cells. Cancer Res 65, 1505–1513.

de Haan, J. B., Gevers, W., and Parker, M. I. (1986). Effects of sodium butyrate on the synthesis and methylation of DNA in normal cells and their transformed counterparts. Cancer Res 46, 713–716.

de Ruijter, A. J., van Gennip, A. H., Caron, H. N., Kemp, S., and van Kuilenburg, A. B. (2003). Histone deacetylases (HDACs): characterization of the classical HDAC family. Biochem J 370, 737–749.

Debatin, K. M. and Krammer, P. H. (2004). Death receptors in chemotherapy and cancer. Oncogene 23, 2950–2966.

Degenhardt, K., Chen, G., Lindsten, T., and White, E. (2002a). BAX and BAK mediate p53-independent suppression of tumorigenesis. Cancer Cell 2, 193–203.

Degenhardt, K., Sundararajan, R., Lindsten, T., Thompson, C., and White, E. (2002b). Bax and Bak independently promote cytochrome C release from mitochondria. J Biol Chem 277, 14127–14134.

Dehm, S. M. and Bonham, K. (2004). SRC gene expression in human cancer: the role of transcriptional activation. Biochem. Cell Biol 82, 263–274.

Deroanne, C. F., Bonjean, K., Servotte, S., Devy, L., Colige, A., Clausse, N., Blacher, S., Verdin, E., Foidart, J. M., Nusgens, B. V., and Castronovo, V. (2002). Histone deacetylases inhibitors as anti-angiogenic agents altering vascular endothelial growth factor signaling. Oncogene 21, 427–436.

Deveraux, Q. L. and Reed, J. C. (1999). IAP family proteins–suppressors of apoptosis. Genes Dev 13, 239–252.

Deveraux, Q. L., Stennicke, H. R., Salvesen, G. S., and Reed, J. C. (1999). Endogenous inhibitors of caspases. J Clin Immunol 19, 388–398.

Doi, S., Soda, H., Oka, M., Tsurutani, J., Kitazaki, T., Nakamura, Y., Fukuda, M., Yamada, Y., Kamihira, S., and Kohno, S. (2004). The histone deacetylase inhibitor FR901228 induces caspase-dependent apoptosis via the mitochondrial pathway in small cell lung cancer cells. Mol Cancer Ther 3, 1397–1402.

Donadelli, M., Costanzo, C., Faggioli, L., Scupoli, M. T., Moore, P. S., Bassi, C., Scarpa, A., and Palmieri, M. (2003). Trichostatin A, an inhibitor of histone deacetylases, strongly suppresses growth of pancreatic adenocarcinoma cells. Mol Carcinog 38, 59–69.

Druesne, N., Pagniez, A., Mayeur, C., Thomas, M., Cherbuy, C., Duee, P. H., Martel, P., and Chaumontet, C. (2004). Diallyl disulfide (DADS) increases histone acetylation and p21(waf1/cip1) expression in human colon tumor cell lines. Carcinogenesis 25, 1227–1236.

Du, C., Fang, M., Li, Y., Li, L., and Wang, X. (2000). Smac, a mitochondrial protein that promotes cytochrome c-dependent caspase activation by eliminating IAP inhibition. Cell 102, 33–42.

Duan, H., Heckman, C. A., and Boxer, L. M. (2005). Histone deacetylase inhibitors down-regulate bcl-2 expression and induce apoptosis in t(14;18) lymphomas. Mol Cell Biol 25, 1608–1619.

Emiliani, S., Fischle, W., Van Lint, C., Al-Abed, Y., and Verdin, E. (1998). Characterization of a human RPD3 ortholog, HDAC3. Proc Natl Acad Sci USA 95, 2795–2800.

Facchetti, F., Previdi, S., Ballarini, M., Minucci, S., Perego, P., and La Porta, C. A. (2004). Modulation of pro- and anti-apoptotic factors in human melanoma cells exposed to histone deacetylase inhibitors. Apoptosis 9, 573–582.

Fandy, T. E. and Srivastava, R. K. (2006). Trichostatin A sensitizes TRAIL-resistant myeloma cells by downregulation of the antiapoptotic Bcl-2 proteins. Cancer Chemother Pharmacol 58, 471–477.

Fandy, T. E., Shankar, S., Ross, D. D., Sausville, E., and Srivastava, R. K. (2005). Interactive effects of HDAC inhibitors and TRAIL on apoptosis are associated with changes in mitochondrial functions and expressions of cell cycle regulatory genes in multiple myeloma. Neoplasia 7, 646–657.

Fang, J. Y. (2005). Histone deacetylase inhibitors, anticancerous mechanism and therapy for gastrointestinal cancers. J Gastroenterol Hepatol 20, 988–994.

Fenic, I., Sonnack, V., Failing, K., Bergmann, M., and Steger, K. (2004). In vivo effects of histone-deacetylase inhibitor trichostatin-A on murine spermatogenesis. J Androl 25, 811–818.

Fischle, W., Emiliani, S., Hendzel, M. J., Nagase, T., Nomura, N., Voelter, W., and Verdin, E. (1999). A new family of human histone deacetylases related to Saccharomyces cerevisiae HDA1p. J Biol Chem 274, 11713–11720.

Folkman, J. (2002). Role of angiogenesis in tumor growth and metastasis. Semin Oncol 29, 15–18.

Folkman, J. (2003a). Angiogenesis and apoptosis. Semin Cancer Biol 13, 159–167.

Folkman, J. (2003b). Angiogenesis and proteins of the hemostatic system. J Thromb Haemost 1, 1681–1682.

Folkman, J. (2003c). Antiangiogenic activity of a matrix protein. Cancer Biol Ther 2, 53–54.

Folkman, J. (2003d). Fundamental concepts of the angiogenic process. Curr Mol Med 3, 643–651.

Folkman, J. and Kalluri, R. (2004). Cancer without disease. Nature 427, 787.

Ford, J., Jiang, M., and Milner, J. (2005). Cancer-specific functions of SIRT1 enable human epithelial cancer cell growth and survival. Cancer Res 65, 10457–10463.

Fronsdal, K. and Saatcioglu, F. (2005). Histone deacetylase inhibitors differentially mediate apoptosis in prostate cancer cells. Prostate 62, 299–306.

Frye, R. A. (1999). Characterization of five human cDNAs with homology to the yeast SIR2 gene: Sir2-like proteins (sirtuins) metabolize NAD and may have protein ADP-ribosyltransferase activity. Biochem Biophys Res Commun 260, 273–279.

Fu, M., Rao, M., Wang, C., Sakamaki, T., Wang, J., Di Vizio, D., Zhang, X., Albanese, C., Balk, S., Chang, C., et al. (2003). Acetylation of androgen receptor enhances coactivator binding and promotes prostate cancer cell growth. Mol Cell Biol 23, 8563–8575.

Fuino, L., Bali, P., Wittmann, S., Donapaty, S., Guo, F., Yamaguchi, H., Wang, H. G., Atadja, P., and Bhalla, K. (2003). Histone deacetylase inhibitor LAQ824 down-regulates Her-2 and sensitizes human breast cancer cells to trastuzumab, taxotere, gemcitabine, and epothilone B. Mol Cancer Ther 2, 971–984.

Gan, Y., Shen, Y. H., Wang, J., Wang, X., Utama, B., Wang, J., and Wang, X. L. (2005). Role of histone deacetylation in cell-specific expression of endothelial nitric-oxide synthase. J Biol Chem 280, 16467–16475.

Gao, L., Cueto, M. A., Asselbergs, F., and Atadja, P. (2002). Cloning and functional characterization of HDAC11, a novel member of the human histone deacetylase family. J Biol Chem 277, 25748–25755.

Gaughan, L., Logan, I. R., Cook, S., Neal, D. E., and Robson, C. N. (2002). Tip60 and histone deacetylase 1 regulate androgen receptor activity through changes to the acetylation status of the receptor. J Biol Chem 277, 25904–25913.

Gaughan, L., Logan, I. R., Neal, D. E., and Robson, C. N. (2005). Regulation of androgen receptor and histone deacetylase 1 by Mdm2-mediated ubiquitylation. Nucleic Acids Res 33, 13–26.

George, P., Bali, P., Annavarapu, S., Scuto, A., Fiskus, W., Guo, F., Sigua, C., Sondarva, G., Moscinski, L., Atadja, P., and Bhalla, K. (2005). Combination of the histone deacetylase inhibitor LBH589 and the hsp90 inhibitor 17-AAG is highly active against human CML-BC cells and AML cells with activating mutation of FLT-3. Blood 105, 1768–1776.

Glass, C. K. and Rosenfeld, M. G. (2000). The coregulator exchange in transcriptional functions of nuclear receptors. Genes Dev 14, 121–141.

Goldsmith, K. C. and Hogarty, M. D. (2005). Targeting programmed cell death pathways with experimental therapeutics: opportunities in high-risk neuroblastoma. Cancer Lett 228, 133–141.

Gozzini, A. and Santini, V. (2005). Butyrates and decitabine cooperate to induce histone acetylation and granulocytic maturation of t(8;21) acute myeloid leukemia blasts. Ann Hematol 1–7.

Gray, S. G. and Ekstrom, T. J. (2001). The human histone deacetylase family. Exp Cell Res 262, 75–83.

Green, D. R. and Reed, J. C. (1998). Mitochondria and apoptosis. Science 281, 1309–1312.

Greene, W. C. and Chen, L. F. (2004). Regulation of NF-kappaB action by reversible acetylation. Novartis Found. Symp 259, 208–217; discussion 218–225.

Grignani, F., De Matteis, S., Nervi, C., Tomassoni, L., Gelmetti, V., Cioce, M., Fanelli, M., Ruthardt, M., Ferrara, F. F., Zamir, I., et al. (1998). Fusion proteins of the retinoic acid receptor-alpha recruit histone deacetylase in promyelocytic leukaemia. Nature 391, 815–818.

Grimm, S., Bauer, M. K., Baeuerle, P. A., and Schulze-Osthoff, K. (1996). Bcl-2 down-regulates the activity of transcription factor NF-kappaB induced upon apoptosis. J Cell Biol 134, 13–23.

Gross, A. (2001). BCL-2 proteins: regulators of the mitochondrial apoptotic program. IUBMB Life 52, 231–236.

Gross, A., McDonnell, J. M., and Korsmeyer, S. J. (1999). BCL-2 family members and the mitochondria in apoptosis. Genes Dev 13, 1899–1911.

Grozinger, C. M. and Schreiber, S. L. (2000). Regulation of histone deacetylase 4 and 5 and transcriptional activity by 14-3-3-dependent cellular localization. Proc Natl Acad Sci USA 97, 7835–7840.

Grozinger, C. M., Hassig, C. A., and Schreiber, S. L. (1999). Three proteins define a class of human histone deacetylases related to yeast Hda1p. Proc Natl Acad Sci USA 96, 4868–4873.

Grunstein, M. (1997). Histone acetylation in chromatin structure and transcription. Nature 389, 349–352.

Han, J. W., Ahn, S. H., Park, S. H., Wang, S. Y., Bae, G. U., Seo, D. W., Kwon, H. K., Hong, S., Lee, H. Y., Lee, Y. W., and Lee, H. W. (2000). Apicidin, a histone deacetylase inhibitor, inhibits proliferation of tumor cells via induction of p21WAF1/Cip1 and gelsolin. Cancer Res 60, 6068–6074.

Han, J. W., Ahn, S. H., Kim, Y. K., Bae, G. U., Yoon, J. W., Hong, S., Lee, H. Y., Lee, Y. W., and Lee, H. W. (2001). Activation of p21(WAF1/Cip1) transcription through Sp1 sites by histone deacetylase inhibitor apicidin: involvement of protein kinase C. J Biol Chem 276, 42084–42090.

He, G. H., Helbing, C. C., Wagner, M. J., Sensen, C. W., and Riabowol, K. (2005). Phylogenetic analysis of the ING family of PHD finger proteins. Mol Biol Evol 22, 104–116.

Henderson, C. and Brancolini, C. (2003). Apoptotic pathways activated by histone deacetylase inhibitors: implications for the drug-resistant phenotype. Drug Resist Updat 6, 247–256.

Heruth, D. P., Zirnstein, G. W., Bradley, J. F., and Rothberg, P. G. (1993). Sodium butyrate causes an increase in the block to transcriptional elongation in the c-myc gene in SW837 rectal carcinoma cells. J Biol Chem 268, 20466–20472.

Hess-Stumpp, H. (2005). Histone deacetylase inhibitors and cancer: from cell biology to the clinic. Eur J Cell Biol 84, 109–121.

Hitomi, T., Matsuzaki, Y., Yokota, T., Takaoka, Y., and Sakai, T. (2003). p15(INK4b) in HDAC inhibitor-induced growth arrest. FEBS Lett 554, 347–350.

Hoffmann, K., Brosch, G., Loidl, P., and Jung, M. (2000). First non-radioactive assay for in vitro screening of histone deacetylase inhibitors. Pharmazie 55, 601–606.

Hong, J., Ishihara, K., Yamaki, K., Hiraizumi, K., Ohno, T., Ahn, J. W., Zee, O., and Ohuchi, K. (2003). Apicidin, a histone deacetylase inhibitor, induces differentiation of HL-60 cells. Cancer Lett 189, 197–206.

Hook, S. S., Orian, A., Cowley, S. M., and Eisenman, R. N. (2002). Histone deacetylase 6 binds polyubiquitin through its zinc finger (PAZ domain) and copurifies with deubiquitinating enzymes. Proc Natl Acad Sci USA 99, 13425–13430.

Howe, G. R., Benito, E., Castelleto, R., Cornee, J., Esteve, J., Gallagher, R. P., Iscovich, J. M., Deng-ao, J., Kaaks, R., Kune, G. A., et al. (1992). Dietary intake of fiber and decreased risk of cancers of the colon and rectum: evidence from the combined analysis of 13 case-control studies. J Natl Cancer Inst 84, 1887–1896.

Hu, J. and Colburn, N. H. (2005). Histone deacetylase inhibition down-regulates cyclin D1 transcription by inhibiting nuclear factor-kappaB/p65 DNA binding. Mol Cancer Res 3, 100–109.

Hylla, S., Gostner, A., Dusel, G., Anger, H., Bartram, H. P., Christl, S. U., Kasper, H., and Scheppach, W. (1998). Effects of resistant starch on the colon in healthy volunteers: possible implications for cancer prevention. Am J Clin Nutr 67, 136–142.

Imai, S., Armstrong, C. M., Kaeberlein, M., and Guarente, L. (2000). Transcriptional silencing and longevity protein Sir2 is an NAD-dependent histone deacetylase. Nature 403, 795–800.

Imai, T., Adachi, S., Nishijo, K., Ohgushi, M., Okada, M., Yasumi, T., Watanabe, K., Nishikomori, R., Nakayama, T., Yonehara, S., et al. (2003). FR901228 induces tumor regression associated with induction of Fas ligand and activation of Fas signaling in human osteosarcoma cells. Oncogene 22, 9231–9242.

Inoue, H., Shiraki, K., Ohmori, S., Sakai, T., Deguchi, M., Yamanaka, T., Okano, H., and Nakano, T. (2002). Histone deacetylase inhibitors sensitize human colonic adenocarcinoma cell lines to TNF-related apoptosis inducing ligand-mediated apoptosis. Int J Mol Med 9, 521–525.

Inoue, S., MacFarlane, M., Harper, N., Wheat, L. M., Dyer, M. J., and Cohen, G. M. (2004). Histone deacetylase inhibitors potentiate TNF-related apoptosis-inducing ligand (TRAIL)-induced apoptosis in lymphoid malignancies. Cell Death Differ 11 (Suppl 2), S193–S206.

Insinga, A., Minucci, S., and Pelicci, P. G. (2005a). Mechanisms of selective anticancer action of histone deacetylase inhibitors. Cell Cycle 4, 741–743.

Insinga, A., Monestiroli, S., Ronzoni, S., Gelmetti, V., Marchesi, F., Viale, A., Altucci, L., Nervi, C., Minucci, S., and Pelicci, P. G. (2005b). Inhibitors of histone deacetylases induce tumor-selective apoptosis through activation of the death receptor pathway. Nat Med 11, 71–76.

Janne, P. A. and Mayer, R. J. (2000). Chemoprevention of colorectal cancer. N Engl J Med 342, 1960–1968.

Jee, C. D., Lee, H. S., Bae, S. I., Yang, H. K., Lee, Y. M., Rho, M. S., and Kim, W. H. (2005). Loss of caspase-1 gene expression in human gastric carcinomas and cell lines. Int J Oncol 26, 1265–1271.

Jiang, Y., Ahn, E. Y., Ryu, S. H., Kim, D. K., Park, J. S., Yoon, H. J., You, S., Lee, B. J., Lee, D. S., and Jung, J. H. (2004). Cytotoxicity of psammaplin A from a two-sponge association may correlate with the inhibition of DNA replication. BMC Cancer 4, 70.

Johnson, J., Hunter, R., McElveen, R., Qian, X. H., Baliga, B. S., and Pace, B. S. (2005). Fetal hemoglobin induction by the histone deacetylase inhibitor, scriptaid. Cell Mol Biol (Noisy-le-grand) 51, 229–238.

Johnstone, R. W. (2002). Histone-deacetylase inhibitors: novel drugs for the treatment of cancer. Nat Rev Drug Discov 1, 287–299.

Johnstone, R. W. and Licht, J. D. (2003). Histone deacetylase inhibitors in cancer therapy: is transcription the primary target? Cancer Cell 4, 13–18.

Johnstone, R. W., Ruefli, A. A., and Lowe, S. W. (2002). Apoptosis: a link between cancer genetics and chemotherapy. Cell 108, 153–164.

Jones, P. A. (2002). DNA methylation and cancer. Oncogene 21, 5358–5360.

Juan, L. J., Shia, W. J., Chen, M. H., Yang, W. M., Seto, E., Lin, Y. S., and Wu, C. W. (2000). Histone deacetylases specifically down-regulate p53-dependent gene activation. J Biol Chem 275, 20436–20443.

Jung, J. H., Sim, C. J., and Lee, C. O. (1995). Cytotoxic compounds from a two-sponge association. J Nat Prod 58, 1722–1726.

Jung, M., Brosch, G., Kolle, D., Scherf, H., Gerhauser, C., and Loidl, P. (1999). Amide analogues of trichostatin A as inhibitors of histone deacetylase and inducers of terminal cell differentiation. J Med Chem 42, 4669–4679.

Kaeberlein, M., McVey, M., and Guarente, L. (1999). The SIR2/3/4 complex and SIR2 alone promote longevity in Saccharomyces cerevisiae by two different mechanisms. Genes Dev 13, 2570–2580.

Kandasamy, K., Srinivasula, S. M., Alnemri, E. S., Thompson, C. B., Korsmeyer, S. J., Bryant, J. L., and Srivastava, R. K. (2003). Involvement of proapoptotic molecules Bax and Bak in tumor necrosis factor-related apoptosis-inducing ligand (TRAIL)-induced mitochondrial disruption and apoptosis: differential regulation of cytochrome c and Smac/DIABLO release. Cancer Res 63, 1712–1721.

Kawaguchi, Y., Kovacs, J. J., McLaurin, A., Vance, J. M., Ito, A., and Yao, T. P. (2003). The deacetylase HDAC6 regulates aggresome formation and cell viability in response to misfolded protein stress. Cell 115, 727–738.

Keane, M. M., Ettenberg, S. A., Nau, M. M., Russell, E. K., and Lipkowitz, S. (1999). Chemotherapy augments TRAIL-induced apoptosis in breast cell lines. Cancer Res 59, 734–741.

Kelly, W. K., O'Connor, O. A., Krug, L. M., Chiao, J. H., Heaney, M., Curley, T., MacGregore-Cortelli, B., Tong, W., Secrist, J. P., Schwartz, L., et al. (2005). Phase I study of an oral histone deacetylase inhibitor, suberoylanilide hydroxamic acid, in patients with advanced cancer. J Clin Oncol 23, 3923–3931.

Khan, S. B., Maududi, T., Barton, K., Ayers, J., and Alkan, S. (2004). Analysis of histone deacetylase inhibitor, depsipeptide (FR901228), effect on multiple myeloma. Br J Haematol 125, 156–161.

Kijima, M., Yoshida, M., Sugita, K., Horinouchi, S., and Beppu, T. (1993). Trapoxin, an antitumor cyclic tetrapeptide, is an irreversible inhibitor of mammalian histone deacetylase. J Biol Chem 268, 22429–22435.

Kim, D., Lee, I. S., Jung, J. H., Lee, C. O., and Choi, S. U. (1999a). Psammaplin A, a natural phenolic compound, has inhibitory effect on human topoisomerase II and is cytotoxic to cancer cells. Anticancer Res 19, 4085–4090.

Kim, D., Lee, I. S., Jung, J. H., and Yang, S. I. (1999b). Psammaplin A, a natural bromotyrosine derivative from a sponge, possesses the antibacterial activity against methicillin-resistant Staphylococcus aureus and the DNA gyrase-inhibitory activity. Arch Pharm Res 22, 25–29.

Kim, D. H., Kim, M., and Kwon, H. J. (2003). Histone deacetylase in carcinogenesis and its inhibitors as anti-cancer agents. J Biochem Mol Biol 36, 110–119.

Kim, J. H., Shin, J. H., and Kim, I. H. (2004a). Susceptibility and radiosensitization of human glioblastoma cells to trichostatin A, a histone deacetylase inhibitor. Int J Radiat Oncol Biol Phys 59, 1174–1180.

Kim, J. S., Jeung, H. K., Cheong, J. W., Maeng, H., Lee, S. T., Hahn, J. S., Ko, Y. W,. and Min, Y. H. (2004b). Apicidin potentiates the imatinib-induced apoptosis of Bcr-Abl-positive human leukaemia cells by enhancing the activation of mitochondria-dependent caspase cascades. Br J Haematol 124, 166–178.

Kim, J. S., Lee, S., Lee, T., Lee, Y. W., and Trepel, J. B. (2001a). Transcriptional activation of p21(WAF1/CIP1) by apicidin, a novel histone deacetylase inhibitor. Biochem Biophys Res Commun 281, 866–871.

Kim, M. S., Kwon, H. J., Lee, Y. M., Baek, J. H., Jang, J. E., Lee, S. W., Moon, E. J., Kim, H. S., Lee, S. K., Chung, H. Y., et al. (2001b). Histone deacetylases induce angiogenesis by negative regulation of tumor suppressor genes. Nat Med 7, 437–443.

Kim, S. H., Ahn, S., Han, J. W., Lee, H. W., Lee, H. Y., Lee, Y. W., Kim, M. R., Kim, K. W., Kim, W. B., and Hong, S. (2004c). Apicidin is a histone deacetylase inhibitor with anti-invasive and anti-angiogenic potentials. Biochem Biophys Res Commun 315, 964–970.

Kim, Y. B., Lee, K. H., Sugita, K., Yoshida, M., and Horinouchi, S. (1999c). Oxamflatin is a novel antitumor compound that inhibits mammalian histone deacetylase. Oncogene 18, 2461–2470.

Klisovic, D. D., Katz, S. E., Effron, D., Klisovic, M. I., Wickham, J., Parthun, M. R., Guimond, M., and Marcucci, G. (2003a). Depsipeptide (FR901228) inhibits proliferation and induces apoptosis in primary and metastatic human uveal melanoma cell lines. Invest Ophthalmol Vis Sci 44, 2390–2398.

Klisovic, D. D., Klisovic, M. I., Effron, D., Liu, S., Marcucci, G., and Katz, S. E. (2005). Depsipeptide inhibits migration of primary and metastatic uveal melanoma cell lines in vitro: a potential strategy for uveal melanoma. Melanoma Res 15, 147–153.

Klisovic, M. I., Maghraby, E. A., Parthun, M. R., Guimond, M., Sklenar, A. R., Whitman, S. P., Chan, K. K., Murphy, T., Anon, J., Archer, K. J., et al. (2003b). Depsipeptide (FR 901228) promotes histone acetylation, gene transcription, apoptosis and its activity is enhanced by DNA methyltransferase inhibitors in AML1/ETO-positive leukemic cells. Leukemia 17, 350–358.

Kouraklis, G. and Theocharis, S. (2002). Histone deacetylase inhibitors and anticancer therapy. Curr Med Chem Anti-Canc Agents 2, 477–484.

Kouzarides, T. (1999). Histone acetylases and deacetylases in cell proliferation. Curr Opin Genet Dev 9, 40–48.

Kovacs, J. J., Murphy, P. J., Gaillard, S., Zhao, X., Wu, J. T., Nicchitta, C. V., Yoshida, M., Toft, D. O., Pratt, W. B., and Yao, T. P. (2005). HDAC6 regulates Hsp90 acetylation and chaperone-dependent activation of glucocorticoid receptor. Mol Cell 18, 601–607.

Kurdistani, S. K. and Grunstein, M. (2003). Histone acetylation and deacetylation in yeast. Nat Rev Mol Cell Biol 4, 276–284.

Kwon, H. J., Owa, T., Hassig, C. A., Shimada, J., and Schreiber, S. L. (1998). Depudecin induces morphological reversion of transformed fibroblasts via the inhibition of histone deacetylase. Proc Natl Acad Sci USA 95, 3356–3361.

Kwon, H. J., Kim, M. S., Kim, M. J., Nakajima, H., and Kim, K. W. (2002a). Histone deacetylase inhibitor FK228 inhibits tumor angiogenesis. Int J Cancer 97, 290–296.

Kwon, S. H., Ahn, S. H., Kim, Y. K., Bae, G. U., Yoon, J. W., Hong, S., Lee, H. Y., Lee, Y. W., Lee, H. W., and Han, J. W. (2002b). Apicidin, a histone deacetylase inhibitor, induces apoptosis and Fas/Fas ligand expression in human acute promyelocytic leukemia cells. J Biol Chem 277, 2073–2080.

Lagger, G., O'Carroll, D., Rembold, M., Khier, H., Tischler, J., Weitzer, G., Schuettengruber, B., Hauser, C., Brunmeir, R., Jenuwein, T., and Seiser, C. (2002). Essential function of histone deacetylase 1 in proliferation control and CDK inhibitor repression. EMBO J 21, 2672–2681.

Lallemand, F., Courilleau, D., Sabbah, M., Redeuilh, G., and Mester, J. (1996). Direct inhibition of the expression of cyclin D1 gene by sodium butyrate. Biochem Biophys Res Commun 229, 163–169.

Langley, E., Pearson, M., Faretta, M., Bauer, U. M., Frye, R. A., Minucci, S., Pelicci, P. G., and Kouzarides, T. (2002). Human SIR2 deacetylates p53 and antagonizes PML/p53-induced cellular senescence. EMBO J 21, 2383–2396.

Lavelle, D., Chen, Y. H., Hankewych, M., and DeSimone, J. (2001). Histone deacetylase inhibitors increase p21(WAF1) and induce apoptosis of human myeloma cell lines independent of decreased IL-6 receptor expression. Am J Hematol 68, 170–178.

Lea, M. A., Randolph, V. M., and Patel, M. (1999). Increased acetylation of histones induced by diallyl disulfide and structurally related molecules. Int J Oncol 15, 347–352.

Lee, B. I., Park, S. H., Kim, J. W., Sausville, E. A., Kim, H. T., Nakanishi, O., Trepel, J. B., and Kim, S. J. (2001). MS-275, a histone deacetylase inhibitor, selectively induces transforming growth factor beta type II receptor expression in human breast cancer cells. Cancer Res 61, 931–934.

Lin, R. J., Nagy, L., Inoue, S., Shao, W., Miller, W. H., Jr., and Evans, R. M. (1998). Role of the histone deacetylase complex in acute promyelocytic leukaemia. Nature 391, 811–814.

Liotta, L. A., Steeg, P. S., and Stetler-Stevenson, W. G. (1991). Cancer metastasis and angiogenesis: an imbalance of positive and negative regulation. Cell 64, 327–336.

Liu, F., Dowling, M., Yang, X. J., and Kao, G. D. (2004). Caspase-mediated specific cleavage of human histone deacetylase 4. J Biol Chem 279, 34537–34546.

Liu, L. T., Chang, H. C., Chiang, L. C., and Hung, W. C. (2003). Histone deacetylase inhibitor up-regulates RECK to inhibit MMP-2 activation and cancer cell invasion. Cancer Res 63, 3069–3072.

Luo, J., Su, F., Chen, D., Shiloh, A., and Gu, W. (2000). Deacetylation of p53 modulates its effect on cell growth and apoptosis. Nature 408, 377–381.

Luo, J., Nikolaev, A. Y., Imai, S., Chen, D., Su, F., Shiloh, A., Guarente, L., and Gu, W. (2001). Negative control of p53 by Sir2alpha promotes cell survival under stress. Cell 107, 137–148.

Luo, X., Budihardjo, I., Zou, H., Slaughter, C., and Wang, X. (1998). Bid, a Bcl2 interacting protein, mediates cytochrome c release from mitochondria in response to activation of cell surface death receptors. Cell 94, 481–490.

Maggio, S. C., Rosato, R. R., Kramer, L. B., Dai, Y., Rahmani, M., Paik, D. S., Czarnik, A. C., Payne, S. G., Spiegel, S., and Grant, S. (2004). The histone deacetylase inhibitor MS-275 interacts synergistically with fludarabine to induce apoptosis in human leukemia cells. Cancer Res 64, 2590–2600.

Mai, A., Massa, S., Rotili, D., Cerbara, I., Valente, S., Pezzi, R., Simeoni, S., and Ragno, R. (2005). Histone deacetylation in epigenetics: an attractive target for anticancer therapy. Med Res Rev 25, 261–309.

Marks, P., Rifkind, R. A., Richon, V. M., Breslow, R., Miller, T., and Kelly, W. K. (2001a). Histone deacetylases and cancer: causes and therapies. Nat Rev Cancer 1, 194–202.

Marks, P. A. and Jiang, X. (2005). Histone deacetylase inhibitors in programmed cell death and cancer therapy. Cell Cycle 4, 549–551.

Marks, P. A. and Rifkind, R. A. (1988). Hexamethylene bisacetamide-induced differentiation of transformed cells: molecular and cellular effects and therapeutic application. Int J Cell Cloning 6, 230–240.

Marks, P. A., Richon, V. M., and Rifkind, R. A. (1996). Cell cycle regulatory proteins are targets for induced differentiation of transformed cells: molecular and clinical studies employing hybrid polar compounds. Int J Hematol 63, 1–17.

Marks, P. A., Richon, V. M., Breslow, R., and Rifkind, R. A. (2001b). Histone deacetylase inhibitors as new cancer drugs. Curr Opin Oncol 13, 477–483.

Marks, P. A., Miller, T., and Richon, V. M. (2003). Histone deacetylases. Curr Opin Pharmacol 3, 344–351.

Marks, P. A., Richon, V. M., Miller, T., and Kelly, W. K. (2004). Histone deacetylase inhibitors. Adv Cancer Res 91, 137–168.

McIntyre, A., Gibson, P. R., and Young, G. P. (1993). Butyrate production from dietary fibre and protection against large bowel cancer in a rat model. Gut 34, 386–391.

McKinsey, T. A. and Olson, E. N. (2005). Toward transcriptional therapies for the failing heart: chemical screens to modulate genes. J Clin Invest 115, 538–546.

McLaughlin, F. and La Thangue, N. B. (2004). Histone deacetylase inhibitors open new doors in cancer therapy. Biochem Pharmacol 68, 1139–1144.

Meinke, P. T. and Liberator, P. (2001). Histone deacetylase: a target for antiproliferative and antiprotozoal agents. Curr Med Chem 8, 211–235.

Meinke, P. T., Colletti, S. L., Doss, G., Myers, R. W., Gurnett, A. M., Dulski, P. M., Darkin-Rattray, S. J., Allocco, J. J., Galuska, S., Schmatz, D. M., et al. (2000). Synthesis of apicidin-derived quinolone derivatives: parasite-selective histone deacetylase inhibitors and antiproliferative agents. J Med Chem 43, 4919–4922.

Michaelis, M., Michaelis, U. R., Fleming, I., Suhan, T., Cinatl, J., Blaheta, R. A., Hoffmann, K., Kotchetkov, R., Busse, R., Nau, H., and Cinatl, J., Jr. (2004). Valproic acid inhibits angiogenesis in vitro and in vivo. Mol Pharmacol 65, 520–527.

Michaelis, M., Suhan, T., Michaelis, U. R., Beek, K., Rothweiler, F., Tausch, L., Werz, O., Eikel, D., Zornig, M., Nau, H., et al. (2005). Valproic acid induces extracellular signal-regulated kinase 1/2 activation and inhibits apoptosis in endothelial cells. Cell Death Differ 13(3), 446–453.

Mie Lee, Y., Kim, S. H., Kim, H. S., Jin Son, M., Nakajima, H., Jeong Kwon, H., and Kim, K. W. (2003). Inhibition of hypoxia-induced angiogenesis by FK228, a specific histone deacetylase inhibitor, via suppression of HIF-1alpha activity. Biochem Biophys Res Commun 300, 241–246.

Mitsiades, C. S., Poulaki, V., McMullan, C., Negri, J., Fanourakis, G., Goudopoulou, A., Richon, V. M., Marks, P. A., and Mitsiades, N. (2005). Novel histone deacetylase inhibitors in the treatment of thyroid cancer. Clin Cancer Res 11, 3958–3965.

Mitsiades, N., Mitsiades, C. S., Richardson, P. G., McMullan, C., Poulaki, V., Fanourakis, G., Schlossman, R., Chauhan, D., Munshi, N. C., Hideshima, T., et al. (2003). Molecular sequelae of histone deacetylase inhibition in human malignant B cells. Blood 101, 4055–4062.

Momparler, R. L. (2003). Cancer epigenetics. Oncogene 22, 6479–6483.

Monneret, C. (2005). Histone deacetylase inhibitors. Eur J Med Chem 40, 1–13.

Mori, N., Matsuda, T., Tadano, M., Kinjo, T., Yamada, Y., Tsukasaki, K., Ikeda, S., Yamasaki, Y., Tanaka, Y., Ohta, T., et al. (2004). Apoptosis induced by the histone deacetylase inhibitor FR901228 in human T-cell leukemia virus type 1-infected T-cell lines and primary adult T-cell leukemia cells. J Virol 78, 4582–4590.

Murakami, J., Asaumi, J., Maki, Y., Tsujigiwa, H., Kuroda, M., Nagai, N., Yanagi, Y., Inoue, T., Kawasaki, S., Tanaka, N., et al. (2004). Effects of demethylating agent 5-aza-2(′)-deoxycytidine and histone deacetylase inhibitor FR901228 on maspin gene expression in oral cancer cell lines. Oral Oncol 40, 597–603.

Myzak, M. C., Karplus, P. A., Chung, F. L., and Dashwood, R. H. (2004). A novel mechanism of chemoprotection by sulforaphane: inhibition of histone deacetylase. Cancer Res 64, 5767–5774.

Myzak, M. C., Hardin, K., Wang, R., Dashwood, R. H., and Ho, E. (2005). Sulforaphane inhibits histone deacetylase activity in BPH-1, LnCaP, and PC-3 prostate epithelial cells. Carcinogenesis 27, 811–819.

Nagane, M., Pan, G., Weddle, J. J., Dixit, V. M., Cavenee, W. K., and Huang, H. J. (2000). Increased death receptor 5 expression by chemotherapeutic agents in human gliomas causes synergistic cytotoxicity with tumor necrosis factor-related apoptosis-inducing ligand in vitro and in vivo. Cancer Res 60, 847–853.

Nam, N. H. and Parang, K. (2003). Current targets for anticancer drug discovery. Curr Drug Targets 4, 159–179.

Nebbioso, A., Clarke, N., Voltz, E., Germain, E., Ambrosino, C., Bontempo, P., Alvarez, R., Schiavone, E. M., Ferrara, F., Bresciani, F., et al. (2005). Tumor-selective action of HDAC inhibitors involves TRAIL induction in acute myeloid leukemia cells. Nat Med 11, 77–84.

Neuzil, J., Swettenham, E., and Gellert, N. (2004). Sensitization of mesothelioma to TRAIL apoptosis by inhibition of histone deacetylase: role of Bcl-xL down-regulation. Biochem Biophys Res Commun 314, 186–191.

Nicolaou, K. C., Hughes, R., Pfefferkorn, J. A., and Barluenga, S. (2001). Optimization and mechanistic studies of psammaplin A type antibacterial agents active against methicillin-resistant Staphylococcus aureus (MRSA). Chemistry 7, 4296–4310.

Nimmanapalli, R., Fuino, L., Stobaugh, C., Richon, V., and Bhalla, K. (2003). Cotreatment with the histone deacetylase inhibitor suberoylanilide hydroxamic acid (SAHA) enhances imatinib-induced apoptosis of Bcr-Abl-positive human acute leukemia cells. Blood 101, 3236–3239.

Nome, R. V., Bratland, A., Harman, G., Fodstad, O., Andersson, Y., and Ree, A. H. (2005). Cell cycle checkpoint signaling involved in histone deacetylase inhibition and radiation-induced cell death. Mol Cancer Ther 4, 1231–1238.

O'Brien, T., Cranston, D., Fuggle, S., Bicknell, R., and Harris, A. L. (1995). Different angiogenic pathways characterize superficial and invasive bladder cancer. Cancer Res 55, 510–513.

O'Shea, J. J., Kanno, Y., Chen, X., and Levy, D. E. (2005). Cell signaling. Stat acetylation–a key facet of cytokine signaling? Science 307, 217–218.

Oikawa, T., Onozawa, C., Inose, M., and Sasaki, M. (1995). Depudecin, a microbial metabolite containing two epoxide groups, exhibits anti-angiogenic activity in vivo. Biol Pharm Bull 18, 1305–1307.

Papeleu, P., Vanhaecke, T., Elaut, G., Vinken, M., Henkens, T., Snykers, S., and Rogiers, V. (2005). Differential effects of histone deacetylase inhibitors in tumor and normal cells-what is the toxicological relevance? Crit Rev Toxicol 35, 363–378.

Park, J. H., Jung, Y., Kim, T. Y., Kim, S. G., Jong, H. S., Lee, J. W., Kim, D. K., Lee, J. S., Kim, N. K., Kim, T. Y., and Bang, Y. J. (2004). Class I histone deacetylase-selective novel synthetic inhibitors potently inhibit human tumor proliferation. Clin Cancer Res 10, 5271–5281.

Park, S. H., Lee, S. R., Kim, B. C., Cho, E. A., Patel, S. P., Kang, H. B., Sausville, E. A., Nakanishi, O., Trepel, J. B., Lee, B. I., and Kim, S. J. (2002). Transcriptional regulation of the transforming growth factor beta type II receptor gene by histone acetyltransferase and deacetylase is mediated by NF-Y in human breast cancer cells. J Biol Chem 277, 5168–5174.

Park, Y., Liu, Y., Hong, J., Lee, C. O., Cho, H., Kim, D. K., Im, K. S., and Jung, J. H. (2003). New bromotyrosine derivatives from an association of two sponges, Jaspis wondoensis and Poecillastra wondoensis. J Nat Prod 66, 1495–1498.

Parker, S. B., Eichele, G., Zhang, P., Rawls, A., Sands, A. T., Bradley, A., Olson, E. N., Harper, J. W., and Elledge, S. J. (1995). p53-independent expression of p21Cip1 in muscle and other terminally differentiating cells. Science 267, 1024–1027.

Paroni, G., Mizzau, M., Henderson, C., Del Sal, G., Schneider, C. and Brancolini, C. (2004). Caspase-dependent regulation of histone deacetylase 4 nuclear-cytoplasmic shuttling promotes apoptosis. Mol Biol Cell 15, 2804–2818.

Pham, N. B., Butler, M. S., and Quinn, R. J. (2000). Isolation of psammaplin A 11′-sulfate and bisaprasin 11′-sulfate from the marine sponge Aplysinella rhax. J Nat Prod 63, 393–395.

Pili, R., Kruszewski, M. P., Hager, B. W., Lantz, J., and Carducci, M. A. (2001). Combination of phenylbutyrate and 13-cis retinoic acid inhibits prostate tumor growth and angiogenesis. Cancer Res 61, 1477–1485.

Pina, I. C., Gautschi, J. T., Wang, G. Y., Sanders, M. L., Schmitz, F. J., France, D., Cornell-Kennon, S., Sambucetti, L. C., Remiszewski, S. W., Perez, L. B., et al. (2003). Psammaplins from the sponge Pseudoceratina purpurea: inhibition of both histone deacetylase and DNA methyltransferase. J Org Chem 68, 3866–3873.

Plass, C. (2002). Cancer epigenomics. Hum Mol Genet 11, 2479–2488.

Qian, D. Z., Wang, X., Kachhap, S. K., Kato, Y., Wei, Y., Zhang, L., Atadja, P., and Pili, R. (2004). The histone deacetylase inhibitor NVP-LAQ824 inhibits angiogenesis and has a greater antitumor effect in combination with the vascular endothelial growth factor receptor tyrosine kinase inhibitor PTK787/ZK222584. Cancer Res 64, 6626–6634.

Qian, D. Z., Kato, Y., Shabbeer, S., Wei, Y., Verheul, H. M., Salumbides, B., Sanni, T., Atadja, P., and Pili, R. (2006). Targeting tumor angiogenesis with histone deacetylase inhibitors: the hydroxamic acid derivative LBH589. Clin Cancer Res 12, 634–642.

Qiu, L., Burgess, A., Fairlie, D. P., Leonard, H., Parsons, P. G., and Gabrielli, B. G. (2000). Histone deacetylase inhibitors trigger a G2 checkpoint in normal cells that is defective in tumor cells. Mol Biol Cell 11, 2069–2083.

Quivy, V. and Van Lint, C. (2004). Regulation at multiple levels of NF-kappaB-mediated trans-activation by protein acetylation. Biochem Pharmacol 68, 1221–1229.

Rahmani, M., Reese, E., Dai, Y., Bauer, C., Payne, S. G., Dent, P., Spiegel, S., and Grant, S. (2005). Coadministration of histone deacetylase inhibitors and perifosine synergistically induces apoptosis in human leukemia cells through Akt and ERK1/2 inactivation and the generation of ceramide and reactive oxygen species. Cancer Res 65, 2422–2432.

Ravery, V., Jouanneau, J., Gil Diez, S., Abbou, C. C., Caruelle, J. P., Barritault, D., and Chopin, D. K. (1992). Immunohistochemical detection of acidic fibroblast growth factor in bladder transitional cell carcinoma. Urol Res 20, 211–214.

Ravi, R., Bedi, G. C., Engstrom, L. W., Zeng, Q., Mookerjee, B., Gelinas, C., Fuchs, E. J., and Bedi, A. (2001). Regulation of death receptor expression and TRAIL/Apo2L-induced apoptosis by NF-kappaB. Nat Cell Biol 3, 409–416.

Richon, V. M., Emiliani, S., Verdin, E., Webb, Y., Breslow, R., Rifkind, R. A., and Marks, P. A. (1998). A class of hybrid polar inducers of transformed cell differentiation inhibits histone deacetylases. Proc Natl Acad Sci USA 95, 3003–3007.

Riggs, M. G., Whittaker, R. G., Neumann, J. R., and Ingram, V. M. (1977). n-Butyrate causes histone modification in HeLa and Friend erythroleukaemia cells. Nature 268, 462–464.

Rocchi, P., Tonelli, R., Camerin, C., Purgato, S., Fronza, R., Bianucci, F., Guerra, F., Pession, A., and Ferreri, A. M. (2005). p21Waf1/Cip1 is a common target induced by short-chain fatty acid HDAC inhibitors (valproic acid, tributyrin and sodium butyrate) in neuroblastoma cells. Oncol Rep 13, 1139–1144.

Rosato, R. R., Wang, Z., Gopalkrishnan, R. V., Fisher, P. B., and Grant, S. (2001). Evidence of a functional role for the cyclin-dependent kinase-inhibitor p21WAF1/CIP1/MDA6 in promoting differentiation and preventing mitochondrial dysfunction and apoptosis induced by sodium butyrate in human myelomonocytic leukemia cells (U937). Int J Oncol 19, 181–191.

Rosato, R. R., Almenara, J. A., Dai, Y., and Grant, S. (2003a). Simultaneous activation of the intrinsic and extrinsic pathways by histone deacetylase (HDAC) inhibitors and tumor necrosis factor-related apoptosis-inducing ligand (TRAIL) synergistically induces mitochondrial damage and apoptosis in human leukemia cells. Mol Cancer Ther 2, 1273–1284.

Rosato, R. R., Almenara, J. A., and Grant, S. (2003b). The histone deacetylase inhibitor MS-275 promotes differentiation or apoptosis in human leukemia cells through a process regulated by generation of reactive oxygen species and induction of p21CIP1/WAF1 1. Cancer Res 63, 3637–3645.

Rossig, L., Li, H., Fisslthaler, B., Urbich, C., Fleming, I., Forstermann, U., Zeiher, A. M., and Dimmeler, S. (2002). Inhibitors of histone deacetylation downregulate the expression of endothelial nitric oxide synthase and compromise endothelial cell function in vasorelaxation and angiogenesis. Circ Res 91, 837–844.

Ryan, Q. C., Headlee, D., Acharya, M., Sparreboom, A., Trepel, J. B., Ye, J., Figg, W. D., Hwang, K., Chung, E. J., Murgo, A., et al. (2005). Phase I and pharmacokinetic study of MS-275, a histone deacetylase inhibitor, in patients with advanced and refractory solid tumors or lymphoma. J Clin Oncol 23, 3912–3922.

Sakajiri, S., Kumagai, T., Kawamata, N., Saitoh, T., Said, J. W., and Koeffler, H. P. (2005). Histone deacetylase inhibitors profoundly decrease proliferation of human lymphoid cancer cell lines. Exp Hematol 33, 53–61.

Sandor, V., Bakke, S., Robey, R. W., Kang, M. H., Blagosklonny, M. V., Bender, J., Brooks, R., Piekarz, R. L., Tucker, E., Figg, W. D., et al. (2002). Phase I trial of the histone deacetylase inhibitor, depsipeptide (FR901228, NSC 630176), in patients with refractory neoplasms. Clin Cancer Res 8, 718–728.

Sartorius, U., Schmitz, I., and Krammer, P. H. (2001). Molecular mechanisms of death-receptor-mediated apoptosis. Chembiochem 2, 20–29.

Sasakawa, Y., Naoe, Y., Inoue, T., Sasakawa, T., Matsuo, M., Manda, T., and Mutoh, S. (2002). Effects of FK228, a novel histone deacetylase inhibitor, on human lymphoma U-937 cells in vitro and in vivo. Biochem Pharmacol 64, 1079–1090.

Sasakawa, Y., Naoe, Y., Noto, T., Inoue, T., Sasakawa, T., Matsuo, M., Manda, T., and Mutoh, S. (2003). Antitumor efficacy of FK228, a novel histone deacetylase inhibitor, depends on the effect on expression of angiogenesis factors. Biochem Pharmacol 66, 897–906.

Sato, N., Ohta, T., Kitagawa, H., Kayahara, M., Ninomiya, I., Fushida, S., Fujimura, T., Nishimura, G., Shimizu, K., and Miwa, K. (2004). FR901228, a novel histone deacetylase inhibitor, induces cell cycle arrest and subsequent apoptosis in refractory human pancreatic cancer cells. Int J Oncol 24, 679–685.

Savickiene, J., Treigyte, G., Pivoriunas, A., Navakauskiene, R., and Magnusson, K. E. (2004). Sp1 and NF-kappaB transcription factor activity in the regulation of the p21 and FasL promoters during promyelocytic leukemia cell monocytic differentiation and its associated apoptosis. Ann NY Acad Sci 1030, 569–577.

Sawa, H., Murakami, H., Ohshima, Y., Murakami, M., Yamazaki, I., Tamura, Y., Mima, T., Satone, A., Ide, W., Hashimoto, I., and Kamada, H. (2002). Histone deacetylase inhibitors such as sodium butyrate and trichostatin A inhibit vascular endothelial growth factor (VEGF) secretion from human glioblastoma cells. Brain Tumor Pathol 19, 77–81.

Sawa, H., Murakami, H., Kumagai, M., Nakasato, M., Yamauchi, S., Matsuyama, N., Tamura, Y., Satone, A., Ide, W., Hashimoto, I., and Kamada, H. (2004). Histone deacetylase inhibitor, FK228, induces apoptosis and suppresses cell proliferation of human glioblastoma cells in vitro and in vivo. Acta Neuropathol (Berl) 107, 523–531.

Shankar, S., Singh, T. R., Chen, X., Thakkar, H., Firnin, J., and Srivastava, R. K. (2004a). The sequential treatment with ionizing radiation followed by TRAIL/Apo-2L reduces tumor growth and induces apoptosis of breast tumor xenografts in nude mice. Int J Oncol 24, 1133–1140.

Shankar, S., Singh, T. R., and Srivastava, R. K. (2004b). Ionizing radiation enhances the therapeutic potential of TRAIL in prostate cancer in vitro and in vivo: intracellular mechanisms. Prostate 61, 35–49.

Shankar, S., Chen, X., and Srivastava, R. K. (2005a). Effects of sequential treatments with chemotherapeutic drugs followed by TRAIL on prostate cancer in vitro and in vivo. Prostate 62, 165–186.

Shankar, S., Singh, T. R., Fandy, T. E., Luetrakul, T., Ross, D. D., and Srivastava, R. K. (2005b). Interactive effects of histone deacetylase inhibitors and TRAIL on apoptosis in human leukemia cells: Involvement of both death receptor and mitochondrial pathways. Int J Mol Med 16, 1125–1138.

Shao, Y., Gao, Z., Marks, P. A., and Jiang, X. (2004). Apoptotic and autophagic cell death induced by histone deacetylase inhibitors. Proc Natl Acad Sci USA 101, 18030–18035.

Shen, S., Li, J., and Casaccia-Bonnefil, P. (2005). Histone modifications affect timing of oligodendrocyte progenitor differentiation in the developing rat brain. J Cell Biol 169, 577–589.

Shetty, S., Graham, B. A., Brown, J. G., Hu, X., Vegh-Yarema, N., Harding, G., Paul, J. T., and Gibson, S. B. (2005). Transcription factor NF-kappaB differentially regulates death receptor 5 expression involving histone deacetylase 1. Mol Cell Biol 25, 5404–5416.

Shim, J. S., Lee, H. S., Shin, J., and Kwon, H. J. (2004). Psammaplin A, a marine natural product, inhibits aminopeptidase N and suppresses angiogenesis in vitro. Cancer Lett 203, 163–169.

Simonsson, M., Heldin, C. H., Ericsson, J., and Gronroos, E. (2005). The balance between acetylation and deacetylation controls Smad7 stability. J Biol Chem 280, 21797–21803.

Singh, T. R., Shankar, S., Chen, X., Asim, M., and Srivastava, R. K. (2003). Synergistic interactions of chemotherapeutic drugs and tumor necrosis factor-related apoptosis-inducing ligand/Apo-2 ligand on apoptosis and on regression of breast carcinoma in vivo. Cancer Res 63, 5390–5400.

Singh, T. R., Shankar, S., and Srivastava, R. K. (2005). HDAC inhibitors enhance the apoptosis-inducing potential of TRAIL in breast carcinoma. Oncogene 24, 4609–4623.

Souleimani, A. and Asselin, C. (1993). Regulation of c-myc expression by sodium butyrate in the colon carcinoma cell line Caco-2. FEBS Lett 326, 45–50.

Sowa, Y. and Sakai, T. (2000). Butyrate as a model for "gene-regulating chemoprevention and chemotherapy." Biofactors 12, 283–287.

Sowa, Y., Orita, T., Minamikawa-Hiranabe, S., Mizuno, T., Nomura, H., and Sakai, T. (1999). Sp3, but not Sp1, mediates the transcriptional activation of the p21/WAF1/Cip1 gene promoter by histone deacetylase inhibitor. Cancer Res 59, 4266–4270.

Strahl, B. D. and Allis, C. D. (2000). The language of covalent histone modifications. Nature 403, 41–45.

Strait, K. A., Warnick, C. T., Ford, C. D., Dabbas, B., Hammond, E. H., and Ilstrup, S. J. (2005). Histone deacetylase inhibitors induce G2-checkpoint arrest and apoptosis in cisplatinum-resistant ovarian cancer cells associated with overexpression of the Bcl-2-related protein Bad. Mol Cancer Ther 4, 603–611.

Su, G. H., Sohn, T. A., Ryu, B., and Kern, S. E. (2000). A novel histone deacetylase inhibitor identified by high-throughput transcriptional screening of a compound library. Cancer Res 60, 3137–3142.

Subramanian, C., Opipari, A. W., Jr., Bian, X., Castle, V. P., and Kwok, R. P. (2005). Ku70 acetylation mediates neuroblastoma cell death induced by histone deacetylase inhibitors. Proc Natl Acad Sci USA 102, 4842–4847.

Suliman, A., Lam, A., Datta, R., and Srivastava, R. K. (2001). Intracellular mechanisms of TRAIL: apoptosis through mitochondrial-dependent and -independent pathways. Oncogene 20, 2122–2133.

Suzuki, T., Ando, T., Tsuchiya, K., Fukazawa, N., Saito, A., Mariko, Y., Yamashita, T., and Nakanishi, O. (1999). Synthesis and histone deacetylase inhibitory activity of new benzamide derivatives. J Med Chem 42, 3001–3003.

Taghiyev, A. F., Guseva, N. V., Sturm, M. T., Rokhlin, O. W., and Cohen, M. B. (2005). Trichostatin A (TSA) Sensitizes the Human Prostatic Cancer Cell Line DU145 to Death Receptor Ligands Treatment. Cancer Biol Ther 4, 382–390.

Takai, N., Desmond, J. C., Kumagai, T., Gui, D., Said, J. W., Whittaker, S., Miyakawa, I., and Koeffler, H. P. (2004a). Histone deacetylase inhibitors have a profound antigrowth activity in endometrial cancer cells. Clin Cancer Res 10, 1141–1149.

Takai, N., Kawamata, N., Gui, D., Said, J. W., Miyakawa, I., and Koeffler, H. P. (2004b). Human ovarian carcinoma cells: histone deacetylase inhibitors exhibit antiproliferative activity and potently induce apoptosis. Cancer 101, 2760–2770.

Takimoto, R., Kato, J., Terui, T., Takada, K., Kuroiwa, G., Wu, J., Ohnuma, H., Takahari, D., Kobune, M., Sato, Y., et al. (2005). Augmentation of Antitumor Effects of p53 Gene Therapy by Combination with HDAC Inhibitor. Cancer Biol Ther 4, 421–428.

Tang, X. X., Robinson, M. E., Riceberg, J. S., Kim, D. Y., Kung, B., Titus, T. B., Hayashi, S., Flake, A. W., Carpentieri, D., and Ikegaki, N. (2004). Favorable neuroblastoma genes and molecular therapeutics of neuroblastoma. Clin Cancer Res 10, 5837–5844.

Taunton, J., Hassig, C. A., and Schreiber, S. L. (1996). A mammalian histone deacetylase related to the yeast transcriptional regulator Rpd3p. Science 272, 408–411.

Terui, T., Murakami, K., Takimoto, R., Takahashi, M., Takada, K., Murakami, T., Minami, S., Matsunaga, T., Takayama, T., Kato, J., and Niitsu, Y. (2003). Induction of PIG3 and NOXA through acetylation of p53 at 320 and 373 lysine residues as a mechanism for apoptotic cell death by histone deacetylase inhibitors. Cancer Res 63, 8948–8954.

Thevenet, L., Mejean, C., Moniot, B., Bonneaud, N., Galeotti, N., Aldrian-Herrada, G., Poulat, F., Berta, P., Benkirane, M., and Boizet-Bonhoure, B. (2004). Regulation of human SRY subcellular distribution by its acetylation/deacetylation. EMBO J 23, 3336–3345.

Tissenbaum, H. A. and Guarente, L. (2001). Increased dosage of a sir-2 gene extends lifespan in Caenorhabditis elegans. Nature 410, 227–230.

Toth, K. F., Knoch, T. A., Wachsmuth, M., Frank-Stohr, M., Stohr, M., Bacher, C. P., Muller, G., and Rippe, K. (2004). Trichostatin A-induced histone acetylation causes decondensation of interphase chromatin. J Cell Sci 117, 4277–4287.

Touma, S. E., Goldberg, J. S., Moench, P., Guo, X., Tickoo, S. K., Gudas, L. J., and Nanus, D. M. (2005). Retinoic acid and the histone deacetylase inhibitor trichostatin a inhibit the proliferation of human renal cell carcinoma in a xenograft tumor model. Clin. Cancer Res 11, 3558–3566.

Trock, B., Lanza, E., and Greenwald, P. (1990). Dietary fiber, vegetables, and colon cancer: critical review and meta-analyses of the epidemiologic evidence. J Natl Cancer Inst 82, 650–661.

Tsatsoulis, A. (2002). The role of apoptosis in thyroid disease. Minerva Med 93, 169–180.

Tsuji, N., Kobayashi, M., Nagashima, K., Wakisaka, Y., and Koizumi, K. (1976). A new antifungal antibiotic, trichostatin. J Antibiot (Tokyo) 29, 1–6.

Van Lint, C., Emiliani, S., and Verdin, E. (1996). The expression of a small fraction of cellular genes is changed in response to histone hyperacetylation. Gene Expr 5, 245–253.

Vanhaecke, T., Henkens, T., Kass, G. E., and Rogiers, V. (2004a). Effect of the histone deacetylase inhibitor trichostatin A on spontaneous apoptosis in various types of adult rat hepatocyte cultures. Biochem Pharmacol 68, 753–760.

Vanhaecke, T., Papeleu, P., Elaut, G., and Rogiers, V. (2004b). Trichostatin A-like hydroxamate histone deacetylase inhibitors as therapeutic agents: toxicological point of view. Curr Med Chem 11, 1629–1643.

Vanoosten, R. L., Moore, J. M., Karacay, B., and Griffith, T. S. (2005). Histone Deacetylase Inhibitors Modulate Renal Cell Carcinoma Sensitivity to TRAIL/Apo-2L-induced Apoptosis by Enhancing TRAIL-R2 Expression. Cancer Biol Ther 4, 1104–1112.

Varshochi, R., Halim, F., Sunters, A., Alao, J. P., Madureira, P. A., Hart, S. M., Ali, S., Vigushin, D. M., Coombes, R. C., and Lam, E. W. (2005). ICI182,780 induces p21Waf1 gene transcription through releasing histone deacetylase 1 and estrogen receptor alpha from Sp1 sites to induce cell cycle arrest in MCF-7 breast cancer cell line. J Biol Chem 280, 3185–3196.

Vaziri, H., Dessain, S. K., Ng Eaton, E., Imai, S. I., Frye, R. A., Pandita, T. K., Guarente, L., and Weinberg, R. A. (2001). hSIR2(SIRT1) functions as an NAD-dependent p53 deacetylase. Cell 107, 149–159.

Verdin, E., Dequiedt, F., and Kasler, H. G. (2003). Class II histone deacetylases: versatile regulators. Trends Genet 19, 286–293.

Verhagen, A. M., Ekert, P. G., Pakusch, M., Silke, J., Connolly, L. M., Reid, G. E., Moritz, R. L., Simpson, R. J., and Vaux, D. L. (2000). Identification of DIABLO, a mammalian protein that promotes apoptosis by binding to and antagonizing IAP proteins. Cell 102, 43–53.

Vigushin, D. M. and Coombes, R. C. (2002). Histone deacetylase inhibitors in cancer treatment. Anticancer Drugs 13, 1–13.

Wang, S., Yan-Neale, Y., Fischer, D., Zeremski, M., Cai, R., Zhu, J., Asselbergs, F., Hampton, G., and Cohen, D. (2003). Histone deacetylase 1 represses the small GTPase RhoB expression in human nonsmall lung carcinoma cell line. Oncogene 22, 6204–6213.

Wang, X. F., Qian, D. Z., Ren, M., Kato, Y., Wei, Y., Zhang, L., Fansler, Z., Clark, D., Nakanishi, O., and Pili, R. (2005). Epigenetic modulation of retinoic acid receptor beta2 by the histone deacetylase inhibitor MS-275 in human renal cell carcinoma. Clin Cancer Res 11, 3535–3542.

Wang, Z. M., Hu, J., Zhou, D., Xu, Z. Y., Panasci, L. C., and Chen, Z. P. (2002). Trichostatin A inhibits proliferation and induces expression of p21WAF and p27 in human brain tumor cell lines. Ai Zheng 21, 1100–1105.

Warrell, R. P., Jr., He, L. Z., Richon, V., Calleja, E., and Pandolfi, P. P. (1998). Therapeutic targeting of transcription in acute promyelocytic leukemia by use of an inhibitor of histone deacetylase. J Natl Cancer Inst 90, 1621–1625.

Wei, M. C., Lindsten, T., Mootha, V. K., Weiler, S., Gross, A., Ashiya, M., Thompson, C. B., and Korsmeyer, S. J. (2000). tBID, a membrane-targeted death ligand, oligomerizes BAK to release cytochrome c. Genes Dev 14, 2060–2071.

Wiedmann, M. W. and Caca, K. (2005). Molecularly targeted therapy for gastrointestinal cancer. Curr Cancer Drug Targets 5, 171–193.

Williams, R. J. (2001). Trichostatin A, an inhibitor of histone deacetylase, inhibits hypoxia-induced angiogenesis. Expert Opin Investig Drugs 10, 1571–1573.

Xiao, H., Hasegawa, T., Miyaishi, O., Ohkusu, K., and Isobe, K. (1997). Sodium butyrate induces NIH3T3 cells to senescence-like state and enhances promoter activity of p21WAF/CIP1 in p53-independent manner. Biochem Biophys Res Commun 237, 457–460.

Xiong, Y., Dowdy, S. C., Podratz, K. C., Jin, F., Attewell, J. R., Eberhardt, N. L., and Jiang, S. W. (2005). Histone deacetylase inhibitors decrease DNA methyltransferase-3B messenger RNA stability and down-regulate de novo DNA methyltransferase activity in human endometrial cells. Cancer Res 65, 2684–2689.

Xu, W. S., Perez, G., Ngo, L., Gui, C. Y., and Marks, P. A. (2005a). Induction of polyploidy by histone deacetylase inhibitor: a pathway for antitumor effects. Cancer Res 65, 7832–7839.

Xu, Y., Voelter-Mahlknecht, S., and Mahlknecht, U. (2005b). The histone deacetylase inhibitor suberoylanilide hydroxamic acid down-regulates expression levels of Bcr-abl, c-Myc and HDAC3 in chronic myeloid leukemia cell lines. Int J Mol Med 15, 169–172.

Yamashita, Y., Shimada, M., Harimoto, N., Rikimaru, T., Shirabe, K., Tanaka, S., and Sugimachi, K. (2003). Histone deacetylase inhibitor trichostatin A induces cell-cycle arrest/apoptosis and hepatocyte differentiation in human hepatoma cells. Int J Cancer 103, 572–576.

Yang, W. M., Inouye, C., Zeng, Y., Bearss, D., and Seto, E. (1996). Transcriptional repression by YY1 is mediated by interaction with a mammalian homolog of the yeast global regulator RPD3. Proc Natl Acad Sci USA 93, 12845–12850.

Yang, X. J. and Gregoire, S. (2005). Class II histone deacetylases: from sequence to function, regulation, and clinical implication. Mol Cell Biol 25, 2873–2884.

Yang, Y., Hou, H., Haller, E. M., Nicosia, S. V., and Bai, W. (2005). Suppression of FOXO1 activity by FHL2 through SIRT1-mediated deacetylation. EMBO J 24, 1021–1032.

Yeung, F., Hoberg, J. E., Ramsey, C. S., Keller, M. D., Jones, D. R., Frye, R. A., and Mayo, M. W. (2004). Modulation of NF-kappaB-dependent transcription and cell survival by the SIRT1 deacetylase. EMBO J 23, 2369–2380.

Yokota, T., Matsuzaki, Y., Miyazawa, K., Zindy, F., Roussel, M. F., and Sakai, T. (2004). Histone deacetylase inhibitors activate INK4d gene through Sp1 site in its promoter. Oncogene 23, 5340–5349.

Yoshida, M., Hoshikawa, Y., Koseki, K., Mori, K., and Beppu, T. (1990a). Structural specificity for biological activity of trichostatin A, a specific inhibitor of mammalian cell cycle with potent differentiation-inducing activity in Friend leukemia cells. J Antibiot (Tokyo) 43, 1101–1106.

Yoshida, M., Kijima, M., Akita, M., and Beppu, T. (1990b). Potent and specific inhibition of mammalian histone deacetylase both in vivo and in vitro by trichostatin A. J Biol Chem 265, 17174–17179.

Yoshida, M., Shimazu, T., and Matsuyama, A. (2003). Protein deacetylases: enzymes with functional diversity as novel therapeutic targets. Prog Cell Cycle Res 5, 269–278.

Yu, C., Rahmani, M., Conrad, D., Subler, M., Dent, P., and Grant, S. (2003). The proteasome inhibitor bortezomib interacts synergistically with histone deacetylase inhibitors to induce apoptosis in Bcr/Abl+ cells sensitive and resistant to STI571. Blood 102, 3765–3774.

Yuan, Z. L., Guan, Y. J., Chatterjee, D., and Chin, Y. E. (2005). Stat3 dimerization regulated by reversible acetylation of a single lysine residue. Science 307, 269–273.

Zgouras, D., Becker, U., Loitsch, S. and Stein, J. (2004). Modulation of angiogenesis-related protein synthesis by valproic acid. Biochem Biophys Res Commun 316, 693–697.

Zhang, X., Wei, L., Yang, Y. and Yu, Q. (2004a). Sodium 4-phenylbutyrate induces apoptosis of human lung carcinoma cells through activating JNK pathway. J. Cell Biochem. 93, 819–829.

Zhang, X. D., Gillespie, S. K., Borrow, J. M., and Hersey, P. (2003). The histone deacetylase inhibitor suberic bishydroxamate: a potential sensitizer of melanoma to TNF-related apoptosis-inducing ligand (TRAIL) induced apoptosis. Biochem Pharmacol 66, 1537–1545.

Zhang, X. D., Gillespie, S. K., Borrow, J. M., and Hersey, P. (2004b). The histone deacetylase inhibitor suberic bishydroxamate regulates the expression of multiple apoptotic mediators and induces mitochondria-dependent apoptosis of melanoma cells. Mol Cancer Ther 3, 425–435.

Zhang, Y., Adachi, M., Zhao, X., Kawamura, R., and Imai, K. (2004c). Histone deacetylase inhibitors FK228, N-(2-aminophenyl)-4-[N-(pyridin-3-yl-methoxycarbonyl)amino- methyl]benzamide and m-carboxycinnamic acid bis-hydroxamide augment radiation-induced cell death in gastrointestinal adenocarcinoma cells. Int J Cancer 110, 301–308.

Zhang, Y., Jung, M., Dritschilo, A., and Jung, M. (2004d). Enhancement of radiation sensitivity of human squamous carcinoma cells by histone deacetylase inhibitors. Radiat Res 161, 667–674.

Zhu, W. G., Lakshmanan, R. R., Beal, M. D., and Otterson, G. A. (2001). DNA methyltransferase inhibition enhances apoptosis induced by histone deacetylase inhibitors. Cancer Res 61, 1327–1333.

Chapter 14
RNA Interference and Cancer: Endogenous Pathways and Therapeutic Approaches

Derek M. Dykxhoorn, Dipanjan Chowdhury, and Judy Lieberman*

Abstract The endogenous RNA interference (RNAi) pathway regulates cellular differentiation and development using small noncoding hairpin RNAs, called microRNAs. This chapter will review the link between mammalian microRNAs and genes involved in cellular proliferation, differentiation, and apoptosis. Some microRNAs act as oncogenes or tumor suppressor genes, but the target gene networks they regulate are just beginning to be described. Cancer cells have altered patterns of microRNA expression, which can be used to identify the cell of origin and to subtype cancers. RNAi has also been used to identify novel genes involved in cellular transformation using forward genetic screening methods previously only possible in invertebrates. Possible strategies and obstacles to harnessing RNAi for cancer therapy will also be discussed.

Keywords RNA interference, microRNA, cancer, microarray, tumor profile, siRNA, therapy, prognosis

1 Introduction

RNA interference (RNAi) is an endogenous, ubiquitous, and evolutionarily conserved pathway for regulating gene expression. Noncoding stem-loop RNAs, encoded within exons or in intergenic regions, are processed by specialized intracellular RNase III enzymes into small RNAs, called microRNAs or miRNAs.[1-4] The microRNAs are taken up by a multiprotein cytoplasmic complex, called the RNA-induced silencing complex (RISC), which directs the posttranscriptional silencing of a partially complementary mRNA target. Silencing of highly complementary mRNAs can occur through mRNA degradation, but for less complementary targets, gene silencing occurs by inhibiting translation. Most mammalian microRNAs work by the latter

Derek M. Dykxhoorn, Dipanjan Chowdhury, and Judy Lieberman
CBR Institute for Biomedical Research and Department of Pediatrics, Harvard Medical School, 200 Longwood Avenue, Boston, MA 02115, USA

* To whom correspondence should be addressed: e-mail: lieberman@cbr.med.harvard.edu

R. Khosravi-Far and E. White (eds.), *Programmed Cell Death in Cancer Progression and Therapy*.
© Springer 2008

pathway. The rules for identifying silenced target genes are still poorly defined; therefore, only a handful of mammalian genes have been clearly shown to be regulated by the endogenous RNAi pathway. However, current bioinformatic estimates suggest that the expression of a third or more of all genes may be regulated by microRNAs. In other species, such as plants, worms, and flies, RNAi regulates critical genes involved in cellular differentiation and survival. In fact, the first identified endogenous microRNAs regulated the progression from one larval state to another in *Caenorhabditis elegans* development.

There is increasing evidence of a role for microRNAs in cancer.[5–8] This should not be surprising since malignant transformation results from abnormally regulated cell differentiation and survival – processes regulated by microRNAs in other organisms. Here, we review how microRNAs are processed within mammalian cells and then describe the evidence for microRNA regulation of genes implicated in cancer and apoptosis. Recent studies provide examples of emerging networks that regulate the expression of microRNAs and transcription factors to control terminal differentiation in a variety of cell types, a step that is aberrant in cancer. We will discuss how microRNA expression profiles are altered in cancer and might be used for diagnosis and prognosis. We will also discuss recent examples of RNAi-based screens to identify tumor-promoting and suppressor-coding genes and micro-RNAs. Lastly, we will discuss the therapeutic prospects for harnessing RNAi to silence oncogenes or other genes involved in cell proliferation and survival or for interfering with microRNAs that play a role in tumorigenesis.

2   microRNA Biogenesis and the Endogenous RNAi Pathway

Most microRNAs are transcribed within coding mRNAs or as independent transcripts by RNA polymerase II as long precursor primary transcripts that are capped and polyadenylated (Fig. 14.1).[4,9–11] microRNA transcripts are highly structured with an elongated hairpin that contains frequent mismatches, bulges, and non-Watson–Crick base-pairings. In some cases, several microRNAs are coordinately expressed as polycistrons from the same primary transcript.[12–16] The microRNA precursors, called pri-miRNAs, have a characteristic fold-back structure that is recognized in the nucleus by an RNase III-type enzyme, Drosha, and its binding partner, variously called DiGeorge syndrome critical region gene 8 (DGCR8) protein in mammals and partner of Drosha (Pasha) in *Drosophila* and *C. elegans*.[15,17–21] Drosha cleaves the pri-miRNA into a ~70 nt fold-back structure, termed the pre-miRNA, which is exported into the cytoplasm by exportin 5.[22–25] The pre-miRNA is then recognized by Dicer and cleaved into a small dsRNA intermediate that contains both the mature microRNA and the accompanying complementary strand.[26–31] The strand whose 5′-end is less tightly bound to its complementary strand is incorporated into the effector RISC or miRISC.[32,33] The complementary strand is rapidly lost when the microRNA is taken up into RISC.[34,35] In some cases, presumably when both ends are comparably paired, microRNAs can be found that correspond

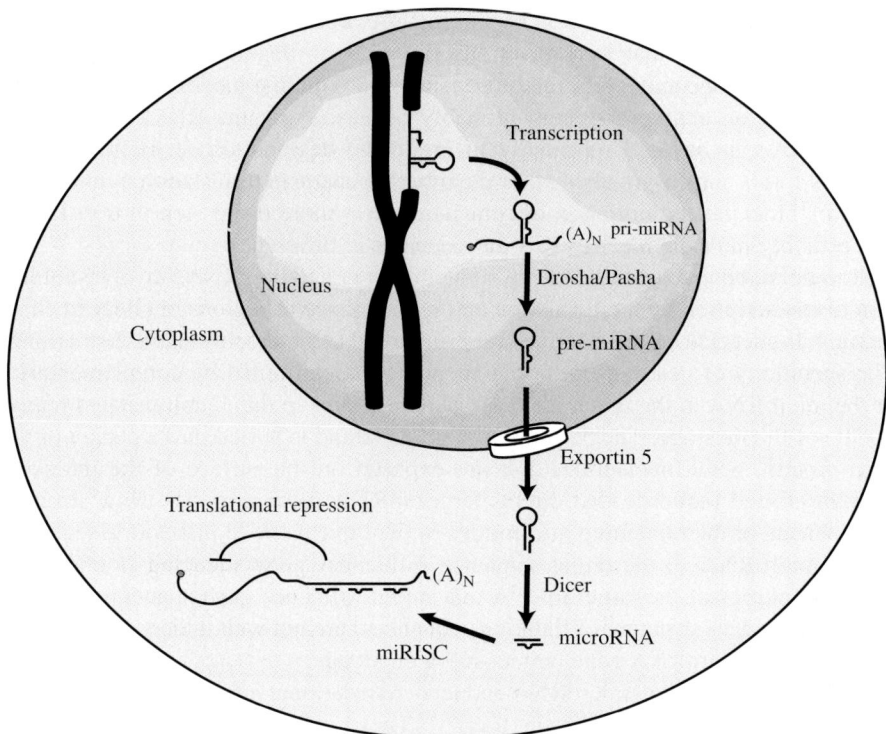

Fig. 14.1 RNA interference pathway. microRNAs that direct the posttranscriptional silencing of gene expression are derived from longer primary transcripts that are expressed from RNA polymerase II promoters.[4,9–11] These primary transcripts, termed pri-miRNAs, can range from several hundred to thousands of nucleotides long with the microRNA sequence encoded in a highly structured RNA hairpin that contains frequent bulges and mismatches.[4] These long hairpins are recognized and cleaved into shorter (~70 nt) hairpin RNAs, pre-miRNAs, in the nucleus by Drosha in conjunction with the double-stranded RNA recognition protein, termed Pasha in *Drosophila* and *Caenorhabditis elegans* and DGCR8 in mammalian cells.[15,17–21] pre-miRNAs are exported from the nucleus into the cytoplasm by Exportin 5 where they are recognized and cleaved into the ~22 nt microRNA by Dicer in conjunction with another dsRNA-binding protein, called Loquacious in *Drosophila* and TRBP in mammals.[22–25,28–30] The miRNA is taken up by the effector complex, miRISC, and the passenger strand is lost, leaving the mature microRNA to guide the recognition of the microRNA-binding sites on the target mRNA, leading to silencing of target gene expression.[34,35] Originally, mammalian microRNAs were thought to mediate target gene silencing by binding to sites on the mRNA that had incomplete complementarity with the microRNA and inducing translational repression, in contrast to small interfering (si)RNAs which have complete (or nearly complete) homology and direct mRNA cleavage.[2,4] However, microRNAs with partial complementarity can facilitate some mRNA degradation, in addition to inducing translational repression[177]

to both strands of the microRNA precursor. The exact composition of the RISC is still unknown. (In fact, this term probably refers to several complexes that may have some core components in common but have additional factors that determine their individual function.) A key component of the RISC is an Argonaute family protein, often Ago 2, which is the RISC endonuclease.[36]

Regulation of gene expression by microRNAs operates through several mechanisms including two that work posttranscriptionally – degradation of the targeted mRNA by cleavage and inhibition of translation.[2,4] The first mechanism has a more potent effect on gene expression, probably because the same RISC-incorporated small RNA can be used repeatedly to guide the degradation of multiple target mRNAs.[37] It is controversial whether the other mechanism (translational inhibition) involves blocking the initiation of translation or a more distal step in translation, and possibly multiple mechanisms may operate in different circumstances. A less well-understood mechanism of gene silencing by noncoding RNAs involves inhibition of transcription by the formation and maintenance of regions of silenced chromatin.[38] In fact, Dicer-deficient cells are impaired in heterochromatin formation.[39] The specificity of posttranscriptional silencing is determined by complementarity of the microRNA to the target mRNA, usually at sites in the 3′ untranslated region (UTR) of the message. The 5′-end of the guide strand is buried into a pocket of the Ago protein, while nucleotides 2–8 are exposed on the surface of the molecule forming a seed sequence that directs target mRNA recognition.[40,41] How strongly base-pairing of the remaining nucleotides of the typical 19–23 nucleotide sequence of the microRNA to the target sequence influences gene silencing is uncertain. Other properties of the target mRNA that might influence gene silencing (such as lack of secondary structure of flanking sequences) are not well understood, making prediction of microRNA gene targets still a challenge.

mRNAs undergoing microRNA-induced translational inhibition appear to be sequestered in distinct cytoplasmic foci.[42–47] These sites, referred to by a variety of names including processing (P)-, cytoplasmic-, GW-, Dcp-, or Lsm-bodies, serve as foci for the accumulation of mRNAs that are destined for degradation.[48–50] In addition to the mRNA, these sites contain essential components of the mRNA degradation pathway, the mRNA decapping enzymes (Dcp1/Dcp2), as well as the 5′-3′ exonuclease Xrn1, Dhh1p, and Pat1p, and in mammalian cells, GW182.[51] The first hint of the interaction of the microRNA machinery with these sites of mRNA turnover was the demonstration that the mammalian Ago proteins implicated in RNAi colocalize with components of mammalian P-bodies.[43,46] Another family of Argonaute proteins, the Piwi family, that have not been found to be associated with microRNAs, do not colocalize.[46] A direct physical interaction between Ago1 and Ago2 with Dcp1 and Dcp2 was also shown by co-immunoprecipitation, even in the absence of RNA or using Ago2 protein, mutated in the Piwi Argonaute Zwille (PAZ) domain required for small RNA binding.[46] However, Ago2 localization to P-bodies is a microRNA/siRNA-dependent process.[45] This was further confirmed by following the fate of reporter mRNAs containing multiple MS2-binding sites that can be visualized with a fluorescently tagged MS2 protein and sites for either an endogenous microRNA or an exogenously introduced siRNA that mimics microRNA function by binding to multiple imperfectly complementary sites on the mRNA.[45] The tagged mRNAs, but not reporter mRNAs that lack the microRNA-binding sites, localize in P-bodies only in the presence of their respective microRNA. A functional link between P-bodies and RNAi-mediated silencing was shown by silencing GW182, which disrupts the formation of P-bodies and significantly impairs gene silencing by both translational repression and mRNA cleavage.[44,45]

The physical and functional link between the sites of mRNA turnover and microRNA/siRNA-mediated silencing raises questions about the potential role of the RNAi machinery in other translational regulation mechanisms. P-bodies are increasingly thought to be sites for the storage of translationally repressed mRNAs, with mRNAs being able to move between the active and inactive pool as needed. One hypothesis put forward by the Parker and Hannon groups is that microRNAs may mediate their repressive function by selectively transporting and possibly even maintaining their mRNA targets in these sites of translational repression, segregated from the translational machinery. It is possible to envision a variety of potential mechanisms by which the RISC could inhibit translation, including impairing various steps in translation (e.g., blocking the processivity or binding of ribosomes along the mRNA) or "tagging" newly formed proteins for degradation, in addition to sequestering the target mRNA from the translational apparatus. The interaction of RNAi components with other sites of mRNA storage and translational regulation, such as stress granules, which are distinct structures that interact with P-bodies, remains to be clarified.

3   Changes in microRNA Expression in Cancer

Mammalian microRNAs were first predicted using RNA-folding algorithms that identified evolutionarily conserved sequences that form into energetically favorable short hairpins that are structurally similar to microRNAs identified in other organisms.[52,53] These algorithms identified about 200 predicted microRNAs in mammalian genomes. A substantial subset of the predicted microRNAs was then verified by cloning small RNAs from a variety of cells. However, when the requirement for evolutionary conservation was relaxed, additional microRNAs were predicted (and the actual number may well exceed 1,000) and some of these have been cloned.[54] These less conserved microRNAs may regulate specialized functions (such as immune responses) that have evolved recently. The makeup and size of the universe of functional mammalian microRNAs is still uncertain, but will soon be more accurately defined using recently available methods for efficiently cloning small RNAs.

microRNAs are expressed in temporally regulated patterns during cell differentiation with distinct expression patterns in different cell types and tissues.[55–62] The total number of microRNAs in a cell can also vary during differentiation and typically constitutes about 1% of the total cellular RNA. Highly expressed microRNAs can be present at as many as 10^4 copies/cell. Highly efficient cloning has enabled researchers to identify microRNAs in rare cell types that are expressed at fewer than 100 copies/cell.[54,63] The functional significance of these rare microRNAs on gene expression is unclear. Figuring this out will be challenging because the gene targets of most microRNAs are unknown, and current target gene prediction algorithms are poor at identifying them. Moreover, the effect on gene expression of a single microRNA binding to an mRNA may be small, particularly when silencing is via translational inhibition. In fact, when single microRNAs are genetically deleted or inhibited, it is rare to find any significant difference in cellular function or fate.[64] However, binding of multiple microRNAs to different sites in the 3' UTR of a gene can coordinately have an impact on its expression.[65] This model of cooperative regulation is reminiscent of models of transcriptional regulation by groups of transcription factors binding to promoter sites on the DNA.

microRNAs have been associated with the regulation of a variety of biological processes from fat metabolism and insulin secretion to cell proliferation, apoptosis, and developmental timing.[66–70] Since microRNAs play such an important role in the regulation of invertebrate development and differentiation, it is not surprising that dysregulation of microRNA expression would be associated with oncogenic transformation in mammals. microRNAs might function as either tumor suppressors or oncogenes depending on their target genes and could contribute to cancer either by enhanced or reduced expression in tumor cells (Fig. 14.2). The first hint that microRNAs might be associated with the development of cancer was the identification of two microRNAs, miR15 and miR16, encoded in a small region of chromosome 13 that is frequently deleted in B-cell chronic lymphocytic leukemia (CLL).[71] These two microRNAs were later found to suppress the expression of bcl-2, an antiapoptotic protein that is frequently overexpressed in B-cell lymphomas and other malignancies. Similarly, expression of miR143 and miR145 is significantly decreased in colorectal cancer specimens compared to matched normal tissue.[72] Expression of these microRNAs is also reduced in a variety of colorectal, breast, prostate, lymphoid, and cervical cancers. In addition, miR26a and mir99a, expressed from regions associated with loss of heterozygosity in lung tumors, have reduced expressed in lung tumors and lung cancer cell lines. Bioinformatic analysis of the regions encoding microRNAs found that most microRNA genes (98 of 186 microRNAs examined) are encoded in regions of the genome associated with cancer, including regions associated with loss of heterozygosity, gene amplification, common break point regions, and fragile sites.[73] Importantly, one of these breakpoint region translocations (t(8,17)) associated with aggressive B-cell lymphoma places the MYC oncogene downstream of the miR142s gene promoter leading to MYC overexpression.[74] Although these studies correlate decreased microRNA expression with the development of cancer, they do not generally identify targets of the microRNAs that can explain their role in tumorigenesis.

Other microRNAs are overexpressed in specific malignancies. A conserved noncoding RNA termed BIC was first identified as a site of insertion of avian leukosis retroviruses, and enhanced expression in chicken B-cell lymphomas.[75] Recently, BIC was found to encode for miR155[74] and was found to be upregulated in human diffuse large B-cell lymphomas (DLBCL) with an activated B-cell phenotype. miR155 overexpressing tumors have poorer prognosis than B-cell lymphomas of the germinal center phenotype.[76–78]

Cancers result from the accumulation of multiple spontaneous and/or inherited mutations that lead to dramatic changes in the pattern of gene expression, particularly

miRNA functioning as an oncogene

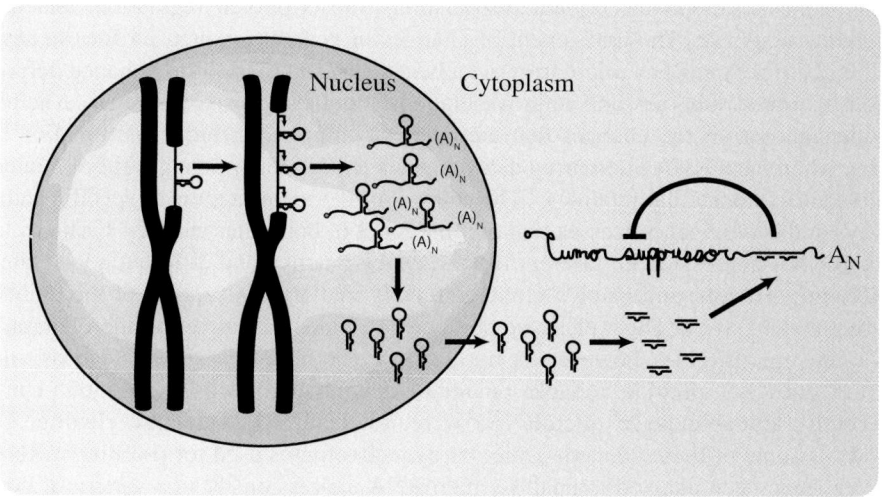

miRNA functioning as a tumor suppressor

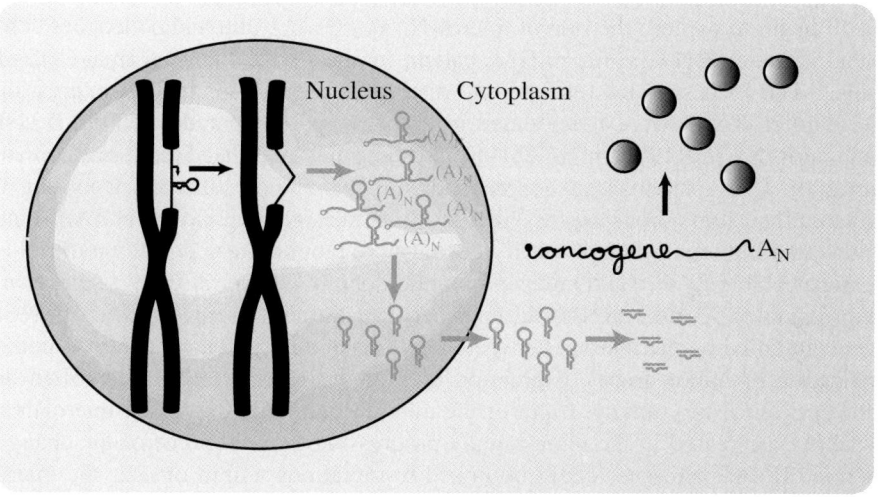

Fig. 14.2 microRNAs can act as either tumor suppressors or oncogenes depending on their targets. (A) *microRNAs as oncogenes*. microRNAs that target a tumor suppressor gene and are overexpressed because of gene amplification (e.g., the miR17–92 polycistron),[86] inappropriate expression of factors that upregulate transcription of the miRNA (e.g., c-Myc upregulation of the miR17–92 cluster)[94] or translocation into a genome locus that alters microRNA expression[73] can lead to cellular transformation, dysregulated proliferation, and tumor formation. (B) *microRNAs as tumor suppressors*. Tumor formation can be induced by the loss or decreased expression of a microRNA whose normal function would be to suppress expression of an oncogene. Inappropriate expression of the oncogene would then lead to cellular transformation. In either case, tumor formation could be a result of increased proliferation, angiogenesis or invasiveness, decreased levels of apoptosis, or alteration of the state of cellular differentiation

in pathways that control cell proliferation and cell cycle regulation, cell signaling, angiogenesis, apoptosis, protein degradation, transcriptional regulation, and the immune response. The assessment of changes in gene expression profiles in specific cancers by mRNA microarrays can be used to some extent to enhance definition of tumor subtypes and improve diagnosis and prognosis. However, a better understanding of the changes that are necessary for oncogenic transformation is seen when the mRNA microarray data are analyzed for changes in groups of related molecules, "molecular modules." This allows for the identification of specific pathways and biological processes that are disrupted in particular cancers. Changes in microRNA expression in different tissues, developmental and differentiation states were initially assessed using a cloning strategy that took advantage of the unique structure and size of Dicer cleavage products to isolate and sequence microRNAs.[13] Cloning microRNAs, however, is not suitable for high-throughput analysis and often does not provide reliable quantitative comparisons of expression. Until recently, low-abundance microRNAs were not readily detected by cloning. To address some of these concerns, microarray technologies used for profiling mRNA levels have been adapted to analyze microRNA expression[79–84] in a variety of normal and cancerous tissues, including CLL[85,86] and solid tumors, including lung, breast, stomach, prostate, colon, and pancreatic cancer.[87,88]

To begin to explore the role of microRNAs in CLL, Calin and colleagues compared the microRNA profile of CLL patient samples with that of normal CD5+B cells.[85] The CLL samples fell into two distinct clusters of microRNA expression. Some microRNAs were upregulated in both groups compared to CD5+B cells (e.g., miR183, miR190, and miR24–1) and some downregulated (e.g., miR213 and miR220). CLL patients can be grouped into two major subtypes according to whether their tumor cells express high levels of the signaling molecule ZAP70 and unmutated immunologlobulin heavy chain (more rapid disease progression) or low or undetectable ZAP70 and mutated immunoglobulin (slower disease progression). Expression of 13 microRNAs differed between the two groups.[89] When patients were classified by the interval between diagnosis and initiation of therapy (another indication of tumor grade), expression of 9 of the 13 microRNAs identified the slower progressing tumors. Eight of the nine differentially expressed microRNAs were overexpressed in the more rapidly progressing tumors. Some of the changes in microRNA expression could be linked to mutations within or near the microRNA sequences.

Microarray analysis also found that microRNA expression differed between normal and cancerous tissue in solid tumors, as well as between solid tumors arising from different organs.[90] microRNA expression by prostate, colon, stomach, and pancreatic adenocarcinomas tend to cluster together. On the other hand, lung and breast cancer samples have distinct patterns of microRNA expression. A few microRNAs (miR21, miR17–5p, and miR191) are overexpressed in a majority of solid tumors. One would expect that these common microRNAs might be involved in dysregulating cellular processes, such as cell proliferation, that are aberrant in all malignancies, while the tissue-specific microRNAs might be involved in oncogenic or differentiation events relevant to specific tissues.[91]

To facilitate microRNA profiling in human cancers, Lu et al. developed a highly effective and specific bead-based solution hybridization procedure.[90] This technique uses oligonucleotide-capture probes linked to polystyrene beads impregnated with a variable combination of fluorescent dyes, a specific combination for each microRNA that is being tested, to analyze rapidly the microRNA composition of large numbers of samples. By binding the oligomer-capture probes to the microRNAs in solution, as opposed to on a solid support (e.g., glass slides), microRNA family members that differ from one another by only a single nucleotide can be distinguished without much cross-reactivity. This method has a robust dynamic range with linear detection over a 100-fold range of microRNA expression. It was used to analyze the microRNA profile from 334 primary tumors representing a variety of tumor types and tissues of origin. Tumor samples showed decreased overall microRNA expression. Tumors of related lineage (i.e., epithelial, endodermal, and hematopoietic) clustered together, and expression patterns differed between tumors and their normal cellular counterparts. Moreover, tumors whose histology was not diagnostic could be assigned based on their microRNA expression profile with much more assurance than would be possible from mRNA profiling.

Early indications suggest that microRNA expression patterns will be more informative than mRNA microarrays in characterizing cancer cells.[90] It is likely that microRNA profiling will soon be used to refine diagnosis and subtype tumors to improve prognostic information and guide the choice of therapies. As the targets of the microRNAs whose expression is altered in various cancers are elucidated, this information will hopefully shed light also on the key events that contribute to the development and progression of cancers. It would not be surprising, for example, if alterations in microRNA genes might underlie poorly understood processes, such as metastasis.

4   microRNAs as Oncogenes or Tumor Suppressor Genes

Recently, a few pathways for microRNA regulation of genes implicated in cellular transformation have begun to be uncovered, but it is clear that this is just the beginning (Table 14.1). let-7, one of the first identified and most well-conserved microRNAs, regulates developmental timing in *C. elegans*. Upregulation of let-7 is necessary for the terminal differentiation of seam cells in adult animals by facilitating their exit from the cell cycle. In worms that lack let-7 expression, seam cells continue to divide and fail to differentiate, similar to cancer cells. In fact, let-7 is downregulated in lung cancer cells and cell growth of a lung cancer cell line is inhibited by overexpression of let-7.[92] These results suggest that let-7 acts as a tumor suppressor. This was demonstrated to be the case when the RAS oncogene was identified as a let-7 target in mammalian cells.[93] In *C. elegans* as well, the RAS homolog, let-60/RAS, is inhibited by the let-7 family members, let-7 and miR84, which bind to multiple target sites in the 3′ UTR of the let-60/RAS mRNA. Overexpression of miR84 in vulval cells

Table 14.1 Validated microRNA targets in mammalian cell proliferation, differentiation, and apoptosis

microRNA	Target gene	Function	Reference
let-7 family	Ras and its homologues	Cell proliferation	93
miR17–5p	E2F1	Transcription and cell proliferation	94
miR20a	E2F1	Transcription and cell proliferation	94
miR181	Hox A11	Hematopoiesis	99
		Skeletal myoblast differentiation	108
miR223	Nuclear factor I-A (NFI-A)	Granulopoiesis	100
miR221	c-kit receptor	Erythropoiesis	101
miR222	c-kit receptor	Erythropoiesis	101
miR130a	Transcription factor MAFB	Platelet physiology	102
miR10a	HoxA1	Megakaryocyte differentiation	102
miR196a	HoxB8	Limb development	106,107
miR1	Histone deacetylase 4 (HDAC 4)	Skeletal myogenesis	112
	Hand2	Cardiac development	111
miR133	Serum response factor (SRF)	Myoblast proliferation	112
miR134	Lim-domain-containing protein kinase 1 (Limk1)	Dendritic spine development	115
miR375	Myotrophin	Insulin secretion	66
miR143	ERK5/BMK1	Adipocyte differentiation	117
miR15a	bcl-2	Antiapoptosis	120
miR16–1	bcl-2	Antiapoptosis	120
miR372	LATS2	Tumor suppressor (germ cells)	135
miR373	LATS2	Tumor suppressor (germ cells)	135

(vulval development being a good model for let-60/RAS function) leads to abnormal vulval development and precocious seam cell terminal differentiation. In addition, miR84 overexpression suppresses the effects of activating mutations of let-60/ RAS. Similar to the *C. elegans* let-60/RAS, the human RAS homologues, HRAS, KRAS, and NRAS, contain multiple putative let-7 family member binding sites in their 3′ UTRs. The introduction of a let-7a siRNA that mimics the let-7a microRNA suppresses RAS expression in a liver cancer cell line. Reciprocally, inhibiting let-7a in HeLa cells by transfection of complementary 2′-*O*-methyl antisense oligomers increases RAS expression (Fig. 14.3). Of note, several of the human let-7 family members, let-7a, let-7c, and let-7g, are encoded in chromosomal locations that are commonly deleted in lung cancer samples,[73] and let-7 expression is reduced in lung tumor samples relative to normal adjacent tissue. In fact, the extent of let-7 reduction is an important independent prognostic indicator; patients with the most drastic reductions in let-7 have the poorest prognosis after potentially curative tumor resection. Moreover, reduced let-7 expression is inversely correlated with the level of NRAS protein. These experiments suggest that let-7 family members are tumor suppressors.

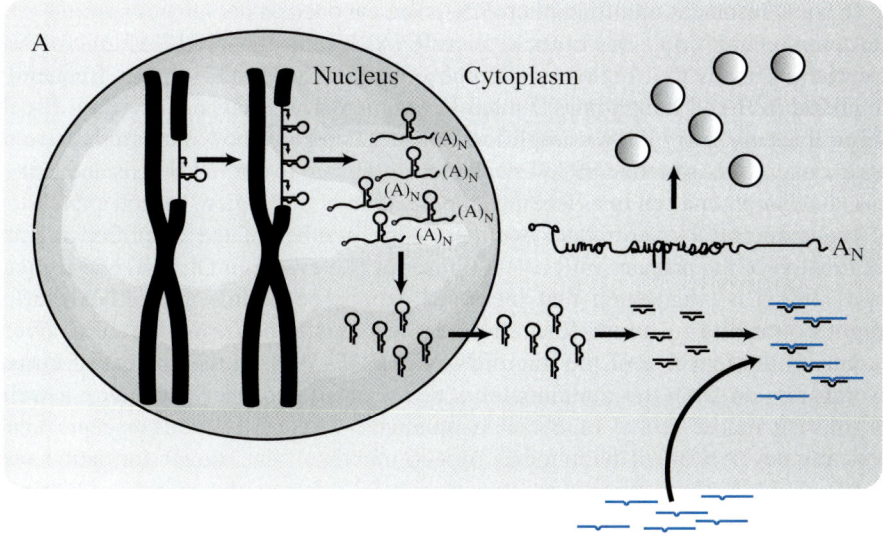

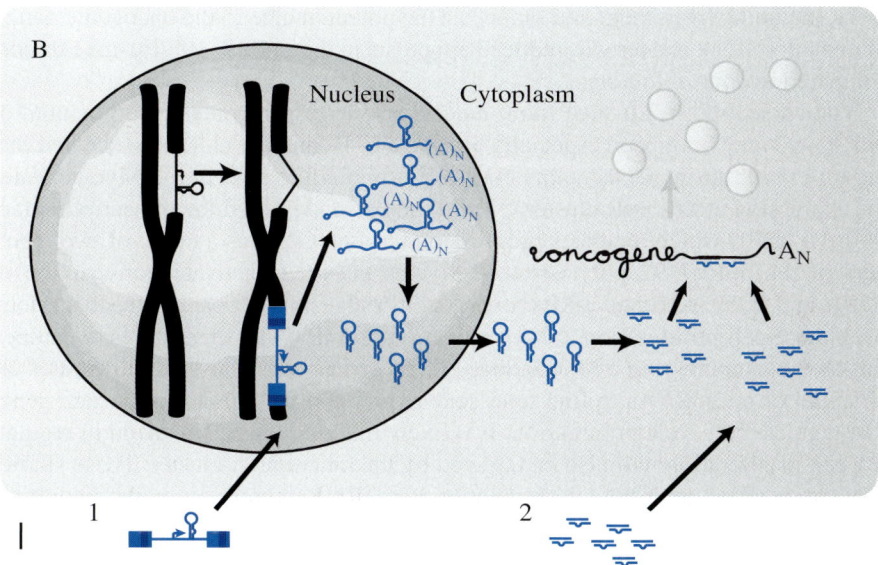

Fig. 14.3 Potential therapeutic approaches to inhibit tumorigenesis associated with altered microRNA expression patterns. (A) Tumor cells resulting from the overexpression of a micro-RNA that functions as an oncogene can be treated with cleavage-resistant single-stranded RNA molecules (e.g., chemically modifying the RNA by replacing the 2′-hydroxy groups on the sugar backbone with 2′-*O*-methyl groups)[178] that are complementary to the mature miRNA. These RNA molecules can effectively bind to the microRNA, preventing the association of the microRNA with its target gene(s) and thereby restore expression of the tumor suppressor gene and inhibit tumor growth. (B) Tumors that result from the loss of expression of a microRNA that acts as a tumor suppressor can be treated by reintroducing the microRNA into the cells. This can be

(continued)

In some instances, multiple microRNAs are encoded as polycistrons from a single common transcript. One of these microRNA clusters, the miR17–92 microRNA polycistron, maps to a region of chromosome 13 (13q31–q32) that is frequently amplified in B-cell lymphomas.[86] microRNA microarray analysis of several B-cell tumor lines that carry known amplifications of this region show increases in five of the six microRNAs in the miR17–92 cluster, compared to normal B cells and leukemia, and lymphoma cell lines lacking amplification of this region. In fact, expression of these microRNAs correlates with the copy number of the amplified region. Expression of the primary miR17–92 transcript is elevated in DLBCL and follicular lymphomas, suggesting that increased expression of this microRNA cluster might contribute to tumor formation. To test this hypothesis, He et al. overexpressed the first five of the microRNAs (miR17–19b), in the context of c-myc overexpression from the immunoglobin heavy chain enhancer (Eμ-myc), a well-established mouse model of B-cell lymphomas. While Eμ-myc transgenic mice typically develop B-cell lymphomas by 4–6 months of age, tumor formation was accelerated in Eμ-myc mice overexpressing miR17–19b, with a mean age of tumor formation of 51 days. Overexpression of each of the microRNAs in the miR17–19b cluster separately failed to enhance the rate of tumor formation. These tumors were particularly aggressive, invading visceral organs outside the lymphoid compartment, including liver, lung, and kidneys. One potential clue to the oncogenic nature of this microRNA cluster was reduced apoptosis in the miR17–19b/Eμ-myc tumors compared to control tumors.

The choice of the Eμ-myc transgenic mouse to test the oncogenic potential of miR17–92 may have been especially apt since O'Donnell et al. (2005) showed that the miR17–92 promoter contains c-Myc E-box binding sites and c-Myc activates the expression of this miR cluster.[94] This study also identified the transcription factor gene E2F1, which regulates progression through G1/S, as a target of two members of the miR17–92 polycistron. c-Myc is known to activate transcription of E2F1, and E2F1 activates c-Myc expression, suggesting a positive feedback loop to enhance cell proliferation. c-Myc induction of miR17–92 then serves to dampen this loop by suppressing c-Myc-induced E2F1 expression. This study illustrates the potential of microRNAs to fine-tune gene expression patterns for important genes that regulate cell cycle progression. It is likely that other genes involved in regulating cell proliferation will also be targeted by this microRNA cluster. These studies provide an example of how dysregulating microRNA expression might disrupt the

---

Fig. 14.3 (continued) achieved either by (1) introducing a DNA-based microRNA expression construct (e.g., using an oncoretroviral or lentiviral vector) that stably expresses the microRNA or (2) by directly introducing a chemically synthesized duplexed form of the microRNA that can enter the microRNA pathway and direct the silencing of the target oncogene. Alternatively, therapeutic benefit could be achieved by introducing siRNAs that silence expression of the dysregulated oncogene or any gene that will inhibit tumor growth (e.g., genes involved in cell cycle progression or angiogenesis) or make the tumor more sensitive to radiation or chemotherapy

fine balance that regulates cell growth, disrupted during oncogenic transformation. However, these regulatory networks may be more complicated than this story suggests; although the miR17–92 cluster is amplified in some B-cell lymphomas and clearly promotes tumor formation in the Eµ-myc mouse, the same region is associated with loss of heterozygosity in some hepatocellular carcinomas.

5   microRNAs and Differentiation in Mammalian Cells

Cancer cells are sometimes considered to be "frozen" in an undifferentiated or partially differentiated state. Until recently, differentiation research primarily focused on transcriptional regulation by regulatory DNA sequences (promoters, enhancers, and locus control regions) that are proximal to protein-coding sequences, paying little attention to the "noncoding" genomic DNA. The discovery of microRNAs has focused attention on mechanisms of posttranscriptional regulation of differentiation. As a general rule, total microRNA expression is higher in terminally differentiated cells than in less-differentiated cells and is higher in adult tissues than in embryos. Moreover, microRNAs have well-defined and distinct expression patterns in different tissues, particularly in cells of different developmental lineages. These findings suggest that microRNAs might play an important role in regulating terminal differentiation in different lineages. In fact, several recent studies provide compelling examples of regulatory networks (or the beginnings of networks) involving microRNAs, discussed below, that hint at an important role for microRNAs in controlling terminal differentiation, a step that is aberrant in cancer. These emerging regulatory circuits often involve intimate connections between transcription factors and microRNAs with, on the one hand, microRNA gene expression being regulated by transcription factors known to be important in lineage determination and, on the other, microRNAs suppressing the expression of key transcription factors.

The first evidence that microRNAs play a role in the differentiation of mammalian cells came from the conditional deletion of Dicer1. (Loss of Dicer1 is lethal early in development.[95]) Conditional deletion of Dicer1 in embryonic stem cells,[39] T cells,[96] limb mesoderm,[97] and skin[98] showed gross defects in differentiation in all these lineages. The logical inference is that impaired production of microRNAs in the absence of Dicer1 interferes with cellular differentiation.

microRNAs in hematopoiesis Much of the initial work in studying the role of microRNAs in mammalian cell differentiation has been elucidated in hematopoiesis, probably the best-studied system of mammalian cellular differentiation. The first example implicated miR181, whose expression is increased in thymus, lymphoid tissues, and bone marrow, in promoting B-cell differentiation. Ectopic expression of miR181 in mouse hematopoietic precursor cells leads to a dramatic increase in B lineage cells.[99] Another microRNA, miR223, expressed in the bone marrow, is important in granulopoiesis.[100] An elegant network involving miR223 and two competing transcription factors, C/EBPα and NFI-A, appears to control the differentiation of promyelocytes into granulocytes. During in vitro and in vivo retinoic acid-induced differentiation of leukemic promyleocytes, miR223 expression is upregulated. The miR223 promoter contains overlapping sites for C/EBPα and NFI-A binding. C/EBPα upregulates and NFI-A inhibits miR223 expression. Upon retinoic acid treatment, NFI-A expression declines, while C/EBPα is upregulated. C/EBPα then binds and displaces NFI-A from the miR223 promoter to enhance miR223 expression. This molecular circuit is complete when miR223 binds the 3′ UTR of NFI-A transcripts, blocking further expression of NFI-A. The importance of miR223 in regulating granulocyte differentiation was shown by inducing differentiation of promyelocytes by ectopic expression of miR223 without retinoic acid and by blocking retinoic acid-induced differentiation by inhibiting miR223.

Two clustered microRNAs, miR221 and miR222, abundantly expressed in CD34 + hematopoietic precursor cells, are downregulated upon in vitro differentiation into the erythroid lineage.[101] One likely target of these microRNAs is the kit receptor, required for proliferation and erythroid differentiation in response to kit ligand. Overexpressing either of these microRNAs reduces kit expression, cell proliferation under erythroid-promoting conditions, and engraftment of CD34+ cord blood cells into immunodeficient mice. Because constitutively activated c-kit has been implicated in leukemias and gastrointestinal stromal tumors, inducing expression or transducing cells with these microRNAs (or their siRNA analogues) might have therapeutic benefit.

Another study that looked at in vitro differentiation of CD34+ progenitor cells into megakaryocytes found a group of downregulated microRNAs, one of which might be involved in targeting the transcription factor *MAFB*, upregulated during megakaryopoiesis and involved in activating transcription of the megakaryocyte-specific gene GPIIB.[102] Another downregulated microRNA miR10a is embedded in the HOX gene cluster and potentially targets HoxA1.

HOX gene microRNAs miR10 and miR196 microRNA families are embedded within the four HOX clusters of mammalian homeobox transcription factor genes,[103] which play an important conserved role in determining the identity of cells in the developing embryo. The intricate expression pattern of HOX genes persists in adult tissues, but their roles are modified according to specific cellular needs (reviewed in[104]). The embryonic expression of the HOX-embedded microRNAs closely follows that of their "host" HOX cluster genes.[105] Moreover, the HOX-embedded microRNAs have been shown in a few examples to regulate the expression of HOX genes. miR196a binds to the HOXB8 3′ UTR and inhibits HOXB8 expression by cleaving the HOXB8 transcript.[106] HOXB8 and miR196a also have complementary expression patterns during embryogenesis, supporting the idea that miR196a regulates the expression of HOXB8 during development.[103] miR196a is overexpressed in embryonic mouse hindlimbs compared to forelimbs, while the expression pattern of HOXB8 is the opposite. At least in chickens, miR196a appears to impede the retinoic acid-induced expression of HOXB8 and sonic hedgehog (Shh) in forelimb development to establish anterior–posterior patterning.[107] However, loss of microRNAs in Dicer-deficient hindlimbs does not induce

HOXB8 expression. This result suggests that the primary regulation of HOX gene expression might not be via microRNAs or that multiple microRNAs might be involved in more complicated regulatory networks.[107]

Another illustration of HOX gene regulation by microRNAs is the regulation of HOXA11 in differentiating myoblasts by miR181.[108] miR181 is upregulated in differentiating muscle cells during development or during regeneration in response to injury, but is not expressed in undifferentiated myoblasts or fully differentiated muscle cells. Its target HoxA11, which inhibits myoblast differentiation,[109,110] is reciprocally expressed in myoblasts and (at low levels) in adult muscle cells, but turned off during differentiation. Inhibiting miR181 interferes with myoblast differentiation, but does not completely restore HoxA11 expression, suggesting that multiple microRNAs or other pathways contribute to this process.[108] Moreover, the ectopic expression of miR181 does not induce myoblast differentiation, again suggesting a more complex regulatory network. Nonetheless, the involvement of miR181 in both B-cell and myoblast differentiation suggests that miR181 might be involved in regulating common pathways activated during the terminal differentiation of cells of mesodermal origin.

microRNAs and muscle development The miR1 family and miR133 genes are specifically and highly expressed in adult skeletal and cardiac muscle tissues and to a lesser extent during development of these tissues. Expression of miR1 genes is activated in the heart by the serum response factor (SRF) transcription factor and its cofactor myocardin and in skeletal muscle by the Mef2 and MyoD transcription factors. One of the targets of miR1 is the Hand2 transcription factor that promotes proliferation of cardiac muscle precursor cells. Cardiac embryonic development is activated when Hand2 begins to be expressed. Although Hand2 mRNA persists in adult cardiac tissue, Hand2 protein is downregulated coincident with miR1 expression. Precocious expression of miR1 in the developing heart leads to severe defects in heart formation because of decreased cell division.[111] Therefore, miR1 controls terminal differentiation of myocardiocytes.

Another microRNA (miR133) is clustered with miR1-1, and they are transcribed as a single transcript beginning late in embryonic development.[112] However, miR1 and miR133 have opposing effects on myoblast fate – as it does for the heart, miR1 promotes skeletal myoblast differentiation, whereas miR133 promotes myoblast proliferation and inhibits differentiation.[112] One of the targets of miR1 in skeletal muscle is HDAC4, which globally represses transcription, including transcription of the muscle-specific transcription factor MEF2C.[113] One way that miR133 inhibits differentiation is by suppressing expression of SRF, which activates myoblast differentiation.[114] Recall that SRF activates expression of miR1 (and thus miR133). This negative feedback loop indicates a complicated microRNA-transcription factor-regulated mechanism for controlling muscle cell differentiation. Likely, there will be more to this story.

microRNAs also regulate the function of terminally differentiated cells microRNAs are especially abundant in terminally differentiated cells compared to their precursors, suggesting that they may not only suppress the genes required for proliferation and progenitor cell pluripotency, but may also regulate their effector functions.

Regulating the function of terminally differentiated cells might not be directly related to cellular transformation and cancer. However, a few instructive examples of tissue-specific microRNAs and their role in differentiated cell function will be briefly described. miR375 appears to be exclusively expressed in pancreatic β cells and regulates insulin response to glucose.[66] Increasing miR375 in β cells suppresses glucose-induced insulin secretion, while inhibiting miR375 has the opposite effect. This effect is mediated by the effect of miR375 on Myotrophin (Mtpn), a protein previously not know to play a role in insulin secretion. Silencing Mtpn by siRNA reproduces the miR375-suppressive effect on insulin secretion.

Another interesting example involves the role of miR134, highly expressed in brain, in regulating dendritic spine development of neurons in response to synaptic stimulation.[115] miR134 inhibits translation of Lim-domain-containing protein kinase 1 (Limk1) which regulates dendritic spine formation.[116] miR134 is localized near synapses where Limk1 synthesis takes place. The authors speculate that miR134 might bind to Limk1 mRNA as it is being transported from the cell body to the dendrites and be responsible for suppressing Limk1 translation during transport and before synaptic stimulation. In response to activating stimuli, such as brain-derived neurotrophic factor (BDNF), the inhibitory effect of miR134 on Limk1 translation is reversed.[115] A surprising observation is that even after BDNF stimulation when Limk1 mRNA is being translated, miR134 continues to associate with the Limk1 transcript. How BDNF stimulation might interfere with miR134-mediated silencing of Limk1 translation or bypass it remains a puzzle.[117]

6 microRNAs and Apoptosis

Deregulation of cell death is an important feature of many cancers (reviewed in[118]). Highly conserved caspase-dependent pathways are often inactivated in transformed cells, principally by overexpression of inhibitors of apoptosis, including antiapoptotic bcl-2 family members, survivin, and other IAP family members. The first example of a role for microRNAs inhibiting apoptosis was in *Drosophila*, where expression of the proapoptotic factor hid is repressed by the microRNA bantam.[67] Bantam not only blocks apoptosis, but also directly increases cell proliferation.[67] In flies, miR14, the miR2 gene family and miR278 also act as potent cell death suppressors.[68,119,120]

In mammals the first evidence for a role of microRNAs in regulating apoptosis comes from conditional deletion of Dicer, which in embryonic limbs causes extensive apoptosis.[97] Deletion of Dicer in the T-cell lineage reduces the numbers of mature T cells, which both proliferate more slowly and are more prone to apoptosis in response to stimulation.[96] Expression of antiapoptotic Bcl-2 in B-cell lymphomas is a likely target of miR15a and 16-1, which are deleted in many high-grade B-cell malignancies.[121] Not only is bcl-2 expression tightly correlated with expression of these microRNAs, but transfection of bcl-2+ leukemia cells with an expression plasmid for either or for both of these microRNAs leads to downregulation of bcl-2 and induction of apoptosis. Glioblastoma cells and some other tumors strongly overexpress miR21.[122] Depletion of miR21 in cultured glioblastoma cells activates caspases and leads to increased apoptosis through an unknown mechanism. Interestingly, another study using antisense microRNA inhibitors in cervical adeno-carcinoma HeLa cells identified miR21 as an inhibitor of cell growth with no direct effect on apoptosis.[123] The biological effects of any particular microRNA, including miR21, in different cells are likely to vary depending on the cell-specific repertoire of expressed target genes. Although these studies support a role for microRNA regulation of apoptosis, understanding the target genes and pathways in mammalian cells awaits further research.

7 RNA Interference-Based Screens to Identify Novel Tumor Suppressor Genes and Oncogenes

RNAi has provided new opportunities to identify novel genes implicated in a variety of diseases by forward genetic screens. Before the discovery that RNAi worked in mammalian cells, the power of unbiased screens to identify unexpected participants in biologically important pathways was only available in invertebrates. Libraries of retroviruses encoding short hairpin RNAs (shRNAs) or arrays of siRNAs mixed with a transfection reagent, designed to silence a large proportion of human-expressed genes or functionally related subsets of genes (i.e., all kinases and phosphatases, all known ubiquitin ligases), can be used to identify genes involved in cellular transformation, susceptibility to apoptosis, or drug resistance. Similarly, libraries of retroviruses encoding microRNAs can be used to identify microRNAs involved in cancer. Identifying a gene candidate in any screen is only the first step to validating its role in a biological pathway or disease. Some illustrative examples of RNAi-based cancer screens are given below.

Cancer cells are especially sensitive to apoptosis induced by the tumor necrosis factor (TNF)-related apoptosis-inducing ligand (TRAIL). To identify genes that might enhance or suppress TRAIL-mediated apoptosis, TRAIL was added to HeLa cells transfected in microtiter plates with a panel of siRNAs targeting 510 genes, including 380 kinases.[124] This screen was able to identify several unknown genes whose silencing either sensitized or desensitized cells to TRAIL-induced apoptosis and to identify several signaling pathways (WNT, MYC) required for maintaining TRAIL sensitivity, since silencing multiple genes in these pathways differentially affected cell survival in response to TRAIL. Another siRNA-based loss-of-function screen surveyed hundreds of kinase and phosphatase genes to identify those that enhance or suppress apoptosis of HeLa cells either on their own or in conjunction with chemotherapeutic drugs.[125] A large proportion of these enzymes (i.e., more than one third of the phosphatases) affected cell survival by at least twofold. The large number of "hits" suggests that more refined screens or biological verification would be needed to winnow through these leads to identify attractive targets for drug development. For example, identifying kinases or phosphatases that are preferentially needed for survival of different types of cancer cells vs normal cells or are only required for survival in the face of radiation or chemotherapy would be a first step.

Another screening approach uses libraries of DNA or viral vectors to express shRNAs, processed intracellularly into active siRNAs. Most libraries have used retroviral vectors because of the transduction efficiency and stability of gene expression afforded by these vectors. Brummelkamp et al.,[126] examining what effect silencing the expression of 50 deubiquitylating enzymes had on TNFα activation of NF-κB, singled out the cylindromatosis (CYLD) tumor suppressor gene, which is linked to a familial proliferative skin disease. With this lead they were then able to pinpoint a role for CYLD in deubiquitinating TRAF2, which activates IKK and consequently NF-κB. By interfering with NF-κB activation using sodium salicylate, they could enhance apoptosis of CYLD-silenced cells. This result was rapidly translated to show that topical aspirin derivatives could be used to treat this rare disfiguring disorder.

Although this study screened a small set of genes, large-scale plasmid and retroviral shRNA expression libraries targeting a large proportion of the human and mouse genome have been constructed and validated by several groups.[127,128–133] Some of these vectors express the shRNA within a microRNA sequence to enhance its processing and increase the efficiency of silencing.[129] In screens for tumor suppressor genes, cells at the brink of transformation because of expression of combinations of oncogenes are transduced to express shRNAs and then selected for outgrowth of transformed cells. One retroviral-based RNAi screen took advantage of a conditionally transformed cell line that expresses the catalytic subunit of telomerase (hTERT) and a temperature-sensitive allele of SV40 large T antigen (tsLT), which allows cells to proliferate at 32°C (the temperature at which tsLT is functional and can inactivate pRb and p53), but not at 39°C at which growth arrest occurs, to identify novel factors that modulate p53-dependent proliferation arrest.[133] After infection with the library, positive colonies, containing cells able to proliferate at 39°C, were selected and sequenced to identify the gene being silenced. shRNAs targeting six genes were pulled out of the screen, including the p53 gene, as well as five novel genes – RPS6KA6 (ribosomal S6 kinase 4, RSK4), Tip60 (histone acetyltransferase), HDAC4 (histone deacetylase), KIAA0828 (putatitive S-adenosyl-L-homocysteine hydrolase, SAH3), and CCT2 (T-complex protein 1, β-subunit). These novel genes were validated by showing that shRNAs targeting each of the genes selected in the screen were able to inhibit growth arrest induced by ionizing irradiation or p19[ARF] overexpression. In a similar manner, Westbrook et al.[134] used another shRNA expression library to look for potential tumor suppressor genes that inhibit transformation of human mammary epithelial cells (HMECs) expressing hTERT and SV40 large T antigen. Colonies of cells that demonstrated anchorage-independent growth after infection with the shRNA library were isolated. The silenced genes were identified by DNA sequencing and bar code (a sequence identifier specific for each shRNA construct) microarray analysis. This approach identified several previously known tumor suppressor genes, including TGFBR2 and PTEN, as well as a gene that had not been previously shown to have tumor suppressing properties, REST/NRSF (RE1-silencing transcription factor/neuron-restrictive silencing factor). The role of REST as a tumor suppressor was confirmed by expressing a dominant negative REST gene. The tumor suppressor activity of REST was then found to be mediated by its ability to suppress PI(3)K-dependent signaling. REST is often deleted in colorectal cancer cell lines.

In a similar assay system, a retroviral shRNA-expressing library was used to screen for potential tumor suppressor genes whose silencing could substitute for overexpression of a constitutively active form of RAS (RAS^{V12}) and permit anchorage-independent growth in fibroblasts that overexpressed the catalytic subunit of telomerase (hTERT), SV40 small t antigen, and had silenced p53 and p16^{INK1A130}. The homeodomain pituitary transcription factor PITX1 was identified and confirmed as a tumor suppressor gene by showing that inhibiting PITX1 expression activates the RAS pathway by activating the promoter of RASAL1, a RAS-GTPase activating protein that connects Ca^{2+} signaling to RAS activity.

These previously described screens could identify tumor suppressor genes whose silencing promotes cellular proliferation or anchorage-independent growth, but they could not be used to identify potential oncogenes, whose silencing would cause growth arrest or cell death. To identify putative oncogenes, Staudt and colleagues[135] used an inducible shRNA retroviral library to identify by microarray analysis shRNAs that were depleted in abundance when transduced and induced DLBCL lines were cultured for 3 weeks. All of the depleted shRNAs silenced NF-κB pathway components, including IKBKB, CARD11, MALT1, and BCL10. Interestingly, these genes were required for the proliferation of only activated-type DLBCL and not germinal center-type DLBCL, suggesting that they might be good selective drug or siRNA targets.

Another type of screen was used to identify microRNAs that act as oncogenes. Using a retroviral library to express many of the known human microRNAs, Voorhoeve and colleagues[136] identified two microRNAs, miR372, and 373, which share the same seed sequence, that cooperate with a constitutively active form of RAS (RAS^{V12}) to transform primary human fibroblasts that express wild-type p53. miR372 and 373 expression was elevated in testicular germ cell tumors, which mostly contain functional p53, but not in normal testes or in samples from breast, colon, lung, and brain tumors. Expression of a putative tumor suppressor gene, large tumor suppressor homolog 2 (LATS2), predicted to contain two potential miR372/373-binding sites, is decreased in cells overexpressing these microRNAs, and its silencing may be contributing to the oncogenic effect of these microRNAs.

8 Harnessing RNA Interference for Cancer Target Validation and Therapeutics

Although we are just beginning to understand the role of microRNAs in cancer, many investigators are already exploring the possibility of exploiting the power of RNAi for cancer therapy and drug target validation in animal models. (A discussion of this extensive body of work is beyond the scope of this review [see, e.g.,[137,138]]). RNAi has become a standard tool to identify the importance of any particular gene in diverse biological pathways, including those implicated in cellular transformation. siRNAs can be highly (but not completely) specific and can distinguish a single nucleotide polymorphism, as was first demonstrated by targeting the point mutation that constitutively activates a RAS oncogene, leaving wildtype RAS unaffected. RNAi can be used effectively to silence the expression of any gene in any cell in vitro.[139] Transgenic mice expressing shRNAs can also be used to identify the importance of particular genes or microRNAs in cancer formation in vivo.[140]

Cleavage of the target mRNA is likely the most potent RNAi mechanism to harness for therapy, because the same RISC-incorporated small RNA can direct the cleavage of many transcripts and because the transcript is eliminated, not merely repressed. (The relative effectiveness of mechanisms in which chromatin is silenced to inhibit transcription is unknown. This mechanism is too poorly understood to use as the basis for therapy at present.) The RISC-stabilized small RNA is highly stable within the cell – probably with a half-life of 1 week or more. The major determinant of durability of silencing is the rate of cell proliferation, where small RNAs are diluted with each cell division.[141] In terminally differentiated nondividing cells silencing can last for weeks, while in rapidly dividing cell lines silencing peaks 3 days after transduction and is gone by 1 week. For cancer cells, frequent and repetitive dosing will likely be required for siRNA-based drugs. However, less rapidly dividing precancerous lesions or potentially cancer stem cells might be particularly effective targets requiring infrequent treatments. In addition to silencing transcripts for oncogenes, RNAi could be used either to mimic microRNAs identified in promoting differentiation, inducing apoptosis or reducing proliferation or to inhibit cancer-promoting microRNAs. If the studies reviewed here that suggest that microRNAs may be master regulatory switches for terminal differentiation hold up, then transducing cells with siRNAs that mimic such microRNAs may be a highly attractive strategy for cancer therapy.

RNAi-based therapy for cancer could be used to target more than oncogenes. Genes implicated in cell cycle progression[142–146] and angiogenesis[147–150] would be good targets. Particularly, if siRNAs can be targeted preferentially to tumor cells, then any gene required for viability is a potential target, although genes needed only for cell division are particularly attractive since they will cause less toxicity to the majority of nondividing cells. Growth factors or their receptors required for tumor growth are also possible targets.[151–154] Targeting viral oncogenes encoded by EBV,[155–157] HPV,[158,159] and other oncogenic viruses also provides an opportunity for specificity. RNAi-based therapy could also be used in conjunction with chemotherapy or radiation to make cells more susceptible to these agents, particularly by silencing genes involved in drug resistance (i.e., transporters that efflux drugs), DNA repair, or metabolic pathways targeted by these drugs.[160–166] It may also be possible to target cells involved in either supporting the growth of the tumor or eliminating it rather than the tumor itself. For example, tumor infiltrating lymphocytes are largely incapacitated in their ability to destroy tumor cells; by targeting inhibitory receptors or regulatory cells, these tumor-specific immune cells might be activated to eliminate residual tumor cells.[138]

Two strategies can be used to harness RNAi for therapy – one is gene therapy (transducing cells with viral vectors that encode for shRNA precursors processed intracellularly like endogenous microRNAs); the other is to develop siRNAs as small molecule drugs.[167,168] The latter is more suitable for cancer therapy and closer to clinical application. siRNAs can be chemically modified to enhance their pharmacokinetics and reduce potential off-target effects caused by binding to Toll-like receptors, immune sensors of pathogenic double-stranded RNAs. The main obstacle to using siRNAs is delivering them into the cytoplasm of cells, where they work. Cancer cells can be transfected in vitro, but except for superficial sites, this is not a viable strategy for treating most cancers, particularly micrometastases and macrometastases. Although most cells do not readily internalize siRNAs, mucosal surfaces appear to be especially susceptible to topically applied siRNAs.[168] Initial siRNA phase I and II studies targeting the eye and lung (to treat age-related macular degeneration and respiratory syncytial virus infection, respectively) have not met with any unexpected toxicity. Therefore, malignancies that are located at these sites or spread locally are good initial targets. Attractive examples for initial studies might include HPV-related cervical cancer (targeting the E6 and E7 oncogenes), EBV-related nasopharyngeal cancer, lung squamous cell carcinomas, retinoblastoma, or head and neck cancer.

However, for most cancers a method for effective systemic administration is needed and is the major obstacle to using siRNAs for cancer therapy. Recently, several systemic siRNA delivery strategies have begun to be described in animal models. These involve covalently coupling the passenger strand of the siRNA to a targeting molecule (e.g., cholesterol),[169] incorporating the siRNA into liposomes,[170,171] lipoplexes,[172–174] or nanoparticles,[149,175] or mixing the siRNA with fusion proteins, capable of specific targeting by binding to cell surface receptors.[176] This latter approach was used to target and inhibit the outgrowth of a subcutaneous mouse melanoma cell line by intravenous injection of 1 mg/kg of a cocktail of siRNAs. siRNA delivery was highly specific since adjacent normal tissues did not take up the siRNAs.

As for other cancer therapies, drug resistance caused by mutating the target site sequence is an anticipated problem. This may be more of an obstacle for siRNAs than for other types of drugs, since conservative mutations that do not alter the encoded protein may interfere with gene silencing. However, dealing with drug resistance to siRNAs is a much simpler problem than for other small-molecule drugs, which usually work by targeting a single active site on a protein. Since multiple sequences can be used to target any gene, alternate siRNAs can readily be designed. Combinations of siRNAs that target more than one sequence in a gene or multiple genes at once are likely to work synergistically to enhance tumor suppression and reduce the likelihood of emerging drug resistance.[141]

References

1. Ambros, V. (2004). The functions of animal microRNAs. Nature 431(7006), 350–355.
2. Bartel, D. P. (2004). MicroRNAs: genomics, biogenesis, mechanism, and function. Cell 116(2), 281–297.
3. Wienholds, E. and Plasterk, R. H. (2005). MicroRNA function in animal development. FEBS Lett 579(26), 5911–5922.
4. Du, T. and Zamore, P. D. (2005). microPrimer: the biogenesis and function of microRNA. Development 132(21), 4645–4652.
5. Croce, C. M. and Calin, G. A. (2005). miRNAs, cancer, and stem cell division. Cell 122(1), 6–7.
6. S. M. Hammond, S. M. (2006). MicroRNAs as oncogenes. Curr Opin Genet Dev 16(1), 4–9.
7. Gregory, R. I. and Shiekhattar, R. (2005). MicroRNA biogenesis and cancer. Cancer Res 65(9), 3509–3512.
8. Esquela-Kerscher, A. and Slack, F. J. (2006). Oncomirs – microRNAs with a role in cancer. Nat Rev Cancer 6(4), 259–269.
9. Cai, X., Hagedorn, C. H., and Cullen, B. R. (2004). Human microRNAs are processed from capped, polyadenylated transcripts that can also function as mRNAs. RNA 10(12), 1957–1966.
10. Lee, Y., Kim, M., Han, J., Yeom, K. H., Lee, S., Baek, S. H., and Kim, V. N. (2004). MicroRNA genes are transcribed by RNA polymerase II. EMBO J 23(20), 4051–4060.
11. Parizotto, E. A., Dunoyer, P., Rahm, N., Himber, C., and Voinnet, O. (2004). In vivo investigation of the transcription, processing, endonucleolytic activity, and functional relevance of the spatial distribution of a plant miRNA. Genes Dev 18(18), 2237–2242.
12. Reinhart, B. J., Weinstein, E. G., Rhoades, M. W., Bartel, B., and Bartel, D. P. (2002). MicroRNAs in plants. Genes Dev 16(13), 1616–1626.
13. Lagos-Quintana, M., Rauhut, R., Lendeckel, W., and Tuschl, T. (2001). Identification of novel genes coding for small expressed RNAs. Science 294(5543), 853–858.
14. Lau, N. C., Lim, L. P., Weinstein, E. G., and Bartel, D. P. (2001). An abundant class of tiny RNAs with probable regulatory roles in *Caenorhabditis elegans*. Science 294(5543), 858–862.
15. Lee, Y., Jeon, K., Lee, J. T., Kim, S., and Kim, V. N. (2002).MicroRNA maturation: stepwise processing and subcellular localization. EMBO J 21(17), 4663–4670.
16. Rodriguez, A., Griffiths-Jones, S., Ashurst, J. L., and Bradley, A. (2004). Identification of mammalian microRNA host genes and transcription units. Genome Res 14(10A), 1902–1910.
17. Denli, A. M., Tops, B. B., Plasterk, R. H., Ketting, R. F., and Hannon, G. J. (2004).Processing of primary microRNAs by the Microprocessor complex. Nature 432(7014), 231–235.
18. Lee, Y., Ahn, C., Han, J., Choi, H., Kim, J., Yim, J., Lee, J., Provost, P., Radmark, O., Kim, S., and Kim, V. N. (2003). The nuclear RNase III Drosha initiates microRNA processing. Nature 425(6956), 415–419.
19. Gregory, R. I., Yan, K. P., Amuthan, G., Chendrimada, T., Doratotaj, B., Cooch, N., and Shiekhattar, R. (2004). The Microprocessor complex mediates the genesis of microRNAs. Nature 432(7014), 235–240.
20. Han, J., Lee, Y., Yeom, K. H., Kim, Y. K., Jin, H., and Kim, V. N. (2004).The Drosha-DGCR8 complex in primary microRNA processing. Genes Dev 18(24), 3016–3027.
21. Landthaler, M., Yalcin, A., and Tuschl, T. (2004). The human DiGeorge syndrome critical region gene 8 and Its D. melanogaster homolog are required for miRNA biogenesis. Curr Biol 14(23), 2162–2167.
22. Bohnsack, M. T., Czaplinski, K., and Gorlich, D. (2004).Exportin 5 is a RanGTP-dependent dsRNA-binding protein that mediates nuclear export of pre-miRNAs. RNA 10(2), 185–191.
23. Lund, E., Guttinger, S., Calado, A., Dahlberg, J. E., and Kutay, U. (2004). Nuclear export of microRNA precursors, Science 303(5654), 95–98.

24. Zeng, Y., and Cullen, B. R. (2004).Structural requirements for pre-microRNA binding and nuclear export by Exportin 5. Nucleic Acids Res 32(16), 4776–4785.

25. Yi, R., Qin, Y., Macara, I. G., and Cullen, B. R. (2003). Exportin-5 mediates the nuclear export of pre-microRNAs and short hairpin RNAs. Genes Dev 17(24), 3011–3016 (2003).

26. Bernstein, E., Caudy, A. A., Hammond, S. M., and Hannon, G. J. (2001). Role for a bidentate ribonuclease in the initiation step of RNA interference. Nature 409(6818), 363–366.

27. Hutvagner, G., McLachlan, J., Pasquinelli, A. E., Balint, E., Tuschl, T., and Zamore, P. D. (2001). A cellular function for the RNA-interference enzyme Dicer in the maturation of the let-7 small temporal RNA. Science 293(5531), 834–838.

28. Chendrimada, T. P., Gregory, R. I., Kumaraswamy, E., Norman, J., Cooch, N., Nishikura, K., and Shiekhattar, R. (2005). TRBP recruits the Dicer complex to Ago2 for microRNA processing and gene silencing. Nature 436(7051), 740–744.

29. Forstemann, K., Tomari, Y., Du, T., Vagin, V. V., Denli, A. M., Bratu, D. P., Klattenhoff, C., Theurkauf, W. E., and Zamore, P. D. (2005). Normal microRNA maturation and germ-line stem cell maintenance requires Loquacious, a double-stranded RNA-binding domain protein. PLoS Biol 3(7), e236.

30. Jiang, F., Ye, X., Liu, X., Fincher, L., McKearin, D., and Liu, Q. (2005). Dicer-1 and R3D1-L catalyze microRNA maturation in Drosophila. Genes Dev 19(14), 1674–1679.

31. Saito, K., Ishizuka, A., Siomi, H., and Siomi, M. C. (2005). Processing of pre-microRNAs by the Dicer-1-Loquacious complex in Drosophila cells. PLoS Biol 3(7), e235.

32. Schwarz, D. S., Hutvagner, G., Du, T., Xu, Z., Aronin, N., and Zamore, P. D. (2003). Asymmetry in the assembly of the RNAi enzyme complex, Cell 115(2), 199–208.

33. Khvorova, A., Reynolds, A., and Jayasena, S. D. (2003). Functional siRNAs and miRNAs exhibit strand bias. Cell 115(2), 209–216.

34. Rand, T. A., Petersen, S., Du, F., and Wang, X. (2005). Argonaute2 cleaves the anti-guide strand of siRNA during RISC activation. Cell 123(4), 621–629.

35. Matranga, C., Tomari, Y., Shin, C., Bartel, D. P., and Zamore, P. D. (2005). Passenger-strand cleavage facilitates assembly of siRNA into Ago2-containing RNAi enzyme complexes. Cell 123(4), 607–620.

36. Rand, T. A., Ginalski, K., Grishin, N. V., and Wang, X. (2004). Biochemical identification of Argonaute 2 as the sole protein required for RNA-induced silencing complex activity. Proc Natl Acad Sci USA 101(40), 14385–14389.

37. Hutvagner, G. and Zamore, P. D. (2002). A microRNA in a multiple-turnover RNAi enzyme complex. Science 297(5589), 2056–2060.

38. Lippman, Z. and Martienssen, R. (2004). The role of RNA interference in heterochromatic silencing. Nature 431(7006), 364–370.

39. Kanellopoulou, C., Muljo, S. A., Kung, A. L., Ganesan, S., Drapkin, R., Jenuwein, T., Livingston, D. M., and Rajewsky, K. (2005). Dicer-deficient mouse embryonic stem cells are defective in differentiation and centromeric silencing. Genes Dev 19(4), 489–501.

40. Ma, J. B., Yuan, Y. R., Meister, G., Pei, Y., Tuschl, T., and Patel, D. J. (2005). Structural basis for 5′-end-specific recognition of guide RNA by the A. fulgidus Piwi protein. Nature 434(7033), 666–670 (2005).

41. Lewis, B. P., Burge, C. B., and Bartel, D. P. (2005). Conserved seed pairing, often flanked by adenosines, indicates that thousands of human genes are microRNA targets. Cell 120(1), 15–20.

42. Rehwinkel, J., Behm-Ansmant, I., Gatfield, D., and Izaurralde, E. (2005). A crucial role for GW182 and the DCP1:DCP2 decapping complex in miRNA-mediated gene silencing. RNA 11(11), 1640–1647.

43. Sen, G. L. and Blau, H. M. (2005). Argonaute 2/RISC resides in sites of mammalian mRNA decay known as cytoplasmic bodies. Nat Cell Biol 7(6), 633–636.

44. Jakymiw, A., Lian, S., Eystathioy, T., Li, S., Satoh, M., Hamel, J. C., Fritzler, M. J., and Chan, E. K. (2005). Disruption of GW bodies impairs mammalian RNA interference, Nat Cell Biol 7(12),1267–1274.

45. Liu, J., Rivas, F. V., Wohlschlegel, J., Yates, J. R., III, Parker, R., and Hannon, G. J. (2005). A role for the P-body component GW182 in microRNA function. Nat Cell Biol 7(12), 1161–1166.
46. Liu, J., Valencia-Sanchez, M. A., Hannon, G. J., and Parker, R. (2005). MicroRNA-dependent localization of targeted mRNAs to mammalian P-bodies. Nat Cell Biol 7(7), 719–723.
47. Valencia-Sanchez, M. A., Liu, J., Hannon, G. J., and Parker, R. (2006). Control of translation and mRNA degradation by miRNAs and siRNAs. Genes Dev 20(5), 515–524.
48. Teixeira, D., Sheth, U., Valencia-Sanchez, M. A., Brengues, M., and Parker, R. (2005). Processing bodies require RNA for assembly and contain nontranslating mRNAs. RNA 11(4), 371–382.
49. Brengues, M., Teixeira, D., and Parker, R. (2005). Movement of eukaryotic mRNAs between polysomes and cytoplasmic processing bodies. Science 310(5747), 486–489.
50. Sheth, U. and Parker, R. (2003). Decapping and decay of messenger RNA occur in cytoplasmic processing bodies. Science 300(5620), 805–808.
51. Coller, J. and Parker, R. (2004). Eukaryotic mRNA decapping. Annu Rev Biochem 73, 861–890.
52. Lim, L. P., Lau, N. C., Weinstein, E. G., Abdelhakim, A., Yekta, S., Rhoades, M. W., Burge, C. B., and Bartel, D. P. (2003). The microRNAs of Caenorhabditis elegans. Genes Dev 17(8), 991–1008.
53. Lim, L. P., Glasner, M. E., Yekta, S., Burge, C. B., and Bartel, D. P. (2003). Vertebrate microRNA genes. Science 299(5612), 1540.
54. Bentwich, I., Avniel, A., Karov, Y., Aharonov, R., Gilad, S., Barad, O., Barzilai, A., Einat, P., Einav, U., Meiri, E., Sharon, E., Spector, Y., and Bentwich, Z. (2005). Identification of hundreds of conserved and nonconserved human microRNAs. Nat Genet 37(7), 766–770.
55. Pasquinelli, A. E. and Ruvkun, G. (2002). Control of developmental timing by microRNAs and their targets. Annu Rev Cell Dev Biol 18495–18513.
56. Mineno, J., Okamoto, S., Ando, T., Sato, M., Chono, H., Izu, H., Takayama, M., Asada, K., Mirochnitchenko, O., Inouye, M., and Kato, I. (2006). The expression profile of microRNAs in mouse embryos. Nucleic Acids Res 34(6), 1765–1771.
57. Willmann, M. R., and Poethig, R. S. (2005). Time to grow up: the temporal role of smallRNAs in plants. Curr Opin Plant Biol 8(5), 548–552.
58. Aboobaker, A. A., Tomancak, P., Patel, N., Rubin, G. M., and Lai, E. C. (2005). Drosophila microRNAs exhibit diverse spatial expression patterns during embryonic development. Proc Natl Acad Sci USA 102(50), 18017–18022.
59. Biemar, F., Zinzen, R., Ronshaugen, M., Sementchenko, V., Manak, J. R., and Levine, M. S. (2005). Spatial regulation of microRNA gene expression in the Drosophila embryo. Proc Natl Acad Sci USA 102(44), 15907–15911.
60. Wienholds, E., Kloosterman, W. P., Miska, E., Alvarez-Saavedra, E., Berezikov, E., de Bruijn, E., Horvitz, H. R., Kauppinen, S., and Plasterk, R. H. (2005). MicroRNA expression in zebrafish embryonic development. Science 309(5732), 310–311.
61. Aravin, A. A., Lagos-Quintana, M., Yalcin, A., Zavolan, M., Marks, D., Snyder, B., Gaasterland, T., Meyer, J., and Tuschl, T. (2003). The small RNA profile during Drosophila melanogaster development. Dev Cell 5(2), 337–350.
62. Houbaviy, H. B., Murray, M. F., and Sharp, P. A. Embryonic stem cell-specific MicroRNAs. Dev Cell 5(2), 351–358 (2003).
63. Pasquinelli, A. E. (2002). MicroRNAs: deviants no longer. Trends Genet 18(4), 171–173.
64. Plasterk, R. H. (2006). Micro RNAs in animal development. Cell 124(5), 877–881.
65. Doench, J. G. and Sharp, P. A. (2004). Specificity of microRNA target selection in translational repression. Genes Dev 18(5), 504–511 (2004).
66. Poy, M. N., Eliasson, L., Krutzfeldt, J., Kuwajima, S., Ma, X., Macdonald, P. E., Pfeffer, S., Tuschl, T., Rajewsky, N., Rorsman, P., and Stoffel, M. (2004). A pancreatic islet-specific microRNA regulates insulin secretion. Nature 432(7014), 226–230.

67. Brennecke, J., Hipfner, D. R., Stark, A., Russell, R. B., Cohen, S. M. (2003). bantam encodes a developmentally regulated microRNA that controls cell proliferation and regulates the proapoptotic gene hid in Drosophila. Cell 113(1), 25–36.

68. Xu, P., Vernooy, S. Y., Guo, M., and Hay, B. A. (2003). The Drosophila microRNA Mir-14 suppresses cell death and is required for normal fat metabolism. Curr Biol 13(9), 790–795.

69. Abrahante, J. E., Daul, A. L., Li, M., Volk, M. L., Tennessen, J. M., Miller, E. A., and Rougvie, A. E. (2003). The Caenorhabditis elegans hunchback-like gene lin-57/hbl-1 controls developmental time and is regulated by microRNAs. Dev Cell 4(5), 625–637.

70. Lin, S. Y., Johnson, S. M., Abraham, M., Vella, M. C., Pasquinelli, A., Gamberi, C., Gottlieb, E., and Slack, F. J. (2003). The C elegans hunchback homolog, hbl-1, controls temporal patterning and is a probable microRNA target. Dev Cell 4(5), 639–650.

71. Calin, G. A., Dumitru, C. D., Shimizu, M., Bichi, R., Zupo, S., Noch, E., Aldler, H., Rattan, S., Keating, M., Rai, K., Rassenti, L., Kipps, T., Negrini, M., Bullrich, F., and Croce, C. M. (2002). Frequent deletions and down-regulation of micro- RNA genes miR15 and miR16 at 13q14 in chronic lymphocytic leukemia. Proc Natl Acad Sci USA 99(24), 15524–15529.

72. Michael, M. Z., SM, O. C., van Holst Pellekaan, N. G., Young, G. P., and James, R. J. (2003). Reduced accumulation of specific microRNAs in colorectal neoplasia. Mol Cancer Res 1(12), 882–891.

73. Calin, G. A., Sevignani, C., Dumitru, C. D., Hyslop, T., Noch, E., Yendamuri, S., Shimizu, M., Rattan, S., Bullrich, F., Negrini, M., and Croce, C. M. (2004). Human microRNA genes are frequently located at fragile sites and genomic regions involved in cancers. Proc Natl Acad Sci USA 101(9), 2999–3004.

74. Lagos-Quintana, M., Rauhut, R., Yalcin, A., Meyer, J., Lendeckel, W., and Tuschl, T. (2002). Identification of tissue-specific microRNAs from mouse. Curr Biol 12(9), 735–739.

75. Tam, W., Ben-Yehuda, D., and Hayward, W. S. (1997). bic, a novel gene activated by proviral insertions in avian leukosis virus-induced lymphomas, is likely to function through its noncoding RNA. Mol Cell Biol 17(3), 1490–1502.

76. Kluiver, J., Poppema, S., de Jong, D., Blokzijl, T., Harms, G., Jacobs, S., Kroesen, B. J., and van den Berg, A. (2005). BIC and miR-155 are highly expressed in Hodgkin, primary mediastinal and diffuse large B cell lymphomas. J Pathol 207(2), 243–249.

77. Metzler, M., Wilda, M., Busch, K., Viehmann, S., and Borkhardt, A. (2004). High expression of precursor microRNA-155/BIC RNA in children with Burkitt lymphoma. Genes Chromosomes Cancer 39(2), 167–169.

78. van den Berg, A., Kroesen, B. J., Kooistra, K., de Jong, D., Briggs, J., Blokzijl, T., Jacobs, S., Kluiver, J., Diepstra, A., Maggio, E., and Poppema, S. (2003). High expression of B-cell receptor inducible gene BIC in all subtypes of Hodgkin lymphoma. Genes Chromosomes Cancer 37(1), 20–28 (2003).

79. Thomson, J. M., Parker, J., Perou, C. M., and Hammond, S. M. (2004). A custom microarray platform for analysis of microRNA gene expression. Nat Methods 1(1), 47–53.

80. Nelson, P. T., Baldwin, D. A., Scearce, L. M., Oberholtzer, J. C., Tobias, J. W., and Mourelatos, Z. (2004). Microarray-based, high-throughput gene expression profiling of microRNAs. Nat Methods 1(2), 155–161.

81. Babak, T., Zhang, W., Morris, Q., Blencowe, B. J., and Hughes, T. R. (2004). Probing microRNAs with microarrays: tissue specificity and functional inference. RNA 10(11), 1813–1819.

82. Sun, Y., Koo, S., White, N., Peralta, E., Esau, C., Dean, N. M., and Perera, R. J. (2004). Development of a micro-array to detect human and mouse microRNAs and characterization of expression in human organs. Nucleic Acids Res 32(22), e188.

83. Barad, O., Meiri, E., Avniel, A., Aharonov, R., Barzilai, A., Bentwich, I., Einav, U., Gilad, S., Hurban, P., Karov, Y.,Lobenhofer, E. K., Sharon, E., Shiboleth, Y. M., Shtutman, M., Bentwich, Z., and Einat, P. (2004). MicroRNA expression detected by oligonucleotide microarrays: system establishment and expression profiling in human tissues. Genome Res 14(12), 2486–2494.

84. Liu, C. G., Calin, G. A., Meloon, B., Gamliel, N., Sevignani, C., Ferracin, M., Dumitru, C. D., Shimizu, M., Zupo, S., Dono, M., Alder, H., Bullrich, F., Negrini, M., and Croce, C. M. (2004). An oligonucleotide microchip for genome-wide microRNA profiling in human and mouse tissues. Proc Natl Acad Sci USA 101(26), 9740–9744.

85. Calin, G. A., Liu, C. G., Sevignani, C., Ferracin, M., Felli, N., Dumitru, C. D., Shimizu, M., Cimmino, A., Zupo, S., Dono, M., Dell'Aquila, M. L., Alder, H., Rassenti, L., Kipps, T. J., Bullrich, F., Negrini, M., and Croce, C. M. (2004). MicroRNA profiling reveals distinct signatures in B cell chronic lymphocytic leukemias. Proc Natl Acad Sci USA 101(32), 11755–11760.

86. He, L., Thomson, J. M., Hemann, M. T., Hernando-Monge, E., Mu, D., Goodson, S., Powers, S., Cordon-Cardo, C., Lowe, S. W., Hannon, G. J., and Hammond, S. M. (2005). A microRNA polycistron as a potential human oncogene. Nature 435(7043), 828–833.

87. Volinia, S., Calin, G. A., Liu, C. G., Ambs, S., Cimmino, A., Petrocca, F., Visone, R., Iorio, M., Roldo, C., Ferracin, M., Prueitt, R. L., Yanaihara, N., Lanza, G., Scarpa, A., Vecchione, A., Negrini, M., Harris, C. C., and Croce, C. M. (2006). A microRNA expression signature of human solid tumors defines cancer gene targets. Proc Natl Acad Sci USA 103(7), 2257–2261.

88. Yanaihara, N., Caplen, N., Bowman, E., Seike, M., Kumamoto, K., Yi, M., Stephens, R. M., Okamoto, A., Yokota, J., Tanaka, T., Calin, G. A., Liu, C. G., Croce, C. M., and Harris, C. C. (2006). Unique microRNA molecular profiles in lung cancer diagnosis and prognosis. Cancer Cell 9(3), 189–198.

89. Calin, G. A., Ferracin, M., Cimmino, A., Di Leva, G., Shimizu, M., Wojcik, S. E., Iorio, M. V., Visone, R., Sever, N. I., Fabbri, M., Iuliano, R., Palumbo, T., Pichiorri, F., Roldo, C., Garzon, R., Sevignani, C., Rassenti, L., Alder, H., Volinia, S., Liu, C. G., Kipps, T. J., Negrini, M., and Croce, C. M. (2005). A MicroRNA signature associated with prognosis and progression in chronic lymphocytic leukemia. N Engl J Med 353(17), 1793–1801.

90. Lu, J., Getz, G., Miska, E. A., Alvarez-Saavedra, E., Lamb, J., Peck, D., Sweet-Cordero, A., Ebert, B. L., Mak, R. H., Ferrando, A. A., Downing, J. R., Jacks, T., Horvitz, H. R., and Golub, T. R. (2005). MicroRNA expression profiles classify human cancers. Nature 435(7043), 834–838.

91. Thorgeirsson, S. S., Lee, J. S., and Grisham, J. W. (2006). Functional genomics of hepatocellular carcinoma. Hepatology 43(2 Suppl 1), S145–S150.

92. Takamizawa, J., Konishi, H., Yanagisawa, K., Tomida, S., Osada, H., Endoh, H., Harano, T., Yatabe, Y., Nagino, M., Nimura, Y., Mitsudomi, T., and Takahashi, T. (2004). Reduced expression of the let-7 microRNAs in human lung cancers in association with shortened postoperative survival. Cancer Res 64(11), 3753–3756.

93. Johnson, S. M., Grosshans, H., Shingara, J., Byrom, M., Jarvis, R., Cheng, A., Labourier, E., Reinert, K. L., Brown, D., and Slack, F. J. (2005). RAS is regulated by the let-7 microRNA family. Cell 120(5), 635–647.

94. O'Donnell, K. A., Wentzel, E. A., Zeller, K. I., Dang, C. V., and Mendell, J. T. (2005). c-Myc-regulated microRNAs modulate E2F1 expression. Nature 435(7043), 839–843.

95. Bernstein, E., Kim, S. Y., Carmell, M. A., Murchison, E. P., Alcorn, H., Li, M. Z., Mills, A. A., Elledge, S. J., Anderson, K. V., and Hannon, G. J. (2003). Dicer is essential for mouse development. Nat Genet 35(3), 215–217.

96. Muljo, S. A., Ansel, K. M., Kanellopoulou, C., Livingston, D. M., Rao, A., and Rajewsky, K. (2005). Aberrant T cell differentiation in the absence of Dicer. J Exp Med 202(2), 261–269.

97. Harfe, B. D., McManus, M. T., Mansfield, J. H., Hornstein, E., and Tabin, C. J. (2005). The RNaseIII enzyme Dicer is required for morphogenesis but not patterning of the vertebrate limb. Proc Natl Acad Sci USA 102(31), 10898–10903.

98. Yi, R., O'Carroll, D., Pasolli, H. A., Zhang, Z., Dietrich, F. S., Tarakhovsky, A., and Fuchs, E. (2006). Morphogenesis in skin is governed by discrete sets of differentially expressed microRNAs. Nat Genet 38(3), 356–362.

99. Chen, C. Z., Li, L., Lodish, H. F., and Bartel, D. P. (2004). MicroRNAs modulate hematopoietic lineage differentiation. Science 303(5654), 83–86.

100. Fazi, F., Rosa, A., Fatica, A., Gelmetti, V., De Marchis, M. L., Nervi, C., and Bozzoni, I. (2005). A minicircuitry comprised of microRNA-223 and transcription factors NFI-A and C/EBPalpha regulates human granulopoiesis. Cell 123(5), 819–831.

101. Felli, N., Fontana, L., Pelosi, E., Botta, R., Bonci, D., Facchiano, F., Liuzzi, F., Lulli, V., Morsilli, O., Santoro, S., Valtieri, M., Calin, G. A., Liu, C. G., Sorrentino, A., Croce, C. M., and Peschle, C. (2005). MicroRNAs 221 and 222 inhibit normal erythropoiesis and erythro-leukemic cell growth via kit receptor down-modulation. Proc Natl Acad Sci USA 102(50), 18081–18086.

102. Garzon, R., Pichiorri, F., Palumbo, T., Iuliano, R., Cimmino, A., Aqeilan, R., Volinia, S., Bhatt, D., Alder, H., Marcucci, G., Calin, G. A., Liu, C. G., Bloomfield, C. D., Andreeff, M., and Croce, C. M. (2006). MicroRNA fingerprints during human megakaryocytopoiesis. Proc Natl Acad Sci USA 103(13), 5078–5083.

103. Mansfield, J. H., Harfe, B. D., Nissen, R., Obenauer, J., Srineel, J., Chaudhuri, A., Farzan-Kashani, R., Zuker, M., Pasquinelli, A. E., Ruvkun, G., Sharp, P. A., Tabin, C. J., and McManus, M. T. (2004). MicroRNA-responsive "sensor" transgenes uncover Hox-like and other developmentally regulated patterns of vertebrate microRNA expression. Nat Genet 36(10), 1079–1083.

104. Morgan, R. (2006). Hox genes: a continuation of embryonic patterning? Trends Genet 22(2), 67–69.

105. Tanzer, A., Amemiya, C. T., Kim, C. B., and Stadler, P. F. (2005). Evolution of microRNAs located within Hox gene clusters. J Exp Zoolog B Mol Dev Evol 304(1), 75–85.

106. Yekta, S., Shih, I. H., and Bartel, D. P. (2004). MicroRNA-directed cleavage of HOXB8 mRNA. Science 304(5670), 594–596.

107. Hornstein, E., Mansfield, J. H., Yekta, S., Hu, J. K., Harfe, B. D., McManus, M. T., Baskerville, S., Bartel, D. P., and Tabin, C. J. (2005). The microRNA miR-196 acts upstream of Hoxb8 and Shh in limb development. Nature 438(7068), 671–674.

108. Naguibneva, I., Ameyar-Zazoua, M., Polesskaya, A., Ait-Si-Ali, S., Groisman, R., Souidi, M., Cuvellier, S., and Harel-Bellan, A. (2006). The microRNA miR-181 targets the homeobox protein Hox-A11 during mammalian myoblast differentiation. Nat Cell Biol 8(3), 278–284.

109. Yamamoto, M. and Kuroiwa, A. (2003). Hoxa-11 and Hoxa-13 are involved in repression of MyoD during limb muscle development. Dev Growth Differ 45(5–6), 485–498.

110. Yamamoto, M., Gotoh, Y., Tamura, K., Tanaka, M., Kawakami, A., Ide, H., and Kuroiwa, A. (1998). Coordinated expression of Hoxa-11 and Hoxa-13 during limb muscle patterning. Development 125(7), 1325–1335.

111. Zhao, Y., Samal, E., and Srivastava, D. (2005). Serum response factor regulates a muscle-specific microRNA that targets Hand2 during cardiogenesis. Nature 436(7048), 214–220.

112. Chen, J. F., Mandel, E. M., Thomson, J. M., Wu, Q., Callis, T. E., Hammond, S. M., Conlon, F. L., and Wang, D. Z. (2006). The role of microRNA-1 and microRNA-133 in skeletal muscle proliferation and differentiation. Nat Genet 38(2), 228–233.

113. Lu, J., McKinsey, T. A., Zhang, C. L., and Olson, E. N. (2000). Regulation of skeletal myo-genesis by association of the MEF2 transcription factor with class II histone deacetylases. Mol Cell 6(2), 233–244.

114. Li, S., Czubryt, M. P., McAnally, J., Bassel-Duby, R., Richardson, J. A., Wiebel, F. F., Nordheim, A., and Olson, E. N. (2005). Requirement for serum response factor for skeletal muscle growth and maturation revealed by tissue-specific gene deletion in mice. Proc Natl Acad Sci USA 102(4), 1082–1087.

115. Schratt, G. M., Tuebing, F., Nigh, E. A., Kane, C. G., Sabatini, M. E., Kiebler, M., and Greenberg, M. E. (2006). A brain-specific microRNA regulates dendritic spine development. Nature 439(7074), 283–289.

116. Bamburg, J. R. (1999). Proteins of the ADF/cofilin family: essential regulators of actin dynamics. Annu Rev Cell Dev Biol 15185–230.

117. Esau, C., Kang, X., Peralta, E., Hanson, E., Marcusson, E. G., Ravichandran, L. V., Sun, Y., Koo, S., Perera, R. J., Jain, R., Dean, N. M., Freier, S. M., Bennett, C. F., Lollo, B., and

Griffey, R. (2004). MicroRNA-143 regulates adipocyte differentiation. J Biol Chem 279(50), 52361–52365.

118. Fearnhead, H. O. (2004). Getting back on track, or what to do when apoptosis is de-railed: recoupling oncogenes to the apoptotic machinery. Cancer Biol Ther 3(1), 21–28.

119. Leaman, D., Chen, P. Y., Fak, J., Yalcin, A., Pearce, M., Unnerstall, U., Marks, D. S., Sander, C., Tuschl, T., and Gaul, U. (2005). Antisense-mediated depletion reveals essential and specific functions of microRNAs in Drosophila development. Cell 121(7), 1097–1108.

120. Nairz, K., Rottig, C., Rintelen, F., Zdobnov, E., Moser, M., and Hafen, E. (2006). Overgrowth caused by misexpression of a microRNA with dispensable wild-type function. Dev Biol 291(2), 314–324.

121. Cimmino, A., Calin, G. A., Fabbri, M., Iorio, M. V., Ferracin, M., Shimizu, M., Wojcik, S. E., Aqeilan, R. I., Zupo, S., Dono, M., Rassenti, L., Alder, H., Volinia, S., Liu, C. G., Kipps, T. J., Negrini, M., and Croce, C. M. (2005). miR-15 and miR-16 induce apoptosis by targeting BCL2. Proc Natl Acad Sci USA 102(39), 13944–13949.

122. Chan, J. A., Krichevsky, A. M., and Kosik, K. S. (2005). MicroRNA-21 is an antiapoptotic factor in human glioblastoma cells. Cancer Res 65(14), 6029–6033.

123. Cheng, A. M., Byrom, M. W., Shelton, J., and Ford, L. P. (2005). Antisense inhibition of human miRNAs and indications for an involvement of miRNA in cell growth and apoptosis. Nucleic Acids Res 33(4), 1290–1297.

124. Aza-Blanc, P., Cooper, C. L., Wagner, K., Batalov, S., Deveraux, Q. L., and Cooke, M. P. (2003). Identification of modulators of TRAIL-induced apoptosis via RNAi-based phenotypic screening. Mol Cell 12(3), 627–637.

125. MacKeigan, J. P., Murphy, L. O., and Blenis, J. (2005). Sensitized RNAi screen of human kinases and phosphatases identifies new regulators of apoptosis and chemoresistance. Nat Cell Biol 7(6), 591–600.

126. Brummelkamp, T. R., Nijman, S. M., Dirac, A. M., and Bernards, R. (2003). Loss of the cylindromatosis tumour suppressor inhibits apoptosis by activating NF-kappaB. Nature 424(6950), 797–801.

127. Paddison, P. J., Silva, J. M., Conklin, D. S., Schlabach, M., Li, M., Aruleba, S., Balija, V., O'Shaughnessy, A., Gnoj, L., Scobie, K., Chang, K., Westbrook, T., Cleary, M., Sachidanandam, R., McCombie, W. R., Elledge, S. J., and Hannon, G. J. (2004). A resource for large-scale RNA-interference-based screens in mammals. Nature 428(6981), 427–431.

128. Moffat, J., Grueneberg, D. A., Yang, X., Kim, S. Y., Kloepfer, A. M., Hinkle, G., Piqani, B., Eisenhaure, T. M., Luo, B., Grenier, J. K., Carpenter, A. E., Foo, S. Y., Stewart, S. A., Stockwell, B. R., Hacohen, N., Hahn, W. C., Lander, E. S., Sabatini, D. M., and Root, D. E. (2006). A lentiviral RNAi library for human and mouse genes applied to an arrayed viral high-content screen. Cell 124(6), 1283–1298.

129. Silva, J. M., Li, M. Z., Chang, K., Ge, W., Golding, M. C., Rickles, R. J., Siolas, D., Hu, G., Paddison, P. J., Schlabach, M. R., Sheth, N., Bradshaw, J., Burchard, J., Kulkarni, A., Cavet, G., Sachidanandam, R., McCombie, W. R., Cleary, M. A., Elledge, S. J., and Hannon, G. J. (2005). Second-generation shRNA libraries covering the mouse and human genomes. Nat Genet 37(11), 1281–1288.

130. Kolfschoten, I. G., van Leeuwen, B., Berns, K., Mullenders, J., Beijersbergen, R. L., Bernards, R., Voorhoeve, P. M., and Agami, R. (2005). A genetic screen identifies PITX1 as a suppressor of RAS activity and tumorigenicity. Cell 121(6), 849–858.

131. Brummelkamp, T. R., Fabius, A. W., Mullenders, J., Madiredjo, M., Velds, A., Kerkhoven, R. M., Bernards, R., and Beijersbergen, R. L. (2006). An shRNA barcode screen provides insight into cancer cell vulnerability to MDM2 inhibitors. Nat Chem Biol 2(4), 202–206.

132. Brummelkamp, T. R., Berns, K., Hijmans, E. M., Mullenders, J., Fabius, A., Heimerikx, M., Velds, A., Kerkhoven, R. M., Madiredjo, M., Bernards, R., and Beijersbergen, R. L. (2004). Functional identification of cancer-relevant genes through large-scale RNA interference screens in mammalian cells. Cold Spring Harb Symp Quant Biol 69439–445.

133. Berns, K., Hijmans, E. M., Mullenders, J., Brummelkamp, T. R., Velds, A., Heimerikx, M., Kerkhoven, R. M., Madiredjo, M., Nijkamp, W., Weigelt, B., Agami, R., Ge, W., Cavet, G.,

Linsley, P. S., Beijersbergen, R. L., and Bernards, R. (2004). A large-scale RNAi screen in human cells identifies new components of the p53 pathway. Nature 428(6981), 431–437.

134. Westbrook, T. F., Martin, E. S., Schlabach, M. R., Leng, Y., Liang, A. C., Feng, B., Zhao, J. J., Roberts, T. M., Mandel, G., Hannon, G. J., Depinho, R. A., Chin, L., and Elledge, S. J. (2005). A genetic screen for candidate tumor suppressors identifies REST. Cell 121(6), 837–848.

135. Ngo, V. N., Davis, R. E., Lamy, L., Yu, X., Zhao, H., Lenz, G., Lam, L. T., Dave, S., Yang, L., Powell, J., and Staudt, L. M. (2006). A loss-of-function RNA interference screen for molecular targets in cancer. Nature 441(7089), 106–110.

136. Voorhoeve, P. M., le Sage, C., Schrier, M., Gillis, A. J., Stoop, H., Nagel, R., Liu, Y. P., van Duijse, J., Drost, J., Griekspoor, A., Zlotorynski, E., Yabuta, N., De Vita, G., Nojima, H., Looijenga, L. H., and Agami, R. (2006). A genetic screen implicates miRNA-372 and miRNA-373 as oncogenes in testicular germ cell tumors. Cell 124(6), 1169–1181.

137. Shankar, P., Manjunath, N., and Lieberman, J. (2005). The prospect of silencing disease using RNA interference. JAMA 293(11), 1367–1373.

138. Pai, S. I., Lin, Y. Y., Macaes, B., Meneshian, A., Hung, C. F., and Wu, T. C. (2006). Prospects of RNA interference therapy for cancer. Gene Ther 13(6), 464–477.

139. Brummelkamp, T. R., Bernards, R., and Agami, R. (2002). Stable suppression of tumorigenicity by virus-mediated RNA interference. Cancer Cell 2(3), 243–247.

140. Rubinson, D. A., Dillon, C. P., Kwiatkowski, A. V., Sievers, C., Yang, L., Kopinja, J., Rooney, D. L., Ihrig, M. M., McManus, M. T., Gertler, F. B., Scott, M. L., and Van Parijs, L. (2003). A lentivirus-based system to functionally silence genes in primary mammalian cells, stem cells and transgenic mice by RNA interference. Nat Genet 33(3), 401–406.

141. Song, E., Lee, S. K., Dykxhoorn, D. M., Novina, C., Zhang, D., Crawford, K., Cerny, J., Sharp, P. A., Lieberman, J., Manjunath, N., and Shankar, P. (2003). Sustained small interfering RNA-mediated human immunodeficiency virus type 1 inhibition in primary macrophages. J Virol 77(13), 7174–7181.

142. Purow, B. W., Haque, R. M., Noel, M. W., Su, Q., Burdick, M. J., Lee, J., Sundaresan, T., Pastorino, S., Park, J. K., Mikolaenko, I., Maric, D., Eberhart, C. G., and Fine, H. A. (2005). Expression of Notch-1 and its ligands, Delta-like-1 and Jagged-1, is critical for glioma cell survival and proliferation. Cancer Res 65(6), 2353–2363.

143. Yuan, J., Yan, R., Kramer, A., Eckerdt, F., Roller, M., Kaufmann, M., and Strebhardt, K. Cyclin B1 depletion inhibits proliferation and induces apoptosis in human tumor cells. Oncogene 23(34), 5843–5852 (2004).

144. Roberson, R. S., Kussick, S. J., Vallieres, E., Chen, S. Y., and Wu, D. Y. (2005). Escape from therapy-induced accelerated cellular senescence in p53-null lung cancer cells and in human lung cancers. Cancer Res 65(7), 2795–2803.

145. Zen, Y., Harada, K., Sasaki, M., Chen, T. C., Chen, M. F., Yeh, T. S., Jan, Y. Y., Huang, S. F., Nimura, Y., and Nakanuma, Y. (2005). Intrahepatic cholangiocarcinoma escapes from growth inhibitory effect of transforming growth factor-beta1 by overexpression of cyclin D1. Lab Invest 85(4), 572–581.

146. Xiao, Z., Xue, J., Sowin, T. J., Rosenberg, S. H., and Zhang, H. (2005). A novel mechanism of checkpoint abrogation conferred by Chk1 downregulation. Oncogene 24(8), 1403–1411.

147. Takei, Y., Kadomatsu, K., Yuzawa, Y., Matsuo, S., and Muramatsu, T. (2004). A small interfering RNA targeting vascular endothelial growth factor as cancer therapeutics. Cancer Res 64(10), 3365–3370.

148. Filleur, S., Courtin, A., Ait-Si-Ali, S., Guglielmi, J., Merle, C., Harel-Bellan, A., Clezardin, P., and Cabon, F. (2003). SiRNA-mediated inhibition of vascular endothelial growth factor severely limits tumor resistance to antiangiogenic thrombospondin-1 and slows tumor vascularization and growth. Cancer Res 63(14), 3919–3922.

149. Schiffelers, R. M., Ansari, A., Xu, J., Zhou, Q., Tang, Q., Storm, G., Molema, G., Lu, P. Y., Scaria, P. V., and Woodle, M. C. (2004). Cancer siRNA therapy by tumor selective delivery with ligand-targeted sterically stabilized nanoparticle. Nucleic Acids Res 32(19), e149.

150. Kilic, N., Oliveira-Ferrer, L., Wurmbach, J. H., Loges, S., Chalajour, F., Neshat-Vahid, S., Weil, J., Fernando, M., and Ergun, S. (2005). Pro-angiogenic signaling by the endothelial presence of CEACAM1. J Biol Chem 280(3), 2361–2369.

151. Liu, N., Bi, F., Pan, Y., Sun, L., Xue, Y., Shi, Y., Yao, X., Zheng, Y., and Fan, D. (2004). Reversal of the malignant phenotype of gastric cancer cells by inhibition of RhoA expression and activity. Clin Cancer Res 10(18 Pt 1), 6239–6247.

152. Chen, Y., Stamatoyannopoulos, G., and Song, C. Z. (2003). Down-regulation of CXCR4 by inducible small interfering RNA inhibits breast cancer cell invasion in vitro. Cancer Res 63(16), 4801–4804.

153. Liang, Z., Yoon, Y., Votaw, J., Goodman, M. M., Williams, L., and Shim, H. (2005). Silencing of CXCR4 blocks breast cancer metastasis. Cancer Res 65(3), 967–971.

154. Lee, E. J., Mircean, C., Shmulevich, I., Wang, H., Liu, J., Niemisto, A., Kavanagh, J. J., Lee, J. H., and Zhang, W. (2005). Insulin-like growth factor binding protein 2 promotes ovarian cancer cell invasion. Mol Cancer 4(1), 7.

155. Yin, Q. and Flemington, E. K. (2006). siRNAs against the Epstein Barr virus latency replication factor, EBNA1, inhibit its function and growth of EBV-dependent tumor cells. Virology 346(2), 385–393.

156. Hong, M., Murai, Y., Kutsuna, T., Takahashi, H., Nomoto, K., Cheng, C. M., Ishizawa, S., Zhao, Q. L., Ogawa, R., Harmon, B. V., Tsuneyama, K., and Takano, Y. (2006). Suppression of Epstein-Barr nuclear antigen 1 (EBNA1) by RNA interference inhibits proliferation of EBV-positive Burkitt's lymphoma cells. J Cancer Res Clin Oncol 132(1), 1–8.

157. Li, X. P., Li, G., Peng, Y., Kung, H. F., and Lin, M. C. (2004). Suppression of Epstein-Barr virus-encoded latent membrane protein-1 by RNA interference inhibits the metastatic potential of nasopharyngeal carcinoma cells. Biochem Biophys Res Commun 315(1), 212–218.

158. Jiang, M. and Milner, J. (2005). Selective silencing of viral gene E6 and E7 expression in HPV-positive human cervical carcinoma cells using small interfering RNAs. Methods Mol Biol 292, 401–420.

159. Jiang, M. and Milner, J. (2002). Selective silencing of viral gene expression in HPV-positive human cervical carcinoma cells treated with siRNA, a primer of RNA interference. Oncogene 21(39), 6041–6048.

160. Wu, H., Hait, W. N., and Yang, J. M. (2003). Small interfering RNA-induced suppression of MDR1 (P-glycoprotein) restores sensitivity to multidrug-resistant cancer cells. Cancer Res 63(7), 1515–1519.

161. Stege, A., Priebsch, A., Nieth, C., and Lage, H. (2004). Stable and complete overcoming of MDR1/P-glycoprotein-mediated multidrug resistance in human gastric carcinoma cells by RNA interference Cancer. Gene Ther 11(11), 699–706.

162. Nieth, C., Priebsch, A., Stege, A., and Lage, H. (2003). Modulation of the classical multidrug resistance (MDR) phenotype by RNA interference (RNAi). FEBS Lett 545(2–3), 144–150.

163. Yague, E., Higgins, C. F., and Raguz, S. (2004). Complete reversal of multidrug resistance by stable expression of small interfering RNAs targeting MDR1. Gene Ther 11(14), 1170–1174.

164. Duan, Z., Brakora, K. A., and Seiden, M. V. (2004). Inhibition of ABCB1 (MDR1) and ABCB4 (MDR3) expression by small interfering RNA and reversal of paclitaxel resistance in human ovarian cancer cells. Mol Cancer Ther 3(7), 833–838.

165. Peng, Z., Xiao, Z., Wang, Y., Liu, P., Cai, Y., Lu, S., Feng, W., and Han, Z. C. (2004). Reversal of P-glycoprotein-mediated multidrug resistance with small interference RNA (siRNA) in leukemia cells. Cancer Gene Ther 11(11), 707–712.

166. Chang, I. Y., Kim, M. H., Kim, H. B., Lee do, Y., Kim, S. H., Kim, H. Y., and You, H. J. (2005). Small interfering RNA-induced suppression of ERCC1 enhances sensitivity of human cancer cells to cisplatin. Biochem Biophys Res Commun 327(1), 225–233.

167. Dykxhoorn, D. M. and Lieberman, J. (2005). The silent revolution: RNA interference as basic biology, research tool, and therapeutic. Annu Rev Med 56, 401–423.

168. Dykxhoorn, D. M., Palliser, D., and Lieberman, J. (2006). The silent treatment: siRNAs as small molecule drugs. Gene Ther 13(6), 541–552.

169. Soutschek, J., Akinc, A., Bramlage, B., Charisse, K., Constien, R., Donoghue, M., Elbashir, S., Geick, A., Hadwiger, P., Harborth, J., John, M., Kesavan, V., Lavine, G., Pandey, R. K., Racie, T., Rajeev, K. G., Rohl, I., Toudjarska, I., Wang, G., Wuschko, S., Bumcrot, D., Koteliansky, V., Limmer, S., Manoharan, M., and Vornlocher, H. P. (2004). Therapeutic silencing of an endogenous gene by systemic administration of modified siRNAs. Nature 432(7014), 173–178.
170. Urban-Klein, B., Werth, S., Abuharbeid, S., Czubayko, F., and Aigner, A. (2005). RNAi-mediated gene-targeting through systemic application of polyethylenimine (PEI)-complexed siRNA in vivo. Gene Ther 12(5), 461–466.
171. Ge, Q., Filip, L., Bai, A., Nguyen, T., Eisen, H. N., and Chen, J. (2004). Inhibition of influenza virus production in virus-infected mice by RNA interference. Proc Natl Acad Sci USA 101(23), 8676–8681.
172. Santel, A., Aleku, M., Keil, O., Endruschat, J., Esche, V., Fisch, G., Dames, S., Loffler, K., Fechtner, M., Arnold, W., Giese, K., Klippel, A., and Kaufmann, J. (2006). A novel siRNA-lipoplex technology for RNA interference in the mouse vascular endothelium. Gene Ther 13(16), 1222–1234.
173. Santel, A., Aleku, M., Keil, O., Endruschat, J., Esche, V., Durieux, B., Loffler, K., Fechtner, M., Rohl, T., Fisch, G., Dames, S., Arnold, W., Giese, K., Klippel, A., and Kaufmann, J. (2006). RNA interference in the mouse vascular endothelium by systemic administration of siRNA-lipoplexes for cancer therapy, Gene Ther 13(18), 1360–1370.
174. Yano, J., Hirabayashi, K., Nakagawa, S., Yamaguchi, T., Nogawa, M., Kashimori, I., Naito, H., Kitagawa, H., Ishiyama, K., Ohgi, T., and Irimura, T. (2004). Antitumor activity of small interfering RNA/cationic liposome complex in mouse models of cancer, Clin Cancer Res 10(22), 7721–7726.
175. Morrissey, D. V., Lockridge, J. A., Shaw, L., Blanchard, K., Jensen, K., Breen, W., Hartsough, K., Machemer, L., Radka, S., Jadhav, V., Vaish, N., Zinnen, S., Vargeese, C., Bowman, K., Shaffer, C. S., Jeffs, L. B., Judge, A., MacLachlan, I., and Polisky, B. (2005). Potent and persistent in vivo anti-HBV activity of chemically modified siRNAs. Nat Biotechnol 23(8), 1002–1007.
176. Song, E., Zhu, P., Lee, S. K., Chowdhury, D., Kussman, S., Dykxhoorn, D. M., Feng, Y., Palliser, D., Weiner, D. B., Shankar, P., Marasco, W. A., and Lieberman, J. (2005). Antibody mediated in vivo delivery of small interfering RNAs via cell-surface receptors. Nat Biotechnol 23(6), 709–717.
177. Bagga, S., Bracht, J., Hunter, S., Massirer, K., Holtz, J., Eachus, R., and Pasquinelli, A. E. (2005). Regulation by let-7 and lin-4 miRNAs results in target mRNA degradation. Cell 122(4), 553–563.
178. Krutzfeldt, J., Rajewsky, N., Braich, R., Rajeev, K. G., Tuschl, T., Manoharan, M., and Stoffel, M. (2005). Silencing of microRNAs in vivo with "antagomirs". Nature 438(7068), 685–689.

Chapter 15
Cancer Stem Cells and Impaired Apoptosis

Zainab Jagani* and Roya Khosravi-Far

Abstract For more than 100 years scientists have fervently sought the fundamental origins of tumorigenesis, with the ultimate hope of discovering a cure. Indeed, these efforts have led to a significant understanding that multiple genetic and molecular aberrations, such as increased proliferation and the inhibition of apoptosis, contribute to the canonical characteristics of cancer. Despite these advances in our knowledge, a more thorough understanding, such as the precise cells, which are the targets of neoplastic transformation, especially in solid tumors, is currently lacking. An emerging hypothesis in the field is that cancer arises and is sustained from a rare subpopulation of tumor cells with characteristics that are highly similar to stem cells, such as the ability to self-renew and differentiate. In addition, more recent studies indicate that stem cell self-renewal pathways that are active primarily during embryonic development and adult tissue repair may be aberrantly activated in various cancers. This chapter introduces the cancer stem cell hypothesis; explores evidence for the presence of cancer stem cells, particularly in leukemia; and discusses various classical stem cell self-renewal pathways in relation to cancer. Investigating the role of cancer stem cells in the context of the major characteristics of cancer, especially impaired apoptosis, offers great promise for the design of superior tumor-selective and apoptosis-inducing therapies.

Keywords cancer stem cell, therapy, leukemia, Notch, Hedgehog, Wnt, Bmi1

Zainab Jagani
Novartis Institutes for Biomedical Research, Novartis Oncology, 250 Massachussetts Avenue, Cambridge MA 02139

Roya Khosravi-Far
Department of Pathology, Harvard Medical School, Beth Israel Deaconess Medical Center, 99 Brookline Avenue, Boston, MA 02215, USA

* To whom correspondence should be addressed: e-mail: zainab.jagani@novartis.com

R. Khosravi-Far and E. White (eds.), *Programmed Cell Death in Cancer Progression and Therapy*.
© Springer 2008

1   Introduction

The inability of conventional chemotherapeutic drugs and even various targeted therapies to produce complete remissions demands a more in-depth understanding of the key cellular events underlying tumor formation, maintenance, and progression, and the molecular pathways that dictate such processes. It has become increasingly apparent that the tumor, rather than consisting of a uniform population of rapidly proliferating cells, is actually composed of a heterogeneous population of cells with variable cellular and molecular characteristics (Foulds, 1965; Heppner, 1984). Therefore, one possible explanation for the failure of chemotherapy is that it cannot eliminate this entire mixed composition of tumor cells, thus necessitating multiple treatment approaches. Along these lines, it has been proposed that a rare group of cells with stem cell-like properties lies within the tumor and gives rise to the heterogeneous tumor cell population (Reya et al., 2001). The existence of these cells indicates that while our current anticancer therapeutics may be successful in debulking a tumor, they remain ineffective in targeting the minute, yet crucial, population of tumor cells that ultimately sustains the tumor. While the "cancer stem cell hypothesis" is supported by seminal findings from hematopoietic cancers, especially acute myeloid leukemia (AML) (Warner et al., 2004), its importance and application in other types of cancers are not clearly understood.

1.1   The Cancer Stem Cell Hypothesis

One intriguing and emerging area of cancer research concerns the striking parallels between cancer cells and stem cells. Both of these cell types have the capacity to self-renew and differentiate. Unlike the highly regulated self-renewal and differentiation decisions of normal stem cells, however, it has been proposed that cancer cells undergo uncontrolled self-renewal and abnormal differentiation. Coincidently, the pathways that regulate stem cell self-renewal and differentiation, such as Notch, Hedgehog (Hh), Wnt, and Bmi1 are dysregulated in various cancers (Reya et al., 2001). In addition, key findings revealing the presence of leukemic stem cells and providing evidence for a stem cell origin for AML are in support of the hypothesis that cancers arise from a small population of tumor-initiating cells known as cancer "stem cells" (Bonnet and Dick, 1997; Buick and Pollak, 1984; Jordan and Guzman, 2004; Lapidot et al., 1994; Mackillop et al., 1983; Reya et al., 2001). These cancer stem cells give rise to the clinically observed, phenotypically diverse tumor population consisting of cells displaying varied capacities for abnormal differentiation, uncontrolled proliferation, and a reduced rate of apoptosis. While the precise identity of a cancer stem cell is difficult to pinpoint, it is possible that cancer stem cells can arise either from the malignant transformation of a stem cell, or the abnormal re-activation of self-renewal pathways in a more committed progenitor cell (Al-Hajj et al., 2004; Burkert et al., 2006; Reya et al., 2001).

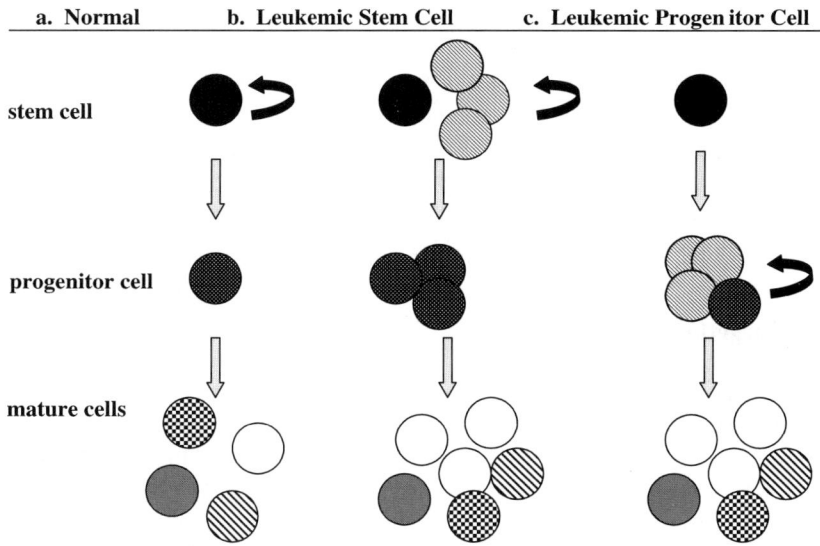

Fig. 15.1 Cancer stem cells and leukemia. (a) A simplified demonstration of normal hematopoietic development in which the self-renewing stem cell is highly regulated leading to normal progenitor and mature cell production. In leukemia however, and according to the cancer stem cell hypothesis; (b) transformation of a stem cell can lead to uncontrolled self-renewal resulting in an abnormal growth and differentiation program; (c) alternatively, transformation of a progenitor cell can abnormally reactivate self-renewal resulting in the abnormal growth and differentiation of hematopoietic cells

1.1.1 Cancer Stem Cells in Leukemia and Other cancers

Since the cellular and developmental biology of the hematopoietic system is well understood, the cancer stem cell hypothesis has been most thoroughly tested in the context of hematopoietic malignancies (Fig. 15.1), such as AML (Dick, 2005). AML is characterized by the uncontrolled growth and accumulation of abnormally differentiated blood cells, or leukemic blasts, which rapidly overwhelm normal blood cell function. Initial studies using various in vitro systems, such as the clonogenic; suspension culture-initiating cells (SC-IC); and long-term culture-initiating cells (LTC-IC) quantitative stem cell assays revealed that only a minor fraction of AML cells are capable of supporting growth in vitro (Warner et al., 2004). These studies were followed by key experiments performed in vivo using the NOD/SCID-leukemia xenotransplantation model. In this model, transplantation of leukemic cells from AML patients into mice can produce leukemic disease resembling human AML (Bonnet and Dick, 1997). It was demonstrated that only a minor percentage (0.1–1%) of AML cells with primitive CD34 + CD38– surface expression was capable of initiating AML in the NOD/SCID mice, thereby providing the first evidence for the presence of cancer stem cells (Bonnet and Dick, 1997; Lapidot et al., 1994). The discovery of leukemic stem cells thus set the groundwork for an investigation of the existence of cancer stem cells in other types of cancers. While the origin of cancer stem cells has not been conclusively defined, recent studies have also identified a subpopulation of tumor-initiating cells in solid tumors, such as breast (Al-Hajj et al., 2003), melanoma (Grichnik et al., 2006), brain (Singh et al., 2003), prostate (Xin et al., 2005), and ovarian (Bapat et al., 2005) cancers. Together, these studies raise important questions regarding the target cells of our current anticancer therapeutics, and the study of cancer signal transduction pathways in the appropriate cellular context.

1.1.2   Targeting Cancer Stem Cells

In the case of the CML-causing oncogene *BCR-ABL*, accumulating evidence suggests that the target cell for transformation is a hematopoietic stem cell (HSC) rather than a committed progenitor cell (Elrick et al., 2005; Huntly and Gilliland, 2005; Huntly et al., 2004). Unfortunately, research has shown that while the Abl kinase inhibitor, Gleevec, can eradicate the majority of proliferating CML progenitors and differentiated granulocytes, it is unable to target the minute population of CML progenitor stem cells that can sustain the disease (Bhatia et al., 2003; Elrick et al., 2005; Graham et al., 2002). In accordance with the cancer stem cell hypothesis, Gleevec treatment can be used continuously to manage chronic phase CML, but not to eliminate leukemic disease, since the remaining cancer stem cells are still able to sustain the disease. Further research must specifically target this cancer stem cell population.

It remains important to determine whether abnormal survival and antiapoptotic signaling, as has been intensively investigated in primary tumor cells, tumor cell lines, and mouse tumor models, actually plays a significant role in the transformation and maintenance of the tumor-initiating cell, or, more specifically, the cancer stem cell population. One goal of such studies is to determine how to selectively induce apoptosis in leukemic stem cells, but not in normal HSCs. Recent studies have shown that the prosurvival pathways, such as NF-κB and PI3-K, are highly activated in the leukemic stem cell population in AML (Guzman et al., 2001; Xu et al., 2003; Zhao et al., 2004). Interestingly, AML leukemic stem cells preferentially undergo apoptosis, unlike normal HSCs, upon combined treatment with the chemotherapeutic agent idarubicin and the proteasome inhibitor MG-132 (Guzman et al., 2002). Such treatments lead to the inhibition of NF-κB activity, along with other currently unidentified mechanisms, and also activate p53, causing the expression of target genes, such as *GADD45, p21*, and the proapoptotic gene *Bax* (Guzman et al., 2002).

1.2   The Role of Stem Cell Regulation Pathways in Tumorigenesis

As early as 1855, the scientist Rudolph Virchow recognized elements of dysregulated embryonic development in tumors, proposing his embryonal-rest hypothesis. In accordance with these earlier findings, there is now evidence for a molecular link between the pathways that regulate stem cell self-renewal during

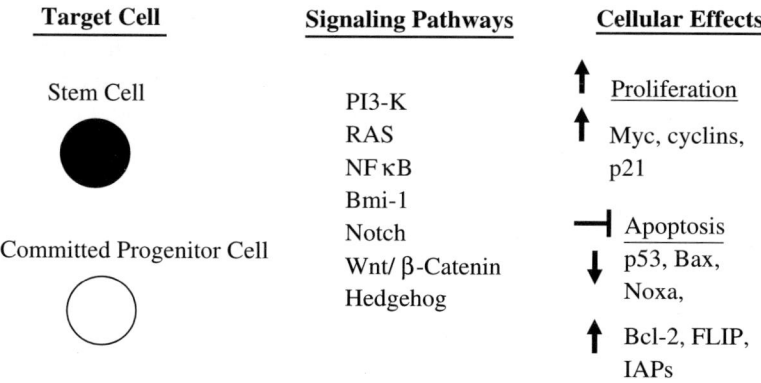

Fig. 15.2 Cancer stem cells and signaling pathways. A summary of the signaling pathways implicated in the survival of cancer stem cells. In general, these signaling pathways could either be aberrantly activated in a stem cell or a committed progenitor cell. Whereas, the outcomes of activating such pathways are numerous, key cellular effects include increase in cellular proliferation, and the inhibition of apoptosis. This figure outlines only a few of the various downstream genes that play important roles in proliferation and apoptosis

development and tumorigenesis (Fig. 15.2) (Burkert et al., 2006; Reya et al., 2001). The major developmental pathways such as Notch, Hh, and Wnt, which intricately control the self-renewal of stem cells during both embryonic development and adult tissue repair and homeostasis, are found to be upregulated in various cancers. These observations have brought forth important questions as to whether these pathways critically contribute to tumor formation and maintenance and whether their inhibition can be utilized in future anticancer therapeutic strategies. Selective inhibition of these developmental pathways in tumor cells may also have the potential to eliminate the elusive population of tumor-initiating cells that share common characteristics with stem cells. Furthermore, determining the direct impact of inappropriate activation of self-renewal pathways on apoptosis in a tumor cell will lead to a better understanding of how to combine therapies that attack upstream self-renewal pathways, with those that unleash downstream apoptotic cascades.

1.2.1 Bmi-1

The *Bmi-1* proto-oncogene was first identified as a target of the Moloney murine leukemia viral insertion in the Eμ-myc lymphoma mouse model (Haupt et al., 1991; van Lohuizen et al., 1991), with further studies suggesting a cooperative role with c-myc in inducing murine lymphogenesis (Haupt et al., 1993). Bmi-1 is a Polycomb-group gene which functions as a transcriptional repressor and plays a role in regulating cellular proliferation and senescence through repression of the INK4A locus (Jacobs et al., 1999). Recently, the *Bmi-1* gene has been shown to play a critical role in the generation of self-renewing adult HSCs, as mice deficient in Bmi-1 show reduced numbers of HSCs (Park et al., 2003). In addition, the *Bmi-1* gene has not only been implicated in regulating the proliferative activity of normal hematopoietic cells, but also of leukemic stem and progenitor cells, in which lack of Bmi-1 leads to proliferation arrest and characteristics of differentiation and apoptosis (Lessard and Sauvageau, 2003).

1.2.2   Notch Signaling

Notch signaling functions in a diverse set of cellular processes during embryonic and postnatal development, including the maintenance of stem cells, cell fate speci- fication, differentiation, and proliferation (Artavanis-Tsakonas et al., 1999; Kadesch, 2004). Interestingly, research points to a role for constitutively active Notch signaling under certain cellular contexts, such as in tumorigenesis (Callahan and Egan, 2004; Hansson et al., 2004; Radtke and Raj, 2003), yet the precise mechanisms underly- ing this effect remain to be determined. In mammalian systems, the Notch signaling pathway consists of four receptors (NOTCH1–4) and five ligands, Delta-like 1, 3, 4 (DLL1, DLL3, and DLL4), Jagged 1 and Jagged 2 (JAG1, JAG2) (reviewed in Artavanis-Tsakonas et al., 1999; Hansson et al., 2004; Kadesch, 2004). Notch receptors are synthesized as precursors, with Notch receptor activation occurring in a series of proteolytic cleavages upon interaction with its ligand. While the first cleavage is facilitated by TACE (tumor-necrosis factor α-converting enzyme/met- alloproteinase) (Brou et al., 2000), the second is mediated by the γ-secretase activ- ity of presenilins, and results in the release of the intracellular cytoplasmic portion of Notch, which then translocates to the nucleus (De Strooper et al., 1999; Mumm et al., 2000; Saxena et al., 2001). The known targets of Notch activation are the HES (hairy/enhancer of split) and HERP (Hes-related repressor protein) families of transcription factors, which regulate the transcription of various genes through development (Bailey and Posakony, 1995; Davis and Turner, 2001). The set of tar- get genes activated by Notch signaling has not been completely defined, and may vary with cellular context. In transformed cells, transcription of the *erbB2* (Chen et al., 1997) and *cyclin D1* (Ronchini and Capobianco, 2001) genes have been reported to be upregulated in response to activated Notch.

The earliest evidence for the involvement of activated Notch in human cancers arose from the identification of a translocation involving the *Notch1* gene in cases of T-cell acute lymphoblastic leukemia (T-ALL) (Ellisen et al., 1991). In particu- lar, the t(7;9) chromosomal translocation fuses a truncated Notch consisting mainly of the intracellular domain (NOTCH1-IC) to the TCRβ promoter/enhancer locus. The oncogenic property of NOTCH1-IC was confirmed by a murine bone marrow transplant model wherein reconstitution with hematopoietic progenitors expressing NOTCH1-IC led to the development of T-cell leukemias (Pear et al., 1996). The presence of activated Notch is not limited to leukemias, as its overex- pression or gain-of-function mutations, resulting in expression of a truncated active Notch, have also been observed in tumors of epithelial origin such as breast, cervical, and colon carcinomas (Callahan and Egan, 2004; Callahan and Raafat, 2001; Gray et al., 1999; Zagouras et al., 1995). A role for constitutive Notch signaling in the development of mammary tumors was first found with the discovery that the *Notch4* gene is a common integration site for the mouse mammary tumor virus (MMTV) in about 18% of virus-induced mouse mammary tumors (Gallahan and Callahan, 1997; Gallahan et al., 1987). MMTV interruption of *Notch4* results in the expression of a transcript that encodes the transmembrane and intracellular regions for Notch4, but that lacks the extracellular regulatory domain. Transgenic mouse models expressing the Notch4 intracellular domain develop mammary tumors (Jhappan et al., 1992; Smith et al., 1995), and therefore support a causative role for activated Notch signaling in mammary tumorigenesis. The relevance of Notch activation in human breast cancers has recently been investigated using tissue microarrays of breast tumor samples from various clinical stages. In these studies, elevated expression of Notch-1 and the Notch ligand, Jag1, was associated with poor survival (Reedijk et al., 2005).

Among the primary mechanisms for Notch-induced tumorigenesis, in addition to increased proliferation, is the inhibition of apoptosis. Activated Notch-1 renders T cells resistant to Fas receptor-mediated signaling, as well as to drugs including dexamethasone and etoposide, via upregulation of antiapoptotic molecules such as Bcl-2, FLIP, and IAPs (Sade et al., 2004). Additional mechanisms for Notch-induced survival include inhibition of p53 tumor suppressor expression, and activation of the RAS, PI3-K, and NF-κB pathways (Leong and Karsan, 2006).

While the precise value of Notch signaling inhibition in cancer therapy remains to be determined, preliminary studies have shown the potential for gamma secretase inhibitors (GSI) (Lanz et al., 2004; Wong et al., 2004), which can block Notch proteolytic processing, to induce apoptosis in various tumor cell lines (Curry et al., 2005; Nickoloff et al., 2005). Treatment of chemoresistant melanoma cells with a small molecule, GSI, induced the expression of the proapoptotic BH3 family member, NOXA, and caused apoptotic cell death (Nickoloff et al., 2005). Future studies will determine which downstream survival or antiapoptotic pathways play a role in the context of Notch activation in leukemias, as well as in solid tumors. In addition, the precise role of each of the four Notch receptors in tumorigenesis, and the development of specific inhibitors and/or antibodies against these receptors, will be crucial for an understanding of the overall role of Notch signaling in cancer and for investigating the potential of Notch inhibition in anticancer therapy. Finally, it will also be important to perform these studies at the cancer stem cell level in order to determine the cellular context in which dysregulated Notch signaling can potentially exert its oncogenic effects.

1.2.3 Hedgehog Signaling

The Hh pathway, first discovered in *Drosophila* (Nusslein-Volhard and Wieschaus, 1980), is highly conserved across vertebrates, with important functions during embryonic development, as well as in adult tissue homeostasis, such as in postembryonic tissue repair and stem cell regulation (Lum and Beachy, 2004; Taipale and Beachy, 2001; Zhang and Kalderon, 2001). The mammalian Hh pathway includes three secreted Hh ligands (Sonic, Indian, and Desert), their 12-pass transmembrane receptors Patched1 (PTCH1) and Patched2 (PTCH2), and the 7-pass transmembrane signal transducer Smoothened (SMO). Hh ligands activate the Hh pathway by inducing the activation of SMO, followed by a signal transduction cascade that causes the nuclear translocation of the GLI family of transcription factors (GLI1, 2, 3), and the subsequent induction of a distinct transcriptional regulatory program (Cohen, 2003; Hooper and Scott, 2005; Kalderon, 2005). The targets of Hh pathway activation include various cell cycle, proliferation, and survival-regulating genes such as the cyclins (Kenney and Rowitch, 2000), c-myc (Kenney et al., 2003), and Bcl-2 (Bigelow et al., 2004; Regl et al., 2004), and also Hh pathway genes themselves, such as Ptch1, Gli1, and Hip (Hh-interacting protein), which in turn regulate pathway activation (Chuang and McMahon, 1999; Goodrich et al., 1996; Lee et al., 1997).

Notably, gene mutations within the Hh pathway have been linked with several human diseases. Mutations resulting in unrestrained Hh pathway activity have been found in Gorlin's syndrome, which is characterized by developmental defects in the brain, spinal cord, and skeleton, and a predisposition for skin and brain cancers, such as basal cell carcinomas (BCCs) and medulloblastomas, respectively (Hahn et al., 1999). Subsequent investigations have substantiated aberrant Hh signaling in BCCs and medulloblastomas (Gailani et al., 1996; Xie et al., 1998). Recent studies have revealed that the Hh pathway is also active in more common tumors such as those of the lung, breast, pancreas, stomach, and prostate (Berman et al., 2003; Karhadkar et al., 2004; Kubo et al., 2004; Pasca di Magliano and Hebrok, 2003; Sheng et al., 2004; Thayer et al., 2003; Watkins et al., 2003). Cyclopamine is a plant-derived steroidal alkaloid that inhibits the Hh pathway by antagonizing SMO (Taipale et al., 2000). Various studies have shown the ability of cyclopamine to induce apoptosis in a variety of tumor cell lines, and to inhibit tumor progression in medulloblastoma, pancreatic, and lung mouse tumor models (Berman et al., 2002; Thayer et al., 2003; Watkins et al., 2003).

1.2.4 Wnt/β-catenin Signaling

Similar to the Notch and Hh pathways, the Wnt signal transduction pathway also plays a critical role during development. Among several functions, Wnt signals regulate the self-renewal of hematopoietic, epidermal, and intestinal stem cells. The canonical Wnt pathway involves signaling through the cytoplasmic protein, β-Catenin. The binding of a Wnt ligand to a complex of a Frizzled receptor and the LRP5/6 receptor leads to a series of signaling events resulting in the inhibition of a destruction complex that promotes the proteasomal degradation of β-Catenin. Therefore, Wnt pathway activity causes the accumulation of β-Catenin and its translocation to the nucleus where it binds to the Lef/Tcf family of transcription factors. This binding elicits the transcriptional activation of various target genes involved in the promotion of cellular proliferation and invasion, and the inhibition of apoptosis (reviewed in Fuchs et al., 2005; Reguart et al., 2005; Reya and Clevers, 2005).

Interestingly, the first *Wnt* gene was identified in mouse mammary tumors induced by the integration of the MMTV (Rijsewijk et al., 1987). Since then, there have been numerous studies on the aberrant activation of Wnt signaling in various cancers, including those of the colon, ovary, prostate, pancreas, breast, and lung, along with melanomas, multiple myeloma, and even leukemias (Fuchs et al., 2005; Janssens et al., 2006; Reguart et al., 2005; Reya and Clevers, 2005). While mutations in the Wnt ligands and receptors have not been identified in cancers thus far, mutations have been identified in downstream effectors of the Wnt pathway, especially in colorectal cancers (CRC). Gain-of-function mutations in oncogenic β-Catenin, and loss-of-function mutations in adenomatous polyposis coli (APC) and Axin, the latter of which are components of the destruction complex, can all lead to uncontrolled β-Catenin-mediated Lef/Tcf target gene expression (Fuchs et al., 2005; Janssens et al., 2006). Wnt pathway target genes involved in the inhibition of apoptosis include *MDR1/PGP, COX-2, PPAR-δ*, and Survivin, each of which has been found to be upregulated in CRCs (Fuchs et al., 2005). Considering the activation of the Wnt pathway in various cancers, inhibition of the Wnt pathway may serve as an attractive and promising therapeutic approach. Recent studies have demonstrated the potential for small-molecule antagonists of the TCF/β-Catenin complex to decrease expression of the Wnt target genes, *Myc* and *Cyclin D*, and to inhibit cellular proliferation in colon carcinoma cell lines (Lepourcelet et al., 2004). In another approach, monoclonal antibodies against Wnt-1 and Wnt-2 ligands have shown promise in inducing apoptosis in a variety of tumor cell lines overexpressing Wnt ligands, both in vitro and in vivo (He et al., 2004; You et al., 2004a–c;). Interestingly, the Wnt-2 antibody was shown to downregulate the expression of Survivin and induce apoptosis in various human non-small-cell lung cancer (NSCLC) cells, while failing to induce apoptosis in normal human airway cells that do not express Wnt-2. In contrast, primary NSCLC tissues showed elevated expression of Wnt-2 (You et al., 2004c).

2 Conclusion and Perspectives

Even though the cellular heterogeneity of tumors has long been recognized, the exact reasons for this feature have not always been clearly understood. The genomic instability that is inherent in cancer cells offers one explanation. Interestingly, recent studies, especially in leukemia, have revealed that the abnormal behavior of a malignant stem cell can give rise to the abnormally differentiated and diverse cellular hierarchy observed in tumors. The cancer stem cell hypothesis proposes that the tumor is actually sustained by a minority of cells, the cancer stem cells. The identification of cancer stem cells in leukemia and some solid cancers has yielded great insight into the cellular underpinnings of cancer, and will greatly affect the consideration of which cells to target critically in future anticancer therapeutics. Together, the study of signal transduction pathways that govern the survival of cancer stem cells, the precise role of cancer stem cells in different cancers, and an analysis of stem cell regulation pathways in cancer offers great promise for the development of more effective treatments in the future.

Acknowledgments   We thank Susan Glueck for assistance with preparation of the manuscript.

References

Al-Hajj, M., Wicha, M. S., Benito-Hernandez, A., Morrison, S. J., and Clarke, M. F. (2003). Prospective identification of tumorigenic breast cancer cells. Proc Natl Acad Sci USA 100, 3983–3988.

Al-Hajj, M., Becker, M. W., Wicha, M., Weissman, I., and Clarke, M. F. (2004). Therapeutic implications of cancer stem cells. Curr Opin Genet Dev 14, 43–47.

Artavanis-Tsakonas, S., Rand, M. D., and Lake, R. J. (1999). Notch signaling: cell fate control and signal integration in development. Science 284, 770–776.

Bailey, A. M. and Posakony, J. W. (1995). Suppressor of hairless directly activates transcription of enhancer of split complex genes in response to Notch receptor activity. Genes Dev 9, 2609–2622.

Bapat, S. A., Mali, A. M., Koppikar, C. B., and Kurrey, N. K. (2005). Stem and progenitor-like cells contribute to the aggressive behavior of human epithelial ovarian cancer. Cancer Res 65, 3025–3029.

Berman, D. M., Karhadkar, S. S., Hallahan, A. R., Pritchard, J. I., Eberhart, C. G., Watkins, D. N., Chen, J. K., Cooper, M. K., Taipale, J., Olson, J. M., and Beachy, P. A. (2002). Medulloblastoma growth inhibition by hedgehog pathway blockade. Science 297, 1559–1561.

Berman, D. M., Karhadkar, S. S., Maitra, A., Montes De Oca, R., Gerstenblith, M. R., Briggs, K., Parker, A. R., Shimada, Y., Eshleman, J. R., Watkins, D. N., and Beachy, P. A. (2003). Widespread requirement for Hedgehog ligand stimulation in growth of digestive tract tumours. Nature 425, 846–851.

Bhatia, R., Holtz, M., Niu, N., Gray, R., Snyder, D. S., Sawyers, C. L., Arber, D. A., Slovak, M. L., and Forman, S. J. (2003). Persistence of malignant hematopoietic progenitors in chronic myelogenous leukemia patients in complete cytogenetic remission following imatinib mesylate treatment. Blood 101, 4701–4707.

Bigelow, R. L., Chari, N. S., Unden, A. B., Spurgers, K. B., Lee, S., Roop, D. R., Toftgard, R., and McDonnell, T. J. (2004). Transcriptional regulation of bcl-2 mediated by the sonic hedgehog signaling pathway through gli-1. J Biol Chem 279, 1197–1205.

Bonnet, D. and Dick, J. E. (1997). Human acute myeloid leukemia is organized as a hierarchy that originates from a primitive hematopoietic cell. Nat Med 3, 730–737.

Brou, C., Logeat, F., Gupta, N., Bessia, C., LeBail, O., Doedens, J. R., Cumano, A., Roux, P., Black, R. A., and Israel, A. (2000). A novel proteolytic cleavage involved in Notch signaling: the role of the disintegrin-metalloprotease TACE. Mol Cell 5, 207–216.

Buick, R. N. and Pollak, M. N. (1984). Perspectives on clonogenic tumor cells, stem cells, and oncogenes. Cancer Res 44, 4909–4918.

Burkert, J., Wright, N., and Alison, M. (2006). Stem cells and cancer: an intimate relationship. J Pathol 209, 287–297.

Callahan, R. and Egan, S. E. (2004). Notch signaling in mammary development and oncogenesis. J Mammary Gland Biol Neoplasia 9, 145–163.

Callahan, R. and Raafat, A. (2001). Notch signaling in mammary gland tumorigenesis. J Mammary Gland Biol Neoplasia 6, 23–36.

Chen, Y., Fischer, W. H., and Gill, G. N. (1997). Regulation of the ERBB-2 promoter by RBPJkappa and NOTCH. J Biol Chem 272, 14110–14114.

Chuang, P. T. and McMahon, A. P. (1999). Vertebrate Hedgehog signalling modulated by induction of a Hedgehog-binding protein. Nature 397, 617–621.

Cohen, M. M., Jr. (2003). The hedgehog signaling network. Am J Med Genet A 123, 5–28.

Curry, C. L., Reed, L. L., Golde, T. E., Miele, L., Nickoloff, B. J., and Foreman, K. E. (2005). Gamma secretase inhibitor blocks Notch activation and induces apoptosis in Kaposi's sarcoma tumor cells. Oncogene 24, 6333–6344.

Davis, R. L. and Turner, D. L. (2001). Vertebrate hairy and Enhancer of split related proteins: transcriptional repressors regulating cellular differentiation and embryonic patterning. Oncogene 20, 8342–8357.

De Strooper, B., Annaert, W., Cupers, P., Saftig, P., Craessaerts, K., Mumm, J. S., Schroeter, E. H., Schrijvers, V., Wolfe, M. S., Ray, W. J., et al. (1999). A presenilin-1-dependent gamma-secretase-like protease mediates release of Notch intracellular domain. Nature 398, 518–522.

Dick, J. E. (2005). Acute myeloid leukemia stem cells. Ann NY Acad Sci 1044, 1–5.

Ellisen, L. W., Bird, J., West, D. C., Soreng, A. L., Reynolds, T. C., Smith, S. D., and Sklar, J. (1991). TAN-1, the human homolog of the Drosophila notch gene, is broken by chromosomal translocations in T lymphoblastic neoplasms. Cell 66, 649–661.

Elrick, L. J., Jorgensen, H. G., Mountford, J. C., and Holyoake, T. L. (2005). Punish the parent not the progeny. Blood 105, 1862–1866.

Foulds, L. (1965). Multiple etiologic factors in neoplastic development. Cancer Res 25, 1339–1347.

Fuchs, S. Y., Ougolkov, A. V., Spiegelman, V. S., and Minamoto, T. (2005). Oncogenic beta-catenin signaling networks in colorectal cancer. Cell Cycle 4, 1522–1539.

Gailani, M. R., Stahle-Backdahl, M., Leffell, D. J., Glynn, M., Zaphiropoulos, P. G., Pressman, C., Unden, A. B., Dean, M., Brash, D. E., Bale, A. E., and Toftgard, R. (1996). The role of the human homologue of Drosophila patched in sporadic basal cell carcinomas. Nat Genet 14, 78–81.

Gallahan, D. and Callahan, R. (1997). The mouse mammary tumor associated gene INT3 is a unique member of the NOTCH gene family (NOTCH4). Oncogene 14, 1883–1890.

Gallahan, D., Kozak, C., and Callahan, R. (1987). A new common integration region (int-3) for mouse mammary tumor virus on mouse chromosome 17. J Virol 61, 218–220.

Goodrich, L. V., Johnson, R. L., Milenkovic, L., McMahon, J. A., and Scott, M. P. (1996). Conservation of the hedgehog/patched signaling pathway from flies to mice: induction of a mouse patched gene by Hedgehog. Genes Dev 10, 301–312.

Graham, S. M., Jorgensen, H. G., Allan, E., Pearson, C., Alcorn, M. J., Richmond, L., and Holyoake, T. L. (2002). Primitive, quiescent, Philadelphia-positive stem cells from patients with chronic myeloid leukemia are insensitive to STI571 in vitro. Blood 99, 319–325.

Gray, G. E., Mann, R. S., Mitsiadis, E., Henrique, D., Carcangiu, M. L., Banks, A., Leiman, J., Ward, D., Ish-Horowitz, D., and Artavanis-Tsakonas, S. (1999). Human ligands of the Notch receptor. Am J Pathol 154, 785–794.

Grichnik, J. M., Burch, J. A., Schulteis, R. D., Shan, S., Liu, J., Darrow, T. L., Vervaert, C. E., and Seigler, H. F. (2006). Melanoma, a tumor based on a mutant stem cell? J Invest Dermatol 126, 142–153.

Guzman, M. L., Neering, S. J., Upchurch, D., Grimes, B., Howard, D. S., Rizzieri, D. A., Luger, S. M., and Jordan, C. T. (2001). Nuclear factor-kappaB is constitutively activated in primitive human acute myelogenous leukemia cells. Blood 98, 2301–2307.

Guzman, M. L., Swiderski, C. F., Howard, D. S., Grimes, B. A., Rossi, R. M., Szilvassy, S. J., and Jordan, C. T. (2002). Preferential induction of apoptosis for primary human leukemic stem cells. Proc Natl Acad Sci USA 99, 16220–16225.

Hahn, H., Wojnowski, L., Miller, G., and Zimmer, A. (1999). The patched signaling pathway in tumorigenesis and development: lessons from animal models. J Mol Med 77, 459–468.

Hansson, E. M., Lendahl, U., and Chapman, G. (2004). Notch signaling in development and disease. Semin Cancer Biol 14, 320–328.

Haupt, Y., Alexander, W. S., Barri, G., Klinken, S. P., and Adams, J. M. (1991). Novel zinc finger gene implicated as myc collaborator by retrovirally accelerated lymphomagenesis in E mu-myc transgenic mice. Cell 65, 753–763.

Haupt, Y., Bath, M. L., Harris, A. W., and Adams, J. M. (1993). bmi-1 transgene induces lymphomas and collaborates with myc in tumorigenesis. Oncogene 8, 3161–3164.

He, B., You, L., Uematsu, K., Xu, Z., Lee, A. Y., Matsangou, M., McCormick, F., and Jablons, D. M. (2004). A monoclonal antibody against Wnt-1 induces apoptosis in human cancer cells. Neoplasia 6, 7–14.

Heppner, G. H. (1984). Tumor heterogeneity. Cancer Res 44, 2259–2265.

Hooper, J. E. and Scott, M. P. (2005). Communicating with Hedgehogs. Nat Rev Mol Cell Biol 6, 306–317.

Huntly, B. J. and Gilliland, D. G. (2005). Cancer biology: summing up cancer stem cells. Nature 435, 1169–1170.

Huntly, B. J., Shigematsu, H., Deguchi, K., Lee, B. H., Mizuno, S., Duclos, N., Rowan, R., Amaral, S., Curley, D., Williams, I. R., et al. (2004). MOZ-TIF2, but not BCR-ABL, confers properties of leukemic stem cells to committed murine hematopoietic progenitors. Cancer Cell 6, 587–596.

Jacobs, J. J., Kieboom, K., Marino, S., DePinho, R. A., and van Lohuizen, M. (1999). The onco-gene and Polycomb-group gene bmi-1 regulates cell proliferation and senescence through the ink4a locus. Nature 397, 164–168.

Janssens, N., Janicot, M., and Perera, T. (2006). The Wnt-dependent signaling pathways as target in oncology drug discovery. Invest New Drugs 24, 263–280.

Jhappan, C., Gallahan, D., Stahle, C., Chu, E., Smith, G. H., Merlino, G., and Callahan, R. (1992). Expression of an activated Notch-related int-3 transgene interferes with cell differ-entiation and induces neoplastic transformation in mammary and salivary glands. Genes Dev 6, 345–355.

Jordan, C. T. and Guzman, M. L. (2004). Mechanisms controlling pathogenesis and survival of leukemic stem cells. Oncogene 23, 7178–7187.

Kadesch, T. (2004). Notch signaling: the demise of elegant simplicity. Curr Opin Genet Dev 14, 506–512.

Kalderon, D. (2005). The mechanism of hedgehog signal transduction. Biochem Soc Trans 33, 1509–1512.

Karhadkar, S. S., Bova, G. S., Abdallah, N., Dhara, S., Gardner, D., Maitra, A., Isaacs, J. T., Berman, D. M., and Beachy, P. A. (2004). Hedgehog signalling in prostate regeneration, neo-plasia and metastasis. Nature 431, 707–712.

Kenney, A. M., Cole, M. D., and Rowitch, D. H. (2003). Nmyc upregulation by sonic hedgehog signaling promotes proliferation in developing cerebellar granule neuron precursors. Development 130, 15–28.

Kenney, A. M. and Rowitch, D. H. (2000). Sonic hedgehog promotes G(1) cyclin expression and sustained cell cycle progression in mammalian neuronal precursors. Mol Cell Biol 20, 9055–9067.

Kubo, M., Nakamura, M., Tasaki, A., Yamanaka, N., Nakashima, H., Nomura, M., Kuroki, S., and Katano, M. (2004). Hedgehog signaling pathway is a new therapeutic target for patients with breast cancer. Cancer Res 64, 6071–6074.

Lanz, T. A., Hosley, J. D., Adams, W. J., and Merchant, K. M. (2004). Studies of Abeta pharma-codynamics in the brain, cerebrospinal fluid, and plasma in young (plaque-free) Tg2576 mice using the gamma-secretase inhibitor N2-[(2S)-2-(3,5-difluorophenyl)-2-hydroxyethanoyl]-N1-[(7S)-5-methyl-6-oxo -6,7-dihydro-5H-dibenzo[b,d]azepin-7-yl]-L-alaninamide (LY-411575). J Pharmacol Exp Ther 309, 49–55.

Lapidot, T., Sirard, C., Vormoor, J., Murdoch, B., Hoang, T., Caceres-Cortes, J., Minden, M., Paterson, B., Caligiuri, M. A., and Dick, J. E. (1994). A cell initiating human acute myeloid leukaemia after transplantation into SCID mice. Nature 367, 645–648.

Lee, J., Platt, K. A., Censullo, P., and Ruiz i Altaba, A. (1997). Gli1 is a target of Sonic hedgehog that induces ventral neural tube development. Development 124, 2537–2552.

Leong, K. G. and Karsan, A. (2006). Recent insights into the role of Notch signaling in tumori-genesis. Blood 107, 2223–2233.

Lepourcelet, M., Chen, Y. N., France, D. S., Wang, H., Crews, P., Petersen, F., Bruseo, C., Wood, A. W., and Shivdasani, R. A. (2004). Small-molecule antagonists of the oncogenic Tcf/beta-catenin protein complex. Cancer Cell 5, 91–102.

Lessard, J. and Sauvageau, G. (2003). Bmi-1 determines the proliferative capacity of normal and leukaemic stem cells. Nature 423, 255–260.

Lum, L. and Beachy, P. A. (2004). The Hedgehog response network: sensors, switches, and routers. Science 304, 1755–1759.

Mackillop, W. J., Ciampi, A., Till, J. E., and Buick, R. N. (1983). A stem cell model of human tumor growth: implications for tumor cell clonogenic assays. J Natl Cancer Inst 70, 9–16.

Mumm, J. S., Schroeter, E. H., Saxena, M. T., Griesemer, A., Tian, X., Pan, D. J., Ray, W. J., and Kopan, R. (2000). A ligand-induced extracellular cleavage regulates gamma-secretase-like proteolytic activation of Notch1. Mol Cell 5, 197–206.

Nickoloff, B. J., Hendrix, M. J., Pollock, P. M., Trent, J. M., Miele, L., and Qin, J. Z. (2005). Notch and NOXA-related pathways in melanoma cells. J Investig Dermatol Symp Proc 10, 95–104.

Nusslein-Volhard, C., and Wieschaus, E. (1980). Mutations affecting segment number and polarity in Drosophila. Nature 287, 795–801.

Park, I. K., Qian, D., Kiel, M., Becker, M. W., Pihalja, M., Weissman, I. L., Morrison, S. J., and Clarke, M. F. (2003). Bmi-1 is required for maintenance of adult self-renewing haematopoietic stem cells. Nature 423, 302–305.

Pasca di Magliano, M., and Hebrok, M. (2003). Hedgehog signalling in cancer formation and maintenance. Nat Rev Cancer 3, 903–911.

Pear, W. S., Aster, J. C., Scott, M. L., Hasserjian, R. P., Soffer, B., Sklar, J., and Baltimore, D. (1996). Exclusive development of T cell neoplasms in mice transplanted with bone marrow expressing activated Notch alleles. J Exp Med 183, 2283–2291.

Radtke, F. and Raj, K. (2003). The role of Notch in tumorigenesis: oncogene or tumour suppressor? Nat Rev Cancer 3, 756–767.

Reedijk, M., Odorcic, S., Chang, L., Zhang, H., Miller, N., McCready, D. R., Lockwood, G., and Egan, S. E. (2005). High-level coexpression of JAG1 and NOTCH1 is observed in human breast cancer and is associated with poor overall survival. Cancer Res 65, 8530–8537.

Regl, G., Kasper, M., Schnidar, H., Eichberger, T., Neill, G. W., Philpott, M. P., Esterbauer, H., Hauser-Kronberger, C., Frischauf, A. M., and Aberger, F. (2004). Activation of the BCL2 promoter in response to Hedgehog/GLI signal transduction is predominantly mediated by GLI2. Cancer Res 64, 7724–7731.

Reguart, N., He, B., Taron, M., You, L., Jablons, D. M., and Rosell, R. (2005). The role of Wnt signaling in cancer and stem cells. Fut Oncol 1, 787–797.

Reya, T. and Clevers, H. (2005). Wnt signalling in stem cells and cancer. Nature 434, 843–850.

Reya, T., Morrison, S. J., Clarke, M. F., and Weissman, I. L. (2001). Stem cells, cancer, and cancer stem cells. Nature 414, 105–111.

Rijsewijk, F., Schuermann, M., Wagenaar, E., Parren, P., Weigel, D., and Nusse, R. (1987). The Drosophila homolog of the mouse mammary oncogene int-1 is identical to the segment polarity gene wingless. Cell 50, 649–657.

Ronchini, C. and Capobianco, A. J. (2001). Induction of cyclin D1 transcription and CDK2 activity by Notch(ic): implication for cell cycle disruption in transformation by Notch(ic). Mol Cell Biol 21, 5925–5934.

Sade, H., Krishna, S., and Sarin, A. (2004). The anti-apoptotic effect of Notch-1 requires p56lck-dependent, Akt/PKB-mediated signaling in T cells. J Biol Chem 279, 2937–2944.

Saxena, M. T., Schroeter, E. H., Mumm, J. S., and Kopan, R. (2001). Murine notch homologs (N1–4) undergo presenilin-dependent proteolysis. J Biol Chem 276, 40268–40273.

Sheng, T., Li, C., Zhang, X., Chi, S., He, N., Chen, K., McCormick, F., Gatalica, Z., and Xie, J. (2004). Activation of the hedgehog pathway in advanced prostate cancer. Mol Cancer 3, 29.

Singh, S. K., Clarke, I. D., Terasaki, M., Bonn, V. E., Hawkins, C., Squire, J., and Dirks, P. B. (2003). Identification of a cancer stem cell in human brain tumors. Cancer Res 63, 5821–5828.

Smith, G. H., Gallahan, D., Diella, F., Jhappan, C., Merlino, G., and Callahan, R. (1995). Constitutive expression of a truncated INT3 gene in mouse mammary epithelium impairs differentiation and functional development. Cell Growth Differ 6, 563–577.

Taipale, J. and Beachy, P. A. (2001). The Hedgehog and Wnt signalling pathways in cancer. Nature 411, 349–354.

Taipale, J., Chen, J. K., Cooper, M. K., Wang, B., Mann, R. K., Milenkovic, L., Scott, M. P., and Beachy, P. A. (2000). Effects of oncogenic mutations in Smoothened and Patched can be reversed by cyclopamine. Nature 406, 1005–1009.

Thayer, S. P., di Magliano, M. P., Heiser, P. W., Nielsen, C. M., Roberts, D. J., Lauwers, G. Y., Qi, Y. P., Gysin, S., Fernandez-del Castillo, C., Yajnik, V., et al. (2003). Hedgehog is an early and late mediator of pancreatic cancer tumorigenesis. Nature 425, 851–856.

van Lohuizen, M., Verbeek, S., Scheijen, B., Wientjens, E., van der Gulden, H., and Berns, A. (1991). Identification of cooperating oncogenes in E mu-myc transgenic mice by provirus tagging. Cell 65, 737–752.

Warner, J. K., Wang, J. C., Hope, K. J., Jin, L., and Dick, J. E. (2004). Concepts of human leukemic development. Oncogene 23, 7164–7177.

Watkins, D. N., Berman, D. M., Burkholder, S. G., Wang, B., Beachy, P. A., and Baylin, S. B. (2003). Hedgehog signalling within airway epithelial progenitors and in small-cell lung cancer. Nature 422, 313–317.

Wong, G. T., Manfra, D., Poulet, F. M., Zhang, Q., Josien, H., Bara, T., Engstrom, L., Pinzon-Ortiz, M., Fine, J. S., Lee, H. J., et al. (2004). Chronic treatment with the gamma-secretase inhibitor LY-411,575 inhibits beta-amyloid peptide production and alters lymphopoiesis and intestinal cell differentiation. J Biol Chem 279, 12876–12882.

Xie, J., Murone, M., Luoh, S. M., Ryan, A., Gu, Q., Zhang, C., Bonifas, J. M., Lam, C. W., Hynes, M., Goddard, A., et al. (1998). Activating Smoothened mutations in sporadic basal-cell carcinoma. Nature 391, 90–92.

Xin, L., Lawson, D. A., and Witte, O. N. (2005). The Sca-1 cell surface marker enriches for a prostate-regenerating cell subpopulation that can initiate prostate tumorigenesis. Proc Natl Acad Sci USA 102, 6942–6947.

Xu, Q., Simpson, S. E., Scialla, T. J., Bagg, A., and Carroll, M. (2003). Survival of acute myeloid leukemia cells requires PI3 kinase activation. Blood 102, 972–980.

You, L., He, B., Uematsu, K., Xu, Z., Mazieres, J., Lee, A., McCormick, F., and Jablons, D. M. (2004a). Inhibition of Wnt-1 signaling induces apoptosis in beta-catenin-deficient mesothelioma cells. Cancer Res 64, 3474–3478.

You, L., He, B., Xu, Z., Uematsu, K., Mazieres, J., Fujii, N., Mikami, I., Reguart, N., McIntosh, J. K., Kashani-Sabet, M., et al. (2004b). An anti-Wnt-2 monoclonal antibody induces apoptosis in malignant melanoma cells and inhibits tumor growth. Cancer Res 64, 5385–5389.

You, L., He, B., Xu, Z., Uematsu, K., Mazieres, J., Mikami, I., Reguart, N., Moody, T. W., Kitajewski, J., McCormick, F., and Jablons, D. M. (2004c). Inhibition of Wnt-2-mediated signaling induces programmed cell death in non-small-cell lung cancer cells. Oncogene 23, 6170–6174.

Zagouras, P., Stifani, S., Blaumueller, C. M., Carcangiu, M. L., and Artavanis-Tsakonas, S. (1995). Alterations in Notch signaling in neoplastic lesions of the human cervix. Proc Natl Acad Sci USA 92, 6414–6418.

Zhang, Y. and Kalderon, D. (2001). Hedgehog acts as a somatic stem cell factor in the Drosophila ovary. Nature 410, 599–604.

Zhao, S., Konopleva, M., Cabreira-Hansen, M., Xie, Z., Hu, W., Milella, M., Estrov, Z., Mills, G. B., and Andreeff, M. (2004). Inhibition of phosphatidylinositol 3-kinase dephosphorylates BAD and promotes apoptosis in myeloid leukemias. Leukemia 18, 267–275.

Index

A

14-3-3 113, 271
17-AAG 64, 115, 256
19S 251
20S 251, 252, 257, 258
26S proteasome complex 64, 251
3-MA 190, 191
4-1-BBL 138
4E-BP1 94, 181
5-AzaC 266, 267
5-FU 57, 58, 205, 209
AKT 92, 94, 96, 135, 137, 182, 186, 187,
        192, 193
A1 82, 84, 86, 87, 91, 113, 230, 233
A20 230, 231
ABC-DLBCL 234, 238
ABT-737 63, 117, 144, 169
acetylation 202, 204, 261, 262, 264,
        266–271, 274, 277, 279
activation 2, 4–8, 13–21, 25–27, 29–32,
        37, 48–59, 61–65, 81–85, 87, 89–97,
        107–116, 128, 130, 133, 135, 136,
        140, 142, 143, 160–163, 167, 169,
        178, 181, 186, 187, 201, 202, 204,
        205, 209, 211, 213, 214, 223–236,
        238, 252–256, 258, 262, 264, 268,
        270–273, 276, 277, 279, 280, 316,
        335–339
activator 14, 17, 28, 30, 58, 62, 63, 83,
        84, 89, 92, 109, 112, 115, 160–165,
        167–169, 186, 187, 231, 273
Adenine nucleotide translocator (ANT)
        32, 38
adenomatous polyposis coli 339
Adenylate cyclase 110
ADP 35, 38
ADP-ribose 113, 263, 276
adriamycin 37, 139, 140, 209–211
AG 35156 144

Ago2 302
agonist 127, 128, 131–133, 135, 138, 140,
        142–145
AI 111
AICD 233
AIDS 4, 7, 8
AIF *see* apoptosis inducing factor
Akt 53, 54, 57, 60, 61, 109, 111, 113–115,
        181, 254, 270, 279
ALL 141, 234
ALLN 252
AML 36, 63, 137, 268, 274, 332–334
androgen receptor 271
angiogenesis 52, 90, 92, 193, 225, 235,
        252, 254, 261, 265, 268–270, 274–277,
        282, 305, 306, 310, 38
anoikis 84, 107
antennapedia 166, 167
anti-apoptotic 49, 117, 128, 160, 225
antimetabolites 8, 54
antiprotozoal 264, 268
antisense 53, 62, 63, 83, 86, 88, 89,
        115–117, 143, 144, 163, 165, 255,
        308, 315
AOE 265, 267, 268
AP-1 55, 112
AP23573 95
APAF1 84, 135–137
Apaf-1 34
apg6 96
API-59-OME 115
apical caspase 13, 15, 16, 18–21, 31, 50
apicidin 265, 268, 275
apicomplexan 268
APN 269
APO-1 48, 84, 204, 205
APO-2 84
APO-2L 276
ApoGossypol 144

apoptosis 1–8, 13, 15, 16, 20, 21, 25–38,
      47–65, 81–97, 105–117, 127, 128,
      130, 134–140, 142–145, 159–162,
      165, 167, 169, 177–179, 183–187,
      189–193, 201, 202, 204–215, 223,
      224, 226–230, 232–239, 251–258,
      261–264, 266–268, 270–274,
      276–282, 299, 300, 304306, 308,
      310, 314–316, 318, 331, 332,
      334–339
spoptosis inducing factor 28–33, 106,
      161, 279
apoptosome 5, 13, 15–18, 21, 29, 31, 50, 84,
      85, 135, 161, 203, 208
apoptotic bodies 26, 48
ARF 58, 109, 182, 185, 186, 231
Argonaute 301, 302
arsenic 38, 95, 190, 238
astrocyte 129
ATF-4 255
Atg 7, 178–181, 183, 186–191
ATL 234
ATM 55, 204, 209
ATP 5, 28, 29, 34, 35, 38, 60, 92, 182, 186,
      235, 251
ATR 55, 204, 231
autoimmune disease 7, 202, 207
autoimmunity 82
autophagy 1, 6–8, 56, 65, 95, 96, 177–184,
      186–194, 229
autoproteolysis 49

B
B23 185
BAD 28, 33, 34, 83, 84, 86, 87, 94, 106,
      107, 111, 137, 143, 160, 162,
      166–168, 209, 273
BAFF 225, 232
Bak 28, 33, 34, 51, 58, 83–85, 87, 106,
      107, 135, 160–162, 164, 165, 167–169,
      189, 190, 193, 203, 208–210, 212, 227,
      255, 272, 273, 276
Basal cell carcinoma 338
Bax 28, 33, 34, 51, 55, 58, 62, 83–86,
      106–108, 111–113, 135–137, 140,
      143, 160–162, 164, 165, 167, 169,
      189, 190, 192, 193, 203, 207–210,
      212, 21, 3, 227, 23, 255, 263, 264,
      272, 273, 276, 280, 334, 335
BAY11-7082 91
BAY11-7085 91
BAY-43-9006 94, 109, 110, 117
bbc3 209

B-cell 5, 36, 52, 138, 140, 159, 225, 226, 230,
      232–234, 304, 306, 310, 311, 313, 314
BCL10 234, 238, 317
Bcl-2 1, 5, 7, 28, 33, 34, 36–38, 50–53, 55,
      56, 58, 62–64, 81–84, 86–89, 92, 94,
      95, 106, 107, 109, 111, 113, 116, 117,
      135–138, 143–145, 159–169, 179, 182,
      183, 191, 193, 202, 203, 208, 209,
      211–213, 230–233, 272, 273, 276, 277,
      282, 304, 308, 314, 335, 337, 338
Bcl-2 Homology 82, 106, 160, 163, 208
BCL-W 33, 82, 84, 87, 160, 169
Bcl-XL 55, 58, 62, 63, 107, 109, 111, 113,
      117, 137, 160, 203
Bcr-Abl 51, 53, 54, 57, 60, 92, 270, 274, 334
BDNF 314
Beclin 1 177–180, 182, 183, 186, 188–191,
      193
benzamide 264, 265, 267, 281
beta-1 252
beta-2 252
beta-5 252
bFGF 110, 275
BFL-1 82, 84, 86, 87, 91, 160, 163, 169,
      230, 233
BH 33, 106, 160, 208, 209, 254, 273
BH1-3 83, 160, 161, 208, 227
BH3 33, 34, 63, 83–87, 90–92, 106, 107, 113,
      114, 117, 159–169, 208, 209, 227, 229,
      255, 273, 337
bid 25, 27, 33, 34, 49, 51, 55, 83, 85,
      86, 106, 108, 112, 117, 135, 136,
      160–162, 166, 167, 192, 203, 205,
      209, 227, 254, 273, 276
BIK 33, 83, 84, 144, 145, 160, 162, 209,
      254, 255, 373
Bim 33, 51, 56, 62, 83, 84, 91, 92, 94,
      106, 107, 109, 111–116, 144, 160,
      161, 165, 209, 254, 255, 272, 273
BimL 107, 111, 114
BimS 167
Bnip3L 23
biochemical 13, 26, 28, 36, 48, 82, 169, 178
biofeedback 48
BIR3 20, 85, 88
BL-193 117
blebbing 26, 48, 178, 184
BLyS 138, 232
BMF 84, 107, 160, 162
Bmi1 331, 332
BMS-345541 238
BNIP3 191, 229
BOK 83
bone marrow stromal cells (BMSC) 253, 254

bortezomib 65, 91, 92, 115, 144, 238,
        251–258, 270
BRAF 108–110
breast 56, 58, 59, 61, 62, 89, 93–95, 130–133,
        139, 143, 162, 185, 188, 192, 206, 207,
        212, 231, 234, 236–238, 269, 272, 274,
        276–278, 304, 306, 317, 334, 336–339
Bryostatin 115
Burkitt's lymphoma 52, 163, 234
butyrate 192, 264, 265, 274–276, 278–280

C
C. elegans 4, 21, 50, 300, 301, 307, 308
Ca^{+2} 84, 128, 254
cAMP 271
cancer 3, 7, 8, 25, 26, 35–38, 47, 48, 51–65,
        81–83, 86–97, 105, 106, 109, 110, 113,
        116–118, 127, 128, 131, 136, 140–144,
        159, 162–165, 169, 177, 178, 182–185,
        188, 190–193, 202, 206–208, 211–214,
        223, 224, 229, 231, 233–239, 252, 253,
        256, 261–264, 266–282, 299, 300, 303,
        304, 306–308, 311, 314–319, 331–335,
        337, 339, 340
cancer stem cell 65, 318, 331–335, 337, 339
cancer therapy 37, 47, 48, 51, 53, 56, 60–62,
        64, 65, 81, 82, 87–89, 91–93, 95–97,
        105, 127, 169, 213, 223, 261, 264, 270,
        272, 276, 282, 299, 317–319, 337
carcinoma 56, 58, 88, 95, 111, 115, 116,
        132–134, 141, 163, 183, 185, 205,
        236, 237, 255, 266, 267, 272, 278,
        280, 339
CARD 17, 29, 136
cardiolipin 25, 34
CARMA 1, 238
casein kinase 1 115
caspase 1, 4–8, 13–21, 25–32, 34, 36, 48–51,
        55, 59, 60, 63, 82–89, 94, 95, 106–108,
        111–113, 117, 134–137, 143, 161, 162,
        178, 184–186, 190–192, 202, 203, 205,
        208, 211–213, 227–230, 252, 254, 255,
        258, 261, 263, 264, 267, 272, 273,
        276–280, 314, 315
castleman disease 163
catalytic 14, 15, 17–21, 92, 94, 128, 135,
        143, 184, 224, 230, 251, 252, 267,
        269, 316, 317
CBHA 266, 279
CCAAT box 271
CCI-779 95, 115
CCT2 316
CD3+ 138

CD34 138, 275, 333
CD38 333
CD4+ 138
CD40 233, 271
CD40L 138, 225
CD80 271
CD86 271
CD95 48, 59, 84, 88, 202, 204, 205
CDDO 144,
CDDO-Imidazolide 144, 256
CDK 108, 109, 237, 254, 266
CDK4 108
CDK6 108
CDKN2a 109
ced 4, 50
cell cycle 53, 55, 56, 83, 90, 91, 93, 95,
        97, 108, 109, 187, 201, 205, 209,
        210, 214, 237, 251, 252, 254, 261,
        263, 264, 266–268, 270–272,
        278–280, 306, 307, 310, 318, 338
cell death 1–8, 16, 25, 26, 31–33, 36, 47,
        50, 53, 56, 57, 59, 60, 62–65, 82–85,
        90, 91, 95, 96, 105, 110, 11, 117, 127,
        128, 133–136, 138, 141–143, 159–161,
        163, 169, 177–179, 183–186, 188–193,
        201–203, 207–212, 214, 215, 223, 226,
        228–234, 254–256, 258, 264, 267, 272,
        273, 279, 314, 317, 337
cell suicide 2, 26, 47
CEP-1347 113
ces 4
cetuximab 93
chemotherapeutic agent 26, 38, 50, 57, 59,
        64, 81, 87, 96, 127, 139–141, 143,
        145, 192, 210, 226, 334
chemotherapy 35–37, 47, 57–59, 64, 81,
        86, 88, 89–95, 97, 106, 115, 117,
        130, 136, 139, 141, 143, 145, 210,
        214, 252, 276, 280–282, 310, 316,
        318, 332
chk1 55, 204, 231
CHK2 55, 204
CHOP 255
chromatin condensation 29, 178, 262
Chronic Myelogenous Leukemia (CML) 53,
        57, 60, 65, 92, 97, 141, 234, 274, 334
CI-1040 94, 114
CI994 264
c-IAP 88, 137, 273
cisplatin 57, 58, 91, 113, 115, 140, 141, 231
CK1 115
c-kit 60, 93, 94, 109, 110, 308, 312
C-L 252, 257
clinical resistance 163

clinical trial 61–64, 86, 88, 90, 93–96,
        115, 116, 133, 145, 164, 169, 214,
        252, 253, 258, 261, 265, 267, 278,
        281, 282
CLL 144, 162, 163, 169, 268, 304, 306
c-myc 51, 52, 108, 109, 111, 137, 185,
        210, 274, 305, 310, 335, 338
colorectal 59, 88, 89, 93, 94, 132, 141,
        143, 162, 211, 212, 231, 256, 304,
        317, 339
corpse 5, 82
COX-2 339
CP-31398 61, 90, 214
CRD 129
CREB 204, 271
CT-L 252, 254, 257, 258
Cyclin B 278
Cyclin D 108, 109, 339
Cyclin D1 108, 111, 115, 336
cycloheximide 6, 168
cyclotetrapeptide 267, 268
CYLD 91, 233, 316
cysteine 13, 19, 82, 128, 129, 138, 181, 204,
        206, 208, 212, 252, 279, 316
Cytochrome C 15, 16, 28–35, 38, 49, 84,
        85, 106, 135, 161, 164, 165, 168,
        186, 192, 203, 208, 210, 211, 227,
        230, 254, 276, 277, 279
cytoplasm 6, 8, 32, 85, 91, 95, 177, 186,
        229, 262, 300, 301, 305, 309, 319
cytoskeletal 26, 48
cytoskeleton 3, 82, 107, 184
cytotoxic T cell (CTL) 106

D
DAPK 177, 182, 184–186, 191
Dcp 302
DcR1 49, 59, 128, 129, 137, 140, 203, 204,
        208, 230
DcR2 49, 59, 128, 140, 203, 204
deacetylation 214, 261, 262, 264, 267, 281
death domain (DD) 17, 49, 59, 85, 112, 128,
        129, 134, 204–206, 208, 226, 227
death receptor (DR) 15, 16, 21, 25, 26,
        34, 48–50, 54–57, 59, 63, 64, 81–85,
        87, 112, 113, 116, 117, 128, 130,
        134, 136, 162, 202–204, 206–208,
        210, 213, 230, 231, 236, 272, 273,
        276, 277, 282
Decoy 59
decoy receptor 49, 59, 128, 129, 137, 140,
        204, 208, 230
DED 49, 85, 134, 205

degradation 14, 31, 49, 55, 56, 61, 62, 85,
        86, 90–92, 94, 96, 107–109, 111,
        163, 178, 179, 191, 202, 204, 213,
        214, 224–226, 230, 234, 238,
        251–253, 256, 257, 264, 299,
        301–303, 306, 338
delivery 82, 86, 88–90, 166, 281, 319
Delta-like 336
DEN 236
dependency receptor 211
depsipeptide 265, 268, 276, 279, 281
Desert 338
Dex 254, 256
dexamethasone 164, 253, 337
Diablo 28, 3–32, 49, 50, 84, 85, 88, 106, 108,
        117, 136, 137, 144, 161, 208, 227, 230,
        273, 279
Dicer 300–302, 306, 311, 312, 314
Dif 224
dimeric 13, 17, 18, 20, 34
DISC 13, 16, 49–51, 85, 134–136, 142,
        162, 205
DLL 336
DNA 6, 26, 36, 37, 54–56, 86, 89, 95, 107,
        191, 202, 204, 205, 214, 224, 225, 231,
        261, 254, 261–263, 266–269, 279, 281,
        304, 310, 311, 316
RNA 6, 88, 131, 179, 181, 226, 228, 234,
        238, 255, 299–304
DNA damage 26, 50, 54, 55, 58, 84, 90, 128,
        135, 160, 187, 201, 204, 206, 207–211,
        213, 273, 306, 309, 315, 317, 318
DNA fragmentation 6, 7, 26, 29, 31,
        48, 178
DNA repair 15, 52, 55–57, 113, 254,
        278, 318
DNA-PK 55, 204, 254
DNMT3B 267
Dorsal 224
doxorubicin 56–64, 87, 113, 115, 140, 207,
        208, 231, 239, 256, 279
DR3 84, 204
DR4 55, 59, 63, 84, 128, 203, 204, 206–208,
        230, 231, 272, 276, 277, 282
DR5 55, 59, 63, 85, 116, 128, 203, 204,
        206–208, 210, 213, 230, 231, 236,
        272, 276, 277, 282
DR6 85, 204, 231
DRAM 182, 187, 188
Drosha 300, 301
DRP-1 184
drug resistance 47, 56–58, 61, 63, 65,
        251–256, 274, 315, 318, 319
Dynein 107, 113

E
E1 ubiquitin enzyme 251
E1A 207
E2 ubiquitin-conjugation enzyme 251
E2F 55, 108, 109, 211, 266
E2F1 108, 185, 308, 310
E3 109
E3 ligase 90, 108, 224, 230
E3 ubiquitin ligase 109, 251
E-box 310
EDAR 204, 233
effector caspase 7, 13, 26, 27, 29, 49, 84, 85,
        108, 135, 162, 202, 203, 205, 207
EGCG 144
EGFR 57, 60, 61, 93, 94, 97
electron transport 32, 168
elongation 178, 180, 181
endonuclease G 28, 31, 106
epigenetic 55, 59, 261, 267, 280–282
epigenic 261
Epoxide 265, 268, 269
epoxyketone 252, 266, 267
Epstein-Barr virus (EBV) 163, 234, 318, 319
ER 106, 160, 178, 180, 182, 212, 237,
        254, 255
ErbB 93, 237, 270, 336
Erbitux 93
ERK 93, 94, 110, 111, 114, 181, 186,
        204, 264, 265
erlotinib 60, 93
E-selectin 253
ETO 268
evasion 38, 51, 54, 65
execution 1, 3, 15, 26, 29, 36, 49, 136, 169,
        177, 186, 205
executioner caspase 15, 16, 18–21, 30, 50,
        85, 88, 136
Exportin 300, 301
extrinsic 26, 28, 33, 38, 47–49, 133
extrinsic pathway 15, 16, 18, 26, 49, 55,
        63–65, 82–84, 88, 105, 116, 133,
        134, 136, 137, 162, 202, 205, 209,
        276, 277

F
FADD 16, 49, 59, 85, 112, 113, 134–136,
        142, 203, 205, 227, 276
Farnesyl transferase inhibitors 61
Fas 1, 8, 48, 49, 51, 55, 57, 59, 62, 63, 84, 85,
        111, 112, 116, 129, 162, 168, 202–207,
        231, 233, 254, 268, 276, 337
FasL 8, 16, 48, 55, 138, 202, 203, 205, 231,
        233, 254, 268, 276

Fc receptors 140, 142
FDA 60, 62, 91, 93, 164, 253
feedback 5, 48, 208, 224, 310, 313
FK228 265, 268, 275
FLICE 136, 227
FLIP 15, 49, 59, 136, 137, 143–145, 227,
        230, 335, 337
Flt-3 94, 270
fluorescence 131, 228
Forkhead 94, 111
FOXO 54, 56, 61, 62, 263
FR901228 268
FRET 228
Frizzled 338
frozen section 131
fusion 51, 53, 92, 159, 178, 179, 181,
        184, 319

G
G2/M 56, 272, 278
G3139 86
GADD45 213, 230, 334
gamma secretase inhibitor (GSI) 337
gastric carcinoma 134, 267
gatekeeper 26, 37, 38
gefitinib 60, 93, 97
geldanamycin 115
Genasense 86, 144, 164
genomic instability 52, 89, 163, 339
GIST 60
Gleevec 60, 65, 92, 334
GLI 338
glioma 56, 62, 140, 190, 192, 278
glycolysis 35, 38, 95, 189, 229, 235
gossypol 87, 117, 168, 169
granulocyte 312, 334
GTP-binding 181
GW182 302
GX01 168, 169
GX015-070 117
GX15-070 63, 117, 144

H
HA14-1 87, 144
hairpin 299–301, 303, 315
HAT 261–263
HCC 236
HDAC 64, 116, 145, 254, 256, 261–282, 308
HDACI 262, 263, 265, 275
HDM2 109, 202
heavy chain 52, 159, 230, 306, 310
Hedgehog 312, 331, 332, 335, 337

hematopoietic 53, 54, 64, 107, 209, 210, 236,
        277, 307, 311, 312, 332, 333, 336, 338
hematopoietic stem cell 334, 335
hepatocyte 129, 139, 140, 223, 232, 236, 237
hepatocyte growth factor 110
Her2 89, 270
herceptin 61, 93
herpes 163, 234
HGS 131, 132
HGS ETR1 88, 141, 144
HGS ETR2 141, 142, 144
HGS-TR2J 142, 144
HHV8 163, 234
HIF1 213
Hip 338
HIPK2 204
Hippel-Lindau tumor suppressor 275
histone 277, 280
Histone deacetylase (HDAC) 231, 262,
        308, 316
histone deacetylase inhibitors 86, 91, 116,
        192, 214, 261, 264
history 1, 2, 4, 6, 8
HIV 163
HMBA 266
HMGB1 228, 235, 237
HNSCC 266
homeostasis 3, 25, 47, 160, 205, 223, 224,
        232, 237, 255, 273, 335, 337
homeostatic 82, 178
homologs 82, 163, 307, 317
homology 33, 82, 106, 130, 160, 180, 184,
        202, 208, 224, 301
hormone 2, 90, 91, 95
HOX gene 312, 313
HPC 266
HPV 318, 319
H-ras 52, 92
HRK 83, 113, 160
HSC see hematopoietic stem cell
Hsp-27 254–256
Hsp-70 254–256
Hsp-90 64, 115, 254, 256, 270, 271
hTERT 316, 317
HtrA2 28, 30, 31, 49, 208, 273
hypoxia 84, 187, 201, 208, 213, 229, 270, 277

I
IAP 13, 20, 30, 31, 59, 83, 85, 88, 89, 107,
        113, 136, 137, 230, 233, 273, 314
ICAM 253, 254
ICAM1 271
IDN-13389 117

IFN-γ 184
IgE 266
IGF-1 254
IgG 134, 140, 142
IkB 317
IKK 62, 91, 94, 113, 224–227, 231, 234,
        236, 238, 253, 316
IKKγ 224–226, 230, 232, 233, 238
IL-15 138
IL-2 138
IL-4 266
IL-5 266
IL-6 110, 253, 254
imatinib 53, 60, 92, 93, 97, 270
immune 4, 7, 25, 50, 81, 84, 87, 95, 117,
        138–140, 142, 166, 193, 205, 224,
        232, 235, 236, 273, 303, 306, 319
immunoglobin 140, 310
immunoglobulin 52, 159, 306
immunohistochemical 130–133
Indian 338
inflammation 82, 95, 139, 224, 236, 237,
        251, 266
ING4 91, 235
inhibition 13, 15, 20, 31, 48, 53, 57,
        59–65, 83, 85, 86, 88–91, 95, 96,
        107, 109–112, 114, 178, 181, 185,
        187–192, 228, 233, 236, 238, 239,
        252–258, 263, 265, 266, 268–270,
        273, 275, 277, 278, 280–282, 302,
        304, 331, 334, 335, 337–339
initiator caspase 13, 17, 85, 107, 135, 143,
        162, 205, 212
INK4a 185, 335
inner membrane (IM) 27, 31, 32, 34,
        35, 50
Interferon 138, 209, 271, 275
intermembrane space (IMS) 29–32, 106
intrinsic 15, 47, 48, 52, 57, 81, 134, 159, 208,
        255, 256
intrinsic pathway 15, 16, 26, 27, 31, 49, 50,
        62, 65, 82, 83, 85, 88, 92, 105, 128,
        135–137, 160, 162, 163, 202, 205, 209,
        276, 277
irinotecan 59, 93, 141, 256
irradiation 54, 62, 96, 183, 206, 210, 264,
        277, 278, 282, 316
ISIS 2181308 144
IκB 14, 62, 83, 91, 224, 234, 235, 253

J
Jag1 336, 337
JAK 92, 94

JNK 55, 112–115, 140, 204, 227–230,
     236–238, 254, 255
Jun 112, 140, 204, 228, 254, 279

K

Kaposi sarcoma 163, 234
keratinocyte 129, 237
ketone 266, 268
KHSV 234
ki-67 108
KIAA0828 316
KILLER 128, 204, 206–208
KMTR2 142
Ku70 271

L

L-Aoe 267
LAQ824 264, 265, 275, 281
LATS2 308, 317
LBH589 275
LC3 178, 180, 181, 191
lactic dehydrogenase (LDH) 116
Lef 338, 339
let-7 307, 308
leukemia 52, 53, 59, 60, 63, 86, 92, 140, 141,
     162, 166–169, 192, 234, 238, 266, 268,
     272, 274, 276, 277, 279, 304, 310, 314,
     331–333, 335, 336, 339
lexatumumab 141
ligand 15, 16, 27, 48, 49, 51, 54, 59, 63,
     64, 83–85, 87, 88, 90, 106, 107,
     111, 116, 127–129, 133, 134, 136,
     138, 139, 142, 144, 145, 162, 190,
     202, 204, 205, 211, 227, 228, 231,
     233, 236, 270, 276, 280, 312, 315,
     336–339
LIGHT 138
LimK 308, 314
lipid 28, 34, 166, 179–181, 279
LRP5/6 338
Lsm-bodies 302
LTC-IC 333
lung 36, 37, 58, 61, 63, 88, 89, 93–95,
     129, 130, 132, 133, 139, 141,
     163, 169, 183, 185, 191, 205,
     206, 210, 234, 238, 272, 276,
     278, 279, 304, 306–308, 310,
     317, 319, 338, 339
lymphoma 5, 36, 52, 58, 87, 88
lymphotoxin b (LT β) 225
Lys-9 266, 267
lysosome 1, 2, 6, 178, 179, 181, 229, 230

M

MAFB 308, 312
maintenance 163, 302, 332, 334–336
MALT 137, 230, 234, 238, 317
mapatumumab 141
MAPKinase 264
mature cells 333
Mcl-1 36, 55, 58, 82, 84, 86, 87, 94, 107,
     109, 111, 115–117, 137, 160, 161,
     163, 169, 208, 272, 276, 279
Mdm2 55, 58, 61, 62, 90, 185, 202, 204,
     213, 214, 230
MDR 56
MDR1 339
medulloblastoma 338
MEF 190, 207, 209–211, 255, 273, 274
MEK 83, 93, 94, 110, 111, 114, 115
MEKK5 229
melanoma 59, 60, 62, 86, 87, 94, 105,
     108–117, 130, 137, 163, 164, 211,
     238, 263, 319, 334, 337, 339
Melphalan 256
membranes 3, 16, 25–38, 48, 50, 56, 84, 106,
     107, 111, 114, 128, 132, 134, 135, 138,
     139, 160, 161, 165–167, 178, 180, 184,
     188, 189, 202, 205, 208, 210, 212, 254,
     255, 269
metabolism 5, 25, 35, 36, 38, 52, 138, 166,
     262, 265, 304
metastasis 185, 225, 236, 268, 270, 275, 307
metastatic 93, 109, 117, 164, 185, 211, 235
methylation 137, 185, 188, 266, 267, 279–281
MG-115 252
MG-132 252, 334
MHC 271
microarray 251, 254, 255, 263, 299, 306, 307,
     310, 316, 317, 337
microRNA 163, 299–319
microtubule 25, 56, 94, 96, 107, 113, 114,
     180, 271
mimics 117, 302, 308
miR 310
miRNA *see* microRNA
MITF 109, 111
mitochondria 6
mitochondrial apoptosis-induced channel
     (MAC) 32–34
mitochondrial DNA (mtDNA) 26, 36–38
mitochondrial membrane premeabilization
     (MMP) 28, 33, 34, 36, 37, 254, 275
mitogenic kinases 54
mitotic 6, 56, 60, 65, 89, 95, 96, 108, 214
mitotic inhibitor 54
mitoxantrone 57, 256

MKP 229
ML-IAP 85, 88, 107, 137, 273
MMTV 337, 339
MnSOD 230, 233
MOMP 161, 162, 165, 167, 210, 211, 213, 255
monoclonal antibody 8, 60, 61, 93, 127, 131,
      132, 140, 141, 145, 339
monomeric 18
morphological 3, 6, 26, 48, 82, 268
MS-275 261, 264–267, 270, 271, 276, 278,
      279, 281
mTOR 60, 61, 83, 94–96, 177, 181, 182,
      186–188, 190, 192, 193
mule 109
multidomain 82–84, 160, 161, 208
multiple myeloma (MM) 36, 54, 91, 92, 141,
      163, 238, 251–258, 266, 272, 339
Myc 51–53, 160, 204, 211, 304, 315, 335, 339
myeloma 36, 64, 91, 92, 139, 141, 163, 164,
      206, 238, 251, 266, 272, 276, 339

N
NAD 262, 263
NBK 83, 273
necrosis 5, 48, 65, 84, 95, 106, 113, 127, 138,
      193, 202, 223, 226–229, 235, 237, 270,
      315, 336
necrotic 3, 6, 56, 95, 193, 223, 228, 235, 237
NEMO 224, 226, 230, 232, 233, 238
Netrin1 211, 212
neu 93
neuroendocrine tumors 95
Nexavar 94
NF-kB see NF-κB
NF-Y 270, 271
NF-κB 14.91, 54, 55, 57, 61, 62, 64, 87,
      91, 92, 111, 113, 115, 140, 190, 205,
      207, 223–239, 252, 253, 258, 271,
      277, 316, 317, 334, 337
NIK 225
NK 138
NMR 87, 168, 169
NOD 333
nonepigenic 271
non-Hodgkin's lymphoma 36, 88, 134,
      262, 206, 207
Notch 331, 332, 335–337
NOXA 55, 83, 84, 90, 106–108, 114, 115,
      160, 162, 203, 209, 210, 254, 264,
      272–274, 335, 337
NPI-0052 92, 256–258
NR13 230
NRSF 316

NSAID 238
NSCL 130, 141
NSCLC 58, 60, 63, 88, 91, 93, 94, 141,
      143, 339
nuclear condensation 26, 48
nucleoplasmin 185
Nur77 270
nutrient 4, 8, 96, 177, 178, 181, 186, 187,
      189, 192, 193, 229

O
oblimersen 62, 63, 86, 117, 143, 164–166
olbimersen sodium 86, 144
oligonucleotides 53, 83, 86, 88, 89, 163,
      164, 251, 254, 255, 307
Omi 28, 30, 31, 49, 106, 108, 208, 273
oncogene 36, 47, 51, 52, 90, 92, 97, 105, 108,
      109, 113, 117, 159, 160, 163, 179, 182,
      185–188, 269, 300, 304, 305, 307, 309,
      310, 315–319, 334, 335
oncoproteins 26, 50, 51, 57, 224, 234
OPG 49, 59, 128
ortholog 14, 96, 179, 183
osteoblast 129, 254
osteosarcoma 139, 206, 268
outer membrane (OM) 26–29, 32, 33, 50, 51,
      106, 160, 161, 165, 180, 210, 255
ovarian 56, 60, 91, 94, 115, 130, 132, 133,
      141, 183, 297, 334
OX40L 138
oxidative 35
oxidative phosphorylation 34, 36, 95, 229
oxidative stress 34, 35, 233
OXO 264

P
P100/p52 NF-κB2 224–226
P105/P50 NF-κB 224
P107 263, 266
p130 263
p21 110, 201, 207, 209, 210, 212, 213, 254,
      280, 334, 335
P21$^{WAF1/CIP1}$ 263, 264, 266, 269, 272, 274, 280
P27^{KIP1} 263, 272, 279
p300 204, 271
p38 204, 254, 256
p53 5, 7, 26, 49, 51, 52, 54–56, 58, 61–64,
      83, 89–91, 95, 106–110, 113, 114, 116,
      117, 135, 136, 161, 179, 182, 184–188,
      190, 201–215, 230, 231, 233, 254, 263,
      271–275, 277, 279, 280, 316, 317, 334,
      335, 337

p53AIP1 203, 204, 211
p53-mSin3a 213
p63 209, 212
p65 91, 224, 225, 227, 231
P70S6kinase 94
p73 55, 209
paclitaxel 38, 56, 57, 60–62, 91, 92, 96, 114,
    141, 169, 290, 239
PAK4 111
pancreatic 93, 94, 132, 133, 192, 238, 256,
    306, 314, 338
pancreatic cancer 93, 94, 192, 238, 306
PARP 113, 254, 272, 277
Pasha 300, 301
patched 338
PAZ 302
P-bodies 302, 303
PCF 270
PD0325901 94
PD184352 94
PD98059 265
PDGFR 60, 93, 190
PDK 111, 182
PDS1 14
PDX101 265, 281
PEL 234
peptide 13, 19, 37, 63, 82, 86–88, 113, 130,
    131, 136, 144, 159, 166–168, 207, 214,
    252, 265, 268
peptidomimetic 63
PERK 254, 255
permeability transition pore (PTP) 32, 37,
    38, 106
PERP 203, 212
PGP 56, 339
Phase 2, 15, 56, 60, 94, 95, 108, 181, 189,
    209, 224, 272, 279–281, 334
Phase I 63, 64, 88, 115, 139, 141, 142, 169,
    252, 253, 268, 281, 282, 319
Phase II 62–64, 88, 93–95, 114, 141, 253,
    281, 282, 319
Phase III 62, 63, 86, 93–95, 116, 164,
    253, 281
Phenyl butyrate 275
Philadelphia chromosome 92, 97
phosphatidylethanolamine (PE) 178, 180
phospholipases 25, 167
PI(3)K 111, 178, 179, 182, 183, 186, 187,
    190, 193, 317
PI3K 55, 57, 61, 92, 94, 115, 254
PI3KC 94
PIDD 107, 108
PIDDOSOME 107
Pin1 230, 231

PITX1 317
piwi 302
PKC 112, 115
PKC412 270
PLAD 129
platinum 139, 141
Plk 214
PMA 112
PMBL 234, 235, 238
polyphenols 168, 237
polyubiquitinated 91, 224
PP1 270, 271
PPAR-γ 86
pre-autophagosomal structure (PAS)
    179, 180
PRIMA-1 90, 214
pri-miRNA 300, 301
pro-apoptotic 15, 49, 160, 203
profiling 256, 306, 307
progenitor cell 52, 138, 209, 312, 313,
    332–336
prognosis 88, 299, 300, 304, 306, 308
programmed cell death (PCD) 1, 3, 7,
    25, 28, 47, 90, 127, 128, 159, 160,
    163, 177, 201, 202, 207, 215, 223,
    226–235, 237
progression 25, 35, 36, 38, 47, 48, 51, 52, 81,
    82, 86, 93, 95, 110, 160, 164, 177, 182,
    185, 187, 201, 207, 213, 223, 232, 234,
    236, 237, 239, 251–253, 275, 279, 300,
    306, 307, 310, 318, 332, 338
protease 4, 13–15, 20, 21, 25, 29–31, 48, 82,
    84, 86, 128, 134, 135, 138, 161, 166,
    167, 178, 181, 182, 212, 230, 252
proteasome 14, 64, 65, 85, 91, 92, 94,
    107–109, 202, 224–226, 238,
    251–254, 256–258, 264, 273
proteasome inhibitor 64, 82, 83, 91, 92, 115,
    191, 238, 251, 252, 256, 334
proteolysis 13–15, 21, 161, 181, 251, 257,
    273
proteolytic 13–16, 20, 21, 32, 82, 89, 112,
    252, 254, 278, 336, 337
proteomic 97, 251, 254
PS-341 91, 115, 252, 254
Psammaplin 269
PSI 252
PTCH1 338
PTD 166, 167
PTEN 57, 62, 94, 111, 137, 179, 182,
    187, 316
PUMA 55, 83, 84, 90, 106–108, 113–115,
    160, 162, 203, 209–213, 264, 272, 273
purpurogallin 168

Q
quinazoline 238

R
RAD001 95
Rad3 204
radiotherapy 88–90, 97, 115, 192, 281
RAF 83, 93, 94, 110, 111, 114, 115, 254, 264
RAGE 228
RAIDD 107
rapamycin 61, 95, 96, 115, 181, 193
Ras 52–54, 57, 60, 61, 92, 93, 108, 110,
        111, 114, 160, 182, 186, 190, 204,
        211, 237, 268, 269, 307, 308, 317,
        318, 335, 337
RASAL1 317
Rb 108, 109, 185, 263, 272
reactive oxygen species (ROS) 28, 34–37,
        191, 212, 227–230, 234, 236, 254, 255,
        264, 273, 279
refractory 91, 93, 96, 143, 192, 238, 251, 253,
        255–257, 263
Rel 223, 224, 232–235, 237
Rel homology domain (RHD) 224
relapsed 36, 91, 238, 251, 253
Relish 224
Relvimid 256
renal cell carcinoma (RCC) 95, 133, 141
resistance 35–38, 47, 48, 51, 56–61, 63–65,
        81, 82, 87–89, 92, 93, 95, 97, 105, 110,
        113, 114, 117, 130, 136, 143, 163, 167,
        204, 206, 210, 213, 235, 237, 238,
        251–257, 263, 274, 315, 318, 319
REST 316, 317
Retinoblastoma 108, 185, 286, 279, 319
Rheb 181, 182, 186
RING 85
RIP1 226–228, 230
RNA interference (RNAi) 179, 184, 189–191,
        234, 238, 299–303, 315–317, 319
RNA-induced silencing (RISC) 299–303, 318
RPS6KA6 316
RSK 186
RTK 60

S
S6K1 94
sarcoma 86, 163, 190, 234
SBHA 266
SCF-βTrCP 224, 226
SC-IC 333
sensitizer 84, 160–164, 166, 169

short-chain fatty acid 265, 280
shrinkage 3, 26, 48
signal transduction 36, 47, 81, 112, 258,
        267, 279, 334, 338, 339
Sir2 262, 263, 272
Sirt1 263, 274
SIRT4 274
SK7041 265
SMAC 28–32, 49, 50, 63, 84, 85, 88, 106,
        108, 117, 136, 137, 144, 161, 208,
        227, 230, 254, 255, 273, 279
SMAC mimetic 83, 144
SMO 338
Smoothened 338
Snk 214
SOCS-1 231
Sonic 338
Sonic hedgehog (shh) 312, 313
Sorafenib 94, 109, 110, 114, 117
Sp1 205, 272, 276, 277
Sp3 280
SP600125 113
SPC839 238
Spi2A 230
Splitomicin 265
Stat 53, 54, 254
Stat3 109, 271
STI571 92
stomach cancer 306, 308
stress 8, 26, 27, 34, 35, 50, 55, 58, 84,
        97, 128, 140, 177, 182, 187, 188,
        193, 201, 208, 224, 228, 233, 254,
        255, 263, 273, 303
stress response 55, 254
strylquinazoline 214
suberoylanilide hydroxamic acid (SAHA) 64,
        91, 192, 261, 264–266, 272, 274, 275,
        278, 281
sulforaphane (SFN) 192, 280
Survivin 60, 85, 88, 89, 107, 109, 137, 143,
        144, 103, 233, 275, 278, 314, 339
surviving 7, 236, 237

T
TAK1 230
Tarceva 60, 93
taxanes 54, 91, 92, 139, 141
taxol 107, 114, 117, 214
tBID 34, 51, 84, 106, 107, 135, 208,
        227, 276
T-cell 87, 138, 140, 185, 205, 225, 233, 234,
        266, 273, 311, 314, 336, 337
Tcf 338, 339

TGFBR2 316
TGF-β 184, 190, 230, 232, 270, 275
Thalidomide 238, 254, 257
therapeutic 26, 36–38, 61, 64, 65, 81, 88,
        90–93, 96, 97, 105, 113, 114, 127,
        129, 140, 144, 145, 159, 163, 164,
        166, 168, 191, 202, 213, 238, 239,
        252–254, 256, 258, 261, 270, 274,
        275, 278, 280, 284, 299, 300, 309,
        310, 312, 335, 339
therapy 25, 26, 35–38, 47, 48, 53, 60–62, 64,
        65, 81–83, 87–90, 92, 93, 95–97, 105,
        117, 128, 130, 145, 163, 191, 213, 223,
        237–239, 251, 256, 258, 261, 264, 270,
        272, 275, 276, 278, 282, 299, 306,
        317–319, 331, 333
Tie-2 275
Tip60 316
T-L 257
TLR 228
TNF see tumor necrosis factor
TNF-R1 56, 84, 204, 206
TNFSFR10a 128
TNFSFR10b 128
TOM 29
topoisomerase 54, 91, 278
toxicity 61, 63, 87, 88, 92, 94, 117, 139,
        140, 167–169, 253, 254, 257, 265,
        281, 318, 319
TP53 136
TRA-8 140, 144
TRADD 49, 56, 226–228
TRAF 2 113, 226, 227, 233, 316
TRAIL 16, 48, 49, 57, 59, 62–64, 87, 88, 107,
        108, 112, 113, 115–117, 127–134, 136,
        138, 139, 143–145, 162, 190, 201–208,
        210-214, 230, 236, 261, 270, 272–277,
        282, 315
TRAIL-R 127, 128, 130, 131, 133–136,
        138–140, 142–145
TRAILR1 85, 88, 116, 128–134, 137,
        140–144, 204, 272, 282
TRAILR2 85, 113, 116, 128–138, 140–145,
        204, 272, 276, 282
transcription factor 51, 55, 56, 61, 62, 108,
        109, 111, 209, 223, 224, 262, 263, 266,
        276, 277, 300, 304, 308, 310–313, 317,
        336, 338
transformation 96, 127, 128, 185, 188, 191,
        234, 271, 281, 299, 300, 304–207, 311,
        314–316, 318, 331–334
translocation 29–31, 33, 52, 91, 92, 111, 113,
        135, 159, 162, 163. 210, 224, 225,
        227, 230, 253, 255, 304, 336, 338

trapoxin A 265, 267, 268
trastuzumab 61, 93, 270
trichostatin A (TSA) 261, 264–267, 274–276,
        278, 279
TRID 204, 206, 208
triterpenoids 63, 144, 254, 256
TRUNDD 204, 206, 208
TSA see trichostatin A
TSC 94, 182, 186–188
tSLT 316
tubulin 271
tumor 27, 53, 95–97, 130–133, 138,
        141, 183, 189, 193, 194, 223,
        258, 275–278, 280, 282, 299–308,
        310, 311, 318, 332, 334,
        337–339
tumor necrosis factor 48, 49, 56, 57, 63, 84,
        87, 91, 106, 112, 127, 128, 138, 162,
        202, 204, 206, 226, 228, 231, 232, 253,
        270, 315, 336
tumor progression 35, 36, 47, 51, 82, 177,
        213, 338
tumor suppressor 26, 50, 55, 57, 89–91,
        94, 110, 177, 182, 183, 185–188,
        201, 215, 230, 231, 233, 235, 237,
        239, 261, 275, 299, 304, 305, 307–309,
        315–317, 319, 337
type I 50, 202, 204–206, 271
type II 6, 27, 50, 51, 138, 202, 208, 210,
        270, 271
tyrosine kinase 53, 55, 57, 60, 61, 83, 92, 108,
        110, 270, 274
TβRII 270, 271

U
ubiquitin 202, 251, 271
ubiquitin ligase 109, 231,
        251, 315
Ubiquitin-proteaosome pathway (UPP) 251,
        252, 254, 273
UTR 302, 304, 307, 312
UVRAG 182, 183

V
vacuoles 95
VCAM 253, 254
VEGFR 93, 94, 277
velcade 64, 91, 238, 251, 254
voltage-dependent anion channel (VDAC)
        32, 33, 36, 38
VPA 265, 275, 278
Vps 179, 183

W
Warburg effect 35, 36
wnt 35, 331, 332, 335, 338, 339

X
xenograft 61, 63, 86–88, 131, 139–141, 144,
    145, 167–169, 207, 213, 214, 256, 258,
    264, 266, 275, 277, 278
XIAP 15, 20, 30, 59, 62, 63, 85, 88, 89, 91,
    107, 111, 137, 143, 144, 230–232, 256,
    273, 278, 279
X-Rell 224

Y
YY1 263

Z
Z-IETD 276
ZIPK 184
Z-VAD 276
zymogen 13–15, 17, 18, 21, 31,
    49, 50
β-catenin 280, 335, 338, 339